10
D1632037
ROM THE LIBRAR

Principles and Techniques of Electromagnetic Compatibility
Second Edition

Principles and Techniques of Electromagnetic Compatibility

Second Edition

Christos Christopoulos

CRC Press
Taylor & Francis Group
Boca Raton London New York

CRC Press is an imprint of the
Taylor & Francis Group, an **informa** business

CRC Press
Taylor & Francis Group
6000 Broken Sound Parkway NW, Suite 300
Boca Raton, FL 33487-2742

CRC Press is an imprint of Taylor & Francis Group, an Informa business

Printed in the United States of America on acid-free paper
10 9 8 7 6 5 4 3 2 1

International Standard Book Number-10: 0-8493-7035-3 (Hardcover)
International Standard Book Number-13: 978-0-8493-7035-9 (Hardcover)

Library of Congress Cataloging-in-Publication Data

Christopoulos, Christos.
Principles and techniques of electromagnetic compatibility / Christos Christopoulos. -- 2nd ed.
p. cm.
Includes bibliographical references and index.
ISBN-13: 978-0-8493-7035-9 (alk. paper)
ISBN-10: 0-8493-7035-3 (alk. paper)
1. Electronic circuits--Noise. 2. Electromagnetic compatibility. 3. Electromagnetic interference. I. Title.

TK7867.5.C47 2007
621.382'24--dc22 2007007235

Visit the Taylor & Francis Web site at
http://www.taylorandfrancis.com

and the CRC Press Web site at
http://www.crcpress.com

Dedication

To Margaret Lydia

Contents

Preface to the Second Edition .... **xiii**

Preface .... **xv**

The Author .... **xvii**

Part I Underlying Concepts and Techniques

1 Introduction to Electromagnetic Compatibility **1**

2 Electromagnetic Fields **5**

2.1 Static Fields 6

2.1.1 Electric Field 6

2.1.2 Magnetic Field 15

2.2 Quasistatic Fields 22

2.2.1 The Relationship between Circuits and Fields 26

2.2.2 Electromagnetic Potentials 31

2.3 High-Frequency Fields 33

2.3.1 Electromagnetic Waves 34

2.3.2 Radiating Systems 39

References 53

3 Electrical Circuit Components **55**

3.1 Lumped Circuit Components 55

3.1.1 Ideal Lumped Components 56

3.1.2 Real Lumped Components 57

3.2 Distributed Circuit Components 64

3.2.1 Time-Domain Analysis of Transmission Lines 67

3.2.2 Frequency-Domain Analysis of Transmission Lines 70

References 76

4 **Electrical Signals and Circuits** 77
4.1 Representation of a Signal in Terms of Simpler Signals 78
4.2 Correlation Properties of Signals 88
4.2.1 General Correlation Properties 89
4.2.2 Random Signals 90
4.3 The Response of Linear Circuits to Deterministic and Random Signals 92
4.3.1 Impulse Response 92
4.3.2 Frequency Response 93
4.3.3 Detection of Signals in Noise 95
4.4 The Response of Nonlinear Circuits 98
4.5 Characterization of Noise 100
References 108

Part II General EMC Concepts and Techniques

5 **Sources of Electromagnetic Interference** 111
5.1 Classification of Electromagnetic Interference Sources 111
5.2 Natural Electromagnetic Interference Sources 112
5.2.1 Low-Frequency Electric and Magnetic Fields 112
5.2.2 Lightning 113
5.2.3 High-Frequency Electromagnetic Fields 117
5.3 Man-Made Electromagnetic Interference Sources 118
5.3.1 Radio Transmitters 118
5.3.2 Electroheat Applications 119
5.3.3 Digital Signal Processing and Transmission 119
5.3.4 Power Conditioning and Transmission 122
5.3.4.1 Low-Frequency Conducted Interference 122
5.3.4.2 Low-Frequency Radiated Interference 123
5.3.4.3 High-Frequency Conducted Interference 124
5.3.4.4 High-Frequency Radiated Interference 124
5.3.5 Switching Transients 125
5.3.5.1 Nature and Origin of Transients 125
5.3.5.2 Circuit Behavior during Switching Assuming an Idealized Switch 126
5.3.5.3 Circuit Behavior during Switching Assuming a Realistic Model of the Switch 131
5.3.6 The Electrostatic Discharge (ESD) 135
5.3.7 The Nuclear Electromagnetic Pulse (NEMP) and High Power Electromagnetics (HPEM) 137
5.4 Surveys of the Electromagnetic Environment 139
References 139

6 Penetration through Shields and Apertures......143
6.1 Introduction......143
6.2 Shielding Theory......145
6.2.1 Shielding Effectiveness......145
6.2.2 Approximate Methods — The Circuit Approach......146
6.2.3 Approximate Methods — The Wave Approach......156
6.2.4 Analytical Solutions to Shielding Problems......161
6.2.5 General Remarks Regarding Shielding Effectiveness at Different Frequencies......161
6.2.6 Surface Transfer Impedance and Cable Shields......163
6.3 Aperture Theory......167
6.4 Rigorous Calculation of the Shielding Effectiveness (SE) of a Conducting Box with an Aperture......174
6.5 Intermediate Level Tools for SE Calculations......176
6.6 Numerical Simulation Methods for Penetration through Shields and Apertures......185
6.6.1 Classification of Numerical Methods......185
6.6.2 The Application of Frequency-Domain Methods......187
6.6.3 The Application of Time-Domain Methods......190
6.7 Treatment of Multiple Apertures through a Digital Filter Interface......194
6.8 Further Work Relevant to Shielding......200
References......201

7 Propagation and Crosstalk......207
7.1 Introduction......207
7.2 Basic Principles......210
7.3 Line Parameter Calculation......223
7.3.1 Analytical Methods......223
7.3.2 Numerical Methods......231
7.4 Representation of EM Coupling from External Fields......232
7.5 Determination of the EM Field Generated by Transmission Lines......246
7.6 Numerical Simulation Methods for Propagation Studies......252
References......253

8 Simulation of the Electromagnetic Coupling between Systems......257
8.1 Overview......257
8.2 Source/External Environment......258
8.3 Penetration and Coupling......259
8.4 Propagation and Crosstalk......270
8.5 Device Susceptibility and Emission......273
8.6 Numerical Simulation Methods......274
8.6.1 The Finite-Difference Time-Domain (FD-TD) Method......275
8.6.2 The Transmission-Line Modeling (TLM) Method......275

8.6.3 The Method of Moments (MM) 276
8.6.4 The Finite-Element (FE) Method 276
References 278

9 Effects of Electromagnetic Interference on Devices and Systems 281
9.1 Immunity of Analogue Circuits 283
9.2 The Immunity of Digital Circuits 284
References 288

Part III Interference Control Techniques

10 Shielding and Grounding 291
10.1 Equipment Screening 291
10.1.1 Practical Levels of Attenuation 292
10.1.2 Screening Materials 292
10.1.3 Conducting Penetrations 297
10.1.4 Slits, Seams, and Gasketing 297
10.1.5 Damping of Resonances 299
10.1.6 Measurement of Screening Effectiveness 299
10.2 Cable Screening 300
10.2.1 Cable Transfer Impedance 300
10.2.2 Earthing of Cable Screens 301
10.2.3 Cable Connectors 302
10.3 Grounding 303
10.3.1 Grounding in Large-Scale Systems 304
10.3.2 Grounding in Self-Contained Equipment 307
10.3.3 Grounding in an Environment of Interconnected Equipment 308
References 309

11 Filtering and Nonlinear Protective Devices 311
11.1 Power-Line Filters 311
11.2 Isolation 316
11.3 Balancing 319
11.4 Signal-Line Filters 320
11.5 Nonlinear Protective Devices 320
References 326

12 General EMC Design Principles 329
12.1 Reduction of Emissions at Source 330
12.2 Reduction of Coupling Paths 331
12.2.1 Operating Frequency and Rise-Time 331
12.2.2 Reflections and Matching 333
12.2.3 Ground Paths and Ground Planes 334
12.2.4 Circuit Segregation and Placement 335
12.2.5 Cable Routing 336

12.3 Improvements in Immunity 336
12.3.1 Immunity by Software Design 338
12.3.2 Spread Spectrum Techniques 339
12.4 The Management of EMC 342
References 344

Part IV EMC Standards and Testing

13 EMC Standards 347
13.1 The Need for Standards 347
13.2 The International Framework 348
13.3 Civilian EMC Standards 349
13.3.1 FCC Standards 349
13.3.2 European Standards 351
13.3.3 Other EMC Standards 354
13.3.4 Sample Calculation for Conducted Emission 354
13.4 Military Standards 357
13.4.1 Military Standard MIL-STD-461D 357
13.4.2 Defence Standard DEF-STAN 59-41 359
13.5 Company Standards 360
13.6 EMC at Frequencies above 1 GHz 360
13.7 Human Exposure Limits to EM Fields 364
References 368

14 EMC Measurements and Testing 371
14.1 EMC Measurement Techniques 371
14.2 Measurement Tools 372
14.2.1 Sources 372
14.2.2 Receivers 373
14.2.3 Field Sensors 375
14.2.4 Antennas 376
14.2.5 Assorted Instrumentation 380
14.3 Test Environments 383
14.3.1 Open-Area Test Sites 384
14.3.2 Screened Rooms 387
14.3.3 Reverberating Chambers 391
14.3.4 Special EMC Test Cells 396
References 397

Part V EMC in Systems Design

15 EMC and Signal Integrity (SI) 403
15.1 Introduction 403
15.2 Transmission Lines as Interconnects 408
15.3 Board and Chip Level EMC 422

15.3.1 Simultaneous Switching Noise (SSN) 422
15.3.2 Physical Models 426
15.3.3 Behavioral Models 433
References 435

16 EMC and Wireless Technologies 437
16.1 The Efficient Use of the Frequency Spectrum 439
16.2 EMC, Interoperability, and Coexistence 442
16.3 Specifications and Alliances 449
16.4 Conclusions 455
References 456

17 EMC and Broadband Technologies 459
17.1 Transmission of High-Frequency Signals over Telephone and Power Networks 460
17.2 EMC and Digital Subscriber Lines (xDSL) 463
17.3 EMC and Power Line Telecommunications (PLT) 465
17.4 Regulatory Framework for Emissions from xDSL/PLT and Related Technologies 466
References 467

18 EMC and Safety 471
References 473

19 Statistical EMC 475
19.1 Introduction 475
19.2 The Basic Stochastic Problem 477
19.3 A Selection of Statistical Approaches to Complex EMC Problems ... 481
References 484

Appendices

A Useful Vector Formulae 485

B Circuit Parameters of Some Conductor Configurations 489

C The sinx/x Function 495

D Spectra of Trapezoidal Waveforms 499

E Calculation of the Electric Field Received by a Short Electric Dipole 503

Index 505

Preface to the Second Edition

In preparing the second edition of this book I have adhered to the same principles as in the first edition, namely, to present the topic of EMC to a wide range of readers and enable as many of them as possible to benefit from the experience and the work of others. A practical approach is adopted whenever appropriate but mathematical analysis and numerical techniques are also presented in connection with the development of EMC predictive tools. I have tried to present the physical basis and analytical models of important interactions in EMC and whenever the mathematical framework became too complex I have guided the interested reader to relevant references. Readers therefore will be able to study the material at a depth appropriate to their skills and interests.

The main thrust of the new edition was to update material, extend the treatment of some topics that are not covered adequately in the first edition, introduce new topics that ten years ago had not emerged as important in EMC, and add more worked examples. Part V is entirely new and covers new and emerging technologies that are having a major impact in EMC theory and practice.

Again, I gratefully acknowledge the support and friendship over the years of professional colleagues at Nottingham and elsewhere.

Christos Christopoulos
Nottingham, October 2006

Preface

Electromagnetic compatibility (EMC) may be approached from two standpoints. In the first approach, one is tempted to address the problem in its generality and seek the development of general tools to predict performance. The complexity of the problem is such that it is not at present possible to make decisive progress along this route and one is left with a collection of mathematical treatises of limited immediate applicability. It is also generally the case that a physical grasp of the problem cannot be easily attained in this way, since mathematical complexity dominates all.

An alternative approach is to view EMC from the multiplicity of practical problems that confront designers on a day-to-day basis. The general appearance of such work is a collection of "EMC recipes and fixes" based on a mixture of theoretical ideas, practical experience, and black magic, which no doubt work under certain circumstances. The student of such works is left, however, without a clear physical grasp of exactly what is happening or an understanding of the range of validity of the formulae he or she is supplied with.

The current text adopts a different viewpoint. Emphasis is placed on understanding the relevant electromagnetic interactions in increasingly complex systems. Mathematical tools are introduced as and when pursuing the physical picture unaided becomes counterproductive.

Approximations and intuitive ideas are also included and an effort is made to define the limitations of each approach. As the reader becomes aware of the physics of EMC interactions and has some mathematical tools at his or her disposal, then the manner in which systems can be engineered to achieve EMC is described and illustrated with practical examples. Many readers will find that the confidence felt when they become familiar with the physical, mathematical, and engineering aspects of EMC is not sufficient to predict the performance of complex systems. In order to handle complexity, numerical tools are developed and the basis and capabilities of these tools are presented. It is hoped that the text will provide useful source material for a serious study of EMC, including references to more advanced work.

The text aims to be comprehensive, although, in a topic of such a wide coverage as EMC, it is difficult to make a selection of material to be included that will appeal to all readers. The text will be useful to all those engaged in EMC analysis and design either as advanced undergraduates, postgraduates, or EMC engineers in industry. The author is, by training and temperament, inclined to take a global view of problems and he hopes that readers will choose to study the entire book at a pace that reflects their own background and interests. However, as guidance for those who may require a selection of topics for a first reading, the following two schemes are suggested:

i. EMC applications-oriented readers — Chapter 1, Sections 2.1, 3.1, 4.1, 4.4; Chapter 5; and all material in Parts III and IV.
ii. EMC analysis-oriented readers — All material in Parts I and II and Chapter 13.

Part I contains material underlying all work in electrical and electronic engineering and it is thus also relevant to EMC. Readers with a degree in electrical engineering will be familiar with a large part of this material. Part II deals with general EMC concepts and techniques and it is thus useful to those engaged in predicting the EMC behavior of systems. More practical techniques used to control electromagnetic interference and the design of EMC into products are presented in Part III. Finally, the main EMC standards and test techniques are described in Part IV.

It is a pleasure to acknowledge the help and support I have received over the years from co-workers at Nottingham University and other higher educational establishments and from those in industry and government laboratories who, by sponsoring work in EMC, have contributed to my understanding of the subject. Miss S. E. Hollingsworth expertly typed large parts of the manuscript and I am grateful to her for doing this with the minimum of fuss.

C. Christopoulos
Nottingham

The Author

Christos Christopoulos was born in Patras, Greece, on September 17, 1946. He received the Diploma in Electrical and Mechanical Engineering from the National Technical University of Athens in 1969 and the MSc and DPhil from the University of Sussex in 1971 and 1974 respectively.

In 1974 he joined the Arc Research Project of the University of Liverpool and spent two years working on vacuum arcs and breakdown while on attachments at the UKAEA Culham Laboratory. In 1976 he was appointed Senior Demonstrator in Electrical Engineering Science at the University of Durham. In October 1978 he joined the Department of Electrical and Electronic Engineering, University of Nottingham, where he is now Professor of Electrical Engineering and Director of the George Green Institute for Electromagnetics Research (GGIEMR).

His research interests are in computational electromagnetics, electromagnetic compatibility, signal integrity, protection and simulation of power networks, and electrical discharges and plasmas. He is the author of over 300 research publications, five books, and several book chapters. He has received the Electronics Letters and the Snell Premiums from the IEE and several conference best paper awards. He is a member of the IEE, IoP, and an IEEE Fellow. He is past Executive Team Chairman of the IEE Professional Network in EMC, member of the CIGRE Working Group 36.04 on EMC, and Associate Editor of the IEEE EMC Transactions. He is Vice-Chairman of URSI Commission E "Noise and Interference" and Associate Editor of the URSI Radio Bulletin.

I

UNDERLYING CONCEPTS AND TECHNIQUES

Any serious study of electromagnetic compatibility (EMC) requires a reasonable background in several areas of electrical engineering. In a topic such as EMC, where the complexity of practical problems is very considerable, it is not always possible to provide complete and accurate answers based on comprehensive quantitative models. In such cases it is imperative that a basic underlying grasp of the nature of electrical components, their interactions through electromagnetic fields, and the nature and characterization of electrical signals is firmly established. Such a foundation provides the analyst and EMC designer with the basic conceptual framework and tools for asking the right questions, for critically evaluating results and assumptions, and for supporting the intuitive search for solutions based on sound scientific understanding. Even those who are not comfortable with mathematical models of physical systems and processes will benefit greatly from studying Part I, which is a distillation of the techniques most relevant to the study of EMC phenomena. Those already familiar with fields circuits and signals may also wish to study Part I, since it presents these topics from an EMC perspective.

The aim of the next four chapters is to provide background material on the nature and importance of EMC, the concept of electromagnetic fields, the description of real components in terms of ideal components, and the relationship between field and network concepts. Finally, the modeling and analysis of signals and circuits in the frequency and time domains will be described.

Chapter 1

Introduction to Electromagnetic Compatibility

Electromagnetic compatibility (EMC) is the branch of science and engineering concerned with the design and operation of equipment in a manner that makes them immune to certain amounts of electromagnetic interference, while at the same time keeping equipment-generated interference within specified limits. The scope of EMC is thus very wide as it encompasses virtually all equipment powered by electrical supplies. Practically, all engineering systems incorporate power conditioning and information processing units and thus fall within the scope of EMC. The frequency range of interest extends from DC to light and in certain parts of this spectrum a strict international regulatory framework has been set up to ensure immunity of equipment to electromagnetic interference (EMI) and to control emission.

Interest in EMC is not new. Since the early days of radio, designers and listeners alike were alerted to noise, interference, and earthing problems. However, the rapid increase in the use of radio communications, digital systems, fast processors, and the introduction of new design practices have brought EMC to the forefront of advanced design. Three technological trends have provided the impetus for these changes. First, modern digital logic and signal processing are based on relatively low-threshold voltages (i.e., a few volts) compared to older technologies based on electronic valves (several hundred volts). The immunity of modern systems is therefore inherently lower. Second, in the process of seeking higher processing speeds, shorter pulse rise-times are used, contributing significant amounts of energy at high frequencies, which is capable of propagating by radiative mechanisms over long distances. Third, the modern physical design of equipment is based increasingly on the use of plastics in preference to metals. This significantly reduces the electromagnetic shielding inherent to an all-metal cabinet. Several other items could be added to this list, such as miniaturization and thus the trend for compact designs, which contribute to EMC problems. Close attention must thus be paid to EMC at all stages of design if equipment is to function properly and meet international EMC regulations.

EMC may be approached from two different directions. First, it may be argued that before anything is designed, a complete EMC study must be performed to predict the electromagnetic signature of equipment and its capacity to withstand externally generated interference. The difficulties in performing such a study are formidable, since many of the necessary predictive analytical and numerical tools are not currently available and the environment in which equipment is installed is not always fully specified. Second, it may be thought that EMC is essentially a fire-righting operation and that any problems that may arise in this area are best dealt with on an ad hoc basis. The dangers of this approach are obvious, since the complexity of modern designs and the nature of EMC preclude easy and inexpensive solutions as an afterthought. A balanced approach to EMC is to bring into this area every tool available to the designer. This includes numerical tools, in-house practical experience, and sound physical grasp of EMC and electromagnetic interactions. No single individual will be gifted in all these areas and it is thus important to establish teams with the right blend of experience and approach to problem solving to form the focus for EMC design. Several years of emphasis in engineering education on digital design have reduced the grasp of radio frequency (RF) design issues among graduates to dangerously low levels and it is therefore of considerable importance to strengthen awareness of EMC and the grasp of fundamental RF design techniques in academia and in industry.

EMC must be regarded as an issue that affects all aspects of design — electrical, electronic, and mechanical. It cannot be adequately addressed in isolation. Typically, a complete design may consist of a number of subsystems that interact with each other through signal and power cables and through reactive (capacitive and/or inductive) or radiative mechanisms. The EMC behavior of a complete system cannot be easily predicted from the known behavior of subsystems, although the purpose of current EMC research is to develop the necessary methodologies and tools to achieve this. The difficulties are more pronounced in large extensive systems where a mixture of new and old technologies is used in the presence of severe external electromagnetic (EM) threats. Clearly, the designer will make every effort to solve EMC problems within a piece of equipment caused by interactions between various subsystems. Nevertheless, the majority of EMC problems seem to be internally generated. In an integrated system in which a number of different types of equipment, often provided by different suppliers, are connected together, EMC problems may develop and they can be difficult to tackle if each unit does not meet specified EMC limits and if insufficient thought has been given to system-level EM interactions.

A designer who has developed a system that operates successfully as far as internally generated electromagnetic interference (EMI) is concerned, has nevertheless to meet, in addition, certain EMC limits specified in national and international standards. Compliance with these limits can be demonstrated by making a number of measurements under specified conditions. In a typical case, these tests will cover emission of EMI from equipment and also the susceptibility or immunity of this equipment to externally generated interference. In an emission test the equipment is placed, depending on the particular standard, inside a screened room or on an open-field site and measurements are taken of the emitted electromagnetic fields at a specified distance using receivers of specified bandwidth over a specified frequency range. The type and polarization of antennas used is also specified to produce, as far as possible, a repeatable measurement. This type of test is called a radiated emission test, to distinguish it from conducted emission tests where the EMI voltage on conductors is measured. In each case, the measured quantity must be below specified limits. A selection of such limits and a more detailed description of test procedures are given in Chapters 13 and 14.

In immunity testing, the equipment is subject to a specified externally generated field or to interference currents injected on conductors and the requirement is that the equipment remains functional. These tests may cover a wide frequency range (typically at least up to 1 GHz) and may also involve pulse-like incident electromagnetic signals to check the response of equipment to transients and electrostatic discharges. For complex equipment with a large number of operating modes it can be difficult to demonstrate immunity.

The designer is obliged to ascertain the nature and significance of EM emissions from different parts of the equipment and the effects of externally generated interference on the functional integrity of an entire system. He or she may, as far as possible, seek to minimize interference generation at its source; reduce or eliminate coupling paths by proper layout, shielding, filtering, and grounding practices; design hardware with an inherent immunity to EMI; and adopt defensive programming practices to develop software that has a high level of immunity to EMI. Very few of these options are cost free and many may have adverse effects on normal operating characteristics, size, appearance, and weight of equipment. It is not an easy matter to seek optimum solutions to such design problems, which depend on so many interdependent parameters. The purpose of research into EMC is to develop the methodologies and tools that allow optimum EMC design procedures to be incorporated into the design process from the very beginning, so that, even at the conceptual stage of a design, EMC issues are taken into account and sensible choices are made with minimum costs.

A glance at the titles of the chapters that follow will convince the reader of the complexity and importance of EMC and of its position at the forefront of teaching and research in advanced system design techniques.

Chapter 2

Electromagnetic Fields

The purpose of this chapter is to restate in broad terms the basic premises of electromagnetic theory, and to establish the fundamental concepts and models of electromagnetic phenomena as applied to electromagnetic compatibility (EMC). It is not intended to give a complete coverage of the topic as this would occupy most of the book. There are many excellent texts on electromagnetics to which the reader may refer.[1–4] The topic is also taken up at a level accessible to first-year university students in my own *Introduction to Applied Electromagnetism* (Wiley, 1990).[5]

No secure understanding of EMC may be attained without at least a physical grasp of what electromagnetics is. The analytical equipment normally associated with the subject is also necessary, if basic understanding is to be followed by the ability to make quantitative predictions of EMC performance. This remains so even when numerical predictive simulation packages are used in EMC studies. In the author's opinion, there is no substitute for a core understanding of electromagnetics. It is therefore proper to devote this chapter to this important topic. It is convenient to examine electromagnetic phenomena under three broadly distinct categories.

First, there are problems where charges are static or they move with constant speed (e.g., direct currents). This type of problem is encountered in DC and low-frequency applications, and the fundamental properties of the field in this limit are that its two aspects, electric and magnetic, can be dealt with separately. Coupling between electric and magnetic field is weak, and it is possible to develop simple models to examine each separately.

Second, there are problems where charges accelerate slowly, so that although coupling between electric and magnetic fields cannot be ignored, certain simplifying assumptions can be made in interpreting what is taking place.

Finally, in cases where charges accelerate rapidly, no approximations can be made and the most general field models must be employed to account for the strong coupling between electric and magnetic fields.

The first two categories are referred to as the static and quasistatic approximations. The task of this chapter is to outline the essential physics applicable in each category, to sketch out the basic models used in making quantitative predictions, and, most importantly, to establish the limits of applicability of each model. In this process the notation, units, and some useful formulae will be given to assist in the more specialized treatment that follows in other chapters.

2.1 Static Fields

2.1.1 Electric Field

Static electric charges set up an influence around them, which is described as an *electric field*. At each point in space, an electric field value is assigned, referred to as the electric field intensity, E, measured in volts per meter (V/m). The amplitude of the electric field does not, however, describe completely the influence of the electric charges. It is also necessary to assign a direction to the field at each point, much like in a weather map where at each point wind speed *and* direction are required. In mathematical jargon, the electric field is a vector quantity **E** described fully when its amplitude and direction are given. The direction of **E** at each point is that in which a positive electric charge placed there would move. The electric field E(r) a distance r away from a point charge q placed in vacuum (or approximately air) is given by the formula

$$E(r) = \frac{q}{4\pi\, \varepsilon_0\, r^2} \tag{2.1}$$

where if q is in coulombs, r is in meters, and ε_0 is a constant. The electric permittivity of free-space is ε_0 = 8.8542 pF/m and then the electric field is obtained in V/m. By the term "point charge" is meant a charge distribution occupying dimensions small compared to r and possessing spherical symmetry so as to look virtually like a charge concentrated at a point a distance r away from the point where the field is observed. The direction of the field is radially outward from a positive charge, as shown in Figure 2.1.

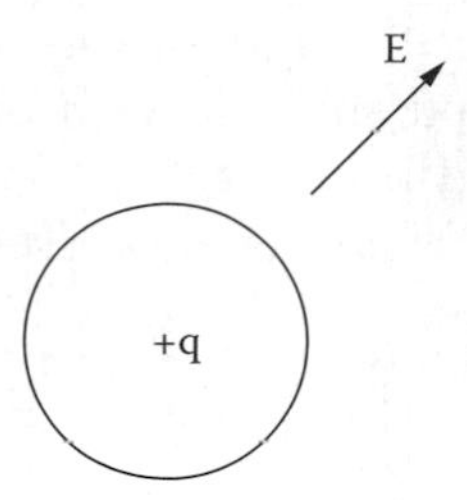

Figure 2.1 The electric field at a point in the proximity of a spherical charge distribution.

If the charge distribution does not meet the above criteria, then Equation 2.1 cannot be used directly. Instead the charge is divided into small volume elements dV, each containing a charge ρdV where ρ is the charge density. Then, since dV is infinitesimally small, charge ρdV is effectively a point charge and Equation 2.1 applies. The field at any point can then be found by *superposition*, i.e., by combining the electric field due to each volume element until all the charges have been accounted for. The individual electric field contributions must be combined as *vectors* to find the total field.

In cases where the surrounding medium is not air but another material, then Equation 2.1 still applies but ε_0 is replaced by ε, the dielectric permittivity of the material used. Materials are normally characterized by their relative dielectric permittivity or dielectric constant ε_r where $\varepsilon = \varepsilon_r \varepsilon_0$. In most, but not all, materials, ε_r is constant and independent of the local value of the electric field or other parameters such as temperature or humidity. The range of ε_r values is remarkably restricted. For most insulating liquids it has a value approximately equal to 3 and in most insulating solids it is equal to about 5. Notable exceptions are water ($\varepsilon_r \simeq 80$) and various titanates ($\varepsilon_r \simeq 10{,}000$).

If different materials surround electric charges, the determination of the field is more complicated, as will be explained shortly. It is also customary to use another quantity to offer an alternative description of the field. This is a vector quantity known as the electric flux density **D** measured in coulombs per m² (C/m²). In most, but not all, materials, **D** is parallel to **E**. The relationship between the two is

$$\mathbf{D} = \varepsilon_r \, \varepsilon_0 \, \mathbf{E} \tag{2.2}$$

It is possible to relate, in a compact and elegant way, the electric field (D or E) to its sources (electric charges). This relationship is expressed by Gauss's Law, which states that if, on any closed surface S, the product is formed on the surface area of a small patch ds and the component of the flux density normal to it D_n, and all the contributions are summed to account for the entire closed surface, then the total sum found is equal to the total charge enclosed by S.

In mathematical form this law is expressed as

$$\int_S D_n \, ds = Q_{tot} \tag{2.3}$$

where the S in the integral sign indicates integration over the closed surface S, and Q_{tot} is the charge enclosed by S. If this charge is distributed in some way (electric charge density ρ) then

$$Q_{tot} = \int_V \rho dv \tag{2.4}$$

where V is the volume enclosed by S.

Term D_n ds may be expressed as a dot product **D** d**s**, where d**s** is a vector of magnitude ds and pointing outward of the volume enclosed by S. Hence, Gauss's Law in its most general form is

$$\int_S \mathbf{D} \, d\mathbf{s} = \int_V \rho dv \tag{2.5}$$

Example: Calculate the electric field at a distance r away from a long cylindrical conductor on which a charge q coulombs per meter length is distributed uniformly.

Solution: A surface S is chosen, as shown in Figure 2.2, on which to apply Gauss's Law. The choice is arbitrary, but a choice that is in tune with the symmetry of the problem, in this case a cylinder, makes the difference between stating the law and obtaining useful results. The symmetry of the problem demands a radial electric field, hence along the flat ends of S the field is parallel to the surface elements and there is zero contribution to the left-hand side of Equation 2.3 or 2.5.

The curved part of S contributes and thus

$$2\pi rl \, D(r) = ql$$

Hence

$$D(r) = \frac{q}{2\pi r}$$

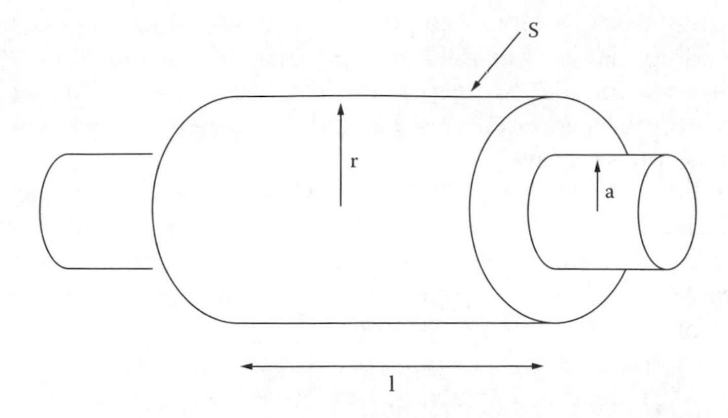

Figure 2.2 Configuration of a coaxial capacitor.

Accordingly

$$E(r) = \frac{q}{2\pi\varepsilon r}$$

Gauss's Law expressed by Equation 2.5 is an example of an integral law. It relates quantities over surfaces and volumes of the problem. Although useful in problems with symmetry and uniform materials, it becomes cumbersome in other more general configurations. It is then necessary to examine the form this law takes when the surface S is reduced until it collapses into a point. If both sides of Equation 2.5 are divided by the volume and the volume is allowed to tend to zero, then the left-hand side of the equation expresses the electric flux per unit volumc and the right-hand side tends to the charge density. The electric flux per unit volume is a quantity known in mathematics as the divergence of the flux density and is designated as div **D**. Gauss's Law in point or differential form is thus

$$\text{div}\, \mathbf{D} = \rho \tag{2.6}$$

In a Cartesian coordinate system

$$\text{div}\, \mathbf{D} = \nabla \cdot \mathbf{D} = \frac{\partial D_x}{\partial x} + \frac{\partial D_y}{\partial y} + \frac{\partial D_z}{\partial z}$$

where D_x, D_y, and D_z are the components of **D**, and $\partial/\partial x$ indicates a derivative with respect to x only. The divergence of a vector is thus a scalar quantity fully described by its amplitude. A detailed treatment of this type of relationship may be found in Reference 6.

Expressions for the divergence in other coordinate systems may be found in Appendix A. Equations 2.5 and 2.6 are equivalent descriptions of the same physical law.

It has been shown how either **E** or **D** may be used to describe the electric field. It is sometimes easier, from the calculation point of view, to use a quantity that is not a vector. Such a quantity is the *electric potential.* At each point in space a potential value is assigned. This is a scalar quantity — there is no direction associated with it. The work done to move a test charge q from point a to point b is q[V(b) – V(a)] where V(a) and V(b) are the potential values at points a and b, respectively. Potentials are always defined with respect to some reference. The potential at a point a distance r away from a point charge q is

$$V(r) = \frac{q}{4\pi\varepsilon}\frac{1}{r} \tag{2.7}$$

The potential due to a distribution of charge occupying volume V is

$$V(r) = \int_v \frac{\rho dv}{4\pi\varepsilon R} \tag{2.8}$$

where ρ is the volume charge density, and R is the distance between the volume element and the observation point at r. The electric field distribution and the potential difference between two points a and b are related by the expression

$$V_{ab} = V(a) - V(b) = \int_a^b E_t \, dl = \int_a^b \mathbf{E}\, d\mathbf{l} \tag{2.9}$$

where E_t indicates the component of the electric field that at each point is parallel (tangent) to a small line segment of length dl on a curve joining points a and b.

Equation 2.9 suggests that the potential difference is the same irrespective of which precise path is chosen to go from a to b. It also follows that if a closed path is followed, the potential difference will be zero. These conclusions are strictly correct for the static field case considered here. They have to be revised when quasistatic and general fields are present. The general form of Equation 2.9 to include nonstatic fields is considered in Section 2.2. The electric field may be obtained if the potential is known. It turns out that

$$E = -\text{grad } V = -\nabla V$$

where, in Cartesian coordinates,

$$\nabla V = \frac{\partial V}{\partial x}\hat{x} + \frac{\partial V}{\partial y}\hat{y} + \frac{\partial V}{\partial z}\hat{z}$$

Thus, the electric field component in the x-direction is $-\partial V/\partial x$ and so on.

Taking the divergence of the expression for **E** gives

$$\nabla \cdot \mathbf{E} = -\nabla \cdot (\nabla V)$$

or

$$\frac{1}{\varepsilon}\nabla \cdot \mathbf{D} = -\nabla^2 V$$

where ∇^2 is the Laplacian operator. In rectangular coordinates

$$\nabla^2 V = \frac{\partial^2 V}{\partial x^2} + \frac{\partial^2 V}{\partial y^2} + \frac{\partial^2 V}{\partial z^2}$$

Combining with Equation 2.6 gives

$$\nabla^2 V = -\rho/\varepsilon \tag{2.10}$$

This is known as Poisson's equation and its solution is given by Equation 2.8. When $\rho = 0$, Equation 2.10 is referred to as Laplace's equation.

At the interface between two different ideal dielectrics, the electric field values on either side are interrelated. Application of Gauss's Law and Equation 2.9 on a closed path gives the following expressions:

$$\begin{aligned} E_{t1} &= E_{t2} \\ D_{n1} &= D_{n2} \end{aligned} \tag{2.11}$$

Simply stated, the component of the electric field tangent to the interface is continuous. Similarly, the flux density component normal to the interface is continuous.

If a battery of voltage V is connected across two conductors, an amount of charge q is transferred from one conductor to the other, thus creating a deficit of electrons in one conductor (charge +q) and a surplus in the other conductor (charge –q). The potential difference V, and the charge q, are related by the *capacitance* C between the two conductors

$$C \equiv q/V \tag{2.12}$$

Whenever charge separation and potential differences are present in a system of conductors, electric fields are also present and some capacitance may be associated with the conductors. The procedure for calculating capacitance is straightforward, but it can only be carried out analytically in a few simple cases. In general terms, in order to calculate capacitance, the following steps are taken:

1. Charges ±q are assumed to be distributed on each conductor forming the two plates of the capacitor. Any symmetries are exploited to simplify further calculation.
2. The electric field **E** between the conductors is obtained using Gauss's Law or any other suitable procedure. The field is only required on points lying on a curve c joining the two conductors.
3. The potential difference between the conductors is then obtained from $V = \int_c \mathbf{E} d\mathbf{l}$, where the curve c is chosen with care to simplify the calculation.
4. The capacitance is then $C \equiv q/V$.

In this procedure, it is implicity assumed that the potential V has the same value irrespective of the path c chosen to evaluate it. Otherwise, a different capacitance would be found depending on the path chosen. As stated earlier, this assumption is true in static fields. However, in rapidly varying fields this cannot be guaranteed, and it is not possible to define a unique value for V (indeed voltages are strongly dependent on the path chosen). In such cases there are conceptual and practical difficulties in defining and calculating capacitance.

A simple example is given below to illustrate the procedure for calculating capacitance.

Example: Calculate the capacitance between two long coaxial cylinders of inner and outer radii r_1 and r_2, respectively.

Solution: Step (1), a charge q_1 is assumed distributed uniformly per meter length of the surface of the inner conductor. From charge conservation, the charge per unit length on the outer conductor is $-q_1$. The assumption of uniform distribution is justified by the cylindrical symmetry of the problem.

Step (2), the electric field in the space between the two conductors may be calculated using Gauss's Law, in exactly the same way as for the case shown in Figure 2.2. It follows that

$$E(r) = \frac{q_1}{2\pi\varepsilon r} \quad r_1 \le r \le r_2$$

Elsewhere, the electric field is zero.

Step (3), the potential difference is calculated by choosing a radial path from the inner to the outer conductor. Along this path, **E** is always tangent to it, and hence

$$V = \int \mathbf{E}\, d\mathbf{l} = \int_{r=r_1}^{r_2} E(r)\, dr = \frac{q_1}{2\pi\varepsilon} \ln \frac{r_2}{r_1}$$

Step (4), the capacitance per unit length is

$$C_1 = \frac{q_1}{V} = \frac{2\pi\varepsilon}{\ln \frac{r_2}{r_1}}, \text{ in F/m} \tag{2.13}$$

If the coaxial conductors are of length l then the total capacitance is $C = C_1 \times l$.

Remarks — The formula given above does not take into account the distortion to the field near the edge of the conductors. In most cases, such secondary considerations are not significant but they may be important in cases, for instance, when r_2/l is large. The capacitance depends on the logarithm of the ratio of outer to inner radii. Hence, it is not a very strong function of these dimensions. Equation 2.13 may be used to estimate capacitance even in cases where there is uncertainty about exact dimensions or when the conductors are not strictly cylindrical.

Formulae for the capacitance of some common geometries are given in Appendix B.

The separation of electric charges and the associated electric field is a store of energy that may be calculated directly from the formula

$$W_e = \frac{q^2}{2C} = \frac{1}{2} CV^2, \text{ in Joule} \tag{2.14}$$

This energy may be viewed, for illustrative purposes, as a form of potential energy found in mechanics. The capacitance is thus a measure of this energy. An alternative way to view this energy is to associate it directly with the electric field rather than the charges. It turns out that in a part of space where the electric field is E, the energy stored per unit volume is

$$w_e = \frac{1}{2} \varepsilon E^2, \text{ in J/m}^3 \tag{2.15}$$

If the total energy stored in the field is calculated by applying Equation 2.15 and all contributions are summed together to account for the entire volume occupied by the field, it is found that it equals that given by Equation 2.14. Thus, these two expressions are equivalent descriptions of the same thing. Equation 2.14 represents the circuit and Equation 2.15 the field point of view. Which formula is used depends on convenience and the nature of the problem.

A powerful tool in understanding electric fields and interactions is to draw on paper or on a computer screen the appearance of the field around charged conductors. *Visualization* removes from the field some of its abstract nature, and aids human reasoning and problem solving.

An obvious technique would be to draw at each point in space an arrow representing the vector of the field **E**, with a length proportional to the magnitude of the field and the same direction as **E**. This approach is adopted by some computer graphics packages, but it is not the best one. A very elegant approach is to plot the field and equipotential lines. Field lines are defined so that they coincide with the direction in which a positive charge would move if placed in the field. They are thus tangent at each point to the direction of the field. They start from positive and terminate in negative charges. This brings out a cardinal feature of the electric field, which is important in EMC work as it helps to identify coupling paths. Electric field lines are shown in Figure 2.3. In such plots a line is shown starting from a conductor for each unit of charge. For example, if a conductor has a total charge of 6 μC and it is decided to plot one line per μC, then, if the charge is distributed uniformly on the conductor, six lines will be shown leaving the conductor from points centered on six equal conductor areas, each containing 1 μC of charge. If the charge distribution is not uniform, the same principle applies, but the areas containing 1 μC of charge are unequal in extent, and thus more

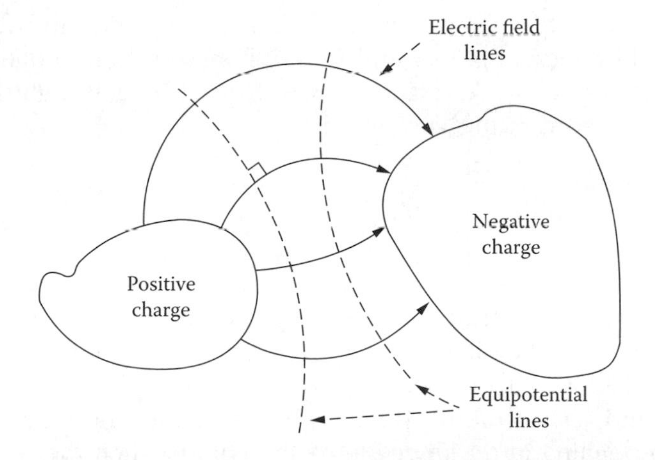

Figure 2.3 Schematic diagram of electric field and equipotential lines.

lines leave from parts of the conductor where charge concentration, and hence the electric field, is stronger. More lines crowded together indicate a strong field. The field lines always impinge on perfect conductors at right angles. In the same diagram, lines of constant potential are drawn (equipotentials). These lines are orthogonal to the field lines and are plotted for fixed potential increments. Hence, in areas of steep potential change, and hence high electric field, more equipotential lines are plotted, as shown schematically in Figure 2.3.

In EMC applications it is common to express electric field values in decibels with respect to some reference. As an illustration, an electric field E expressed in dB μV/m is the value found by calculating

$$20 \log \left(\frac{E}{1 \times 10^{-6}} \right)$$

where E is in units of V/m.

An electric field equal to 10 V/m may thus be expressed as 140 dB μV/m.

2.1.2 Magnetic Field

Treatment so far has been limited to the effects of stationary electric charges. When electric charges are moving with constant velocity to produce an electric current, an additional influence is set up that is described as a *magnetic field*. At each point in space a magnetic field value is assigned, referred to as the magnetic flux density B, measured in teslas (T) or, equivalently, in Wb/m^2. The magnetic flux density is a vector quantity requiring, in a full description, both an amplitude and a direction. Its value at a distance r away from a long straight conductor placed in air and carrying current I is

$$B(r) = \frac{\mu_0 I}{2\pi r} \tag{2.16}$$

where $\mu_0 = 4\pi\ 10^{-7}$ A/m and is the magnetic permeability of vacuum or approximately air. The direction of **B** is perpendicular to the plane formed by the conductor and the observation point and is fully determined by the right-hand rule. If the conductor is grasped with the right-hand, with the thumb pointing in the direction of the current, then the fingers show the magnetic field direction. Another quantity often used to describe the magnetic field is the magnetic field intensity **H** measured in A/m. In most, but not all, materials **B** and **H** are parallel and are related by the expression

$$\mathbf{B} = \mu \mathbf{H} = \mu_r \mu_0\ \mathbf{H} \tag{2.17}$$

where μ_r is the relative permeability of the medium and is a material constant. Most materials have a μ_r of the order of 1 with the exception of ferromagnetic materials, where values approaching a few hundred are common. Some materials, such as mumetal, have values of μ_r exceeding several thousands. The fundamental expression, known as Ampere's Law, relating the magnetic field to its sources (currents), states that if the component of the magnetic field intensity H_t parallel (tangent) at a point on a curve c is multiplied by the length of a small element of the curve around this point, and if all such contributions are summed together to cover the entire closed curve, then the total sum is equal to the current linked with this curve:

$$\int_c H_t\ dl = I_{tot} \tag{2.18}$$

The current I_{tot} may best be identified by calculating the current emerging from a surface S bounded by the curve c, as shown in Figure 2.4. Equation 2.18 may then be written in a more general form as

$$\int_c \mathbf{H}\ d\mathbf{l} = \int_s \mathbf{j}\ d\mathbf{s} \tag{2.19}$$

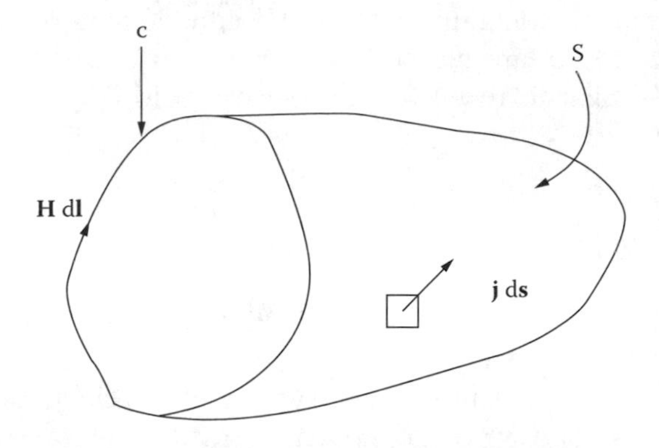

Figure 2.4 Configuration used to apply Ampere's Law.

This is the well-known Ampere's Law. It can be used to calculate the field very easily in cases where the problem has some symmetry. In static fields, the current linked is that due to conductors (conduction current). However, at high frequencies other terms, in addition to conduction currents, must be included in the right-hand side (displacement currents). This generalization is discussed further on in this chapter (see Equation 2.32). Equation 2.19 is an example of an integral law, as it involves a relationship between quantities evaluated over a finite area. It is useful to determine the form of Ampere's Law as the curve c and the associated area S are reduced in size to the extent that in the limit, they collapse to a point. Both sides of Equation 2.19 are divided by the surface area S and the limit is taken as the area tends to zero. The resulting expression on the left-hand side is well known in mathematics and it is called the curl of the vector **H** or $\nabla \times \mathbf{H}$. The expression on the right-hand side reduces simply to the current density. Ampere's Law in differential or point form is then

$$\nabla \times \mathbf{H} = \mathbf{j} \tag{2.20}$$

In Cartesian coordinates

$$\nabla \times \mathbf{H} = \hat{x}\left(\frac{\partial H_z}{\partial y} - \frac{\partial H_y}{\partial z}\right) + \hat{y}\left(\frac{\partial H_x}{\partial z} - \frac{\partial H_z}{\partial x}\right) + \hat{z}\left(\frac{\partial H_y}{\partial x} - \frac{\partial H_x}{\partial y}\right)$$

Similar expressions in other coordinate systems may be found in Appendix A.

In configurations containing different magnetic materials, the magnetic field on either side of an interface between two such materials is subject to conditions similar to those for the electric field (Equation 2.11).

$$\begin{aligned} H_{t1} &= H_{t2} \\ B_{n1} &= B_{n2} \end{aligned} \tag{2.22}$$

In obtaining the first of these two expressions, it was assumed that no current flows at the interface.

Whenever a current I flows in a system of conductors, it establishes a magnetic flux ϕ. These two quantities are related by a parameter known as the inductance L of the conductors.

$$L \equiv \phi / I \tag{2.23}$$

Current flow and an associated field indicate that an inductance may be assigned to the conductors.

The inductance may be calculated by following the procedure outlined below:

1. Current $\pm I$ is assumed flowing in the pair of conductors. Any symmetries are identified since they simplify subsequent calculations.
2. The flux density **B** around the conductors is calculated using Ampere's Law or any other appropriate procedure.
3. The flux ϕ linked with the conductors is then calculated from $\phi = \int_S \mathbf{B} d\mathbf{s}$ where surface S is any surface bounded by the conductors.
4. The inductance is then $L \equiv \phi/I$.

Difficulties with this procedure arise in practical problems where the current paths (return conductor) cannot be easily identified. These problems are dealt with in other chapters.

A simple example will be shown to illustrate the procedure for calculating inductance.

Example: Calculate the inductance between two long coaxial cylinders of inner and outer radii r_1 and r_2, respectively.

Solution: Step (1), a current I on the inner conductor is assumed to flow as shown in Figure 2.5, and returns on the outer conductor. Cylindrical symmetry indicates that the current is distributed uniformly.

The flux density B(r) in the space between the two conductors may be calculated (step 2) by applying Ampere's Law on a curve of radius r concentric with the axis of the conductors. It follows that

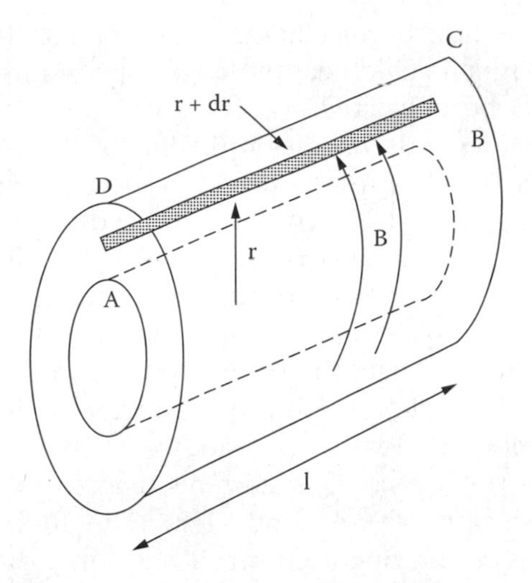

Figure 2.5 Calculation of the inductance of a coaxial line.

$$B(r) = \frac{\mu I}{2\pi r} \quad r_1 \le r \le r_2$$

The field for $r > r_2$ is zero as can be confirmed by applying Equation 2.18 on a curve with radius $r > r_2$ (total current linked is zero).

In step (3) the magnetic flux linked with 1 m length of the conductors is calculated. This is done by finding first the amount of flux $d\phi$ crossing the small strip shown hatched in Figure 2.5. It is clear that

$$d\phi = 1 \, dr \, B(r)$$

Hence the total flux linked is

$$\phi = \int_{r=r_1}^{r_2} d\phi = \frac{\mu l I}{2\pi} \int_{r=r_1}^{r_2} \frac{dr}{r} = \frac{\mu l I}{2\pi} \ln \frac{r_2}{r_1}$$

Finally, the inductance may be calculated (step 4) from Equation 2.23:

$$L \equiv \frac{\phi}{I} = \frac{\mu l}{2\pi} \ln \frac{r_2}{r_1}, \text{ in H} \tag{2.24}$$

The inductance per unit length is clearly $\frac{\mu}{2\pi} \ln \frac{r_2}{r_1}$

Remarks — The above calculation does not take into account the distortion of the magnetic field near the edges of conductors. These effects may in many cases be neglected.

The inductance depends on the logarithm of the ratio of outer and inner radii, and for the dimensions found in most practical systems does not vary greatly. Equation 2.24 may thus be used to estimate inductance even in cases where the conductors are not strictly cylindrical. Formulae for the inductance of some common geometries are given in Appendix B.

Implicit in this example was the assumption that the magnetic flux is entirely confined in the space between the two conductors. It is thus assumed that there is no flux inside the conductors. This is true provided the current is distributed close to the surface of the conductors. At low frequencies, this is not the case, as currents tend to flow uniformly over the entire cross section of the conductors. The flux density inside a conductor of radius a, carrying a current I uniformly distributed, may be found by applying Ampere's Law on a curve of radius r where $r \leq a$. It is easy to show that

$$B(r) = \frac{\mu I}{2\pi a^2} r, \qquad \text{where } r \leq a \tag{2.25}$$

Hence, in this case, the flux density rises linearly from the center of the conductor (Equation 2.25) and then falls in magnitude according to Equation 2.16 outside the conductor, as shown in the example.

The extra flux inside the conductor cross section indicates a higher inductance (called in this case the "internal inductance"). At high frequencies, the current tends to concentrate on conductor surfaces, hence the inductance tends to the value calculated in the example. These matters are discussed in more detail in future chapters. It is stressed at this stage that the inductance, even of a simple conductor, depends on the manner in which current is distributed, which in turn depends, among other things, on the frequency.

Current flow (movement of charges) and the associated magnetic field, represent storage of energy, which may be calculated directly from the formula

$$W_m = \frac{1}{2} L I^2 \tag{2.26}$$

This energy may be viewed, for illustrative purposes, as a form of kinetic energy found in mechanics. The inductance is thus a measure of this energy. An alternative point of view is to associate this energy directly with the magnetic field rather than the current. The energy stored in a part of space where the magnetic field intensity is H is

$$w_m = \frac{1}{2}\mu H^2 = \frac{B^2}{2\mu}, \quad \text{in J/m}^3 \tag{2.27}$$

The energy stored calculated from Equation 2.26 (circuit point of view), or from Equation 2.27 over the entire volume occupied by the field (field point of view), yields the same value. These two approaches are thus equivalent and the one best suited to each problem must be chosen in each case.

Equation 2.27 for energy storage in a magnetic field may be used to estimate the internal inductance of a conductor referred to earlier. Using a thin cylindrical region as the basic element, energy stored using the field and circuit points of view is

$$\int_{r=0}^{a} 2\pi r l \frac{1}{2} \mu \, H(r)^2 \, dr = \frac{1}{2} \left(L_{int} \, l\right) I^2$$

where L_{int} is the internal inductance per unit length of the conductor. Substituting H(r) from Equation 2.25 in the expression above gives

$$L_{int} = \frac{\mu}{8\pi}, \quad \text{in H/m} \tag{2.28}$$

It is seen that for a current distributed uniformly over the conductor cross section, the internal inductance per unit length is independent of the radius.

A complication often arising in magnetic field systems in the presence of ferromagnetic materials is that μ is not a constant but varies with the local value of the field. A similar remark can be made for ε and electric field systems, but the former is a more common problem. In either case, difficulties in interpretation and in obtaining solutions are encountered as the resulting equations are nonlinear.

It is common to *visualize* the magnetic field in a manner similar to that used for the electric field. This is done by plotting flux lines, as shown in Figure 2.6 for a simple solenoid. Flux lines are always closed lines (they start and finish nowhere). Magnetic flux is conserved. Hence, in any volume, the flux entering and leaving must be the same — or, put another way, the total magnetic flux crossing a closed surface is always equal to

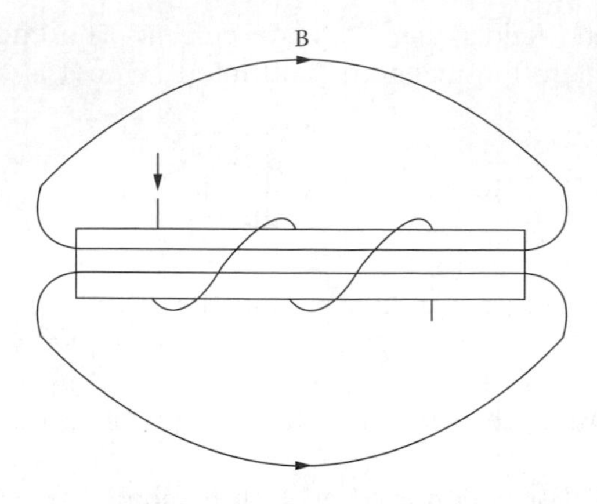

Figure 2.6 Schematic diagram of the magnetic field due to a solenoid.

zero. In a plot such as the one shown in Figure 2.6, areas with many flux lines indicate a strong local magnetic field.

This brief treatment of static fields may be summarized further to bring out more clearly the connection of these ideas with EMC. A conductor with net electric charge and/or current flow sets up influences around it that are understood in terms of electric and magnetic fields. Any other conductors, not necessarily physically connected to it, will come under these influences, with the result that further charge separation and current flow may result. This constitutes interference and its quantification and prevention is the subject of EMC. Such interactions between conductors at low frequencies may be studied in terms of capacitance and inductance or, as it will be shown in future chapters, by mutual capacitance and inductance to stress more clearly the effect of one conductor on another. The same concepts may also be applied with benefit at high frequencies, but care is required in understanding when this approach is permissible and how it should be applied. These issues are explained further in the remainder of this chapter.

2.2 Quasistatic Fields

In the previous section electric and magnetic fields were treated as if they are completely independent of each other. This is acceptable at low frequencies or low rates of change of field quantities. When this assumption becomes invalid, it is found that a changing electric (magnetic) field induces a magnetic (electric) field. The ideas and equations presented earlier must therefore be generalized to include the coupling between these two manifestations of the electromagnetic field. Treatment in this section will not, however, be of the most general kind. The assumption will be made that any interactions between different parts of a system or circuit are practically instantaneous (quasistatic approximation). If the speed of propagation for electromagnetic disturbances is u (the speed of light for propagation in air) and the largest dimension of the system studied is D, then the time it takes for different parts of the system to interact electromagnetically with each other is of the order of D/u. If the period T of the electrical disturbance is much larger than the propagation time D/u, then, in electrical terms, interaction across the system is instantaneous, and the phenomena observed can be classed as quasistatic. Another way of expressing the same thing is to compare the wavelength λ of the disturbance with the largest dimension D. If $\lambda >> D$, quasistatic conditions apply. Otherwise, the phenomena observed belong to the truly high-frequency regime and are described in Section 2.3.

The feature of most importance in the quasistatic regime, where field quantities vary relatively slowly, is the coupling between electric and magnetic fields. A magnetic field changing in time induces an electric field. In its simplest form, this coupling is observed when a time-varying magnetic flux ϕ is coupled with a circuit. It is found then that a voltage is induced on this circuit as follows:

$$V_{ind} = \frac{d\phi}{dt} \tag{2.29}$$

In field terms, the induced voltage is simply $\int_C \mathbf{E}\, d\mathbf{l}$, where c is a curve following the outline of the circuit. The magnetic flux linked is $\int_S \mathbf{B}\, d\mathbf{s}$, where the surface S is any surface bounded by the curve c. Hence Equation 2.29 may be expressed in a more general form as

$$-\frac{\partial}{\partial t}\int_S \mathbf{B}\, d\mathbf{s} = \int_C \mathbf{E}\, d\mathbf{l} \tag{2.30}$$

Equations 2.29 and 2.30 are alternative forms of *Faraday's Law*. The positive reference directions for the quantities shown in these equations are chosen as shown in Figure 2.7 to satisfy Lenz's Law, namely, that the induced voltage must drive currents that tend to oppose any change in the magnetic flux coupled with the circuit. It is straightforward to determine the form of Equation 2.30 when the curve c and surface S collapse toward a point. This is done in exactly the same way as for Equation 2.19 and gives

$$\nabla \times \mathbf{E} = \frac{\partial \mathbf{B}}{\partial t} \tag{2.31}$$

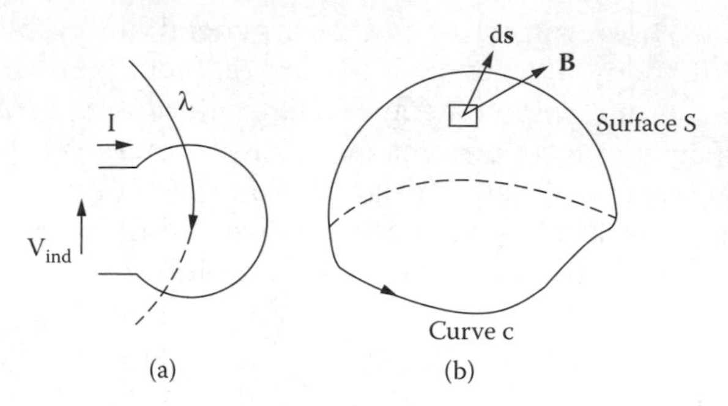

Figure 2.7 Positive reference directions for Faraday's Law in circuit (a) and field (b) problems.

This equation expresses Faraday's Law in a point or differential form. Hence, Equations 2.29, 2.30, and 2.31 all express the physical fact that a changing magnetic field produces an associated electric field. Since the magnetic field produced by a part of a circuit extends over a wide area, it follows that it may induce electric fields and potential differences in other, remote, circuits. This is termed as magnetic or inductive coupling and is a powerful coupling mechanism in EMC problems.

A changing electric field can similarly induce a magnetic field. Since the source of the magnetic field is current, the way in which this feature is introduced into the equations describing fields is to assign to a changing electric field a fictitious current as the source of the induced magnetic field. This current is described as a "displacement current" to distinguish it from normal conduction current due to the motion of free electrons in conductors. The displacement current density is $\partial\mathbf{D}/\partial t$, and must be included in the expression for total current in Ampere's Law (Equation 2.19).

$$\int_C \mathbf{H}\, d\mathbf{l} = \int_S \left(\mathbf{j}_f + \frac{\partial \mathbf{D}}{\partial t} \right) ds \qquad (2.32)$$

The subscript f has been added to indicate clearly the conduction current density (free currents). Equation 2.32 is Ampere's Law in a generalized form and in an integral formulation. It can be reduced to a differential or point form to obtain the equivalent of Equation 2.20.

$$\nabla \times \mathbf{H} = \mathbf{j}_f + \frac{\partial \mathbf{D}}{\partial t} \qquad (2.33)$$

Table 2.1 The Basic Equations Describing Electromagnetic Fields

	Integral Formulation	*Differential Formulation*
Ampere's Law	$\int_c \mathbf{H}\, d\mathbf{l} = \int_s \left(\mathbf{j}_f + \frac{\partial \mathbf{D}}{\partial t} \right) ds$	$\nabla \times \mathbf{H} = \mathbf{j}_f + \frac{\partial \mathbf{D}}{\partial t}$
Faraday's Law	$\int_c \mathbf{E}\, d\mathbf{l} = -\int_s \frac{\partial \mathbf{B}}{\partial t} d\mathbf{s}$	$\nabla \times \mathbf{E} = -\frac{\partial \mathbf{B}}{\partial t}$
Gauss's Law	$\int_S \mathbf{D}\, d\mathbf{s} = \int \rho\, dv$	$\nabla \cdot \mathbf{D} = \rho$
Flux Conservation	$\int_S \mathbf{B}\, d\mathbf{s} = 0$	$\nabla \cdot \mathbf{B} = 0$
Continuity Equation	$\int_S \mathbf{j}_f\, d\mathbf{s} = -\frac{\partial}{\partial t} \int_v \rho dv$	$\nabla \cdot \mathbf{j}_f = -\frac{\partial \rho}{\partial t}$
Constitutive Expressions:	$\mathbf{D} = \varepsilon_r \varepsilon_0\, \mathbf{E},\ \mathbf{B} = \mu_r \mu_0\, \mathbf{H}$	
Ohm's Law:	$\mathbf{j}_f = \sigma \mathbf{E}$	

These equations are summarized in Table 2.1 both in integral and differential forms, together with other useful expressions. The static case may be obtained from the expressions in this table, by demanding that the time derivatives are equal to zero.

In summary, in any situation where the magnetic field varies with time, an electric field is induced as shown schematically in Figure 2.8b. Similarly, a changing electric field induces a magnetic field as shown in Figure 2.8a. The term $\varepsilon\, \partial \mathbf{E}/\partial t$ is the displacement current. It is of particular importance to EMC. We all expect that current flows in a closed "loop." Part of the loop is through good conductors and the current there is almost entirely a conduction current. If no convenient "loop" can be established through good conductors, the current returns as a displacement current and it is this part that is usually imperfectly understood by the designer. In a good design the current loops are kept as small as possible and identified as such at the design stage. Otherwise, the current finds its own path as displacement current with unpredictable impact on emission and coupling. It can be shown that the ratio of displacement to conduction current density is proportional to $\omega\varepsilon/\sigma$.[1–5] Hence, at high frequencies and in good dielectrics the displacement current dominates. In contrast, in good conductors conduction current dominates even at high frequencies.

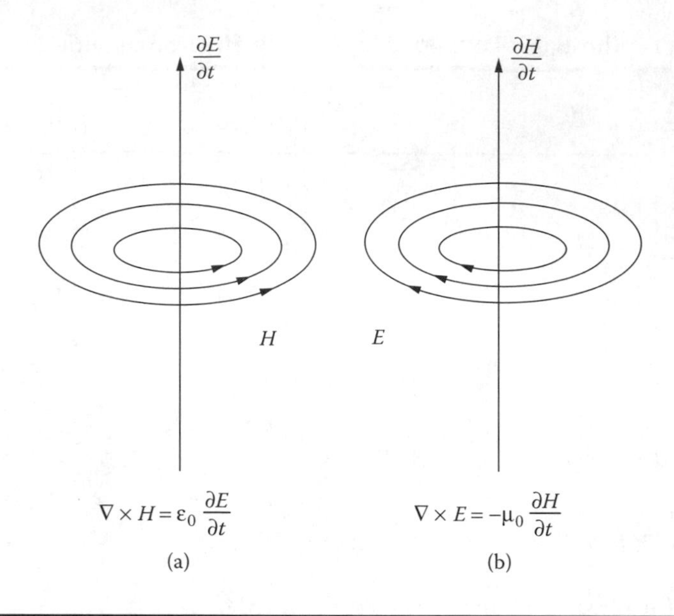

Figure 2.8 A changing electric field generates a magnetic field (a), and a changing magnetic field generates an electric field (b).

2.2.1 The Relationship Between Circuits and Fields

Sufficient material has been presented about the electromagnetic field up to this point to attempt a synthesis of field ideas with the more familiar circuit concepts. This exercise is useful both in order to better understand circuits and fields, and also to help with practical calculations.

The basic question to ask is why is it necessary to resort to field concepts instead of reasoning throughout using circuit (network) concepts? There are two reasons for this. First, it is not possible to calculate parameter values for circuit components, such as inductors and capacitors, without an understanding of fields. This may not seem a serious practical problem, as in most cases component values are marked on components used for circuit construction. In EMC work, however, it is usually the components that are not explicitly indicated (normally referred to as "stray" or "parasitic" components) that are crucial in understanding the behavior of the circuit. Second, there is a whole class of important phenomena that cannot be understood entirely in terms of circuit concepts. High-frequency applications, wave propagation, and radiating systems require the concept of field for an adequate understanding. The very notion of a lumped component representing all the capacitance of a system breaks down when the wavelength of the disturbance approaches the physical dimensions of the system. However, circuit ideas are a very useful tool in understanding

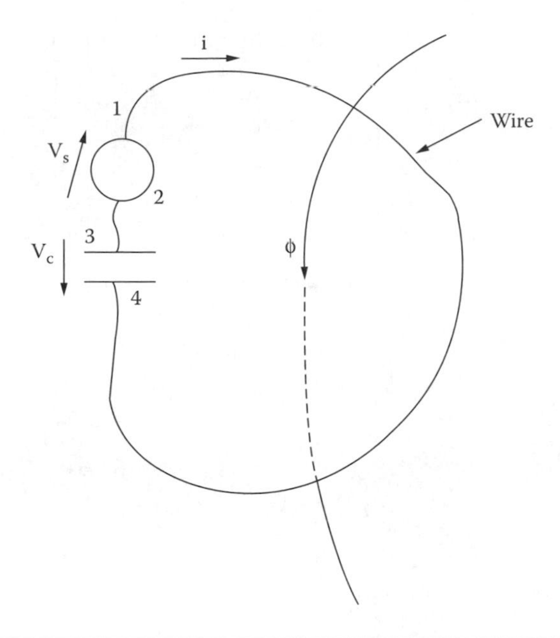

Figure 2.9 Circuit used to illustrate Faraday's Law in circuits and fields.

complex interactions, as the circuit equations (Kirchhoff's Laws) are in the form of simple algebraic or ordinary differential equations, which are relatively easy to manipulate. In contrast, the EM field laws involve, in general, temporal and spatial validations and vector quantities, and result in partial differential equations that are relatively difficult to manipulate. It can be shown that circuit laws are a special case of EM field laws, in the limit when the wavelength is much larger than the largest physical dimension of a system.[7] The relationship between fields and circuits is illustrated below.

Let us first find the circuit equivalent of Faraday's Law (Equation 2.30). A circuit is shown in Figure 2.9 with a time-varying magnetic flux ϕ_i linked with it. Subscript i indicates an externally produced, incident, flux. This flux will induce a current that in turn will produce its own magnetic flux ϕ_s, the subscript s indicating a scattered flux, i.e., the contribution of the circuit to the total flux $\phi = \phi_I + \phi_s$. Applying Faraday's Laws on a curve c coinciding with the circuit loop gives for the right-hand side:

$$\int_c \mathbf{E}dl = \int_1^2 \mathbf{E}dl + \int_3^4 \mathbf{E}dl + \int_2^3 \mathbf{E}dl + \int_4^1 \mathbf{E}dl = v_s - v_c - \int \frac{\mathbf{j}}{\sigma} \mathbf{dl}$$

where the last integral represents the voltage drop due to the resistance of the wire (paths 2 to 3 and 4 to 1).

Hence $\int_c \mathbf{E}d\mathbf{l} = v_s - v_c - iR$. The left-hand side of Equation 2.30 gives

$$-\frac{\partial}{\partial t}\int \mathbf{B}\, d\mathbf{s} = \frac{d\phi}{dt} = \frac{d\phi_i}{dt} + \frac{d\phi_s}{dt}$$

where ϕ_s is the flux produced by the induced current. Hence

$$\frac{d\phi_s}{dt} = L\frac{di}{dt}$$

and substituting

$$-\frac{\partial}{\partial t}\int \mathbf{B}\, d\mathbf{s} = \frac{d\phi_i}{dt} + L\frac{di}{dt}$$

Faraday's Law can now be expressed as

$$v_s - \frac{d\phi_i}{dt} = v_c + iR + L\frac{di}{dt}$$

This expression is none other than Kirchhoff's Voltage Law (KVL) where the total driving term is the voltage source plus the magnetic induction term. It is clear from this expression that any distortion to the path or orientation of the circuit will result in a change to the coupled incident flux ϕ_i and hence a different current. This is the mechanism operating in many situations of interest to EMC.

Let us now find the circuit equivalent of Ampere's Law.[7] The configuration chosen to demonstrate this is one of interest to EMC. It consists of a large ground place (i), a hatch (ii), and a gasket (iii) as shown in Figure 2.10a. Items (i) and (ii) are perfect conductors, and item (iii) is made out of material of electrical conductivity σ, width w, and thickness t. The mean value of the perimeter of the hatch is l. It is assumed that an electric field is incident on the hatch and that its value, when the hatch and gasket are both replaced by a conducting sheet, is E_i. This is referred to as the short-circuit field to indicate that it is obtained on the assumption of a perfect short across the hatch.

First, let us evaluate the term on the left-hand side of Ampere's Law (Equation 2.32). This is done on a curve C consisting of two parts C_t and C_b, running along the top and bottom of the gasket, as shown in Figure 2.10b.

Clearly, there is only free current I_f linked with this curve, and it is equal to GV, where V is the potential difference between the hatch and

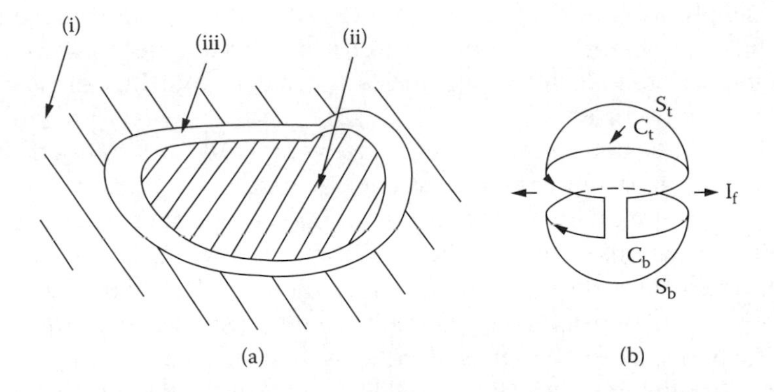

Figure 2.10 Circuit used to illustrate Ampere's Law in circuits and fields.

the surrounding conducting plate, and G is the electrical conductance across the gasket ($\approx t\,\sigma\,1/w$). Taking account of positive reference directions gives

$$\int_{C_t - C_b} \mathbf{H}\, d\mathbf{l} = -GV$$

The expression on the right-hand side of Equation 2.32 is evaluated on a surface bounded by c, namely S_t and S_b as shown in Figure 2.10b. Clearly, no free current flows through this surface, and the only term left to evaluate is $\partial/\partial t \int_S \mathbf{D}\, d\mathbf{s}$, where $\mathbf{D}$ is the superposition of two parts — the field without the hatch shorted out $\mathbf{D}_i$, and the field $\mathbf{D}_s$ due to the charges Q induced on the hatch.

$$\frac{\partial}{\partial t}\int \mathbf{D}\, d\mathbf{s} = \frac{\partial}{\partial t}\int \mathbf{D}_i\, d\mathbf{s} + \frac{\partial}{\partial t}\int \mathbf{D}_s d\mathbf{s}$$

The first term on the right represents a displacement current I_i, which is effectively a source term responsible for the induced charges and voltage on the hatch. The second term represents simply the rate of change of the induced charge on the hatch $\partial Q/\partial t$. The induced charge is equal in turn to the capacitance of the hatch with respect to the surrounding plate, times the induced voltage. Hence, equating both sides of Equation 2.32 gives

$$-I_i = C\frac{dV}{dt} + GV$$

This is simply Kirchhoff's Current Law (KCL) in the circuit representation of the hatch problem, which consists of a current source I_i (source term) producing a potential difference across the hatch and thus displacement and conduction currents.[7]

We have indicated the relationship between fields and lumped circuits. In EMC work it is imperative to recognize under what circumstances a particular circuit may be treated using lumped components (network laws, Kirchoff's equations, etc.) or field theory (Maxwell's equations). The general rule as already stated depends on the electrical size of the circuit. If the physical dimensions of the circuit are much smaller than the shortest wavelength of interest (highest frequency) in all dimensions, then the network paradigm applies (Figure 2.11a). Network topology is paramount and the usual network laws apply. If one of the three dimensions is comparable to the wavelength, then transmission line theory based on Telegrapher's equation is a reasonable model on which to study system response (Figure 2.11b). If, however, all three dimensions are comparable to the wavelength (Figure 2.11c), then the network paradigm is no longer valid and full-field concepts need to be introduced based on Maxwell's equation. Under these circumstances, in addition to topology, the exact geometry of the circuit has a profound impact on the response. It goes without saying that a full-field solution is difficult and time consuming but it cannot be avoided under the circumstances outlined above. Persistence with the network paradigm when it is no longer valid, simply because it is simple and easier to handle, inevitably results in a catastrophic loss of accuracy and profound conceptual difficulties.

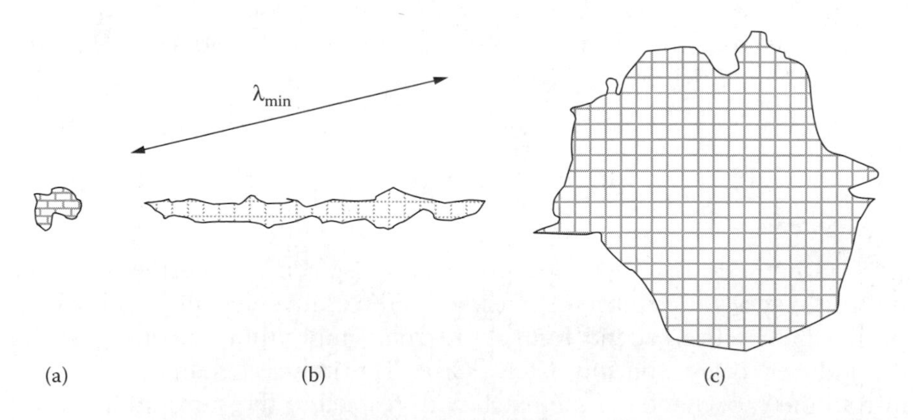

Figure 2.11 Schematic showing circuit size relative to the minimum wavelength of interest: (a) conditions when the network paradigm applies; (b) treatment based on transmission line equations; (c) conditions requiring the application of the full-field paradigm.

2.2.2 Electromagnetic Potentials

It is very convenient in many problems to define potential functions that may then be used to evaluate **E** and **B**. An electrostatic potential has already been defined by Equation 2.7. This potential, however, accounts for electrostatic field components only. Electric fields can also be induced by changing magnetic fields, as indicated in the last section. There is, therefore, a need to define an additional potential function to account for this induced electric field and also the magnetic field. It turns out that a scalar potential ϕ and a vector potential **A** provide complete information about all aspects of the EM field. In particular, the electric and magnetic components are obtained from

$$\mathbf{E} = -\nabla\phi - \frac{\partial \mathbf{A}}{\partial t} \tag{2.34}$$

$$\mathbf{B} = \nabla \times \mathbf{A} \tag{2.35}$$

Taking the divergence of Equation 2.34 and combining with Equation 2.6 gives

$$\nabla \cdot \mathbf{E} = -\nabla^2\phi - \frac{\partial}{\partial t}\left(\nabla \cdot \mathbf{A}\right) = \rho/\varepsilon \tag{2.36}$$

Similarly, from the curl of Equation 2.35

$$\nabla \times \mathbf{B} = \nabla \times \nabla \times \mathbf{A} = -\nabla^2\mathbf{A} + \nabla\left(\nabla \cdot \mathbf{A}\right)$$

Combining Equations 2.33 and 2.34 gives

$$\nabla \times \mathbf{B} = \mu\mathbf{j_f} + \mu\varepsilon\left(-\frac{\partial}{\partial t}\nabla\phi - \frac{\partial^2 \mathbf{A}}{\partial t^2}\right)$$

Equating the right-hand side of the last two equations gives

$$-\nabla^2\mathbf{A} + \mu\varepsilon\frac{\partial^2 \mathbf{A}}{\partial t^2} + \nabla\left[\nabla \cdot \mathbf{A} + \mu\varepsilon\frac{\partial\phi}{\partial t}\right] = \mu\mathbf{j_f}$$

It can be shown that in order to fully define a vector, its divergence and curl must be specified. In Equation 2.35, the curl has been chosen; therefore, an additional constraint is required to define $\nabla \cdot \mathbf{A}$. The particular choice of divergence is made for convenience and commonly the Lorentz gauge is used, namely

$$\nabla \cdot \mathbf{A} = -\mu\varepsilon \frac{\partial \phi}{\partial t} \tag{2.37}$$

Substituting this expression for $\nabla \cdot \mathbf{A}$ gives

$$\nabla^2 \mathbf{A} = -\mu\varepsilon \frac{\partial^2 A}{\partial t^2} = -\mu \mathbf{j}_f \tag{2.38}$$

Similarly, substituting Equation 2.37 into Equation 2.36 gives

$$\nabla^2 \phi - \mu\varepsilon \frac{\partial^2 \phi}{\partial t^2} = -\rho/\varepsilon \tag{2.39}$$

The equations given above are wave equations for the potentials. For static or quasistatic fields the second term on the left-hand side may be neglected, and Equation 2.39 reduces to Poisson's equation as expected. Its solution is given by Equation 2.8. By analogy, the solution of Equation 2.38 for **A** is obtained by substituting **j** for ρ as shown below:

$$\phi(r) = \int_V \frac{\rho(\mathbf{r}')\,dv'}{4\pi\varepsilon R} \tag{2.40}$$

$$\mathbf{A}(r) = \int_V \frac{\mu \mathbf{j}(\mathbf{r}')\,dv'}{4\pi R} \tag{2.41}$$

The configuration to which these equations apply is shown in Figure 2.12 where **r** is the coordinate point P at which the potentials are evaluated, V is the volume occupied by the source charges and currents, and R is the distance between point P and point $\mathbf{r}'$ at the center of the volume element dv' used in the integration. The source terms ρ and **j** are evaluated at point $\mathbf{r}'$.

A useful application of the vector potential **A** is in the calculation of inductance, which is defined as the flux linked with a circuit loop over the current that produced it. The flux linkage is the integral of the flux

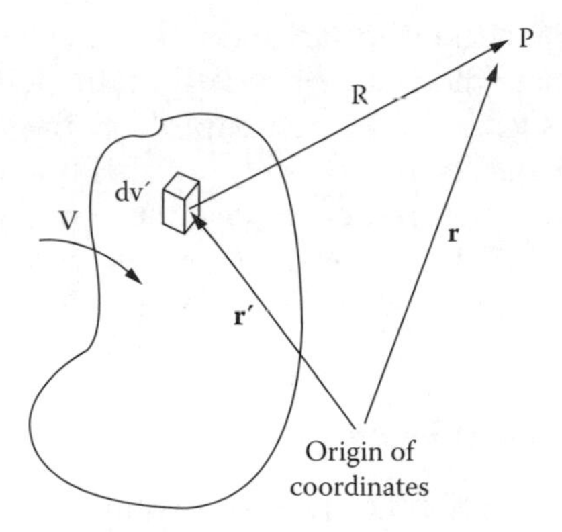

Figure 2.12 Configuration used to calculate potentials due to charge and current distributions.

density over a surface bounded by the circuit loop. In terms of the vector potential, this is equal to the surface integral of $\nabla \times \mathbf{A}$ and therefore from Stake's Theorem

$$L \equiv \frac{\int_c \mathbf{A}\, d\mathbf{l}}{I}$$

where the integral is calculated along a curve coinciding with the circuit loop.

In many problems there is no a priori knowledge of the source distributions $\mathbf{j}$ and ρ, and hence the potentials cannot be determined from Equations 2.40 and 2.41 directly. In such cases, Equations 2.38 and 2.39, subject to the quasistatic approximation (negligible second-time derivatives), must be solved subject to the boundary conditions of the particular problem. This is a difficult task. Analytical solutions to some simple problems are available, and these may be used for EMC studies. It is important, however, to understand the approximations involved, and usually there are many, before using ready-made formulae. For most practical configurations, solutions can only be obtained by numerical simulation methods. This topic will be taken up in more detail in Chapters 6 and 7.

2.3 High-Frequency Fields

In the last two sections, static and quasistatic fields were examined. In both cases, the essence of the approximations made is that whatever changes in charge or current distribution take place, their effects on the system studied are practically instantaneous. When this is not the case, different classes of phenomena are observed. If the period and wavelength of a disturbance are such as to be comparable to the longest propagation time and longest dimension, respectively, of the system studied, then the problem belongs to the high-frequency regime. Two types of phenomena then dominate the response of systems, namely, waves and radiation. These two aspects are examined separately.

2.3.1 Electromagnetic Waves

Wave-like behavior is inherent in systems in which the propagation velocity of disturbances is finite. If the propagation velocity is infinite then any disturbance caused by the source would instantaneously spread throughout the system and the entire system would change in unison. This would represent an oscillation rather than a wave.

To illustrate how waves come about, a simple case of a wave propagating along the x-direction will be considered. The only nonzero electric and magnetic field components are then E_y and H_z. Ampere's and Faraday's Laws in differential form (Table 2.1) reduce to

$$-\frac{\partial H_z}{\partial x} = \varepsilon \frac{\partial E_y}{\partial t}$$

$$\frac{\partial E_y}{\partial x} = -\mu \frac{\partial H_z}{\partial t} \tag{2.42}$$

The conduction current in Ampere's Law was neglected (no losses). Differentiating the first equation with respect to t and the second with respect to x and combining gives

$$\frac{\partial^2 E_y}{\partial x^2} = \mu\varepsilon \frac{\partial^2 E_y}{\partial t^2} \tag{2.43}$$

This equation describes an electromagnetic wave. The solution may be found for harmonic fields of the form

$$E_y = \text{Re}\left\{\bar{E}_y(x)\, e^{j\omega t}\right\}$$

where $\bar{E}_y(x)$ is the phasor describing the electric field component. Substituting in Equation 2.43 gives

$$\frac{d^2\bar{E}_y(x)}{dx^2} = -\omega^2\mu\varepsilon\,\bar{E}_y(x) = -\beta^2\bar{E}_y(x)$$

where β is the phase constant of the medium. The solution to this equation is of the form

$$\bar{E}_y(x) = A\,e^{-j\beta x} + B\,e^{j\beta x}$$

Hence

$$E_y(x,t) = \mathrm{Re}\left\{A\,e^{j(\omega t - \beta x)} + B\,e^{j(\omega t + \beta x)}\right\}$$

$$= A\cos(\omega t - \beta x) + B\cos(\omega t + \beta x)$$

Both terms in this expression represent waves. A point of constant phase on a wave represented by the first term must be such that

$$\omega t - \beta x = \text{constant}$$

Differentiating this expression with respect to t gives

$$\omega - \beta\frac{dx}{dt} = 0$$

or, that the phase velocity is

$$u = \frac{dx}{dt} = \frac{\omega}{\beta} \tag{2.44}$$

Similar reasoning with the second term gives a phase velocity equal to $-\omega/\beta$. Hence, the first term represents a wave propagating along the positive x-direction and the second term represents a wave propagating along the negative x-direction. Attention is now focused on the first wave term alone, as shown below

$$E_y(x,t) = E_o\cos(\omega t - \beta x)$$

An equation identical to Equation 2.43 may be obtained for H_z with a solution

$$H_z(x,t) = H_o \cos(\omega t - \beta x)$$

The magnitudes of the electric and magnetic field components are not independent. Substituting the above two expressions to one of Equation 2.42 gives

$$\frac{E_o}{H_o} = Z_o = \sqrt{\frac{\mu}{\varepsilon}} \tag{2.45}$$

The quantity $Z_0 = \sqrt{\mu/\varepsilon}$ is the intrinsic impedance of the medium and in the case of air it is equal to 377 Ω. It is instructive to rearrange the phase term in the field components as follows:

$$\omega t - \beta x = \omega\left(t - \frac{\beta x}{\omega}\right) = \omega\left(t - \frac{x}{u}\right)$$

Two consecutive points at the same phase on the wave must be a distance λ apart, such that

$$\frac{\omega\lambda}{u} = 2\pi$$

Since $\omega = 2\pi f$, it follows that

$$\lambda = \frac{u}{f} \tag{2.46}$$

The quantity λ is referred to as the wavelength.

Energy transport by the wave may be calculated using Poynting's vector defined as follows:

$$\mathbf{P} = \mathbf{E} \times \mathbf{H}, \text{ in W/m}^2 \tag{2.47}$$

This expression represents an instantaneous value. The power averaged over a period is

$$\mathbf{P}_{ave} = \frac{1}{2}\left(\bar{\mathbf{E}} \times \bar{\mathbf{H}}^*\right)$$

where the asterisk indicates the complex conjugate of the phasor representing the magnetic field intensity.

For the example presented in this section

$$P_z = E_y\ H_z = \frac{E_0}{\sqrt{2}}\frac{H_0}{\sqrt{2}} = \frac{1}{2}\frac{E_0^2}{Z_0},\ \text{in W/m}^2$$

where the $\sqrt{2}$ factor is used to convert peak to rms values.

Losses may be easily accounted for by retaining the conduction term in Ampere's Law ($\mathbf{j}_f = \sigma$ E) and proceeding as in the start of this section to obtain the following expression:

$$\frac{\partial^2 E_y}{\partial x^2} = \mu\varepsilon\frac{\partial^2 E_y}{\partial t^2} + \mu\sigma\frac{\partial E_y}{\partial t} \tag{2.48}$$

The first term on the right-hand side indicates wave-like behavior, whereas the second term is typical of diffusion-like behavior. Which behavior dominates the response of the medium depends on the relative magnitudes of μ, ε, and σ and on the frequency. Assuming harmonic excitation, the wave term is of the order of $\mu\varepsilon\omega^2$, whereas the diffusion term scales as $\mu\sigma\omega$. Therefore, when $\omega\varepsilon >> \sigma$, wave behavior predominates (high frequencies, poor conductors). For low frequencies and good conductors, diffusion becomes the dominant mechanism.

Assuming, as before, harmonic fields, Equation 2.48 reduces to

$$\frac{d^2\bar{E}_y(x)}{dx^2} = j\omega\mu\left(j\omega\varepsilon + \sigma\right)\bar{E}_y\left(x\right) = \gamma^2\bar{E}_y\left(x\right)$$

where

$$\begin{aligned}\gamma &= \sqrt{j\omega\mu\left(\sigma + j\omega\varepsilon\right)}\\ &= j\omega\sqrt{\mu\varepsilon}\left(1 + \frac{\sigma}{j\omega\varepsilon}\right)^{1/2} = \alpha + j\beta\end{aligned} \tag{2.49}$$

is the propagation constant of the medium. As for the lossless case, β is the phase constant but now, in addition, propagation is affected by the attenuation constant α. The wave components for propagation along +x are

$$\bar{E}_y(x) = A\, e^{-\gamma x}$$

$$\bar{H}_z(x) = B\, e^{-\gamma x}$$

The second part of Equation 2.42 reduces for harmonic fields to

$$\frac{\partial \bar{E}_y(x)}{\partial x} = -\, j\omega\mu\, \bar{H}_z(x)$$

Substituting into this equation the expressions for the phasors gives the intrinsic impedance of the medium:

$$Z_0 = \frac{A}{B} = \frac{j\omega\mu}{\gamma} = \sqrt{\frac{\mu}{\varepsilon}}\left(1 + \frac{\sigma}{j\omega\varepsilon}\right)^{-1/2} \tag{2.50}$$

Hence

$$\bar{E}_y(x) = E_0\; e^{-\gamma x}$$

$$\bar{H}_z(x) = \frac{E_0}{Z_0}\, e^{-\gamma x} \tag{2.51}$$

The time variation of the electric field is

$$E_y(x,t) = \mathrm{Re}\,\left[E_0\; e^{-\gamma x} e^{j\omega t}\right] = E_0 e^{-\alpha x} \cos(\omega t - \beta x)$$

Since now Z_0 is a complex number, the electric and magnetic field components are out of phase by an amount dependent on the magnitude of the parameter $\sigma/\omega\varepsilon$.

Two limiting cases may be distinguished. First, there is the case of poor conductors $\sigma << \omega\varepsilon$. Then, $Z_0 \simeq \sqrt{\mu/\varepsilon}$ (since Z_0 appears as a multiplier in Equation 2.51), and using the binomial theorem $\gamma \simeq 0.5\, \sigma$

$\sqrt{\mu/\varepsilon} + j\omega\sqrt{\mu\varepsilon}$ (since γ appears as an exponent). These expressions imply that the wave attenuates without a significant phase shift between the electric and magnetic field components. Second, there is the case of good conductors $\sigma >> \omega\varepsilon$.

Then

$$\gamma \simeq j\omega\sqrt{\mu\varepsilon}\sqrt{\frac{\sigma}{j\omega\varepsilon}} = (1+j)\sqrt{\frac{\omega\mu\sigma}{2}} = \frac{1+j}{\delta} \tag{2.52}$$

The quantity δ is known as the *skin depth* of the material

$$\delta = \sqrt{\frac{2}{\omega\mu\sigma}} \tag{2.53}$$

It has the dimension of length and its physical significance will be explained shortly. Similarly,

$$Z_0 \simeq (1+j)\sqrt{\frac{\omega\mu}{2\sigma}} \tag{2.54}$$

Hence

$$\begin{aligned} E_y(x,t) &= E_0 e^{-x/\delta}\cos(\omega t - x/\delta) \\ H_z(x,t) &= E_0\sqrt{\frac{\sigma}{\omega\mu}}\, e^{-x/\delta}\cos(\omega t - x/\delta - \pi/4) \end{aligned} \tag{2.55}$$

These expressions show that the electric field leads the magnetic field by $\pi/4$ and that both wave components decay by $1/e$ for every skin depth δ that the wave penetrates inside the material. Approximate values of the skin depth for copper are 2 mm at 1 kHz and 100 μm in the megahertz range.

2.3.2 Radiating Systems

It is already established that charge and current distributions on conductors set up around them electric and magnetic fields. These fields were calculated under static or quasistatic conditions. The problem of determining the electromagnetic field around conductors is now considered for the high-frequency case. This means that the physical dimensions considered are comparable to the wavelength and that propagation times are significant when compared to the period of variation of the signals. The potentials are related to the sources by Equations 2.38 and 2.39, but now the second derivatives with respect to time cannot be neglected. The solution to these equations is in the form of Equations 2.40 and 2.41 as for the quasistatic case, but now the source charge and current used to determine ϕ and **A** at time t must be those present at time t – R/u, where R is the distance between the observation and source points, and u is the velocity of propagation. This simply takes account of the fact that the field at the observation point at time t is due to the condition of source charges and currents at the earlier time t – R/u.

The potentials thus calculated are referred to as the retarded potentials and are given by the following expressions

$$\phi(r,t) = \int_v \frac{[\rho]\, dv'}{4\pi\varepsilon R} \tag{2.56}$$

$$\mathbf{A}(r,t) = \int_v \frac{\mu[\mathbf{j}]\, dv'}{4\pi R} \tag{2.57}$$

The brackets around ρ and **j** indicate distribution calculated at the retarded time t – R/u. Hence if $\bar{\rho} = \rho_0\, e^{j\omega t}$, the retarded value is

$$[\bar{\rho}] = \rho_0\, e^{j\omega(t-x/u)} = \rho_0\, e^{j(\omega t-\beta x)}$$

and similarly for [**j**].

Equations 2.56 and 2.57 may be used to evaluate the potentials and hence the EM field due to any specified charge and current distribution. Unfortunately, in many practical situations these distributions are not precisely known, and complex calculations are then necessary, using numerical methods, to obtain solutions to Equations 2.38 and 2.39. However, some simple cases are amenable to calculation by analytical techniques based on reasonable approximations for the distribution of ρ and j. These are studied below, both because they illustrate the basic properties of the field, and also because they can form the first step in understanding the behavior of more complex systems.

Let us now consider a very short piece of wire of length 1 ($1 < \lambda/50$) where a current I is established such that its peak value is constant everywhere on the wire and varies with an angular frequency ω. Charge conservation implies that charges will accumulate at the ends of the wire

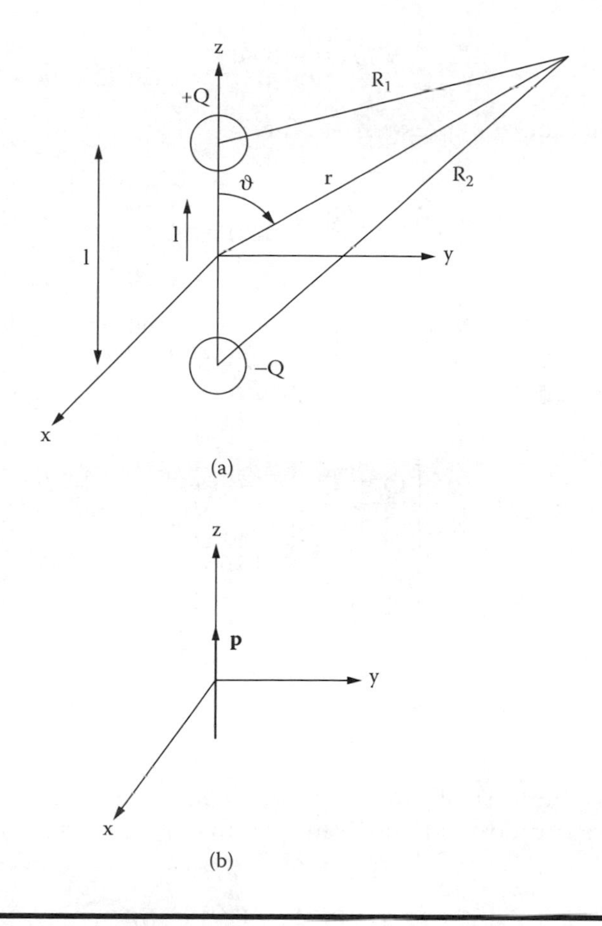

Figure 2.13 A short linear electric dipole (a) and its equivalent electric dipole moment (b).

so that I = dQ/dt. This configuration is shown in Figure 2.13. The current distribution just described is an acceptable approximation provided that the length l is much smaller than the wavelength λ. This structure is then described as a very short (infinitesimal) electric dipole.

The electromagnetic fields surrounding this dipole may be calculated from Equations 2.34 and 2.35. For a harmonic electric field, the following expression is obtained:

$$\bar{\mathbf{E}} = -\nabla\bar{\phi} - j\omega\,\bar{\mathbf{A}} \tag{2.58}$$

Potentials $\bar{\phi}$ and $\bar{\mathbf{A}}$ must be first obtained for the distribution of charges and currents shown in Figure 2.13. The vector and scalar potentials are related by the Lorentz condition (Equation 2.37):

$$\nabla \cdot \bar{\mathbf{A}} = -\mu\varepsilon j\omega\bar{\phi}$$

Hence, an alternative expression for $\bar{\mathbf{E}}$ is

$$\bar{\mathbf{E}} = -j\frac{1}{\mu\omega\varepsilon}\nabla\left(\nabla \cdot \bar{\mathbf{A}}\right) - j\omega\,\bar{\mathbf{A}} \tag{2.59}$$

The electric field may thus be obtained using either of these two expressions (Equations 2.58 and 2.59). For a harmonic variation $\bar{Q} = Q_0\ e^{j\omega t}$, hence from Equation 2.56:

$$\bar{\phi} = \frac{1}{4\pi\varepsilon}\left[\frac{Q_0 e^{j(\omega t - \beta R_1)}}{R_1} - \frac{Q_0 e^{j(\omega t - \beta R_2)}}{R_2}\right]$$

For a short dipole and R >> 1, it follows that

$$R_1 = r - \frac{1}{2}\cos\theta = r - \Delta r \quad \text{and} \quad R_2 = r + \frac{1}{2}\cos\theta = r + \Delta r$$

Since Δr is very small, the expression for $\bar{\phi}$ may be expanded and terms of second order and above in $\Delta r/r$ may be neglected, to give

$$\bar{\phi} \simeq \frac{Q_0 l\cos\theta}{4\pi\varepsilon}\left(\frac{1}{r^2} + j\frac{\beta}{r}\right)e^{j(\omega t - \beta r)} \tag{2.60}$$

It is convenient at this stage to refer to the short dipole shown in Figure 2.13a as an electric dipole of electric dipole moment $p = \hat{a}_z\, p_z = \hat{a}_z\, Q_0\, l$ as shown in Figure 2.13b.

For this example

$$\left[\bar{p}_z\right] = Q_0 l\, e^{j(\omega t - \beta r)} \tag{2.61}$$

Substituting the expression above into Equation 2.60 gives

$$\bar{\phi} \simeq \frac{\left[\bar{p}_z\right]\cos\theta}{4\pi\varepsilon}\left(\frac{1}{r^2} + \frac{j\beta}{r}\right)$$

Similarly, the vector potential may be calculated and is found to be

$$\mathbf{A} = \hat{a}_z\, A_z = \hat{a}_z \frac{\mu}{4\pi r} Il$$

Since I = dQ/dt, it follows that

$$\bar{I}l = j\omega\bar{Q}l$$

and therefore

$$\bar{A}_z = \frac{j\omega\mu}{4\pi r}\left[\bar{p}_z\right]$$

The electric field is obtained by substituting $\bar{\phi}$ and $\bar{\mathbf{A}}$ in Equation 2.58 and is

$$\begin{aligned}
\bar{E}_r &= \frac{\left[\bar{p}_z\right]\cos\theta}{2\pi\varepsilon}\left(\frac{1}{r^3} + j\frac{\beta}{r^2}\right) \\
\bar{E}_\theta &= \frac{\left[\bar{p}_z\right]\sin\theta}{4\pi\varepsilon}\left(\frac{1}{r^3} - \frac{\beta^2}{r} + j\frac{\beta}{r^2}\right) \\
\bar{E}_\phi &= 0
\end{aligned} \tag{2.62}$$

The magnetic field components are obtained from the curl of the vector potential and are

$$\begin{aligned}
\bar{H}_\phi &= \frac{j\omega\left[\bar{p}_z\right]\sin\theta}{4\pi}\left(\frac{1}{r^2} + j\frac{\beta}{r}\right) \\
\bar{H}_r &= \bar{H}_\theta = 0
\end{aligned} \tag{2.63}$$

In these expressions

$$\left[\bar{p}_z\right] = Q_0 l\, e^{j(\omega t - \beta r)} = \frac{I_0 l}{j\omega} e^{j(\omega t - \beta r)}$$

Clearly, as $r \rightarrow \infty$, the only significant terms are

$$\begin{aligned} \bar{E}_\theta &= -\frac{[\bar{p}_z]\ \sin\theta}{4\pi\varepsilon}\frac{\beta^2}{r} = j\frac{I_0 l\beta}{4\pi r} Z_0 \sin\theta\ e^{j(\omega t - \beta r)} \\ \bar{H}_\phi &= -\frac{\omega[\bar{p}_z]\ \sin\theta}{4\pi}\frac{\beta}{r} = j\frac{I_0 l\beta}{4\pi r}\sin\theta\ e^{j(\omega t - \beta r)} \end{aligned} \tag{2.64}$$

The power flux is given by Poynting's vector defined by Equation 2.45 and in the far-field region ($r \rightarrow \infty$) it has a radial component only:

$$\mathbf{P} = \hat{a}_\rho\ E_\theta H_\phi$$

Energy therefore radiates away from the dipole. The power flux is

$$P(r,\theta) = \frac{1}{2}\left(\frac{I_0 l\beta}{4\pi r}\sin\theta\right)^2 Z_0, \text{ in W/m}^2 \tag{2.65}$$

where I_0 is the peak dipole current. The total power radiated by the dipole may be obtained by integrating over a spherical surface surrounding it

$$P_t = \int_{\theta=0}^{\pi} (2\pi r\sin\theta)(r d\theta)\, P(r,\theta) = \frac{\beta^2 I_0^2 l^2 Z_0}{12\pi}, \text{ in W} \tag{2.66}$$

The radiation resistance R_{rad} of the dipole is defined as

$$R_{rad} = \frac{P_t}{\left(I_0/\sqrt{2}\right)^2} \tag{2.67}$$

For the case of air, $Z_0 = 120\pi = 377\ \Omega$, and combining the last two expressions gives

$$R_{rad} = 80\pi^2 (l/\lambda)^2, \text{ in } \Omega \tag{2.68}$$

It should also be noted that in the far-field region, the electric and magnetic field components are related by the intrinsic impedance of the medium — the wave impedance $Z_w = \bar{E}/\bar{H}$ is equal to Z_0. E and H form a transverse EM wave (TEM wave) with respect to the radial propagation direction $\hat{\alpha}_p$.

Let us now return to examine the field for the case when r cannot be regarded as very large, in what is described as the near-field region.

In this region the fields are given by the full set of Equations 2.62 and 2.63. The terms decaying as $1/r^3$ represent electrostatic fields and hence capacitive energy storage. Terms decaying as $1/r^2$ represent a quasistatic field due to a current element and associated energy storage around the dipole. This contribution is described as an inductive near field. The presence of the E_r component shows that in this region not all energy flow is radially out. The dipole does not yet display its far-field radiation properties as regards directivity. Moreover, in the near-field region, the electric and magnetic field components are not related simply by the intrinsic impedance of the medium ($E/H \neq Z_0$). Even if E_r is neglected, it is found that the wave impedance $Z_w = E_\theta/H_\phi \neq Z_0$. To illustrate this, the E_θ and H_ϕ components in Equations 2.62 and 2.63 are expressed in terms of the dimensionless parameter (βr) to give

$$\begin{aligned} \bar{E}_\theta &= j\frac{\beta I_0 l \sin\theta}{4\pi r} Z_0 \left(1 + \frac{1}{j(\beta r)} - \frac{1}{(\beta r)^2}\right) \\ \bar{H}_\phi &= j\frac{\beta I_0 l \sin\theta}{4\pi r} \left(1 + \frac{1}{j(\beta r)}\right) \end{aligned} \tag{2.69}$$

Hence

$$\begin{aligned} Z_\omega &= \frac{\bar{E}_\theta}{\bar{H}_\phi} = Z_0 \frac{1 - j\dfrac{1}{(\beta r)} - \dfrac{1}{(\beta r)^2}}{1 - j\left(\dfrac{1}{\beta r}\right)} \\ &= Z_0 \frac{1 - j\dfrac{1}{(\beta r)^3}}{1 + \dfrac{1}{(\beta r)^2}} \end{aligned} \tag{2.70}$$

The magnitude of the wave impedance is

$$\left|Z_w\right| = Z_0 \frac{\sqrt{1+\dfrac{1}{(\beta r)^6}}}{1+\dfrac{1}{(\beta r)^2}} \tag{2.71}$$

It is clear that Z_w differs from the intrinsic impedance of the medium Z_0 both in magnitude and in phase. Since $\beta r = \omega\sqrt{\mu\varepsilon}r = 2\pi fr/u = 2\pi r/\lambda$, it follows that when $r >> \lambda$ (far-field region), $1/(\beta r) \rightarrow 0$ and $|Z_w| \rightarrow Z_0$ as expected. Very near to the dipole ($r << \lambda$), the wave impedance is $Z_w \simeq Z_0\, \lambda/2\pi r$ and hence much larger than Z_0. This implies that the electric field component is larger than would be expected for a radiation field. At a distance $r = \lambda/2\pi$, the wave impedance has equal real and reactive (capacitive) parts. The reactive part is due to the storage of energy in the near field. This is reasonable since the dipole described operates primarily by separating electric charges and by setting up potential differences. Near this electric dipole, the character of the source is reflected in the EM wave properties. In the far-field region, however, there is nothing in the field properties to identify the character of its source. The wave impedance is important as wave reflection and penetration are both determined by impedance differences along the path of propagation.

Antennas are characterized by their directivity or gain and by their effective aperture. These quantities are defined below and are evaluated for the case of the short electric dipole just described.

The *directive gain* $D(\theta,\phi)$ of an antenna is a measure of how the radiated power is distributed in different directions, and is defined as the ratio of its power flux [Poynting's vector $P(\theta,\phi)$] over the power flux of an isotropic antenna radiating the same total power. For the short electric dipole, $D(r,\theta,\phi)$ may be calculated by dividing Equation 2.60 by the flux of an isotropic dipole radiating the power given by Equation 2.65 uniformly over an area equal to $4\pi r^2$. Hence, for the short electric dipole

$$D(\theta,\phi) = 1.5 \sin^2\theta \tag{2.72}$$

This radiation pattern is shown in Figure 2.14. The *directivity* of an antenna is equal to the maximum value of its directive gain. Hence, the directivity of the dipole considered here is equal to 1.5. The *gain* G of an antenna is equal to its directive gain times its radiation efficiency. In most cases, radiation efficiency is very high and the gain is approximately equal to the directivity. In many applications the gain is expressed in decibels:

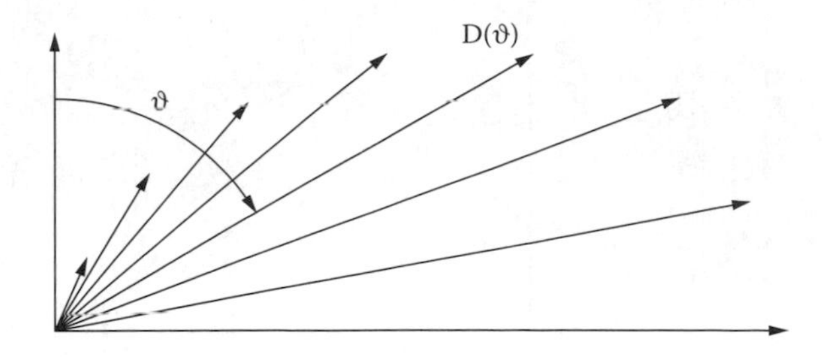

Figure 2.14 The directive gain of a short dipole.

$$G(\mathrm{dB}) = 10\ \log G$$

Hence the gain in dB of the short electric dipole is

$$G(\mathrm{db}) = 10\ \log(1.5) = 1.76\ \mathrm{dB}$$

Another useful parameter is the antenna *effective aperture* A_e, defined as the ratio of the power absorbed by the antenna load to the power density of the incident wave. Assuming that the short dipole is polarized in the same direction as the incident electric field and that the antenna load is equal to its radiation resistance (maximum power transfer conditions) then

$$P_{load} = \left(\frac{V_{pk}}{2\sqrt{2}}\right)^2 \frac{1}{R_{rad}}$$

The peak voltage induced on the antenna is $V_{pk} = 1 \cdot E_{pk}$, where l is the dipole length and E_{pk} the peak value of the incident field. Therefore, $P_{load} = E_{pk}^2\ l^2/8R_{rad}$. The wave power density is equal to $(E_{pk}/\sqrt{2})^2/Z_0$ where Z_0 is the medium impedance. The effective aperture for the short dipole is therefore equal to

$$A_e = \frac{P_{load}}{\left(E_{pk}/\sqrt{2}\right)^2 / Z_0} = \frac{Z_0 l^2}{4\,R_{rad}}$$

Substituting for the radiation resistance from Equation 2.68 gives

$$A_e = \frac{\lambda^2}{4\pi}\frac{3}{2} \tag{2.73}$$

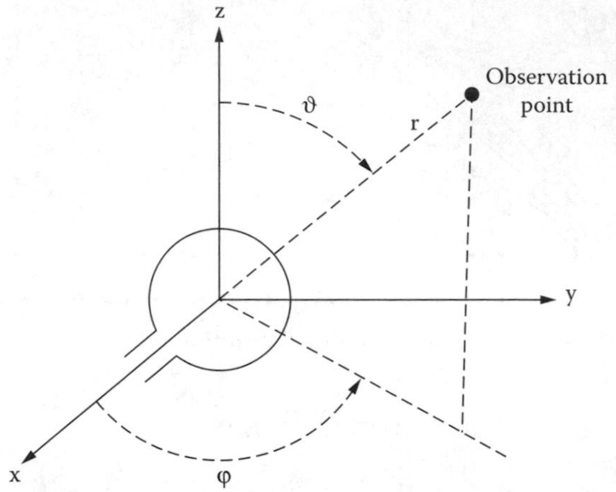

Figure 2.15 Coordinate system for the calculation of fields due to a small loop on the x-y plane.

The EM field established by the small *loop antenna* shown in Figure 2.15 may be similarly obtained and is

$$
\begin{aligned}
&E_r = E_\theta = H_\rho = 0 \\
&\bar{E}_\varphi = \frac{-j\omega\mu\left(\pi a^2\right) I_0\, e^{j(\omega i - \beta r)} \sin\theta}{4\pi}\left(\frac{1}{r^2} + j\frac{\beta}{r}\right) \\
&\bar{H}_r = \frac{\left(\pi a^2\right) I_0\, e^{j(\omega t - \beta r)} \cos\theta}{2\pi}\left(\frac{1}{r^3} + j\frac{\beta}{r^2}\right) \\
&\bar{H}_\theta = \frac{\left(\pi a^2\right) I_0\, e^{j(\omega t - \beta r)} \sin\theta}{4\pi}\left(\frac{1}{r^3} - \frac{\beta^2}{r} + j\frac{\beta}{r^2}\right)
\end{aligned}
\tag{2.74}
$$

In the far-field region (r >>) the only significant components are

$$
\begin{aligned}
&\bar{E}\varphi = \frac{\omega\mu\beta\left(\pi a^2\right) I_0\, e^{j(\omega t - \beta r)} \sin\theta}{4\pi r} \\
&\bar{H}_\theta = \frac{-\beta^2\left(\pi a^2\right) I_0\, e^{j(\omega t - \beta r)} \sin\theta}{4\pi r}
\end{aligned}
\tag{2.75}
$$

The quantity I_0 (πa^2) is called the magnetic moment m of the loop, which in this example is directed along the z-axis. Thus, the equations above show the radiated field due to a magnetic dipole. These are the dual of those obtained for the electric dipole (Equations 2.62 to 2.64). The magnetic dipole equations are obtained from those for the electric dipole by replacing E by $-H/\varepsilon$ and H by E/μ.

The radiation resistance of the small magnctic dipole is given by the formula

$$R_{rad} = Z_0 \left(\frac{2\pi}{3}\right)\left(\frac{\beta}{\lambda}\right)^2 \left(\pi a^2\right)^2, \text{ in } \Omega \tag{2.76}$$

If the loop has N turns, then the radiation resistance in Equation 2.76 must be multiplied by N^2.

The same comments for near and far fields apply as for the electric dipole. As before, the wave impedance near the loop is not equal to the medium impedance Z_0. Expressing as before $\bar{E}_\phi$ and $\bar{H}_\theta$ as functions of the parameter βr gives

$$\begin{aligned} \bar{E}_\phi &= \frac{-j\omega\mu[m]\sin\theta}{4\pi r}\beta\left(\frac{1}{(\beta r)}+j\right) \\ \bar{H}_\theta &= \frac{[m]\sin\theta}{4\pi r}\beta^2\left[-1+\frac{1}{(\beta r)^2}+j\frac{1}{(\beta r)}\right] \end{aligned} \tag{2.77}$$

Hence

$$Z_w = \frac{\bar{E}_\varphi}{H_\theta} = Z_0 \frac{1-j\dfrac{1}{(\beta r)}}{\left(-1+\dfrac{1}{(\beta r)^2}\right)+j\left(\dfrac{1}{\beta r}\right)} \tag{2.78}$$

Thus, Z_w differs from Z_0 both in magnitude and phase. In the limit as $r \gg \lambda$ it is found that $|Z_w| \to Z_0$. In the near field, $r \ll \lambda$, $|Z_w| \to Z_0\, 2\pi r/\lambda$. Thus, in the near field the wave impedance is much smaller than Z_0. This is not surprising, since near the antenna the character of the field reflects the manner of its production (i.e., the magnetic field is predominant). In the far region, however, its character is independent of the source.

Example: Calculate how much power is needed to drive enough current through a short wire length to exceed Class A limits at 30 m. We assume that the frequency is f = 1 GHz and that the length of the wire is $\lambda/30$ at this frequency.

A possible practical configuration is of monopole as shown in Figure 2.16a. We exploit the formulae already derived for a short dipole with a few adjustments. Electromagnetically a short monopole of length ℓ is equivalent to a dipole in free space of length 2ℓ (method of images). Hence the total radiated power is obtained from Equation 2.66 by setting the length to 2ℓ and multiplying by ½ since the monopole radiates only in half space. Equation 2.64 is similarly adjusted by changing the length to 2ℓ to obtain the magnitude of the electric field. We have therefore for a monopole of length ℓ

$$|E_\vartheta| = \left(\frac{I_0(2\ell)\beta}{4\pi r}\right) Z_0 \sin\vartheta, \quad W/m^2 \tag{2.79}$$

$$P_t = \frac{1}{2}\frac{\beta^2 I_0^2 (2\ell)^2 Z_0}{12\pi}, \quad W \tag{2.80}$$

The maximum electric field is obtained for an angle of 90 degrees. We have in free space $Z_0 = 377\ \Omega$ and $\beta = \omega/c = 2\pi f/c = 20.94$ rad/m. The Class A limit is 37 dBμV/m, which in V/m is equivalent to 70.8×10^{-6} V/m. Substituting in Equation 2.79 we have that the required current to obtain this field is obtained from

$$70.8 \times 10^{-6} = \frac{I_0 \times 10^{-2} \times 20.94}{4\pi 30} 377 \sin(90^\circ)$$

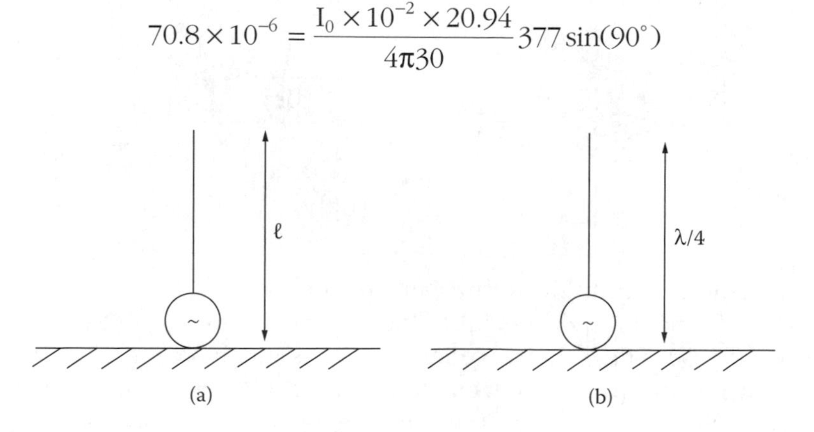

Figure 2.16 Monopole antenna configurations.

Hence, a current of $I_0 = 337$ μA is sufficient to exceed the limits. Substituting this current into Equation 2.80, the total radiated power can be calculated:

$$P_t = \frac{1}{2}\frac{(20.94)^2(337\times 10^{-6})^2(0.01)^2 377}{12\pi} \approx 25\text{nW}$$

In conclusion, a short piece of wire 0.5 cm in length carrying 337 μA at 1 GHz and radiating in total 25 nW radiates enough to exceed the Class A limit at 30 m distance.

Example: Radiation from a half-wavelength dipole.

A useful example is the calculation of the maximum radiated electric field at a frequency such that the dipole length is equal to half the wavelength. This case represents an efficient radiator, which is used intentionally to construct antennas. From the EMC point of view, however, if a particular object meets this criterion, significant (size comparable to half wavelength), strong emissions and hence EMI problems may result. The starting point for making a quantitative assessment is the formulae for the far-field radiation from the half-wavelength dipole, which are given below:[8]

$$E_\vartheta = \frac{j60I_0}{r}e^{-j\beta r}\frac{\cos\left[\frac{\pi}{2}\cos\vartheta\right]}{\sin\vartheta}$$
$$H_\varphi = \frac{jI_0}{2\pi r}e^{-j\beta r}\frac{\cos\left[\frac{\pi}{2}\cos\vartheta\right]}{\sin\vartheta} \tag{2.81}$$

The time-averaged Poynting's vector is then

$$P_{ave} = \frac{1}{2}E_\vartheta H_\varphi^* = \frac{15\times I_0^2}{\pi r^2}\left[\frac{\cos\left[\frac{\pi}{2}\cos\vartheta\right]}{\sin\vartheta}\right]^2 \tag{2.82}$$

and the total radiated power is obtained by integrating Equation 2.82 over a spherical surface of radius r:

$$P_r = \int_0^{2\pi}\int_0^{\pi} P_{ave} r^2 \sin\vartheta d\vartheta d\varphi = 36.54 I_0^2, \quad \text{W} \tag{2.83}$$

The radiation resistance is therefore

$$R_r = \frac{2P_r}{I_0^2} = 73.1\Omega \tag{2.84}$$

The radiation intensity U (power radiated in a particular direction per solid angle) has a maximum value at 90°. Hence, the directivity D of this antenna and its gain G (assuming no losses) are

$$U_{max} = r^2 P_{ave}\Big|_{\vartheta=90^\circ} = 15\frac{I_0^2}{\pi}$$

$$D = \frac{4\pi U_{max}}{P_r} = 1.64 \tag{2.85}$$

$$G = 10\log 1.64 = 2.15 \text{ dB}$$

The half-power beam-width is obtained from Equation 2.82 by setting

$$\frac{\cos\left[\dfrac{\pi}{2}\cos\vartheta\right]}{\sin\vartheta} = \frac{1}{\sqrt{2}}$$

Hence, the beam-width is 78°. These are standard results from antenna theory. Let us now look at the EMI implications. As for the case depicted in Figure 2.16a, the configuration is identical but the length is $\lambda/4$ as shown in Figure 2.16b. By the methods of images the quarter wavelength monopole is equivalent to a half-wavelength dipole. Since the monopole radiates only in half space, the total radiated power is half that in Equation 2.83 and hence the radiation resistance of the monopole is equal to 36.54Ω. The maximum electric field is at 90° and its magnitude from Equation 2.81 is $|E_\vartheta| = 60I_0/r$. Equating this with the Class A limit calculated earlier (70.8×10^{-6} V/m) we obtain the current at the feed point $I_0 = 35.4$ μA. Under these circumstances, the total radiated power is $P_{rad} = (I_0/\sqrt{2})^2 R_r = (35.4)^2 36.54/2 = 22.9$ nW. A quarter wavelength at 1 GHz corresponds to monopole length equal to 7.5 cm. The conclusion, therefore, is that a 7.5-cm-long wire carrying a current of 35.4 μA at 1 GHz generates sufficient radiation to exceed the Class A limit at 30 m.

References

1. Krauss, J D, "Electromagnetics," Fourth Edition, McGraw-Hill, NY, 1992.
2. Zahn, M, "Electromagnetic Field Theory: a Problem Solving Approach," Krieger Publ., Malabar, Florida, 1987.
3. Ramo, S, Whinnery, J R, and Van Duzer, T, "Fields and Waves in Communication Electronics", Second Edition, John Wiley, NY, 1984.
4. Haus, H A and Melcher, J R, "Electromagnetic Fields and Energy," Prentice Hall, 1989.
5. Christopoulos, C, "An Introduction to Applied Electromagnetism," John Wiley, 1990.
6. Skilling, H H, "Fundamentals of Electric Waves," Krieger Publ, Malabar, Florida, 1974.
7. Lee, K S H (editor), "EMP Handbook — Principles, Techniques and Reference Data," Report AFWL-TR-80-402, Dec 1980.
8. Balanis, C A, "Antenna Theory — Analysis and Design," Harper and Row, NY, 1982.
9. Balanis, C A, "Advanced Engineering Electromagnetics," John Wiley, NY, 1989.

Chapter 3

Electrical Circuit Components

In trying to interpret the world as it is, humans endeavor to construct *models* that in some way approximate the behavior of real things. These models are then manipulated by analytical or computational means to predict and interpret the behavior of more complex systems. In electrical work, a very powerful set of models used for this purpose consists of what are described as electrical circuit components. It goes without saying that any model is inherently an approximation to the real thing. A skillful investigator selects the simplest possible model consistent with acceptable accuracy for the type of problem under study. The purpose of this chapter is to outline the basic models used in electrical circuit work and describe their advantages and limitations. It would perhaps have been justifiable to start from general models and reduce to more basic ones. Since, however, most readers are more familiar with the basic models, these are chosen to form the starting point of this treatment. Their relationship to more general models will be established gradually in subsequent sections.

The basic classification of circuit components is divided into lumped and distributed. These are examined in detail in the sections that follow.

3.1 Lumped Circuit Components

Electrical phenomena require for their description models to represent energy dissipation (losses) and energy supply and storage. From the discussion in Chapter 2 it is evident that, in general, two modes of energy storage are required, namely, storage associated with the electric and magnetic fields. In real situations, all energy supply, dissipation, and storage processes are mixed together and distributed over finite parts of space. The lumped circuit representation is based on the approximation that it is acceptable to assign each of these processes to individual components, concentrated virtually at a point (lumped) in space. Thus, there is a lumped circuit component with the sole function of representing energy storage in the electric field and so on. These basic circuit components are examined below.

3.1.1 Ideal Lumped Components

Models of complex systems may be constructed using a small number of ideal basic components. The term ideal should be understood as meaning that such components cannot actually be constructed in practice. They are idealizations of real things.

There is an ideal energy dissipation component known as a *resistor*. The value of voltage and current across a resistor R is given by the following expressions in the time and frequency domains:

$$\begin{aligned} v(t) &= i(t)R \\ \bar{V} &= \bar{I}\,R \end{aligned} \tag{3.1}$$

In obtaining Equation 3.1 it was assumed that positive reference current direction is from the higher to the lower potential across the component.

There is an ideal energy storage component, to represent storage associated with electric fields, described as a *capacitor*. The voltage and current across a capacitor C, using the same conventions as for Equation 3.1, are related as follows:

$$\begin{aligned} i(t) &= C\frac{dv(t)}{dt} \\ \bar{V} &= \bar{I}\frac{1}{j\omega C} \end{aligned} \tag{3.2}$$

Since voltage and current are out of phase by $\pi/2$, there is no net exchange of energy with this component — it simply deals in borrowing energy from and lending energy to the rest of the network.

A similar ideal component exists, described as an *inductor*, to represent energy storage in the magnetic field. The voltage-current relationships across an inductance L are

$$v(t) = L\frac{di(t)}{dt}$$
$$\bar{V} = \bar{I}\, j\omega L \tag{3.3}$$

The same comments as regards energy exchange apply for inductors as for capacitors. To this list a component representing magnetic coupling between circuits may be added (mutual inductance). Similarly, voltage and current sources are required to represent energy supply. A voltage source is a device that can supply any current, without any change in voltage across its terminals. Clearly, in practice, this can only be approximately correct over a limited range of currents. Hence, a voltage source is an idealization. It may appear that ideal components are of limited use when studying real problems. This, however, is not the case. There are many situations in which these idealizations are good enough in gaining an adequate understanding of real systems. The art of modeling is not to produce the most comprehensive description of a component, but merely to produce one that is good enough for the intended application. The simplicity of ideal components makes them a powerful tool in reasoning about complex problems. More importantly, they may be combined to approximate to a very high degree of accuracy the behavior of real components. This is described in more detail in the next section.

3.1.2 Real Lumped Components

In many cases, the behavior of real components is difficult to evaluate and quantify accurately. Thus, emphasis will be placed in this section on identifying the circumstances in which real behavior deviates substantially from the ideal. Trends then will be established and quantitative results given whenever possible. Each component is examined separately below.

Resistors — The question posed here is how a resistor deviates from its ideal behavior as a dissipative element. This question may be answered by considering the construction of a resistor. Taking a wire-wound resistor with short leads as an example, it is straightforward to identify behavior other than that which could be described as dissipation. Since current flows through the leads and the dissipative element, it follows that a magnetic field will be established and, thus, any equivalent circuit of this component must contain an inductance associated with the leads (L_{lead})

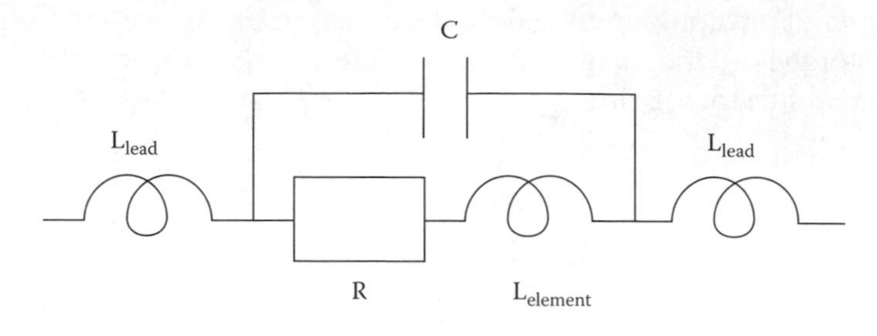

Figure 3.1 Equivalent circuit of a real resistor.

and the resistive element ($L_{element}$), as shown in Figure 3.1. Potential differences between different points of such a structure are unavoidable, and hence an electric field will be established around the component. This in turn implies that a capacitance C must be associated with the resistor R as shown in Figure 3.1. All components except R may be regarded as stray components. The way that the stray components have been introduced in Figure 3.1 is somewhat arbitrary. Other configurations are possible that can adequately represent the real resistor. At low frequencies, the reactance of the inductive elements is very low; hence, since they are placed in series with R, their contribution is negligible. Similarly, the reactance of the capacitor is high and since it appears in parallel with R, its effects are negligible. Thus, at low frequencies the influence of the stray components is minimal and the component behaves, approximately, as an ideal resistor. At high frequencies, however, the reactance of the stray components dominates the behavior of the real component. Resonant behavior is also possible due to the stray inductance and capacitance. It should also be borne in mind that the values of the stray components are in general frequency dependent.

Although an exact calculation of these values is a major task, it should be anticipated that, even for a simple piece of wire, capacitances of the order of a picofarad and inductances of the order of a few nanohenry per centimeter length of wire can appear as stray components. Thus, for frequencies exceeding a few megahertz, simple resistors used in circuit construction start to exhibit behavior influenced by stray components. In EMC work, where circuits and components have to meet a demanding specification over a wide frequency range, normally exceeding the normal operating range, the distribution and magnitude of stray components is of paramount interest. It should also be noted that even the value of the resistance R cannot be regarded as constant independent of frequency. It was established in Chapter 2 that current and magnetic field distribute nonuniformly over the cross section of a conductor. It was found that current concentrates near the surfaces over a thickness of a few skin depths δ (Equation 2.53), where

$$\delta = \sqrt{\frac{1}{\pi f \mu \sigma}}$$

In copper ($\varepsilon_r = \mu_r = 1$, $\sigma = 5.8 \times 10^7$ S/m) the skin depth at several frequencies is shown below.

f	δ
50 Hz	9.3 mm
60 Hz	8.5 mm
1 kHz	2.09 mm
1 MHz	66 μm
1 GHz	2.1 μm

If the skin depth is much larger than the conductor radius r, then the resistance per unit length is

$$R = \frac{1}{\sigma \pi r^2}$$

If, however, $\delta << r$, then only a small layer near the surface is available for conduction. Hence, approximately, the resistance per unit length is

$$R = \frac{1}{\sigma 2\pi r \delta} \tag{3.4}$$

It follows from this expression that the resistance increases as the square root of the frequency. For conductors with a noncircular cross section, the same principles apply and Equation 3.4 may be used with suitable modifications. Another factor that affects the distribution of current is the proximity of other conductors or components. Any change to the distribution of current in a conductor cross section due to proximity effects will inevitably change the resistance and inductance associated with the conductor.

Capacitors — Capacitors exhibit nonideal behavior for similar reasons as described above. A large value electrolytic capacitor cannot be regarded as being lossless and the capacitor leads contribute their own losses and stray inductance and capacitance, as shown in Figure 3.2. This circuit may look unnecessarily complicated, but it brings out the various processes

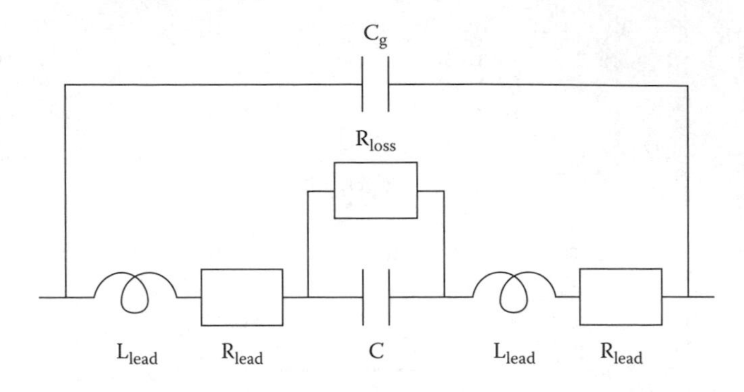

Figure 3.2 Equivalent circuit of a real capacitor.

that affect the high-frequency behavior of a capacitor. The lead inductance and resistance are not dissimilar to those associated with a resistive component of similar physical dimensions. Similarly, the stray capacitance C_s is that associated with the electric field established between the capacitor leads. The loss resistance is due to dissipative mechanisms inside the dielectric. If low-loss dielectrics are used, R_{loss} is very large and can normally be neglected. Its value is related to the loss tangent of the dielectric by the expression

$$R_{loss} = \frac{1}{\omega\, C \tan \delta} \tag{3.5}$$

The combined effect of all these factors is to change the character of the impedance presented by a real capacitor. While at low frequencies the impedance decreases with increasing frequency as expected; this situation gradually reverses and, beyond a frequency described as the resonant frequency of the capacitor, the impedance increases with frequency. This indicates inductive rather than capacitive behavior. As expected, the value of the resonant frequency depends, among other factors, on the length of the leads. Common-type capacitors have resonant frequencies of the order of a few megahertz. Lead-through capacitors approach a few tens of megahertz. Typical behavior for a range of capacitors is shown in Figure 3.3, showing how the nature of a nominally capacitive component alters at modest high frequencies. Surface mount capacitors, with much shorter leads, have resonance frequencies that are much higher, but nevertheless the same limitations apply. Some manufacturers of capacitors give an equivalent series inductance (ESL) to account partially for the complex real capacitor behavior at high frequencies.

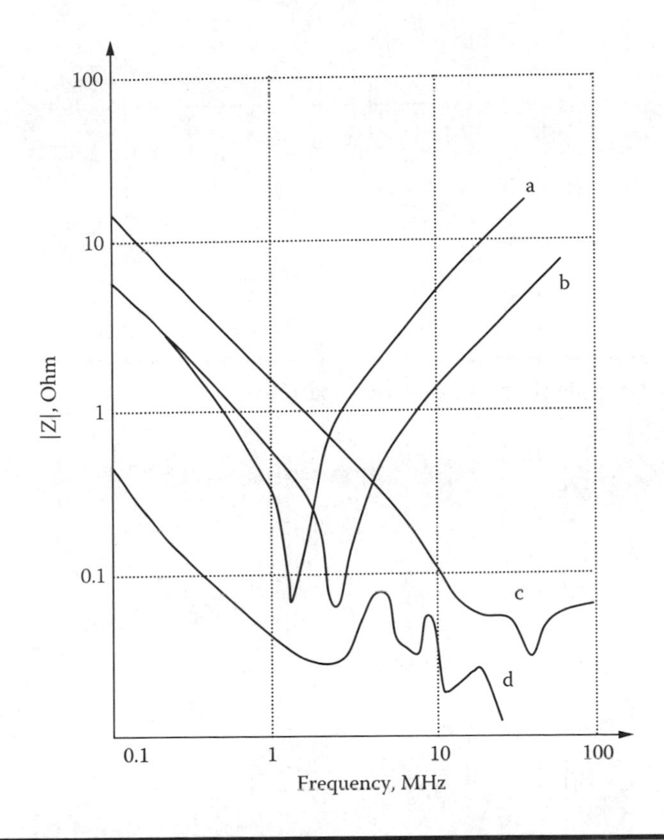

Figure 3.3 Magnitude of capacitor impedance versus frequency: (a) 0.25 μF, 10-cm leads; (b) 0.25 μF, 2-cm leads; (c) 0.1 μF, feed-through capacitor; (d) 4 μF, bushing capacitor.

It is important to ascertain carefully the high-frequency performance of capacitors used in EMC applications.

Inductors — Inductors are perhaps the most difficult component to design. A possible equivalent circuit is shown in Figure 3.4. Resistor R representing losses is usually significant, especially for inductors wound on magnetic cores. It represents several processes, i.e., ohmic losses in the wire and leads, eddy, and hysteresis losses. The latter normally predominate. At low frequencies the impedance increases with frequency. However, beyond a frequency referred to as the resonant frequency of the inductor, the impedance decreases with increasing frequency. This behavior is characteristic of a capacitor rather than an inductor. Typical behavior for inductors is depicted in Figure 3.5. For high-value inductances, the resonant frequency can be quite low (less than 1 MHz). Inductors of a few microhenry in value have resonant frequencies in the 100 MHz range. The actual value of the inductance depends on the

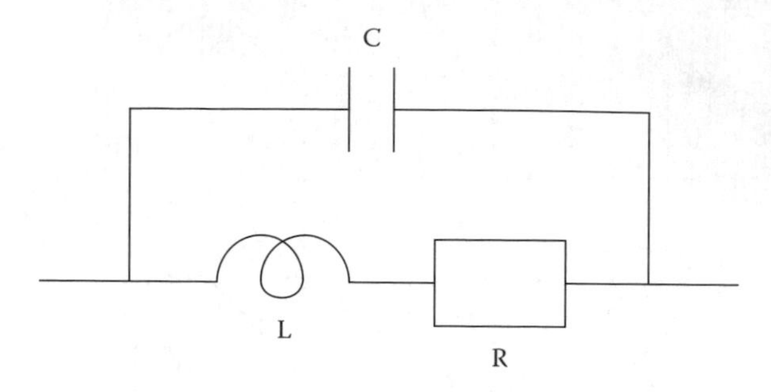

Figure 3.4 Equivalent circuit of a real inductor.

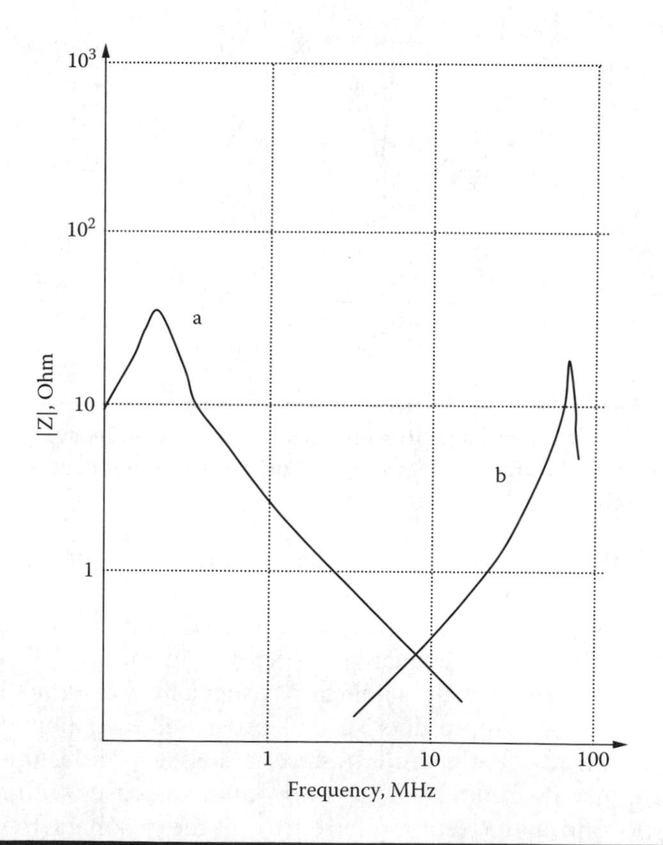

Figure 3.5 Magnitude of inductor impedance versus frequency: (a) 10 mH; (b) 6 μH.

distribution of current. This can be seen even in the simple case of the inductance of a wire. From the discussion leading to the derivation of Equation 2.28 for the internal inductance, it can be seen that it was based

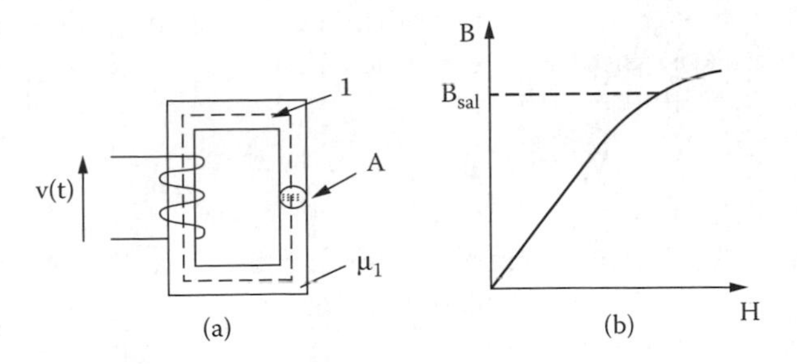

Figure 3.6 A nonlinear inductor (a) and the B-H characteristic of the iron (b).

on the assumption that the current was distributed uniformly on the wire cross section. This is only one contribution to the total inductance, but it illustrates the range of factors that influence inductance values at different frequencies.

Another factor that cannot normally be ignored in inductance calculations is the influence of nonlinear effects. In cases where wires are wound on magnetic cores, or are in the proximity of magnetic materials, nonlinear effects may significantly affect inductance values. This may best be illustrated with reference to the simple system shown in Figure 3.6. A wire of N turns is wound on an iron former as shown. Actual magnetic cores may be different in shape from that shown. In any case, an effective cross section A and magnetic length 1 may be defined, so that the analysis presented here is of general value in understanding how the inductance of the coil depends on the parameters of the magnetic circuit.

Two fundamental laws may be applied to this circuit to obtain an expression for the coil inductance and a relationship between the applied voltage and the magnetic flux density in the magnetic core. Applying Ampere's Law on a curve coinciding with a mean magnetic line gives:

$$H1 = NI \tag{3.6}$$

Hence, $\phi = B\,A = \mu_r\,\mu_0 \dfrac{NI}{1}$ A. The inductance of the coil is thus

$$L = \frac{N\phi}{I} = \mu_r \mu_0 \frac{N^2 A}{1} \tag{3.7}$$

Applying Faraday's Law gives

$$v(t) = N\frac{d\phi(t)}{dt}$$

Assuming that v(t) represents a harmonic voltage of frequency ω and peak value V_{pk}, the expression above implies that

$$V_{pk} = N\ \omega\ \phi_{pk} = N\ \omega\ A\ B_{pk} \tag{3.8}$$

From Equation 3.8 it is clear that the peak flux density in the core depends on the peak applied voltage and its frequency. Magnetic materials have a B–H characteristic of the shape shown in Figure 3.6b. If the flux density exceeds the value B_{sat}, the magnetic field intensity increases disproportionately. This has two implications. First, it means that a much higher current will be drawn by the coil (Equation 3.6) with undesirable effects on the power supply and circuits feeding the coil. Second, if $B_{pk} > B_{sat}$ the magnetic permeability of the iron will decrease substantially. This is clear from Figure 3.6b and the relationship $\mu_r\ \mu_0 = \Delta B/\Delta H$. In turn, the coil inductance decreases substantially. Equation 3.8 offers guidance as to what is likely to happen if the frequency is much lower than that originally intended for the coil. Assuming that the peak voltage remains the same, the magnetic core will operate at a much higher peak magnetic flux density. If this value exceeds B_{sat}, the coil inductance will be much lower than expected.

It is therefore necessary to examine carefully the interplay of the various factors mentioned above when a system is evaluated under conditions (such as frequency, peak voltage, pulse duration) that are different to those applicable under normal operating conditions. This is important in many EMC assessments.

The development of realistic models for real components is thus a nontrivial task that must be approached with care. In most cases, simple equivalents can give acceptable semiquantitative results. Further details may be found in References 1, 2, and 3.

3.2 Distributed Circuit Components

In real physical systems energy storage and dissipation are mixed together and distributed over relatively large areas. It is therefore necessary to investigate under what circumstances it is permissible to separate resistive, capacitive, and inductive behavior and model it by lumped components. It turns out that in answering this question, the relationship between the physical size of the system and the wavelength of the disturbance to be studied is crucial. Let the size of the system be D and the wavelength of the disturbance be λ. In cases where disturbances of several wavelengths are present, then the shortest wavelength is compared with D. If $D \ll \lambda$, then conditions across the network are established practically instantaneously — the propagation time of disturbances is negligible compared to their period. In such cases, it is a good approximation to lump together various components to model circuit behavior in the most acceptable way. If, however, $D \geq \lambda$, propagation effects are dominant and a lumped component representation fails to represent accurately a whole class of phenomena observed under these circumstances. It is still possible to take small sections of the system, of a size that is much smaller than λ, and thus study these subsections using a lumped parameter representation. However, when examining the behavior of the entire system, a new way of reasoning is necessary to take full account of propagation effects. The purpose of this section is to develop an understanding of, and techniques for, the solution of distributed networks, i.e., networks for which $D \geq \lambda$.

Let us start first with a few examples to illustrate the circumstances under which a distributed model is necessary. A high-voltage power transmission line of length D = 100 km needs to be modeled (Figure 3.7a). A typical model used for this purpose is a π-equivalent as shown in Figure 3.7b. The inductance L is equal to the inductance per unit length of the line times the line length. The capacitance C is calculated similarly. Under what circumstances is this model an acceptable representation of the line? Consider first the study of power signals (50 or 60 Hz). The wavelength for a 50-Hz signal is $\lambda = 3 \times 10^8/50 = 6 \times 10^6$ m. This is much larger than the length of the line and hence the model is quite adequate. Let us now examine whether this model is suitable to study the propagation of lightning-induced transients on this line. These are, typically, unidirectional pulses with a rise-time of 1 μsec. Such a pulse contains a rich spectrum of frequencies, but a substantial amount of energy will be associated with frequencies of the order of $1/\tau_r$, where τ_r is the rise-time of the pulse. For this example, it is reasonable to examine whether the model is adequate for the study of frequencies of the order of $1/\tau_r \simeq 1$ MHz. The

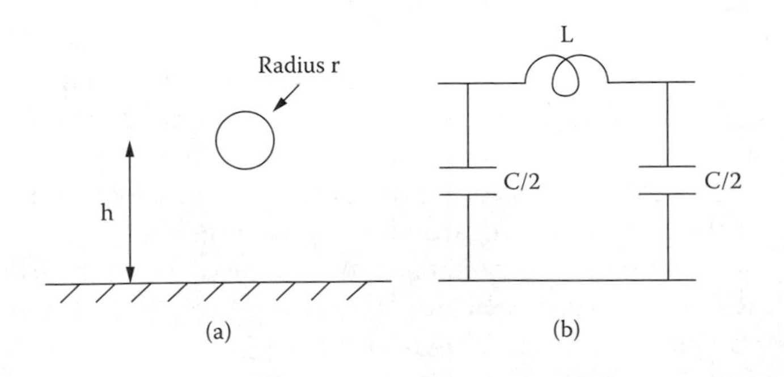

Figure 3.7 A wire-above-ground transmission line (a) and its -equivalent (b).

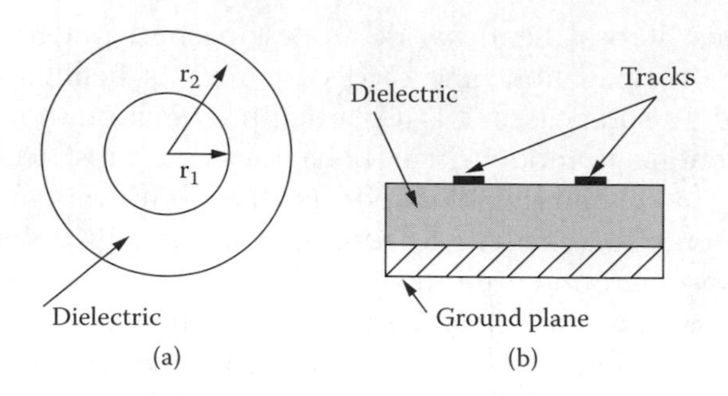

Figure 3.8 A coaxial line (a) and a line consisting of two parallel tracks above a ground plane (b).

wavelength at this frequency is $\lambda = 3 \times 10^8/10^6 = 300$ m. This is smaller than the length of the line, hence the model in Figure 3.7b is not suitable for studying the propagation of this pulse.

A coaxial cable (Figure 3.8a) 1.5 m in length is used for the transmission of a 100-MHz signal. The cable length can be represented, at low frequencies, by a circuit of the types shown in Figure 3.7b. The wavelength at 100 MHz is $\lambda = 2/3 \times 3 \times 10^8/10^8 = 2$ m where the velocity of propagation in the cable was taken to be equal to 2/3 of the speed of light in air. Since the wavelength is comparable to the cable length, the model shown in Figure 3.7b is not suitable for studying this problem.

Finally, let us consider two parallel tracks on a printed circuit board (Figure 3.8b) connecting electronic equipment placed 30 cm apart. Signals with a rise-time of 5 nsec are transmitted along the transmission line formed by these tracks. Treating this problem in the same way as for the lightning pulse and the power line, and taking as the propagation velocity 0.6 times the speed of light in air gives $\lambda = 90$ cm. Hence, this line cannot be represented by a circuit of the type shown in Figure 3.7b.

These examples illustrate that each physical system may be modeled in different ways depending on the type of problem considered. It is the task of the investigator to select, in each case, the appropriate model and to obtain solutions. Models of distributed systems and their solutions are presented in subsequent sections. The examples given, and the analyses that follow, refer to transmission lines. There are, however, other distributed circuits that cannot be regarded as transmission lines, and yet their behavior and analysis are very similar. An example is a transformer coil, which at low frequencies may be adequately represented by an inductance or a more complete circuit as shown in Figure 3.4. However, when steep-fronted pulses are applied to the coil, it behaves very much like a transmission line and may be studied using techniques very similar to those described below.

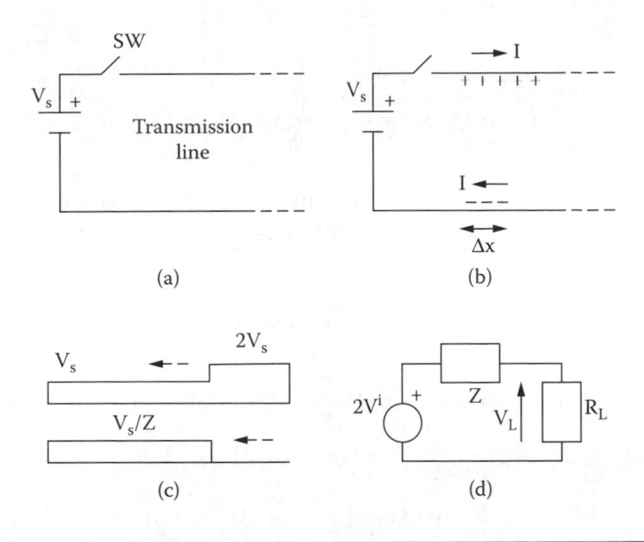

Figure 3.9 A parallel-wire transmission line (a), conditions after time Δt (b), voltage and current distributions after reflection from the load end (c), and the Thevenin equivalent circuit at the load (d).

3.2.1 Time-Domain Analysis of Transmission Lines

Let us consider the parallel-wire transmission line shown schematically in Figure 3.9a. The connection of a voltage source on this line will result in the separation and movement of charges and thus in the establishment of electric and magnetic fields. Energy storage in the field may be represented by capacitance and inductance per unit length, C_d and L_d, respectively. These may be calculated if the dimensions and the medium surrounding the line are known (Appendix B). The termination of the line at the far end is not specified. The following question is posed: Immediately after switch SW is closed, what current will flow into the line? More specifically, is this current affected immediately by the actual line termination?

The answer to these questions can only be given after a systematic examination of conditions at the start of the line has been made following the closure of the switch. The source will transfer electric charge so that the top wire is left with a surplus of positive charge and the bottom wire with a surplus of negative charge. This situation is depicted in Figure 3.9b, where it has been assumed that over a time interval Δt this disturbance has progressed into the line a distance Δx. The charge induced on each wire is

$$\Delta Q = \left(C_d \,\Delta x\right) V_s$$

and hence the current is

$$I = \Delta Q/\Delta t = C_d V_s (\Delta x/\Delta t) = C_d V_s u \tag{3.9}$$

where u is the velocity of propagation of the disturbance.

The magnetic flux linked with this portion of the line is

$$\Delta\phi = (L_d \Delta x) I$$

The voltage V_s must be related to the magnetic flux by Faraday's Law, hence

$$V_s = \Delta\phi/\Delta t = L_d I (\Delta x/\Delta t) = L_d I u \tag{3.10}$$

Combining Equations 3.9 and 3.10 gives

$$V_s = L_d C_d V_s u^2$$

or that the *velocity of propagation* of the disturbance on the line is

$$u = \frac{1}{\sqrt{L_d C_d}} \tag{3.11}$$

Similarly, by substituting from Equation 3.11 into Equation 3.10, the current is found to be related to the voltage by the expression

$$I_s = V_s \Big/ \sqrt{L_d/C_d} = V_s/Z \tag{3.12}$$

where Z is known as the *characteristic or surge impedance* of the line. Hence, the current flowing initially into the line is determined by its characteristic impedance (and source impedance if it is not zero). Only after the disturbance has reached the line termination, has been suitably modified, and traveled back to the source, will the influence of the termination be felt by the source.

Let us now follow this process for a little longer, assuming that the line length is 1 and that its termination is an open circuit. After time $\tau = 1/u$ (the transit time of the line) the voltage and current on the line will be V_s and I_s, respectively. However, this situation cannot persist any longer, since the open-circuit termination demands zero current. The only way to meet this constraint is to assume that a voltage pulse V' will be reflected back with an associated current equal to $-V'/Z$ so that the total current on the line is forced to zero. This demands that $V' = V_s$, i.e., that following the arrival of pulse V_s at the termination the voltage doubles in magnitude and that this disturbance travels toward the source. The current is simultaneously forced to zero, as shown in Figure 3.9c. As this disturbance reaches the source, further reflections will take place, and the process will continue until damped by losses that inevitably are present in real systems.

It is useful to ascertain exactly what reflections take place when a termination other than an open circuit is present. This can be done easily by replacing the line, as seen from its termination, by its Thevenin equivalent circuit. It is assumed that a voltage V^i is traveling along the line and is incident on the termination. In the previous discussion $V^i = V_s$, but the incident voltage would have a different value if the source had a finite impedance, etc. Clearly, the Thevenin equivalent circuit consists of a voltage source equal to the open-circuit voltage of the line $2V^i$ as shown earlier, in series with an impedance Z equal to the characteristic impedance of the line. This circuit is shown in Figure 3.9d connected to the load resistance R_L. It is obvious from Figure 3.9d that the load voltage is

$$V_L = \frac{2R}{R_L + Z} V^i \tag{3.13}$$

The voltage anywhere on the line is equal to the sum of the incident V^i and reflected V^r voltages. Hence, at the load end

$$V^r = V_L - V^i = \left(\frac{R_L - Z}{R_L + Z} \right) V^i \tag{3.14}$$

The quantity in the brackets is known as the *load reflection coefficient.*

$$\Gamma_L = \frac{R_L - Z}{R_L + Z} \tag{3.15}$$

As expected, when the load is an open circuit $R_L \to \infty$, the reflection coefficient is equal to one. For a short circuit ($R_L = 0$), it follows from Equation 3.15 that $\Gamma_L = -1$. For the special case when $R_L = Z$, the reflection coefficient is equal to zero. This means that the entire line charges to voltage V^i in time τ without further reflections. Under these circumstances, the line is described as *matched*.

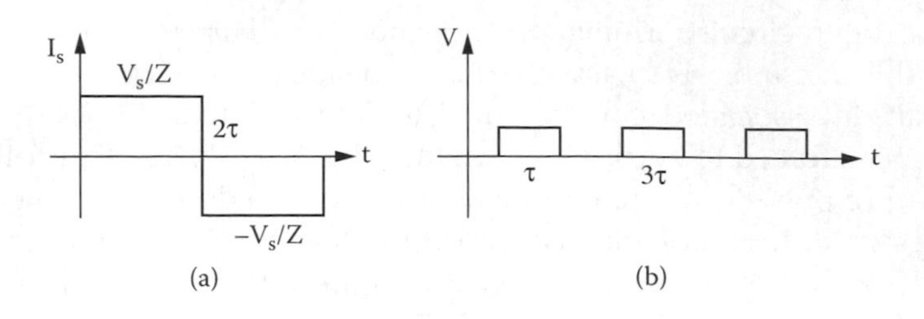

Figure 3.10 Current variation at the source end (a) and voltage in the middle of the line (b).

The circuit in Figure 3.9d may be generalized to deal with more general terminations. For instance, if the load is a capacitor C_L, then R_L may be replaced in this circuit by C_L and the voltage V_L determined in the normal way for a capacitor-charging circuit. Clearly, these analyses are valid as long as no further pulses arrive from further reflections. These more general situations will be dealt with in more detail in Chapter 8.

Several observations relevant to EMC can be made based on this simple model of propagation and reflection. A line with an open-circuit termination, as shown in Figure 3.9a, will carry current pulses of magnitude V_s/Z and duration related to its transit time and the position on the line where the current is measured. For example, the current at the source end will have the waveform shown in Figure 3.10a. If losses are neglected then this waveform persists for ever. Otherwise, the current eventually decays to zero. A line with a short-circuit termination does not have a zero voltage across its entire length. The voltage in the middle of such a line is shown in Figure 3.10b. If losses are added to this system, the waveform shown decays in an exponential manner. Both these examples indicate the presence of pulses even when the supply is a step voltage. Only in the case of a matched line would a train of pulses be avoided. Clearly, this situation has EMC implications. Further details regarding the propagation of pulses on transmission lines may be found in References 4 and 5.

3.2.2 Frequency-Domain Analysis of Transmission Lines

In the previous section, the response of a line to a step voltage was examined. Clearly any time-varying input could have been employed instead and the same principles used to obtain a solution. The complexity of such calculations is considerable, however, and they are not normally attempted without computational aids, except for the simplest waveforms.

In many applications, however, the response of a line to a harmonic excitation only is required. In such cases, it is possible to obtain analytical solutions as shown below.

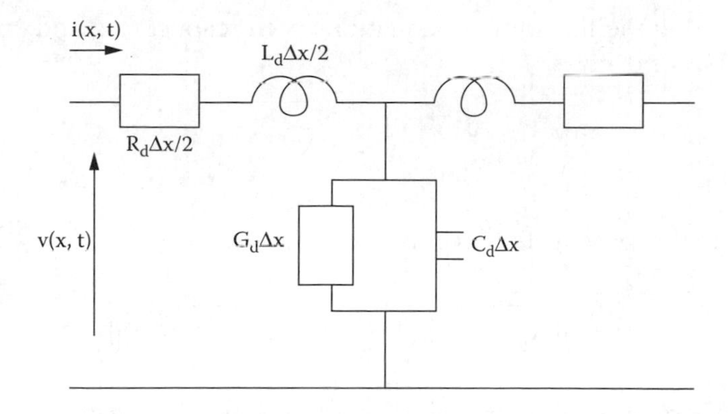

Figure 3.11 Electrical model of a line segment.

A segment of a line, length Δx, is shown in Figure 3.11. Kirchhoff's voltage (KVL) and current (KCL) laws give

$$R_d\, i(x,t) + L_d \frac{\partial i(x,t)}{\partial t} = -\frac{\partial v(x,t)}{\partial x} \tag{3.16}$$

$$G_d\, v(x,t) + C_d \frac{\partial v(x,t)}{\partial t} = -\frac{\partial i(x,t)}{\partial x} \tag{3.17}$$

For harmonic signals, phasors $\bar{V}(x)$ and $\bar{I}(x)$ representing voltage and current, respectively, are defined such that

$$\begin{aligned} v(x,t) &= \mathrm{Re}\left[\bar{V}(x)e^{j\omega t}\right] \\ i(x,t) &= \mathrm{Re}\left[\bar{I}(x)e^{j\omega t}\right] \end{aligned} \tag{3.18}$$

Equations 3.16 and 3.17 may be expressed in terms of the phasor quantities to eliminate time dependence, giving

$$\begin{aligned} \frac{d\bar{V}(x)}{dx} &= -\left(R_d + j\omega L_d\right)\bar{I}(x) \\ \frac{d\bar{I}(x)}{dx} &= -\left(G_d + j\omega C_d\right)\bar{V}(x) \end{aligned} \tag{3.19}$$

Differentiating the first of these equations with respect to x and combining with the second gives

$$\frac{d^2\overline{V}(x)}{dx^2} = \gamma^2\overline{V}(x) \tag{3.20}$$

where γ is the propagation constant of the line

$$\gamma = \sqrt{(R_d + j\omega L_d)(G_d + j\omega C_d)} = \alpha + j\beta \tag{3.21}$$

The real part α is the attenuation constant, and the imaginary part β is the phase constant of the line. An equation similar to Equation 3.20 is obtained for the current. The solution of these equations is in the general form

$$\begin{aligned} \overline{V}(x) &= V_A\, e^{\gamma x} + V_B e^{-\gamma x} \\ \overline{I}(x) &= -\frac{V_A}{Z} e^{\gamma x} + \frac{V_B}{Z} e^{-\gamma x} \end{aligned} \tag{3.22}$$

where Z is the characteristic impedance of this line

$$Z = \sqrt{\frac{R_d + j\omega L_d}{G_d + j\omega C_d}} \tag{3.23}$$

For a lossless line ($R_d = G_d = 0$) this expression reduces to that obtained in Equation 3.12.

Two types of term appear in Equation 3.22, namely those varying as $e^{-\gamma x}$ and $e^{\gamma x}$. The instantaneous value of the voltage due to a term of the first type is

$$\sim \mathrm{Re}\left[e^{-(\alpha + j\beta)x} e^{j\omega t}\right]$$

or proportional to $\cos(\omega t - \beta x)$. This expression represents a wavelike term where a point of constant phase is subject to the constraint $\omega t - \beta x = \text{const}$. This implies that $\omega dt - \beta dx = 0$, and therefore that the phase velocity is $u = \omega/\beta$. Disturbances represented by this term propagate in the forward direction. Terms of the type $e^{\gamma x}$ may be similarly studied and it is found that they describe waves moving in the backward direction ($u = -\omega/\beta$). In conclusion, conditions on the line are the result of a succession of waves traveling in the forward and backward directions.

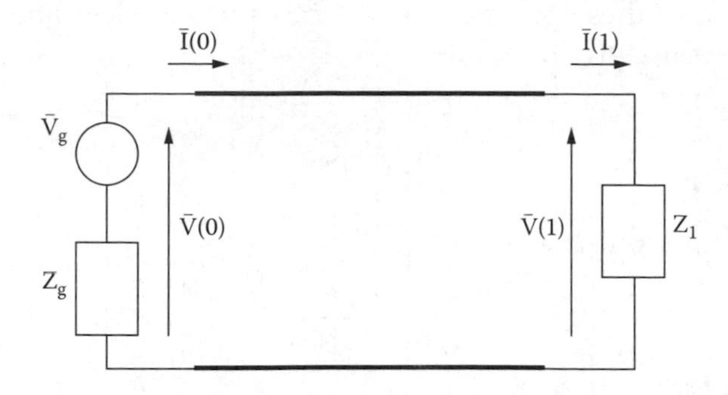

Figure 3.12 A model of a complete line in the frequency-domain.

In Equation 3.22, V_A and V_B are constants determined from the source and load conditions of the line. Assuming a line terminated as shown in Figure 3.12, V_A and V_B may be expressed in terms of $\bar{V}(0)$ and $\bar{I}(0)$, the voltage and current at the start of the line, to give

$$\begin{aligned} \bar{V}(x) &= \bar{V}(0)\cosh(\gamma x) - \bar{I}(0)\,Z\sinh(\gamma x) \\ \bar{I}(x) &= \frac{-\bar{V}(0)}{Z}\sinh(\gamma x) + \bar{I}(0)\cosh(\gamma x) \end{aligned} \tag{3.24}$$

The voltage and current at the load end of the line may be obtained by setting x = 1 in Equation 3.24. The resulting equations are then solved to express voltage and current at the source end in terms of the corresponding quantities at the load end, i.e.,

$$\begin{bmatrix} \bar{V}(0) \\ \bar{I}(0) \end{bmatrix} = \begin{bmatrix} A & B \\ C & D \end{bmatrix} \begin{bmatrix} \bar{V}(1) \\ \bar{I}(1) \end{bmatrix} \tag{3.25}$$

The matrix of coefficients A, B, C, and D is known as the transmission matrix [T] and, in terms of line parameters its elements are

$$\begin{aligned} A &= D = \cosh(\gamma 1) \\ B &= Z\sinh(\gamma 1) \\ C &= \frac{\sinh(\gamma 1)}{Z} \end{aligned} \tag{3.26}$$

A special case of these general equations is for a lossless line, where the propagation constant reduces to

$$\gamma = j\,\omega\,\sqrt{L_d C_d} = j\beta$$

and the characteristic impedance to

$$Z = \sqrt{L_d / C_d}$$

The transmission parameters in Equation 3.26 are then reduced to

$$A = D = \cos(\beta 1)$$
$$B = j\,Z \sin(\beta 1) \tag{3.27}$$
$$C = \frac{j\,\sin(\beta 1)}{Z}$$

At this stage it is interesting to investigate the value of the input impedance of this line, i.e., $Z_{in} = \bar{V}(0)/\bar{I}(0)$. Using the transmission parameters in Equation 3.27

$$Z_{in} = \frac{AZ_1 + B}{C\,Z_1 + D}$$

and therefore

$$\frac{Z_{in}}{Z} = \frac{\frac{Z_1}{Z} + j\tan(\beta 1)}{1 + j\frac{Z_1}{Z}\tan(\beta 1)} \tag{3.28}$$

Let us study in some detail this expression. It is clear, without much further consideration, that the input impedance Z_{in} is not always equal to the load impedance Z_1. Only as ($\beta 1$) tends to zero does the input impedance approach Z_1. The quantity ($\beta 1$) is described as the "electrical length" of the line. Since

$$\beta = \omega\,\sqrt{L_d C_d} = \frac{\omega}{u} = \frac{2\pi}{\lambda}$$

it follows that a line where $l/\lambda << 1$ is "electrically small," and normal intuition suggests that $Z_{in} \simeq Z_1$ holds. However, for electrically long lines, Equation 3.28 paints a much more complex picture. Let us choose as an example the case when $Z_1 \rightarrow \infty$. Then from Equation 3.28

$$\frac{Z_{in}}{Z} = \frac{1}{j\tan(\beta 1)} \tag{3.29}$$

The input impedance of this open-circuit line assumes the following values depending on the value of $(\beta 1)$

(β1)	Z_{in}
0	Very large capacitive
π/2	Zero
π	Very large inductive

This shows that a line that is a quarter of a wavelength short $(\beta 1 = \pi/2)$ appears at its input as a short circuit, even though its load end is an open circuit. In fact, a vast range of input impedance values is possible depending on the value of $(\beta 1)$. The input impedance of a short-circuit line is similarly obtained from Equation 3.28 and is

$$\frac{Z_{in}}{Z} = j\tan(\beta 1) \tag{3.30}$$

This expression implies that a short-circuit line one-quarter of a wavelength long appears at its input as an open circuit. The presence of losses on the line modifies somewhat these conclusions, but the essence remains the same.

From the EMC point of view, it is clear that the line termination alone is a poor guide as to the effective impedance seen at the start of a line. A connection that at low frequencies appears as a short circuit (DC short) looks very different at high frequencies. The material that has been presented in this section is not limited to systems specifically designed as transmission lines. Most wire connections and components behave like distributed components at sufficiently high frequencies, and thus exhibit similar behavior. The importance of these effects may be further illustrated by the following example. A line is formed on a substrate so that the velocity of propagation is $u = 0.6 \times 3 \times 10^8$ m/s. It is required to establish its behavior at a frequency of 300 MHz. The wavelength under these circumstances is $\lambda = 0.6 \times 3 \times 10^8/3 \times 10^8 = 60$ cm. Hence a section of this line one-quarter of a wavelength long (15 cm) will exhibit very strongly the behavior described above and due consideration must be given to transmission-line effects. It should be noted that distances of the order of tens of centimeters are typical of track lengths in many electronic systems. This topic will be examined again in future chapters.

More detailed descriptions of transmission line effects may be found in References 5 and 6.

References

1. Ott, H W, "Noise Reduction Techniques in Electronic Systems," Second Edition, John Wiley, NY, 1990.
2. Chatterton, P A and Houghton, M A, "EMC — Electromagnetic Theory to Practical Design," John Wiley, NY, 1992.
3. Paul, C R, "Introduction to Electromagnetic Compatibility," John Wiley, NY, 1992.
4. Greenwood, A, "Electrical Transients in Power Systems," Second Edition, John Wiley, NY, 1991.
5. Ramo, S, Whinnery, J R, and Van Duzer, T, "Fields and Waves in Communication Electronics," John Wiley, NY, 1984.
6. Bylanski, P and Ingram D G W, "Digital Transmission Systems," Revised Edition, Peter Peregrinus, IEE, 1980.

Chapter 4

Electrical Signals and Circuits

The purpose of this chapter is to introduce the reader to the basic techniques used to describe signals, their detection by simple circuits, and their manipulation using signal processing techniques.

Electrical signals are usually voltages that may, however, represent a wide variety of physical processes, e.g., electric field or temperature variations. Similarly, the term "circuits" may be taken to mean a physical system which, when excited in a particular way, responds appropriately. Hence the material presented in this chapter is of general value in understanding signals of all types.

A signal is described by a mathematical function, in which the variable is normally time. For example, a voltage source may be described as $v(t) = V_{pk} \cos(\omega t)$. This, however, is not the only kind of functional relationship found in EMC work. The function v(t) may take many analytical forms, or may be defined by a set of experimental data obtained at different times. It is necessary, therefore, to separate the exact functional form of v(t) from the more general properties of the signal, and its further processing by hardware (circuits) and software (computer programs) means. This mathematical description of signals offers great flexibility to the investigator, as it makes it possible to isolate and study specific attributes of the signal and its processing.

Signals may be broadly divided into deterministic and random or stochastic signals. Deterministic signals are those for which the instantaneous value can be specified accurately at each instant of time. These are signals that are used to transmit information and are therefore of primary interest in all applications. There are, however, signals for which even approximate predictions of variation with time cannot be made. These are signals that are due to chaotic thermal fluctuations in circuits and are described as random or stochastic signals. The term "noise" is also used to express the manifestation of these phenomena. A distinction may be drawn between noise, which is generated internally by a circuit, and interference, which is coupled to it from external sources. In cases where the interfering signal is low level, it can sometimes be difficult to distinguish between internally and externally generated unwanted signals and the term "noise" is used indiscriminately to indicate an unwanted signal. Study of the basic properties of random signals is useful in EMC as it determines the limits of detectability of signals in noisy environments.

In many cases signals may be measured and specified at any instant. They are then described as continuous (or analogue) signals. There are, however, circumstances in which signals are known only at specific instants, multiples of a basic sampling time interval Δt. They are then known as sampled (or discrete) signals. The magnitude of each sample of a discrete signal may be represented by a binary number. A common representation is to use 8 bits, and thus a pulse train of eight pulses of magnitude 1 or 0 to represent each sample. This signal is then described as a digital signal. Whichever signal representation is chosen, certain general principles apply that permit the description of any complex signal in terms of simpler signals. Examples of such representations are given below.

4.1 Representation of a Signal in Terms of Simpler Signals

The unit step function is shown in Figure 4.1a and is defined as

$$u(t) = \begin{cases} 0 & t < 0 \\ 0.5 & t = 0 \\ 1 & t > 0 \end{cases} \tag{4.1}$$

It may be shifted in time to obtain the function $u(t - t_0)$ shown in Figure 4.1b. The signal $v(t)$ shown in Figure 4.1c may be represented by a succession of step voltages, as shown in Figure 4.1d. Mathematically, $v(t)$ is then given by the formula

$$v(t) = V_0\, u(t) + \sum_{k=1}^{\infty} (V_k - V_{k-1})\, u(t - k\Delta t) \tag{4.2}$$

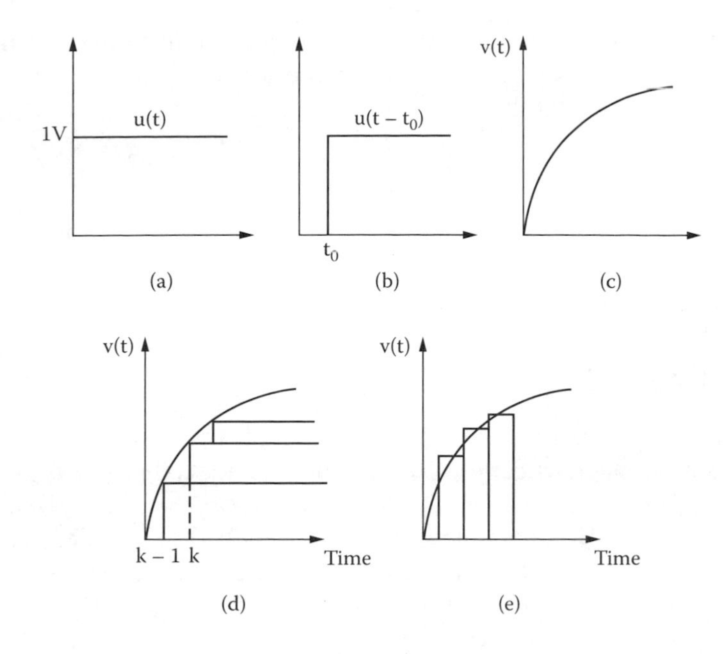

Figure 4.1 (a) A unit step voltage waveform, (b) as in (a) but shifted by t_0, (c) a general waveform and its representation in terms of steps (d) and impulses (e).

If the sampling interval is made very small and in the limit reduced to zero, this expression reduces to

$$v(t) = V_0\, u(t) + \int_0^{\infty} \frac{dv(\tau)}{d\tau}\, u(t-\tau)\, d\tau \tag{4.3}$$

An alternative approach is to represent the signal in Figure 4.1c as a succession of pulses, as shown in Figure 4.1e. In the limit, the basic constituent of this representation is the delta function defined by the following properties:

$$\delta(t - t_0) = \begin{cases} 1 & t = t_0 \\ 0 & t \neq t_0 \end{cases} \quad \text{and} \quad \int_{-\infty}^{\infty} \delta(t)\, dt = 1$$

In mathematical form the signal v(t) is then represented by the expression

$$v(t) = \int_0^{\infty} v(\tau)\delta(t-\tau)\, d\tau \tag{4.4}$$

Another way of representing a signal is to express it in terms of a number of basis functions, such as sinusoids, in the form

$$v(t) = \sum_{i=1}^{\infty} C_i \, \phi_i(t) \tag{4.5}$$

It should be noted that only certain signals are suitable as basis signals in Equation 4.5.

The selection of basis functions $\phi_i(t)$ and the calculation of the coefficients C_i will be explained shortly. However, it is worth pausing to reflect on the significance of Equations 4.3 to 4.5. In all cases, the original signal is described in terms of other simpler signals. If the response of a circuit to these simpler signals can be found, and provided that the principle of superposition holds (linear circuit), then the response to the original signal is easily obtained by combining the responses to the simpler signals. A particularly useful formulation is when harmonic signals are used as the basis functions.[1,2] A *periodic signal* v(t) of period T may then be represented in different ways as a collection of harmonic signals, as shown below.

$$\begin{aligned} v(t) &= a_0 + \sum_{n=1}^{\infty} \left[\left(a_n \cos(n\omega_0 t) + b_n \sin(n\omega_0 t) \right) \right] \\ &= a_0 + \sum_{n=1}^{\infty} A_n \cos(n\omega_0 t - \phi_n) \\ &= \sum_{n=-\infty}^{\infty} C_n \, e^{jn\omega_0 t} \end{aligned} \tag{4.6}$$

The harmonic functions are all multiples of the fundamental frequency $\omega_0 = 2\pi/T$ and this representation of the signal is referred to as an expansion into a Fourier series. The three expressions above are equivalent and the particular choice is a matter of convenience. The unknown coefficients a_n, b_n, A_n, and C_n may be obtained from the following formulae:

$$a_0 = \frac{1}{T} \int_0^T v(t) \, dt$$

$$a_n = \frac{2}{T} \int_0^T v(t) \cos(n\omega_0 t) \, dt, \qquad n = 1, 2, \ldots$$

$$b_n = \frac{2}{T}\int_0^T v(t)\sin(n\omega_0 t)\,dt, \qquad n = 1, 2, \ldots$$

$$A_n = \sqrt{a_n^2 + b_n^2}, \quad \phi_n = \tan^{-1}(b_n/a_n)$$

$$C_n = \frac{1}{T}\int_0^T v(t)\,e^{-jn\omega_0 t}\,dt, \qquad n = 0, \pm 1, \pm 2, \ldots \qquad (4.7)$$

Any periodic signal of period T may thus be represented by a collection of harmonic signals (infinite in number), which have a frequency an integer multiple of $\omega_0 = 2\pi/T$. This is known as a spectral representation of the signal.

To illustrate the application of these formulae, the square wave signal shown in Figure 4.2a will be studied. Its period is T and the signal is supposed to exist forever! A signal that is zero for $t < 0$ is clearly not periodic and cannot be represented accurately by a Fourier series. Applying Equation 4.7 gives

$$a_0 = \frac{1}{T}\int_0^T v(t)\,dt = V_0\frac{\tau}{T}$$

$$a_n = \frac{2}{T}\int_0^T V_0\cos(n\omega_0 t)\,dt = \frac{V_0}{n\pi}\sin\left(2\pi n\frac{\tau}{T}\right), \quad n \neq 0$$

$$b_n = \frac{2}{T}\int_0^T V_0\sin(n\omega_0 t)\,dt = \frac{V_0}{n\pi}\left[1 - \cos\left(2\pi n\frac{\tau}{T}\right)\right], \quad n \neq 0$$

For the special case when $\tau/T = 0.5$ these formulae reduce to

$$a_0 = 0.5\,V_0$$

$$a_n = 0$$

$$b_n = \begin{cases} \dfrac{2\,V_0}{n\pi} & \text{n: odd} \\ 0 & \text{n: even} \end{cases}$$

Thus, for this special case the signal shown in Figure 4.2a may be represented as follows:

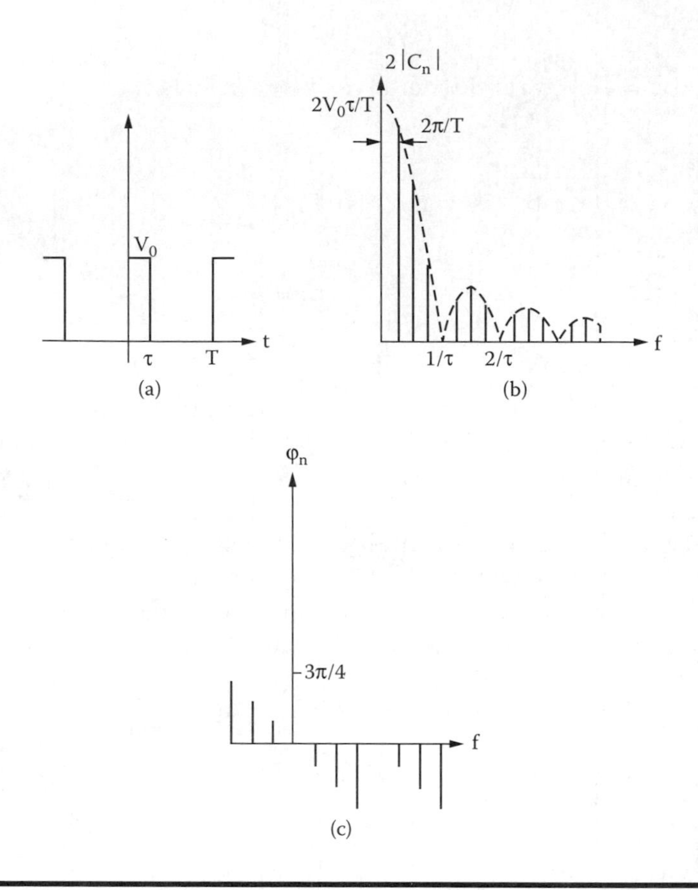

Figure 4.2 A repetitive pulse waveform (a), its amplitude (b), and phase (c) spectra.

$$v(t) = \frac{V_0}{2} + \frac{2V_0}{\pi} \sin\left(2\pi \frac{t}{T}\right) + \frac{2\,V_0}{3\pi} \sin\left(2\pi \frac{3t}{T}\right) + \cdots$$

It contains a DC term, first, third, fifth, and so on, harmonics. It is interesting to speculate how many terms are required to describe accurately this signal in terms of its spectral content. Clearly, for higher harmonics (n large) the amplitude decreases and hence contributions are small. The reader may plot the formula above retaining the first three terms only and compare the result with the waveform in Figure 4.2a. It will be seen that a substantial error remains at the transition between low and high values. This error persists even when many more terms are included in the Fourier series (Gibbs phenomenon).

Let us now visualize in more detail the spectrum of this signal for the general case. The coefficient C_n is easily calculated and is

$$C_n = \left[V_0 \frac{\tau}{T} \frac{\sin(n\omega_0\tau/2)}{(n\omega_0\tau/2)} \right] e^{-jn\omega_0\tau/2} \tag{4.8}$$

Hence,

$$v(t) = C_0 + \sum_{n=1}^{\infty} \left[C_n \, e^{jn\omega_0 t} + C_{-n} \, e^{-jn\omega_0 t} \right]$$

The coefficient C_n is equal to $|C_n| \, e^{j\phi_n}$ where $|C_n|$ is the magnitude of the term in brackets (Equation 4.8) and $\phi_n = \pm n \, \omega_0 \, \tau/2$. The negative sign is selected when $n \, \omega_0 \, \tau/2 < \pi$. In addition, C_{-n} is equal to the complex conjugate of C_n, hence

$$v(t) = C_0 + \sum_{n=1}^{\infty} \left[|C_n| \, e^{j(n\omega_0 t + \phi_n)} + |C_n| \, e^{-j(n\omega_0 t - \phi_n)} \right]$$

$$= C_0 + \sum_{n=1}^{\infty} 2|C_n| \, \cos(n\omega_0 t + \phi_n)$$

Coefficients C_0, $2|C_n|$, and ϕ_n constitute the magnitude and phase spectrum of this signal. The first few terms are plotted in Figure 4.2b and c. The shape of the envelope of the magnitude spectrum is that of the function sinx/x, which is tabulated in Appendix C. Naturally, real square-wave pulses do not have zero rise- and fall-times as assumed here. The influence of finite rise- and fall-times on the spectra is explained in Appendix D. However, it is already possible to gain an insight into the spectral content of this signal. Let us assume that the repetition rate is 50 MHz ($T = 1/50\ 10^6 = 20$ nsec) and that the duty cycle is 50% ($\tau = 0.5\ T = 10$ nsec). Then the fundamental frequency is 50 MHz and the first zero in the magnitude spectrum is at $f = 1/\tau = 100$ MHz. It can be seen that at the ninth harmonic ($f = 450$ MHz) the amplitude is still in excess of 10% of its value at the fundamental frequency. This type of pulse thus contains many frequencies covering the entire range of interest to EMC.

Let us now consider how to deal with signals that are not periodic. In mathematical terms, an *aperiodic signal* may be regarded as a periodic signal with very large period ($T \to \infty$). Hence, the last expression in

Equation 4.6 can be generalized to give the *Fourier transform* of a signal v(t).

$$v(t) = \frac{1}{2\pi}\int_{-\infty}^{\infty} V(j\omega)\, e^{j\omega t}\, d\omega \tag{4.9}$$

where

$$V(j\omega) = \int_{-\infty}^{\infty} v(t)\, e^{-j\omega t}\, dt \tag{4.10}$$

The essential difference between the spectra of periodic and non-periodic waveforms is that in the former the spectrum consists of contributions at distinct frequencies multiples of the fundamental, while in the latter there are contributions at all frequencies, i.e., the spectrum is continuous. As an example, the spectrum of the pulse shown in Figure 4.3a is calculated. Using Equation 4.10 and substituting for v(t) gives

$$\begin{aligned} V(j\omega) &= \int_{0}^{\tau} V_0\, e^{-j\omega t} = \frac{V_0}{-j\omega}\left(e^{-j\omega\tau} - 1\right) \\ &= \left[V_0\tau \frac{\sin\left(\frac{\omega\tau}{2}\right)}{\left(\frac{\omega\tau}{2}\right)} \right] e^{-j\omega\tau/2} = \left|V(j\omega)\right|\, e^{j\phi(j\omega)} \end{aligned} \tag{4.11}$$

where $\phi(j\omega) = \pm\omega\tau/2$ and the negative sign is selected when $(\omega\tau/2) < \pi$. The amplitude and phase spectra are plotted in Figure 4.3b and c.

Let us now examine the spectrum of a very-short-duration impulse, such that the product of its amplitude and duration is equal to A. This is referred to as a *delta function pulse*, and mathematically is described by the following formula

$$v(t) = A\delta(t)$$

Substituting into Equation 4.10 gives

$$V(j\omega) = A\int_{-\infty}^{\infty} e^{-j\omega t}\delta(t)\, dt = A$$

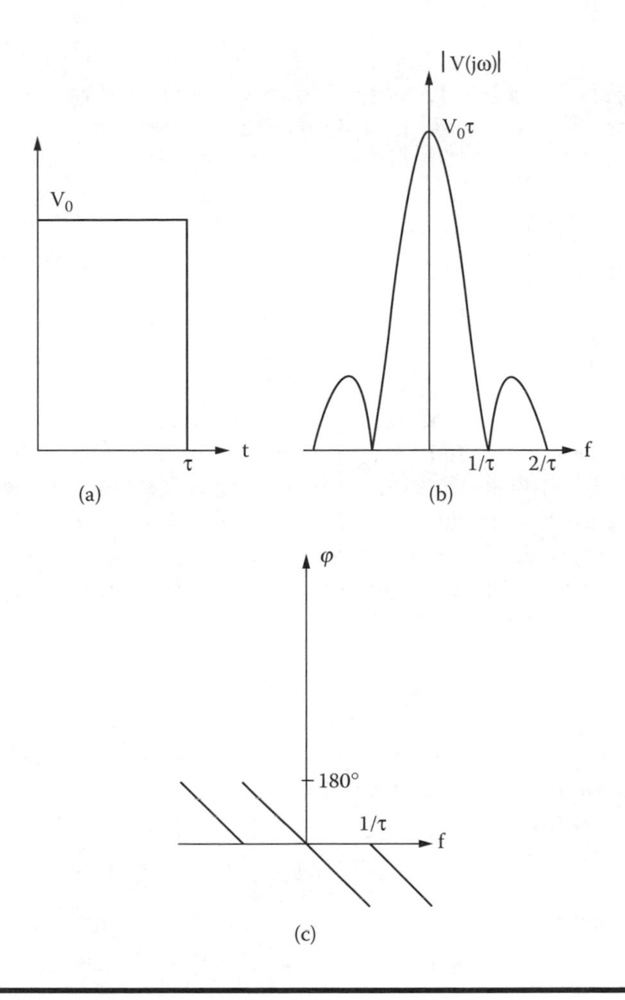

Figure 4.3 A pulse (a), its amplitude (b), and phase (c) spectra shown up to a frequency 2/τ.

Thus, the delta impulse has a flat spectrum. Applying a very-short-duration pulse is like injecting signals at all frequencies having the same magnitude and phase. The value of this so-called *impulse excitation* is that it contains within it the entire spectrum of signals.

In practical situations, it is difficult to generate pulses of zero rise- and fall-times. It is therefore useful to study the spectrum of signals with a smooth transition from zero to the peak voltage value. A particularly useful signal of this type is the Gaussian pulse defined as

$$v(t) = V_0\, e^{-at^2} \tag{4.12}$$

The effective duration of this pulse τ_d is defined as the time interval in which the pulse exceeds 10% of its peak value. It can be shown that $\tau_d = 3/\sqrt{a}$. The spectrum of this pulse is found to be

$$V(j\omega) = V_0 \int_{-\infty}^{\infty} e^{-at^2}\, e^{-j\omega t}\, dt = V_0 \sqrt{\pi/a}\, e^{-\omega^2/(4a)} \qquad (4.13)$$

Thus, the spectrum is also a smooth function of frequency of the same shape as the pulse.

The pair of transforms described by Equations 4.9 and 4.10 offers two alternatives for the study of signals. First, a signal may be described as a function of time or in what is referred to as the *time domain*. If the time variation of the signal is a complicated one, it is normally very difficult to proceed directly by calculation to find the response of circuits to it. Second, an alternative approach may be adopted whereby the spectral content $V(j\omega)$ of the original signal $v(t)$ is found using Equation 4.10. The response of a circuit to $v(t)$ can then be studied by reference to its response to the spectral components $V(j\omega)$. This approach is normally easier to pursue, especially for signals with complicated time waveforms. The spectrum $V(j\omega)$ of the signal contains complete information about it and may be regarded as its description in the *frequency domain*. Conversion between the time and frequency domains is achieved by using the direct (Equation 4.10) and inverse (Equation 4.9) Fourier transforms. It can be shown that

$$\int_{-\infty}^{\infty} v^2(t)\, dt = \frac{1}{2\pi} \int_{-\infty}^{\infty} \left|V(j\omega)\right|^2 d\omega \qquad (4.14)$$

This equation, referred to as Parseval's Theorem, relates measures of energy associated with signal description in the time and frequency domains. A plot of $|V(j\omega)|^2$ vs. frequency is described as the *power spectrum* of the signal.

Examination of Figure 4.3 shows that a decrease in the duration of a pulse (τ small) results in a broad spectrum of frequencies ($1/\tau$ increases), i.e., a narrow pulse implies a wide bandwidth and vice versa. In broad terms it can be stated that the product of signal duration in the time domain τ_d and of its bandwidth (BW) in the frequency domain is of the order of 1, i.e.,

$$\tau_d(BW) \simeq 1 \qquad (4.15)$$

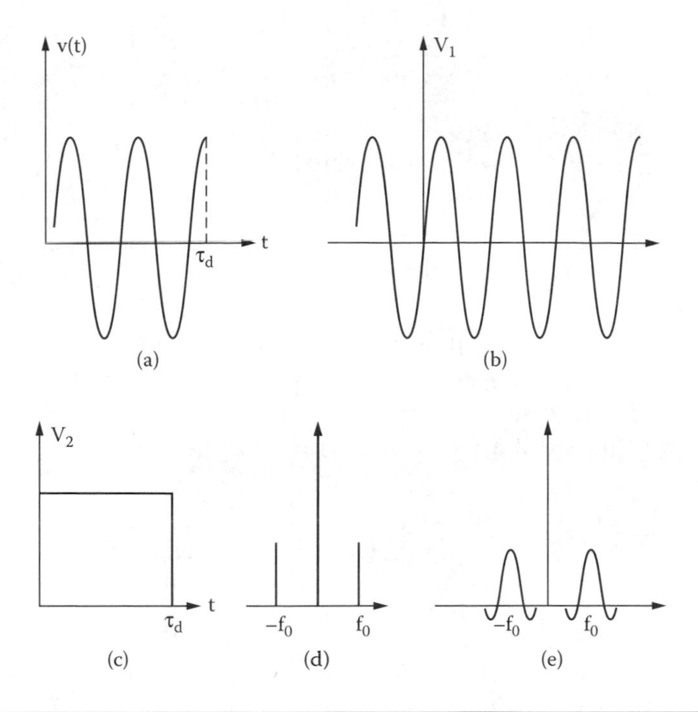

Figure 4.4 A sinusoidal signal of finite duration (a) and its derivation from a periodic signal (b) and pulse (c). The spectrum of V_1 (d) and of v(t) (e).

Let us now study the spectrum of a more complex waveform, such as the one shown in Figure 4.4a. Clearly, this waveform may be regarded as the product of the two waveforms shown in Figure 4.4b and c. The spectrum of the waveform (b) can be obtained easily.

$$v_1(t) = V_0\cos(\omega_0 t) = V_0\frac{e^{j\omega_0 t} + e^{-j\omega_0 t}}{2}$$

Hence

$$V_1(j\omega) = \frac{V_0}{2}\int_{-\infty}^{\infty}\left(e^{j\omega_0 t} + e^{-j\omega_0 t}\right)e^{-j\omega t}\,dt$$

This integral can be evaluated if the sampling property of the delta function is invoked, i.e.,

$$e^{j\omega_0 t} = \frac{1}{2\pi}\int_{-\infty}^{\infty} e^{j\omega t}\,2\pi\delta(\omega - \omega_0)\,d\omega$$

This means that the Fourier transform of $e^{j\omega_0 t}$ is $2\pi\delta(\omega - \omega_0)$. Using this result in the expression for $V_1(j\omega)$ gives

$$V_1(j\omega) = V_0\pi\left[\delta(\omega - \omega_0) + \delta(\omega + \omega_0)\right] \tag{4.16}$$

Thus, the amplitude spectrum of $V_1(t)$ consists of two contributions of $\pm\omega_0$ as shown in Figure 4.4d. The spectrum of $v_2(t)$ has already been obtained (Figure 4.3b).

The question arises whether the spectrum of the waveform shown in Figure 4.4a may be obtained from the spectra of $v_1(t)$ and $v_2(t)$. It turns out that this is the case, but while $v = v_1(t)\, v_2(t)$, the spectrum $V(j\omega) \neq V_1(j\omega)\, V_2(j\omega)$. Instead the following expression holds:

$$V(j\omega) = \frac{1}{2\pi}\int_{-\infty}^{\infty} V_1(j\omega')\, V_2(j\omega - j\omega')\, d\omega' \tag{4.17}$$

The integral on the right-hand side of Equation 4.17 is called the convolution of waveforms $V_1(j\omega)$ and $V_2(j\omega)$ and it is normally indicated using the notation $V_1(j\omega) * V_2(j\omega)$.

Substituting into Equation 4.17 for $V_1(j\omega')$ gives

$$\begin{aligned} V(j\omega) &= \frac{1}{2\pi}\int_{-\infty}^{\infty} V_0\pi\left[\delta(\omega' - \omega_0) + \delta(\omega' + \omega_0)\right] V_2(j\omega - j\omega')\, d\omega' \\ &= \frac{V_0}{2}\left[V_2(j\omega - j\omega_0) + V_2(j\omega + j\omega_0)\right] \end{aligned} \tag{4.18}$$

This expression indicates that the spectrum of v(t) consists of the spectrum of the rectangular pulse shifted by $\pm\omega_0$, as shown in Figure 4.4e.

Several conclusions may be drawn from these spectra. If the duration of the cosinusoidal wave is long (τ_d >>), then its spectrum shown in Figure 4.4e becomes very narrow around $\pm\omega_0$ and tends to that of a waveform of infinitely long duration. If only a few periods of the signal are included (τ_d <<), then the spectrum contains very substantial contributions at frequencies other than ω_0.

4.2 Correlation Properties of Signals

There are many applications that are critically dependent on evaluating the degree of similarity between two signals or, in technical parlance, their correlation. A particular example of this is in radar. The correlation properties of signals are described in this section.

4.2.1 General Correlation Properties

The autocorrelation of a signal v(t) is defined as the quantity

$$R(\tau) = \int_{-\infty}^{\infty} v(t)\, v(t-\tau)\, dt \tag{4.19}$$

where the signal $v(t - \tau)$ has the same functional dependence as signal v(t) but it is shifted in time by an amount τ. For the signal shown in Figure 4.5a, the autocorrelation is shown in Figure 4.5b. Similarly, the autocorrelation of the signal in (c) is given in (d). It can be shown that the power spectrum $|V(j\omega)|^2$ of a signal and its autocorrelation function are Fourier transform pairs (Wiener–Khinchine Theorem). From this it follows that the wider the frequency spectrum of a signal is, the narrower

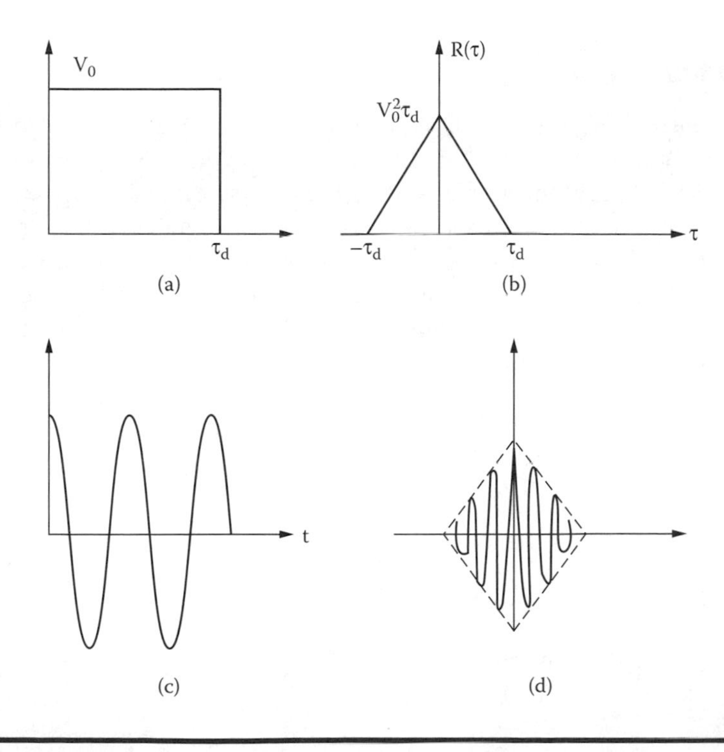

Figure 4.5 Autocorrelation of the signals in (a) and in (c), shown in (b) and (d), respectively.

the width of its autocorrelation function will be and the easier it will be to locate it in time. Under some circumstances it may be easier to obtain the power spectrum of a signal by first finding its autocorrelation and then its Fourier transform.

A measure of the similarity between two signals is provided by their *cross-correlation*, defined as

$$R_{12}(\tau) = \int_{-\infty}^{\infty} v_1(t)\, v_2(t-\tau)\, dt \tag{4.20}$$

It can be shown that the cross-correlation of two signals is related to their spectra by the expression

$$R_{12}(\tau) = \frac{1}{2\pi} \int_{-\infty}^{\infty} V_1(j\omega)\, V_2^*(j\omega)\, e^{j\omega\tau}\, d\omega \tag{4.21}$$

where the star * indicates the complex conjugate quantity.

4.2.2 Random Signals

Random signals and their mathematical description is an extensive subject that cannot be covered fully in this text. However, it is necessary to get a basic idea of random signals, which manifest themselves as noise in many systems. It may be thought that random signals, by their nature, cannot be described in any precise way. Although this is true as far as predicting their precise evolution in time is concerned, it does not, however, prevent us from making precise statements about them, which are true in a statistical sense.

Let us consider a quantity x varying randomly, with a probability P(x) that this quantity has a particular value x. Then the mean or expected value of x, designated as $\langle x \rangle$, is

$$m = \langle x \rangle = \int_{-\infty}^{\infty} x\, P(x)\, dx \tag{4.22}$$

Similarly, the mean square value of x is

$$\langle x^2 \rangle = \int_{-\infty}^{\infty} x^2\, P(x)\, dx \tag{4.23}$$

Another quantity of particular interest is the variance of x, σ_x^2 defined as

$$\sigma_x^2 = \left\langle (x - m)^2 \right\rangle = \left\langle x^2 \right\rangle - \left\langle x \right\rangle^2 \tag{4.24}$$

The square root of the variance is called the standard deviation σ_x of the signal x. For most signals encountered in engineering applications, the averaging indicated by Equations 4.22, 4.23, and 4.24 may be replaced by a time-average over a period of time T, which tends to infinity (ergodic signals), i.e.,

$$m = \lim_{T \to \infty} \frac{1}{T} \int_0^T x(t)\, dt$$

$$\sigma^2 = \lim_{T \to \infty} \frac{1}{T} \int_0^T \left[x(t) - m \right]^2 dt$$

A representation of a random signal in the time domain has a corresponding description in the frequency domain. It will be assumed here that the statistical characteristics of a random signal remain unchanged with respect to time (stationary random signal). The autocorrelation of such a signal is defined as

$$R(\tau) = \left\langle x(t)\, x(t + \tau) \right\rangle \tag{4.25}$$

It will be seen from this expression and from Equation 4.24, that provided the mean value of x(t) is zero, then $\sigma_x^2 = R(0)$.

The autocorrelation of a random signal v(t) and its power spectrum $W_v(j\omega)$ are a Fourier transform pair, i.e.,

$$\begin{aligned} R_v(\tau) &= \frac{1}{2\pi} \int_{-\infty}^{\infty} W_v(j\omega)\, e^{j\omega t}\, d\omega \\ W_v(j\omega) &= \int_{-\infty}^{\infty} R_v(\tau)\, e^{-j\omega\tau}\, d\tau \end{aligned} \tag{4.26}$$

This is known as the Wiener–Khinchin Theorem. It follows that the variance of the signal is

$$\sigma_v^2 = R_v(0) = \frac{1}{2\pi} \int_{-\infty}^{\infty} W_v(j\omega)\, d\omega \tag{4.27}$$

A particularly useful random signal is "white noise," defined as a signal with a constant power spectrum at all frequencies, i.e., $W_{wn}(j\omega) = W_0$. For this signal, the variance may be obtained from Equation 4.27 and is

$$\sigma_{wn}^2 = 2\frac{1}{2\pi}\int_{f=0}^{\infty} W_0 2\pi \, df = 2\int_0^{\infty} W_0 \, df \tag{4.28}$$

The rms value of the signal associated with the white noise is equal to the square root of the variance. From the first of the set of Equation 4.26 it follows that the autocorrelation of white noise is

$$R_{wn}(\tau) = \frac{W_0}{2\pi}\int_{-\infty}^{\infty} e^{j\omega\tau} \, d\omega = W_0\delta(\tau)$$

4.3 The Response of Linear Circuits to Deterministic and Random Signals

The response of linear circuits to a variety of inputs can be described in several ways. Two particular techniques used frequently are to find the circuit's impulse and frequency response. These two cases are examined below.

4.3.1 Impulse Response

The impulse response h(t) of a circuit is the output obtained when the input signal is a delta function. In practice, the impulse applied at the input must have a duration much shorter than the shortest time constant of the circuit. If the impulse response h(t) is known, then the output to any input signal is equal to the convolution of h(t) and the function describing the input signal (Figure 4.6a), i.e.,

$$v_{out}(t) = \int_{-\infty}^{\infty} v_{in}(\tau)\, h(t-\tau)\, d\tau \tag{4.29}$$

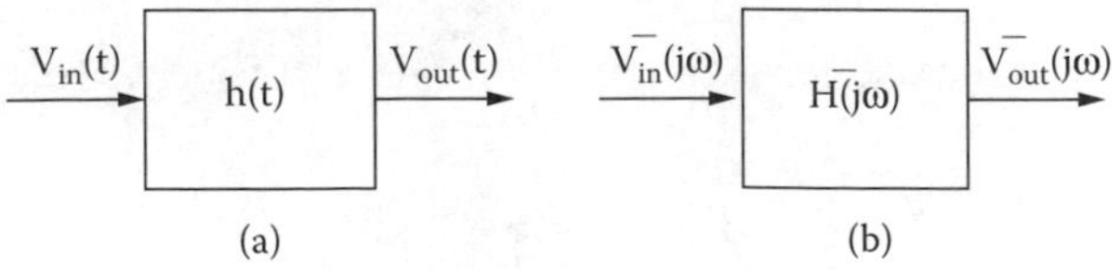

$V_{in}(t)$ → h(t) → $V_{out}(t)$

(a)

$\bar{V}_{in}(j\omega)$ → $\bar{H}(j\omega)$ → $\bar{V}_{out}(j\omega)$

(b)

Figure 4.6 Impulse (a) and frequency (b) response of a system.

4.3.2 Frequency Response

An alternative approach is to find the frequency response of the circuit H(jω) to harmonic inputs covering the entire frequency range of interest. The phasor describing the output signal at a particular frequency is given by the expression

$$\bar{V}_{out}(j\omega) = H(j\omega)\,\bar{V}_{in}(j\omega) \tag{4.30}$$

as shown in Figure 4.6b.

To find the response to an arbitrary input, the spectrum of the input signal is first determined using Fourier transforms. The output to each frequency component is then obtained from Equation 4.30 and thus the spectrum of the output signal is found. The output voltage in the time domain can then be obtained by taking the inverse Fourier transform of the output spectrum.

The impulse and frequency responses h(t) and H(jω), respectively, are Fourier transform pairs. Either of these two functions describes fully the response of the circuit. An alternative method for determining the impulse response of a linear system is to exploit the property that the input-output cross-correlation is equal to the convolution of the impulse response and the input autocorrelation.[3] The input-output cross-correlation is defined as $R_{in/out} = \langle v_{in}(t)\, v_{out}(t + \tau)\rangle$. If the input to the system is supplied with white noise [autocorrelation $W_0\delta(\tau)$] and the input-output cross-correlation is measured, then $h(t) = R_{in\text{-}out}(t)/W_0$. This approach has several advantages.[3]

The following simple example will help to illustrate the techniques described above.

Example 1: Find the frequency and impulse responses of the circuit shown in Figure 4.7 and then obtain an expression for the output voltage when the input is a step voltage V_0.

Solution: The frequency response is easily obtained:

$$\bar{V}_2(j\omega) = \frac{1/j\omega C}{R + 1/j\omega C}\,\bar{V}_1(j\omega) = \frac{1}{1 + j\omega RC}\,\bar{V}_1(j\omega)$$

Hence

$$H(j\omega) = \frac{1}{1 + j\omega RC}$$

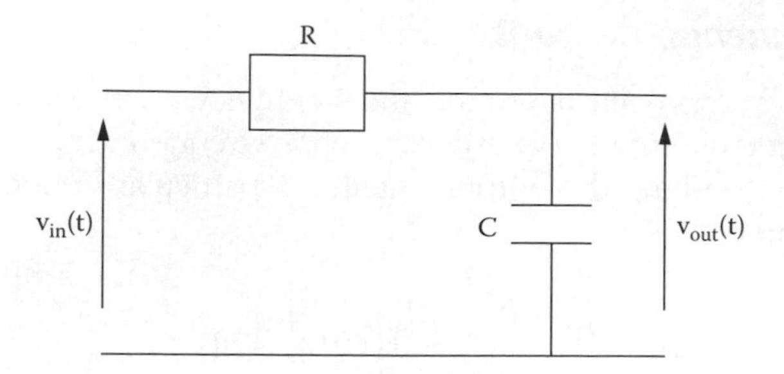

Figure 4.7 Circuit used to study the relationship between frequency and impulse response.

The impulse response is obtained from Fourier transform tables (inverse transform of H(jω)) and is

$$h(t) = \frac{1}{RC} e^{-t/(RC)} u(t)$$

Working in the frequency domain the output voltage is then

$$\bar{V}_2(j\omega) = H(j\omega)\,\bar{V}_1(j\omega)$$

$$= \frac{1}{1 + j\omega RC}\,\frac{V_0}{j\omega} = \left[\frac{-1}{\frac{1}{RC} + j\omega} + \frac{1}{j\omega}\right] V_0$$

Hence, from Fourier transform tables

$$v_2(t) = -e^{-t/(RC)} V_0 u(t) + V_0 u(t) = \left(1 - e^{-t/(RC)}\right) V_0 u(t)$$

Working in the time domain, the output voltage is

$$v_2(t) = \int_{-\infty}^{\infty} v_1(\tau) h(t - \tau)\, d\tau$$

$$= \frac{V_0}{RC} \int_{-\infty}^{\infty} u(\tau) e^{-\frac{t-\tau}{RC}} u(t - \tau)\, d\tau$$

$$= \frac{V_0}{RC} \int_{0}^{t} e^{-\frac{t-\tau}{RC}}\, d\tau = \left(1 - e^{-t/(RC)}\right) V_0$$

As expected, the same expression for the output voltage is obtained whichever appoach is adopted. This general procedure can be easily adapted to work with sampled signals.

Example 2: Find the response of the circuit shown in Figure 4.7 to a white noise input voltage of power spectrum W_0.

Solution: The power transfer function of this circuit is

$$|H(j\omega)|^2 = \frac{1}{1+(\omega RC)^2}$$

Hence, from Equation 4.26

$$R_{v2}(\tau) = \frac{1}{2\pi}\int_{-\infty}^{\infty} W_0 |H(j\omega)|^2 e^{j\omega\tau}\, d\omega$$

The variance is obtained from Equation 4.28 and is

$$\sigma_{v2}^2 = R_{v2}(0) = \frac{W_0}{\pi}\int_0^{\infty}\frac{d\omega}{1+(\omega RC)^2} = \frac{W_0}{2RC}$$

The rms value of the output voltage is equal to the square root of the variance, i.e., $\sqrt{W_0/2RC}$.

4.3.3 Detection of Signals in Noise

Strategies have evolved to detect signals that are mixed with noise. The objective is to design circuits that are capable of detecting signals in an optimum manner. The theory relevant to this task is known as optimum filtering.[5] A particular example will be presented here, whereby characteristics of a filter are desired that are capable of maximizing the ratio of signal to noise at its output. It is assumed that the input signal waveform is known and that the output signal is mixed with white noise.

Let us assume that the input voltage waveform is known, $v_{in}(t)$, and that its frequency spectrum is

$$\bar{V}_1(j\omega) = |V_1(j\omega)|\; e^{j\phi_1(\omega)} \tag{4.31}$$

The white noise at the input has a power spectrum W_0. Let us also assume that the desired frequency response of the filter is

$$H(j\omega) = |H(j\omega)| \; e^{j\phi(\omega)} \tag{4.32}$$

The variance of the noise signal at the output of the filter is then given by Equation 4.33:

$$\sigma_2^2 = \frac{W_0}{\pi} \int_0^{\infty} |H(j\omega)|^2 \; d\omega \tag{4.33}$$

Maximizing the ratio of the output signal at a particular time t_0 to the square root of the variance of the noise at the output requires the following filter frequency response:[5]

$$H(j\omega) = A \; \bar{V}_1^*(j\omega) \; e^{-j\omega t_0} \tag{4.34}$$

where A is a constant.

It can also be shown that the impulse response of the filter is $h(t) = A\,V_1\,(t_0 - t)$, i.e., it is, but for a scale factor, a mirror image of the input signal about the $t = 0$ axis, shifted by an amount of time t_0. This type of filter is known as a "matched filter" as it is clearly matched to perform optimally for a particular input waveform. The output voltage due to $v_1(t)$ is given by the inverse Fourier transform of $V_2(j\omega) = H(j\omega)\,V_1(j\omega)$, hence

$$\begin{aligned} v_2(t) &= \frac{1}{2\pi} \int_{-\infty}^{\infty} A \; \bar{V}_1^*(j\omega) \; e^{-j\omega t_0} \; \bar{V}_1(j\omega) \; e^{j\omega t} \; d\omega \\ &= \frac{A}{2\pi} \int_{-\infty}^{\infty} |\bar{V}_1(j\omega)|^2 \; e^{j\omega(t - t_0)} \; d\omega \end{aligned} \tag{4.35}$$

The quantity on the right may be recognized from Equation 4.26 as the autocorrelation of the input signal, i.e.,

$$v_2(t) = A \; R_{v1}(t - t_0) \tag{4.36}$$

Taking as an example the input waveform shown in Figure 4.8a, its autocorrelation function is then as shown in Figure 4.8b and the output signal as in Figure 4.8c, where it was chosen to maximize the ratio of

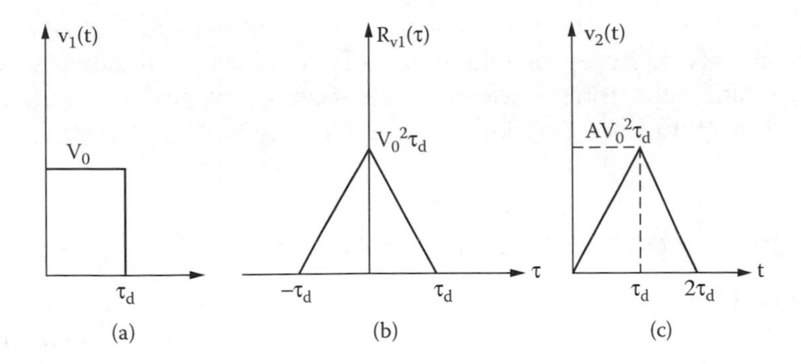

Figure 4.8 Input signal (a), its autocorrelation (b), and output signal (c).

signal to noise at the time the input waveform goes to zero ($t_0 = \tau_d$). Clearly, the output waveform for this matched filter is very different to the input waveform. The objective is maximizing the chances of detecting the signal from the noise and not its faithful reproduction. This situation typically occurs in radar applications. Using Equations 4.33 and 4.34 the variance of the noise at the output is found to be

$$\sigma_2^2 = \frac{W_0 A^2}{\pi} \int_0^\infty \left|\bar{V}_1(j\omega)\right|^2 d\omega$$

The integral on the right is related to the energy associated with the input pulse, hence from Equation 4.14,

$$\sigma_2^2 = W_0 A^2 \int_{-\infty}^{\infty} v_1^2(t)\, dt = W_0 A^2 V_0^2 \tau_d$$

At the moment when the ratio of signal to noise is maximum, the output signal is from Equations 4.35 and 4.14:

$$v_2(t_0) = \frac{A}{2\pi} \int_{-\infty}^{\infty} \left|\bar{V}_1(j\omega)\right|^2 d\omega = A\, V_0^2\, \tau_d$$

Hence, the ratio of signal to noise is

$$\frac{v_2(t_0)}{\sigma_2} = \sqrt{\frac{V_0^2 \tau_d}{W_0}}$$

and it is seen to depend on the energy associated with the signal ($V_0^2\ \tau_d$). It is customary to accept a minimum acceptable ratio of signal to noise for adequate detection of about three. Based on this assumption the required magnitude of the signal at the input is $V_0 \geq 3\ W_0 \cdot \tau_d$.

4.4 The Response of Nonlinear Circuits

Nonlinear circuits are frequently used in engineering applications. Their main utility is to add to input signals other frequency components. Whereas in linear circuits the principle of superposition holds, the same cannot be said of nonlinear circuits. Thus, circuit analysis in nonlinear circuits is more difficult. A simple example is presented here to illustrate some of the solution principles and also introduce a particular circuit used often as a detector.

A circuit configuration used to detect signals is shown in Figure 4.9a. The circuit consists of a diode and charging (R – C) and discharging (R_L – C) circuits. The load resistance is R_L and R represents the diode forward resistance. The detector circuit may be given different characteristics by varying the relative magnitude of charging and discharging time

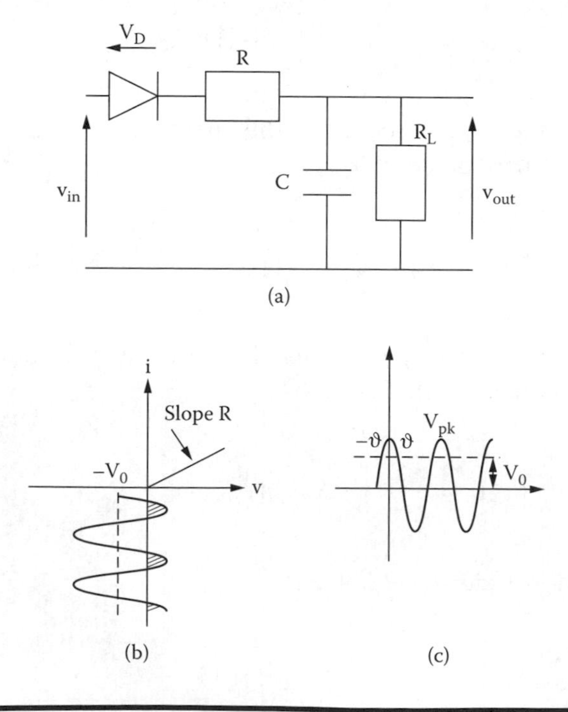

Figure 4.9 A nonlinear detector (a), current (b), and voltage (c) waveforms.

constants. If, for instance, the charging time constant τ_c = RC is much larger than the discharging time constant $\tau_D = R_L C$, then the detector output is sensitive to the peak value of the input signal. If, however, this inequality does not hold, then the detector does not quite respond to the peak and it is described as a quasipeak detector. A low-pass filter after the detector produces an output proportional to the average of the signal. Similarly, modifications may be introduced to produce the rms value of the input signal.

For illustration purposes the response of the circuit shown in Figure 4.9a to a sinusoidal input signal of frequency ω_0 and peak amplitude V_{pk} will be studied. Other cases are presented in detail in Reference 6. Let us assume that the DC component of the output current is I_0. Then the voltage across the diode is $V_D = V_{in} - I_0 R_L$. Assuming an idealized diode characteristic, then charging current flows during the period indicated by the hatched waveform in Figure 4.9b. The DC component of the current is related to the DC component of the hatched voltage waveform V′ by the expression $I_0 = V'/R$. From Figure 4.9c it is clear that $V_0 = V_{pk} \cos\theta$ and hence $\theta = \arccos(V_0/V_{pk})$. Using Equation 4.7, the DC component V′ is obtained as shown below:

$$V' = a_0 = \frac{\omega_0}{2\pi}\int_{-\theta/\omega_0}^{\theta/\omega_0}\left(V_{pk}\cos\omega_0 t - V_0\right)dt = \frac{V_{pk}}{\pi}\left(\sin\theta - \theta\cos\theta\right)$$

Since

$$I_0 = \frac{V_0}{R_L} = \frac{V_{pk}}{\pi R}\left(\sin\theta - \theta\cos\theta\right) \tag{4.37}$$

it follows that

$$\cos\theta = \frac{V_0}{V_{pk}} = \frac{R_L}{\pi R}\left(\sin\theta - \theta\cos\theta\right)$$

and hence

$$\frac{\pi R}{R_L} = \tan\theta - \theta \tag{4.38}$$

If the circuit parameters are such as to give, say, $\tau_D/\tau_c = R_L/R = 2400$, then θ is very small then tan θ may be approximated by the first two terms in its Taylor series expansion

$$\tan\theta \simeq \theta + \theta^3/3$$

Substituting the above expression in Equation 4.38 gives

$$\theta = \left(3\pi R/R_L\right)^{1/3}$$

For the case chosen $R_L/R = 2400$, it follows that $\theta = 0.158$ rad and therefore $V_0/V_{pk} = \cos\theta = 0.987$. The DC component of the output voltage is virtually equal to the peak input voltage. The relationship between the two time constants may be varied to give different characteristics to the detector. In EMC applications, peak and quasipeak detectors are common. The latter has a relatively short discharge time constant to produce a lower output for input pulses occurring infrequently. This is aimed at producing a measure of interference that is related to the relative annoyance caused to listeners of radio signals that are subject to frequent and infrequent pulses of similar magnitude. Descriptions of the different types of detector in use may be found in References 7 and 8.

4.5 Characterization of Noise

Noise may be due to many causes both external (e.g., galactic noise) and internal (e.g., Johnson noise) to a system. More details of the spectral properties of noise sources are given in Chapter 5. Irrespective of the origins of noise, it is necessary to establish a set of parameters that allows quantitative results in regard to noise to be determined for a variety of networks. A more detailed description of noise and its characterization may be found in References 9 to 11. Let us first examine the noise that may be expected at the terminals of an antenna.

An antenna placed in an enclosure of temperature T is surrounded by a low-level electromagnetic field that is the result of random fluctuations.[5] The flux of electromagnetic energy at a given point coming from a solid angle of one steradian is given by Planck's formula and is[12]

$$I(f) = \frac{2hf^3}{u^2\left(e^{hf/K_BT} - 1\right)} \tag{4.39}$$

I(f) is given in units of $Wm^{-2}\,Hz^{-1}\,srad^{-1}$, f is the frequency, u the speed of light, and $h = 6.626 \cdot 10^{-34}$ J/Hz is Planck's constant. In most practical cases $hf \ll K_BT$ and therefore $e^{hf/K_BT} \simeq 1 + hf/K_BT$. Equation 4.39 thus reduces to

$$I(f) \simeq 2\,K_B T/\lambda^2 \tag{4.40}$$

where λ is the wavelength. This is known as *Rayleigh's Law*. Let the mean square value of the electric field strength per hertz be $\langle E^2\rangle = \langle E_x^2\rangle + \langle E_y^2\rangle + \langle E_z^2\rangle$, where the right-hand side indicates the field contributions in each coordinate direction. On average these are expected to be at the same value, hence $\langle E^2\rangle = 3\langle E_z^2\rangle$. The power flux associated with the EM field may be found from Poynting's vector and for air is

$$\frac{\langle E^2\rangle}{120\,\pi} = \frac{\langle E_z^2\rangle}{40\,\pi}.$$

The power flux per steradian is thus

$$\langle E_z^2\rangle / 160\,\pi^2.$$

Equating this with the value obtained from Equation 4.40 gives for the z-directed electric field component

$$\langle E_z^2\rangle = 320\,\pi^2\,K_B T/\lambda^2$$

For a short antenna of length l polarized in the z-direction, the induced voltage will be

$$\langle V^2\rangle = \langle E_z^2\rangle\, l^2 = 320\,\pi^2\,K_B T (1/\lambda)^2$$

Substituting the radiation resistance for this dipole from Equation 2.68 gives

$$\langle V^2\rangle = 4\,K_B T R_{rad},\ \text{in } V^2/H_z \tag{4.41}$$

This formula gives the square of the rms voltage per hertz appearing across the antenna terminals due to EM field fluctuations. The temperature T in this formula represents the regions of space through which EM waves propagate and ranges from a few degrees Kelvin to several thousand degrees in the direction of radio galaxies and is referred to as the *antenna noise temperature*. For an antenna with $R_{rad} = 5\ \Omega$, $T = 20$ K, and system bandwidth 30 MHz, the rms noise voltage is 0.4 μV.

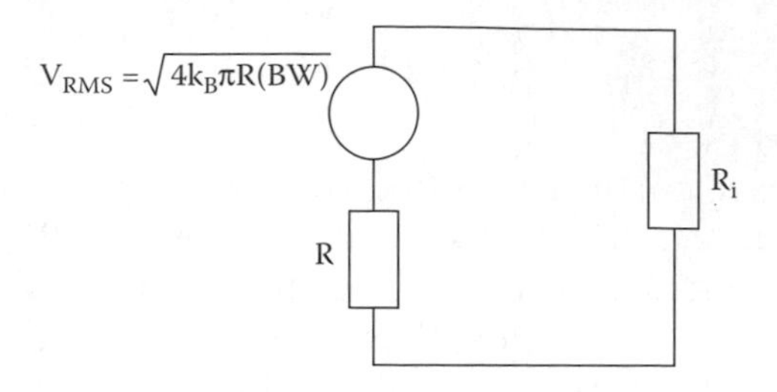

Figure 4.10 Equivalent circuit of a "noisy" resistor connected to a device with input resistance R_i.

It turns out that, irrespective of the physical origin of the resistance, Equation 4.41 applies, where the temperature is that at which the particular resistance is kept (Johnson thermal noise). Although the rms value of thermal white noise is specified in Equation 4.41 and is $V_{rms} = \sqrt{4K_B TR}$, its instantaneous value varies statistically according to a Gaussian probability distribution. The probability that the instantaneous voltage lies between V and V + dV is

$$PdV = \frac{1}{V_{rms}\sqrt{2\pi}} e^{-\frac{V^2}{2V_{rms}^2}} dV \tag{4.42}$$

If a real resistor R is connected to a circuit that presents an input resistance R_i, then some of the noise associated with R will couple to the circuit through R_i. We may then represent the real resistor R by an ideal noiseless resistor R in series with a noise source as shown in Figure 4.10 where we have indicated the RMS value measured over a bandwidth (BW). Maximum power transfer to the load (R_i in this case) occurs when the load impedance is the complex conjugate of the source impedance. In the present case this corresponds to the condition $R_i = R$. Thus the maximum available noise power from the resistor R is

$$P_{max} = \left(\frac{V_{RMS}}{2R}\right)^2 R = \frac{V_{RMS}^2}{4R}$$

Substituting the RMS value from (4.41) we obtain

$$P_{max} = K_B T(BW)$$

where BW is the bandwidth of the circuit (the noise power spectral density is $K_B T$). If as an example we calculate the available noise power over a bandwidth of 1 Hz at room temperature (say T = 290 K) we obtain P_{max} = 4×10^{-21} W. Referring to 1 mW and expressing in decibels gives a maximum available power of

$$P_{max} = 10 \log\left(\frac{4 \times 10^{-21}}{10^{-3}}\right) = -174 \text{ dBm/Hz}$$

This is the minimum achievable noise level at room temperature (noise floor). This noise floor can only be reduced further by cooling equipment below room temperature. It is also pointed out that there are other sources of noise (in addition to thermal noise described above), hence the noise floor is likely to be even higher.

Every device or network used for signal processing receives the desired signal plus noise and produces at its output the modified signal plus noise. At each stage of this process the *signal-to-noise ratio* (SNR) may be defined as

$$\text{SNR} = 10 \log \frac{S}{N} \tag{4.43}$$

where S is the power associated with the signal and N is the power associated with noise. Clearly, a high signal-to-noise ratio guarantees good signal reception and detection. Which value of SNR is acceptable depends to a large extent on the application. For TV reception, values in excess of 40 dB are required for good picture quality. For voice communication of good quality, values in excess of 30 dB are required. It is considered just possible to establish communication between experienced operators at SNR values of 5 dB.

Every device, such as an amplifier, adds its own noise component to the signal and hence the signal-to-noise ratio at its input and output are not the same. A measure of the noise introduced by a device is given by its *noise factor*, defined as follows:

$$\text{Noise factor } F = \frac{(S/N)_{input}}{(S/N)_{output}} \tag{4.44}$$

If N_s is the source noise power and N_a the noise power added by the device per hertz, then

$$F = \frac{S/N_s}{S/(N_s + N_a)} = 1 + \frac{N_a}{N_s} \tag{4.45}$$

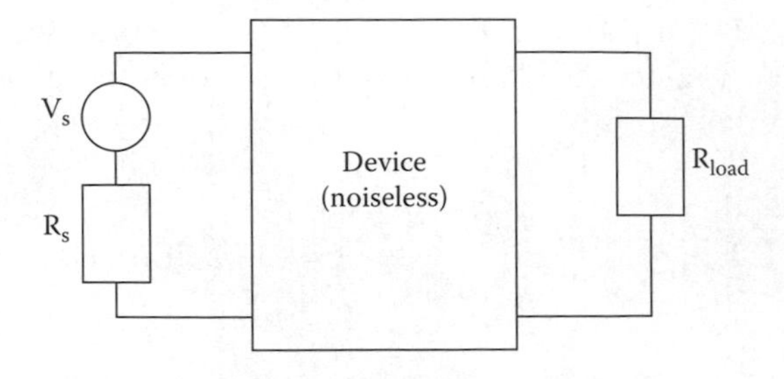

Figure 4.11 Noise equivalent circuit of a device.

The *noise figure* of the device is defined as

$$\text{Noise figure NF} = 10 \log F, \text{ dB} \tag{4.46}$$

A low-noise amplifier has typically a noise figure less than 3 dB.

Another way to describe the noise added by a device is to assign to it an "effective noise temperature" T_e. This is defined as the temperature of a fictitious additional source resistance connected at the input of the device that gives the same noise power at the output. The device is now assumed to be noiseless. Thus, the noise equivalent circuit of a device where the source impedance is R_s is as shown in Figure 4.11. From Equation 4.45 it can be seen that the noise factor and the noise temperature are related by the expression

$$F = 1 + T_e/T_0 \tag{4.47}$$

where T_0 is the temperature of the source resistance (usually the ambient temperature).

Example: An antenna has a noise temperature of 25 K and is connected to a receiver of noise temperature 30 K. Assuming that the minimum acceptable SNR at the output of the receiver is 5 dB, calculate the minimum acceptable signal power at the input of the receiver. A bandwidth of 10 kHz may be used in the calculation.

Solution: The noise factor of the receiver is

$$F = 1 + T_e/T_0 = 1 + 30/290 = 1.1$$

where T_0 was taken as the ambient temperature 20°C. Hence,

$$NF = 10 \log 1.1 = 0.427 \text{ dB}$$

$$SNR_{in} = SNR_{out} + NF = 5 + 0.427 = 5.427 \text{ dB}$$

The antenna noise power is

$$N_{ant} = 4\ K_B T(BW) = 4 \times 1.38\ 10^{-23} \times 25 \times 10^4 = 1.38 \times 10^{-17}$$

Hence,

$$10 \log \frac{\text{antenna signal power}}{1.38 \times 10^{-17}} = 5.427$$

Therefore, the minimum acceptable input signal power is 1.38×10^{-17} W.

It remains now to examine more carefully how the bandwidth of a system may be determined, specifically for noise calculations. If a wide-band noise signal is applied to a frequency-selective network such as a filter or tuned amplifier, the spectrum of the noise at the output follows the shape of the frequency response of the networks. The question often raised is, what is an appropriate bandwidth $(BW)_n$ to use in calculating the total noise power at the output? This *equivalent-noise bandwidth* $(BW)_n$ is defined as the bandwidth of an idealized network with a rectangular response that produces the same noise power at the output as the real network. This is shown in Figure 4.12, where the condition mentioned above is met provided the two hatched areas are equal. It follows that if the input noise spectral power density is W_n (V^2/Hz), then

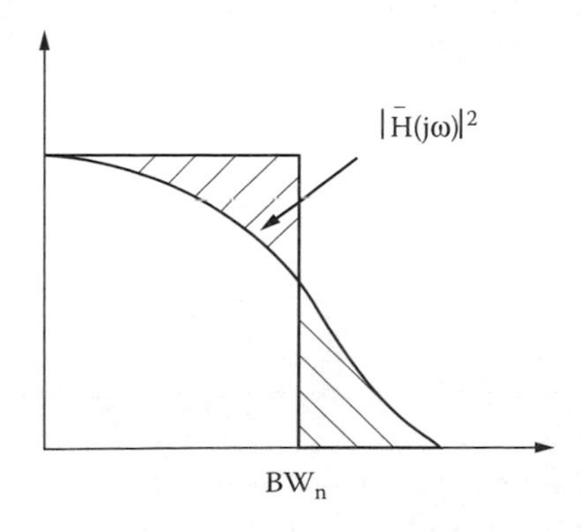

Figure 4.12 Equivalent noise bandwidth.

$$2\,W_n \int_0^\infty \left|H(j\omega)\right|^2 df = 2\,W_n\,H_{max}^2\,(BW)_n$$

or that the equivalent noise bandwidth of a network of frequency response H(jω) (peak value H_{max}) is

$$(BW)_n = \frac{1}{H_{max}^2} \int_0^\infty \left|H(j\omega)\right|^2 df \tag{4.47}$$

For a simple RC network

$$H(j\omega) = \frac{1}{1 + j\omega RC}$$

Hence, $H_{max} = 1$ and

$$\left|H(j\omega)\right|^2 = \frac{1}{1 + 4\,\pi^2 f^2 (RC)^2}$$

Substituting in Equation 4.47 gives the following expression for the equivalent-noise bandwidth of an RC circuit

$$(BW)_n = 1/(4\,RC) \tag{4.48}$$

This should be compared with the bandwidth (BW) of this network defined as the point at which the frequency response falls to $1/\sqrt{2}$ of its peak value (3-dB point), which is (BW) = $1/2\pi RC$. Hence, for this network

$$(BW)_n = \frac{\pi}{2}(BW) \tag{4.49}$$

The relationship between noise and normal bandwidths depends on the exact shape of H(jω). Equation 4.49, however, still holds, approximately, for a range of tuned networks with simple bell-shaped responses.

Example: Calculate the equivalent noise bandwidth and the noise voltage at the output of a tuned amplifier with resonant frequency 1 MHz, quality factor Q = 50, and gain at resonance G = 60. Assume that white noise is injected at the input with $W_0 = 10^{-8}$ V^2/Hz and that Equation 4.49 holds for this network.

Solution: The bandwidth of the amplifier is

$$(BW) = f_{res}/Q = 10^6/50 = 20 \text{ kHz}$$

Hence, the noise bandwidth is

$$(BW)_n \cong \frac{\pi}{2}(BW) = 31.4 \text{ kHz}$$

The variance of the noise at the output is obtained from Equation 4.28 and is

$$\sigma^2 = 2 \times 10^{-8} \times 60^2 \times 31.4 \cdot 10^3 = 2.26 \text{ V}^2$$

It follows that the rms value of the noise voltage at the output is 1.5 V.

Example: Estimate the minimum detectable signal at the output of a system with a noise figure (NF) assuming that we need 3 dB above noise for secure detection.

Solution: Let us assume thermal noise at the input $N_s = K_BT$ and that the system adds noise N_a as shown in Figure 4.13. Then, at the output we have,

$$(N_a + N_s)(BW) = N_s\left(1 + \frac{N_a}{N_s}\right)(BW)$$

or, in decibels,

$$10 \log N_s + 10 \log\left(1 + \frac{N_a}{N_s}\right) + 10 \log(BW) = -174\text{dB} + NF + 10 \log(BW)$$

Hence the minimum detectable signal is the above plus 3 dB.

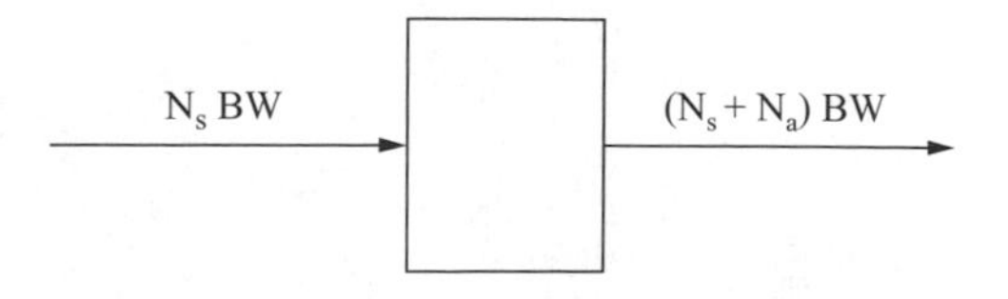

Figure 4.13 Schematic to calculate the noise at the output of a system with noise figure NF.

References

1. Poularikas, A D and Seely, S, "Elements of Signals and Systems," PWS-Kent Publ., Boston, 1988.
2. Candy, J V, "Signal Processing — The Modern Approach," McGraw-Hill, NY, 1988.
3. Lee, Y W, "Statistical Theory of Communication," John Wiley, NY, 1960.
4. Papoulis, A, "Signal Analysis," McGraw-Hill, NY, 1984.
5. Baskalov, S I, "Signals and Circuits," Mir Publ., Moscow, 1986.
6. Geselowitz, D B, "Response of ideal radio noise meter to continuous sine-wave, recurrent impulses and random noise," IRE Trans, on Radio Frequency Interference, pp. 2–11, May 1961.
7. British Standard Specification for Radio-Interference Measuring Apparatus, BS727:1983.
8. CISRP Publ. 16, Specification for Radio Interference Measuring Apparatus and Measurement Methods, 1987.
9. Ott, H W, "Noise Reduction Techniques in Electronic Systems," Second Edition, John Wiley, NY, 1988.
10. Suprynowitz, V A, "Electrical and Electronics Fundamentals," West Publ. Company, 1987.
11. Horowitz, P and Hill, W, "The Art of Electronics," Cambridge University Press, Cambridge, England, 1989.
12. Feynman, R P, Leighton, R B, and Sands, M, "The Feynman Lectures on Physics," Vol. 1, Addison-Wesley, Reading, MA, 1963.

GENERAL EMC CONCEPTS AND TECHNIQUES

In Part I, the basic techniques necessary for studying EMC were described. These may be applied to the study of specific problems that affect the EMC behavior of practical systems. A global view of EMC requires a study of the sources of electromagnetic interference (EMI) in the first instance, so that general and specific threats may be identified for each application. The presence of EMI sources would not be a problem if suitable coupling paths to other circuits were not present. Hence, a study of these paths is necessary and it includes topics such as shielding, apertures, propagation and crosstalk, and coupling at the system level. Finally, the impact of EMI on devices and systems needs to be assessed so that realistic limits and safety margins may be established.

These topics are addressed in Part II using a variety of approaches, including simple intuitive models, analytical formulations, and advanced numerical simulation techniques. The material in the next five chapters is particularly relevant to the EMC analyst.

Chapter 5

Sources of Electromagnetic Interference

The aim of this chapter is to describe the characteristics of the main sources of EM interference (EMI). The reader will be alerted to the great variety of EM threats and will be introduced to the basic qualitative and quantitative models describing their origin, magnitude, and spectral content.

5.1 Classification of Electromagnetic Interference Sources

A large number of sources contribute to the electromagnetic environment. Although in any given application normally only a few of these sources are significant, it is instructive to survey the entire range so that the reader is alerted to various possibilities and is therefore able to ask the right questions when an assessment of EM threats is made. It is undoubtedly difficult to be comprehensive in identifying sources of interference, as each application has its own special features. An astronomical measurement and a domestic electrical appliance are subject to similar EM threats but the dominant design and operational factors are in each case very different. A broad classification of sources of EM interference is useful, however, as it offers a basis for a systematic study. As is natural, the criterion used for classification influences strongly the outcome of this exercise. Sources may be classified according to whether they are continuous or transient in nature. Clearly, in certain applications a transient source of interference may be acceptable, whereas continuous interference would be detrimental to the application. Similarly, sources may be intentional or unintentional in nature. An example of an intentional source is a broadcast radio transmitter, while the radiated interference caused by a computer-to-computer digital signal communication line is clearly unintentional. Another possible classification is into broad- and narrow-band interference. A burst or spike-like disturbance has a broad frequency spectrum unlike, say, interference caused by a mains signal that is at 50 or 60 Hz. A source of interference may be internal or external to a system. Interference caused by switching operation in a power system is internally generated, but phenomena due to lightning are due to sources external to the system. Many other criteria may be used to classify sources, but in this book EMI sources are classified into two broad categories, namely, those due to natural phenomena and those due to human activity.

5.2 Natural Electromagnetic Interference Sources

Human activity, as we understand it, takes place almost exclusively very near the surface of an insignificant planet. It should therefore come as no surprise that this environment is subject to electromagnetic fields that can originate from great distances. We describe these as natural fields. They are grouped in this section under the heading of low-frequency, lightning, and high-frequency fields.

5.2.1 Low-Frequency Electric and Magnetic Fields

It may appear at first sight that low-frequency (LF) fields are of no interest to EMC studies. Undoubtedly, coupling to these fields is generally weak except for particular types of instrument. There are, however, some situations where LF fields may cause significant effects. The geomagnetic field is known to all and may be regarded as due to a magnetic dipole with one pole near Antarctica and the other in North America. Both the position of the poles and the strength of the field are subject to variations. A typical average value is H = 30 A/m. Variations of this field over a period of hours are described as magnetic storms and they generate an associated electric field. It is believed that these storms are related to the sun's activity. Although these fields are small in engineering terms, they may nevertheless cause undesirable effects in large exposed networks. Varying magnetic fields induce potential differences on the earth's surface (typically 1 to 10 V/km) that can drive current through the neutrals of earthed transformers. Currents as high as 100 A have been measured. The effect of these currents is to drive power and current transformers into saturation, thus causing severe difficulties in the protection arrangements of power systems. Further details, and a description of the effects of a particular geomagnetic storm that occurred on March 13, 1989, may be found in Reference 1.

At a distance of approximately ten earth radii major distortions to the dipole magnetic field are observed (earth's magnetosphere). These are the result of the interaction between particles traveling from the sun (solar wind) and the geomagnetic field.[1,2]

Under normal fair weather conditions, an electric field of average value of about 100 V/m can be measured at the earth's surface. It is pointing downward, indicating a negatively charged earth. With increasing height the electric field decreases as air becomes more and more ionized. At a height of approximately 50 km, the electrical conductivity is so high as to permit us to describe the earth's surface as the inner electrode of a giant spherical capacitor with the highly ionized layer at approximately 50 km as the outer electrode. The drift of ions between these electrodes contributes under fair weather conditions a total current of about 1500 A to the earth's surface. Clearly, such a situation, if continued, would result in a change to the electric charge balance on the earth. Charge balance is maintained, however, by an opposite current that flows during electrical storms.[3]

A small electric field of the order of a few millivolts per kilometer is also observed in the earth's crust (tellurian electric field). Values depend on latitude, time, and soil properties and increase substantially during magnetic storms.[4]

5.2.2 Lightning

Lightning is one of the most energetic of all electromagnetic phenomena. During the lightning phenomenon, the potential difference between earth and thunderclouds is of the order of 100 MV and the charge transfer is of the order of 20 C. Hence, energies of the order of 10^{10} J are potentially in transit during electrical storms.[5] Details of lightning may be found in specialist literature on the subject. A simple phenomenological description of lightning is as follows. Through complex processes, charge separation takes place in thunderclouds so that an excess of negative charge is established near their base. Field enhancement and the resulting ionization result in an ionized channel that proceeds in steps (stepped leader) toward the earth as shown schematically in Figure 5.1. As the stepped leader approaches the earth, it causes an upward positive streamer. When the

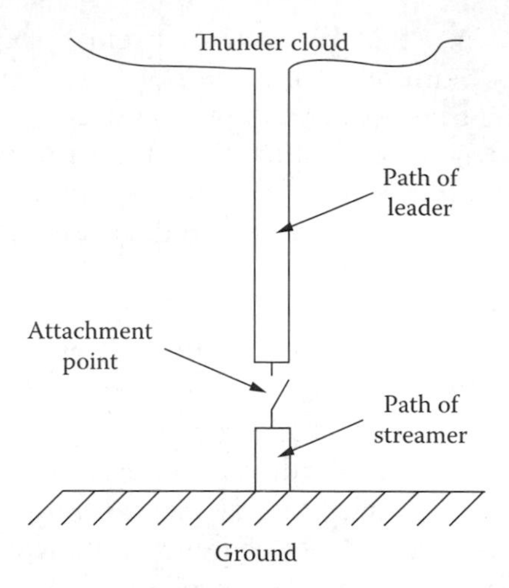

Figure 5.1 Schematic of a lightning discharge.

two meet, a highly conducting ionized channel is formed through which a very high current flows (the return stroke). Return stroke currents can be very high. In approximately 10% of all events, the stroke current exceeds 40 kA, and in about 1% of all cases, currents in excess of 100 kA are observed. The current pulse has a rise-time typically 1 μsec and a decay time of 50 μsec. The number of strokes in any locality may be estimated from the following empirical formula

$$N = 0.15 \times \begin{pmatrix} \text{number of days in a year} \\ \text{when thunder is heard} \end{pmatrix}, \text{ in } \frac{\text{strokes}}{\text{km}^2\text{year}} \tag{5.1}$$

Hence in the United Kingdom, where typically thunder is heard on about 15 days a year, about two strokes per km^2 should be expected in this period. Two types of lightning effects may be distinguished. First, there are direct effects whereby the attachment of lightning to a structure such as a building or aircraft may cause mechanical damage, fires, or explosions. Second, there are indirect effects whereby the rapid current variations during lightning induce significant interference signals on adjacent circuits and structures. It is this latter case that is of most interest to EMC. Of particular concern is the lightning threat to airborne systems. Measurements on aircraft indicate field values of the order of 600 A/m and 50 kV/m and a rate of rise of current of the order of 150 kA/μsec.[6] Recent measurements indicate that rise-times as short as a few nanoseconds are not uncommon. Measurements of the spectrum of the electrical field

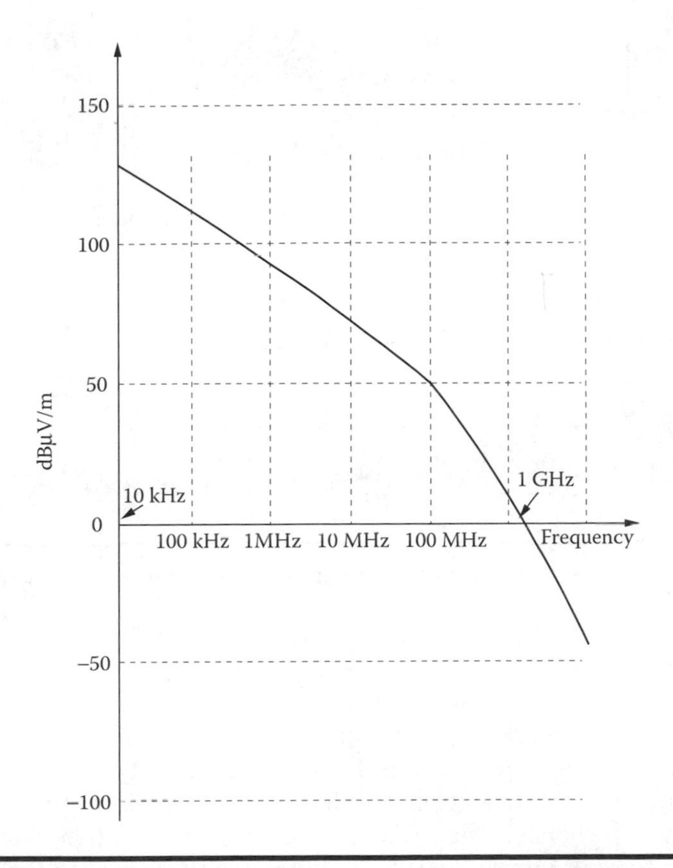

Figure 5.2 Amplitude distribution of sferics — peak electric field at a distance of one mile measured with a bandwidth of 1 kHz. (From Oh, L L, IEEE Trans, EMC-11, 125, 1969. With permission.)

normalized to a distance of 1 mile are shown in Figure 5.2.[7] Further information on the lightning threat may be found in References 8 and 9. Comparisons of lightning with other electromagnetic threats [e.g., nuclear electromagnetic pulse (NEMP) and electrostatic discharge (ESD)] may be found in References 10 and 11.

Lightning is a complex and very severe threat, especially for flying systems, with large statistical variations, and therefore arrangements for testing were developed over the years for the lightning certification of aircraft. These are based on idealized current test waveforms, a classification of effects into direct and indirect, and zoning concepts.

The standard test waveform shown in Figure 5.3 consists of four components, A representing the initial stroke (peak current 200 kA, action integral 2×10^6 A^2s, rise-time 25 μs, and duration 500 μs), B representing an intermediate current (average amplitude 2 kA transferring charge of

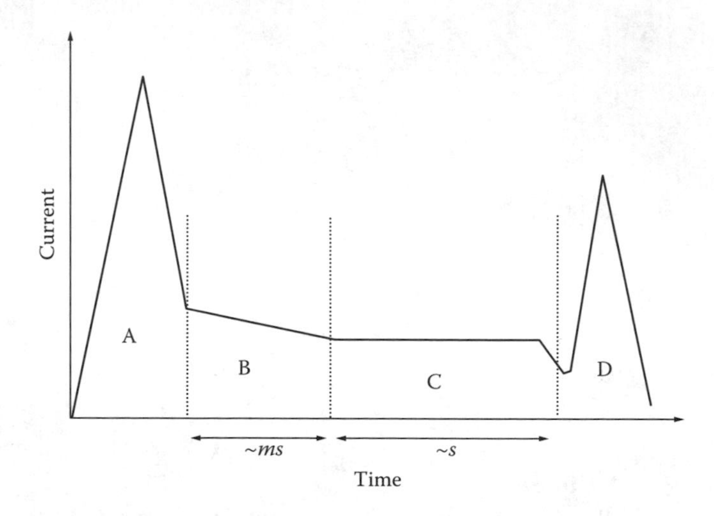

Figure 5.3 Idealized aircraft lightning current waveforms (not to scale).

10 C), C representing a continuing current (amplitude 100s of amperes, charge transfer 200 C), and D representing a restrike current pulse (current peak 100 kA, action integral 0.25×10^6 A^2s). The action integral is defined as $\int i^2\, dt$ and is a measure of the energy dissipated in resistive parts in the path of lightning (responsible for direct current effects). In contract, indirect effects (e.g., inductive coupling) are dependent on the current and its rate of change. The zoning of aircraft is an attempt to identify parts of its structure most likely to be lightning attachment points. Zone 1 refers to parts with a high probability of attachment (e.g., wing tips, nose, engine nacelle, etc.). Following initial attachment and because of aircraft movement the lightning channel is swept along its surface over areas that constitute Zone 2. Finally all other remaining areas are classed as Zone 3. International regulations determine which test current waveforms (A, B, etc.) are applicable to each zone.

The increasing use of carbon fiber composites (CFC) in aircraft manufacture because of their strength and low weight has nevertheless enhanced the importance of lightning as a source of interference with aircraft systems (indirect effects) and also direct effects such as rupture of CFC skins, etc. In a nonmetal skinned aircraft the lightning current path is difficult to establish as it may be significantly affected by local features such as fastenings and other conducting structural members. Similarly, indirect effects can be more severe as the EM shielding afforded by the CFC is inferior to that of aluminum. One can see that significant voltage drops appear across the CFC surface that may contribute to common mode voltage on adjacent circuits. Taking the resistivity of CFC as 3500×10^{-8} Ωm and its thickness as 1 mm we obtain its surface resistivity (the ratio of voltage drop per unit length to current per unit width), ρ/h = 335 mΩ/sq. Hence, a 100-kA peak current will induce a voltage drop of 3500 V/m. The situation will be very different if Al was used ($\rho = 2.8 \times 10^{-8}$ Ωm).

The EM generated by the lightning channel some distance away may be obtained by considering the channel as an antenna where the current distribution is time and space varying. The current need only be known to obtain the vector potential and then the scalar potential is obtained from the Lorentz gauge condition (see Section 2.2.2). With the two potentials known, the EM field is obtained from Equations 2.34 and 2.35. Details of such calculations may be found in References 12–14.

5.2.3 High-Frequency Electromagnetic Fields

High-frequency fields in the range below about 30 MHz are due mainly to terrestrial processes such as electrical storms described earlier and are referred to generally as "static" or "sferics." There are wide seasonal and topographical variations. In general, fields are higher during the night and near the equator, and decrease with frequency. For frequencies higher than about 30 MHz the dominant contribution to the field is extraterrestrial (cosmic) in origin. The earth's magnetosphere, ionosphere, and atmosphere form a partial EM shield to fields originating from outside the earth's environment. At the earth's surface only fields that can penetrate these layers are observed. Penetration is only possible at optical frequencies (infrared to UV) and at radio frequencies between 10 MHz and 37 GHz. The radio frequency passband of interest to EMC is determined by the shielding properties of the ionosphere and absorption by water molecules. The sources of cosmic radiation are in the galaxies (strongest near the center) and in the sun. It has been estimated that each galaxy emits about 10^{35} W in the range 10 MHz to 10 GHz with a large thermal component, but also particular emissions at specific frequencies (e.g., hydrogen at 1.428 GHz). Solar radiation consists of a background level (during a quiet sun), on which are superimposed slow variations (period 27 days), which in the range 1 to 10 GHz may exceed the background value by a factor of three, and radio bursts (flares), which can last from a few seconds to hours with intensities higher by a factor of a thousand compared to background levels.

Due to the statistical nature and wide dependence on location, it is difficult to give specific values for these fields. Surveys of natural background radio noise usually quote results in terms of an equivalent or effective antenna noise factor F_a in decibels. Thus, at 10 MHz a typical value for atmospheric noise is F_a = 40 dB, while at 100 MHz galactic noise dominates and contributes approximately 8 dB.[15] The peak value of the electric field received by a short vertical dipole antenna may be found from F_a using the following formula, which is derived in Appendix E.

$$E(\text{dB}\ \mu\text{V/m}) = F_a(\text{dB}) + 10\log(\text{BW}) + 20\log f(\text{MHz}) - 95.5 \quad (5.2)$$

where BW is the bandwidth of the receiver in hertz and f is the frequency in megahertz. Further details of the spectral content of natural radio noise are presented in Section 5.4.

5.3 Man-Made Electromagnetic Interference Sources

The operation of man-made engineering devices makes an increasingly larger contribution to the electromagnetic environment. This may be regarded as a form of pollution and it is expected that increasingly closer attention will be paid to reducing and controlling electromagnetic emissions. A survey of the main sources of man-made EM emissions is given below.

5.3.1 Radio Transmitters

This is a form of intentional man-made emission. Mobile and fixed transmitters, radar, and computer-to-computer communications are continuously increasing in numbers, power, and geographical distribution. International regulatory bodies allocate fixed frequency bands for the different types of application. An example of allocation in the United Kingdom is shown in Table 5.1.[16] Clearly, in frequency bands allocated to broadcast transmitters, substantial signal strength may be expected, the

Table 5.1 Selected U.K. Frequency Allocations*

148.5–283.5 kHz	Broadcasting
526.5–1606.5 kHz	Broadcasting
13.553–13.567 MHz	ISM
26.957–27.283 MHz	ISM
27.6–28 MHz	Citizens Band
88–108 MHz	Broadcasting (FM)
470–582 MHz	Broadcasting (TV)
886–906 MHz	ISM
934–935 MHz	Citizens Band
2.4–2.5 GHz	ISM

* Different allocations may apply in other regions.

exact value depending on the power of the transmitter, the distance from the transmitter, and the proximity of other bodies (e.g., buildings, transmission lines, etc.). Typical urban environment electric field strength due to broadcast transmitters rarely exceeds 200 mV/m. However, near powerful transmitters field strengths of several tens of V/m are possible. Estimates obtained from 250-kW transmitters at a distance of 100 m range from 87 V/m at 4 MHz to 272 V/m at 26 MHz.[17] Although these emissions are narrowband, interference is also observed at harmonics of the carrier frequency, at sideband frequencies, and there may also be a broadband noise contribution from the various stages inside the transmitter. In a well-designed transmitter, harmonics and noise are well below, typically, 70 dB, the strength at the fundamental frequency.

5.3.2 Electroheat Applications

A valuable means of heating materials is by using high-frequency signals. In cases where the specimen is highly conducting, heating takes place by inducing currents to it (induction heating), whereas for materials that are poor conductors heating takes place by generating dielectric losses (dielectric heating). Typical frequencies used for induction heating are 1 to 100 KHz, 1 MHz, whereas for dielectric heating, operating frequencies are selected from the following list: 13.560 MHz, 27.12 MHz, 40.68 MHz, 433 MHz, 915 MHz, 2.45 GHz, and 5.8 GHz. Power ratings of several kilowatts are typical. Measurements taken at 30 m away from such heaters have shown average electric field values typically 100 (dB μV/m)[18]. Harmonics of the operating frequency of domestic microwave ovens have also been investigated as a source of interference at broadcast satellite frequencies (Table 5.1).[19]

5.3.3 Digital Signal Processing and Transmission

Modern digital signal processing and transmission methods use fast pulses to code information. Processing, storage, input, and output take place continuously synchronized to an electronic clock. The requirement for higher processing speeds and higher rates of information transfer implies high clock rates and therefore shorter transition times (fall- and rise-times). Clock rates of several hundreds of megahertz and transition times of a few nanoseconds are common. The presence of fast pulses on printed circuit boards and transfers across communication cables can give rise to radiation and coupling (crosstalk) to adjacent circuits. As indicated in Appendix D, a trapezoidal pulse having a duration τ and equal rise- and fall-times τ_r has an approximately flat spectrum extending up to a frequency $1/\pi\tau$.

Thereafter the spectrum falls by –20 dB/decade up to a frequency $1/\pi\tau_r$. Beyond this frequency, the spectrum falls by –40 dB/decade. It is therefore evident that substantial power is associated with harmonics of the fundamental frequency and that the highest significant harmonics are determined by the transition time of the pulse waveform. Fast transition times, although desirable from the operation point of view, are undesirable in EMC terms as they make substantial contributions to high-order harmonics. Significant amounts of power are found at harmonics that may be at frequencies twenty times or higher above the fundamental. Estimates of radiated noise for different logic families may be found in Reference 20.

Example: Obtain the envelope of the spectrum for the signal shown in Figure 5.4a and then repeat the calculation by changing the transition time to 200 ns. Duty cycle is 50% and f = 50 kHz. Calculate the interference voltage at 3 MHz for the two values of rise-time.

Solution: Following Appendix D, the envelope of the spectrum is as shown in Figure 5.4b. We need to calculate the values A_1 and A_2 and the frequency breakpoints f_{b1} and f_{b2}. We have

$$T = 1/f = 1/50\times10^3 = 20\ \mu s \quad \text{and} \quad \tau - \tau_r \approx 0.5\times 20\ \mu s = 10\ \mu s$$

Hence,

$$A_1 = 20\log\left[\frac{2V_0(\tau-\tau_r)}{T}10^6\right] \simeq 20\log\left[\frac{2\times100\times10}{20}10^6\right]$$

$$= 120 + 40 = 160\ \text{dB}\mu\text{V}$$

$$f_{b1} = \frac{1}{\pi(\tau-\tau_r)} \simeq \frac{1}{\pi 10\times10^{-6}} = 31.8\ \text{kHz}$$

$$f_{b2} = \frac{1}{\pi\tau_r} = \frac{1}{\pi 300\times10^{-9}} = 1.06\ \text{MHz}$$

A_2 can be found from A_1 and the slope of –20 dB/decade

$$A_2 = A_1 - 20\log\left(\frac{1.06\times10^6}{31.8\times10^3}\right) = 160 - 30.46 = 129.5\ \text{dB}\mu\text{V}$$

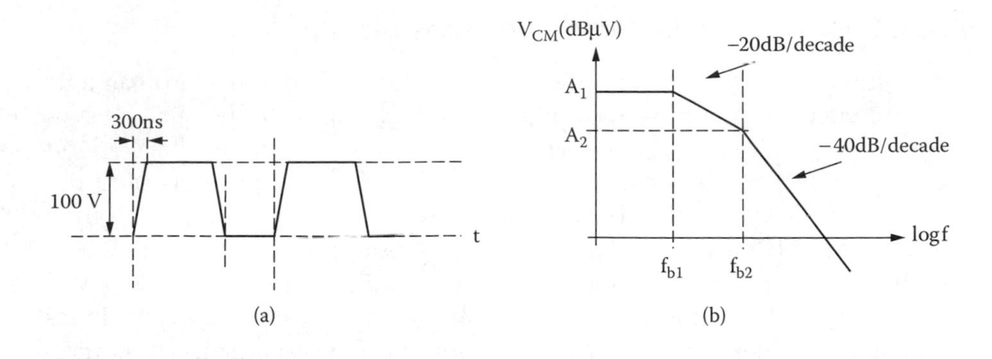

Figure 5.4 A pulse in the time-domain (a) and the shape of its frequency spectrum (b).

The level of the signal at 3 MHz is therefore

$$129.5 - 40\log\left(\frac{3}{1.06}\right) = 111.43 \text{ dB } \mu\text{V}$$

We now repeat the calculation but with a transition time of 200 ns. A_1, f_{b1} are as before:

$$f_{b2} = \frac{1}{\pi\tau_r} = \frac{1}{\pi 200 \times 10^{-9}} = 1.59 \text{ MHz}$$

and

$$A_2 = A_1 - 20\log\left(\frac{1.59 \times 10^6}{31.8 \times 10^3}\right) = 160 - 33 = 127 \text{ dB}\mu\text{V}$$

Hence the level at 3 MHz is

$$127 - 40\log\left(\frac{3}{1.59}\right) = 116 \text{ dB}\mu\text{V}$$

We see that reducing the transition time increases in this case the high-frequency interference level by 5 dB.

5.3.4 Power Conditioning and Transmission

The power supply network is the source and the victim of electromagnetic interference. Although a very high quality of supply is normally maintained, a whole range of signals and disturbances are established on extensive and partially exposed power networks. The presence and the quality of this network, which by its nature is all pervasive, are of great interest to EMC. A broad legal framework exists in most countries, which defines the maximum variation of voltage and frequency (e.g., 240 V ± 6%, 50 Hz ± 1% in the United Kingdom). However, most power utilities maintain a tighter control of their network and impose additional requirements on the quality of the supply. Increasingly, EMC-related regulations impose additional constraints.

A survey of EMC in power systems may be found in Reference 21 and broadly the same classification is used here.

5.3.4.1 Low-Frequency Conducted Interference

Harmonics — Due to the presence of converter equipment,[22] nonlinear devices such as transformers[23] and arc furnaces[24] in power systems, sinusoidal voltages, and currents are generated with frequencies that are multiples (harmonics) of the fundamental mains frequency (50 or 60 Hz). As an example, a converter contributes signals at harmonic frequencies that are a multiple n of the mains frequency

$$n = mp \pm 1$$

where m = 1, 2, 3, ... and p is the converter pulse number p = 6, 12, ... The voltage total harmonic distortion is defined as

$$\mathrm{VTHD} = \frac{100}{V_1}\sqrt{\sum_{n=2}^{\infty} V_n^2}, \text{ percent} \tag{5.3}$$

Some of the undesirable effects of harmonics are interference with communications, overheating, maloperation of instrumentation, control and protection systems,[25] and torque pulsations in motors. The level of harmonics varies throughout the day. Of particular concern is the level of the 5th harmonic where values as high as 6% of the nominal voltage have been observed.[26]

Voltage fluctuations — Small changes in the voltage as the result of connection and disconnection of loads are referred to as fluctuations or flicker. The latter term is to indicate the effect that these changes have on the luminosity of incandescent lamps. A 27% fluctuation is considered acceptable provided it does not occur more than once a minute, whereas a fluctuation occurring ten times per minute should not exceed 1.3%.

Voltage dips — Sudden reductions in the supply voltage (more than 10%) followed by recovery within a short time (typically 300 msec) are common in supply networks. Surveys have shown that the causes of these dips and occasional interruptions originate in the medium- and high-voltage networks.[27] Most of the dips do not exceed 30%. It is estimated that for a consumer in an urban environment, there are about four dips per month that exceed 10% of the nominal voltage. Dips are caused by faults, the switching of large loads, and the starting of induction motors. Dips and interruptions cause problems to data processing equipment and may require restarting and proper sequencing of motors and automatic processes.

Unbalance — The presence of large single-phase loads and the unequal loading of each phase in a three-phase system may cause an inbalance in amplitude and/or phase. Some motors may overheat if supplied by an unbalanced source. In practice, unbalance is kept below 2%.

Mains signaling systems — The power network is primarily intended for the transmission of power to consumers. It is nevertheless possible to transmit information that is used for control (public lighting), load management, and tariff fixing. Mains signaling uses relatively high frequencies and networks not specifically intended for signal transmission. There are, therefore, EMC implications, especially since the use of mains signaling is likely to increase in the future. Three types may be distinguished, namely, ripple control systems (100 Hz to 5 KHz, amplitude 5% of the nominal voltage), power-line carrier systems (up to 100 kHz, amplitude 2.5% of the nominal voltage), and home signaling systems used by individual consumers. The EMC implications of using the power network for digital signal transmission are discussed in Reference 28.

Other lower-frequency effects — Voltages may be induced from adjacent circuits, especially during transient conditions, DC components may be present if half-wave rectifiers are used, and small variations in frequency may be present, normally not exceeding 1 Hz/s.

5.3.4.2 Low-Frequency Radiated Interference

Stray electric and magnetic fields due to power lines and domestic appliances are subject to a variety of limits in various countries. These are normally set with biological effects in mind. At power frequencies electric fields not exceeding 30 kV/m (industrial environment and near power lines) and 0.3 kV/m near domestic appliances may be expected. Magnetic field values rarely exceed 1 mT in industrial environments and 100 μT near domestic appliances and cables.[21,29] Much medical apparatus is sensitive to magnetic fields that are a fraction of a microtesla and video display units exhibit flicker when the magnetic field is of the order of a few microtesla.

Low-frequency electric fields at harmonics of the power frequency have been studied near 500 kV transmission lines and are reported in Reference 30. Values ranging from 13 mV/m at 15 kHz to 130 μV/m at 200 kHz have been measured. Global measurements of natural background radio noise at power frequencies and contributions made by power lines are reported in References 31 and 32.

5.3.4.3 High-Frequency Conducted Interference

Voltage spikes in LV networks — A variety of faults and switching operations and lightning strikes on power networks result in a number of spikes (typically ten per day exceeding 200 V) appearing at the domestic consumer level. A number of surveys are available aiming to establish statistical data for spike amplitude and frequency of occurrence. A thorough discussion appears in Reference 33. About 2% of spikes exceed 500 V and about one in a thousand exceeds 3000 V. The duration of most spikes does not exceed a few microseconds, indicating a bandwidth of a few hundred kilohertz. Less than one in a thousand have a duration exceeding 100 μsec. Direct lightning strokes generate overvoltages of the order of 100 kV. These are attenuated as they propagate through the network but surges of several kilovolts in magnitude can be expected at domestic consumer level.

Voltage surges in high-voltage (HV) substations — A substation is a harsh electromagnetic environment, the most severe problems from the EMC point of view being the operation of disconnect switches. With proper control and protection measures, voltage surges do not normally exceed a few thousand volts. A survey of the substation EM environment appears in Reference 34. In modern gas-insulated substations (GIS) where SF_6 is used as the insulating medium, faster switching times are obtained with the result that steep-fronted transients with rise-times of the order of 10 to 20 nsec and a rate-of-rise as high as 40 MV/μsec are generated.[35] These surges, in addition to the insulation problems they cause, also have undesirable EMC effects.

5.3.4.4 High-Frequency Radiated Interference

Significant radiated fields are measured near substations during switching operations. Electric fields as high as 70 kV/m with frequency components in excess of 200 MHz have been observed. In a conventional substation the data shown in Table 5.2 are regarded as typical.[34,36]

Table 5.2 Radiated Fields Near Substations

	Electric Field		*Magnetic Field*	
	345-kV Substation	*500-kV Substation*	*345 kV*	*500 kV*
Amplitude	5 kV/m	50 kV/m	1.2 A/m	2 A/m
Rise-time	180 nsec	700 nsec	60 nsec	100 nsec
Duration	100 nsec	1500 nsec	2000 nsec	5000 nsec
Bandwidth	20 MHz	20 MHz	5 MHz	20 MHz

Adapted from D. Russell, et al., IEEE Trans, PAS-103, 1863, 1984. With permission.

Field measurements near a HV DC converter station have been reported in Reference 37. A detailed discussion of EMC in power plants and substations may be found in the CIGRE Guide.[38]

5.3.5 Switching Transients

Many of the particular examples of interference-generating phenomena, outlined in the last two sections, can be understood by reference to fundamental principles. Many other interference sources, which have not been explicitly mentioned here, can be similarly understood. It is therefore important to present and illustrate these principles in a manner that allows proper characterization of known interference sources and also permits the identification of new ones. This task is tackled in this section.

5.3.5.1 Nature and Origin of Transients

Although a lot of interference is due to the operation of circuits in steady state, some of the most severe cases of EMI are due to transients that are externally imposed on systems, or those generated internally as a result of normal functional requirements, or in response to abnormal conditions (faults). In an electrical circuit operating under steady-state conditions, energy is stored in particular components such as capacitors and inductors. Any change in circuit conditions, irrespective of its origin or purpose, requires a redistribution of this energy. Capacitors store energy associated with electric fields.

A capacitor C with a potential difference V across its plates stores an amount of energy given by

$$W_e = \frac{1}{2} CV^2 \text{ J} \tag{5.4}$$

Similarly, in a part of space where the electric field is E and the dielectric permittivity is ε the energy storage density is

$$w_e = \frac{1}{2}\varepsilon E^2 \text{ J/m}^3 \tag{5.5}$$

Inductors store energy associated with the magnetic field and the following formulae apply

$$W_m = \frac{1}{2}LI^2 \text{ J} \tag{5.6}$$

$$w_m = \frac{1}{2}\mu H^2 \text{ J/m}^3 \tag{5.7}$$

Any redistribution of energy in a circuit, whether described by lumped components or more generally distributed components, cannot be done instantaneously. This is a fundamental law of physics, and it implies that during a period of time, however short, potentially large amounts of energy are in transit throughout the circuit. During this period the circuit is described as being under transient conditions. It is important to realize that during this time the values of voltage and current in the different parts of the circuit bear little direct relationship to their steady-state values prior to or after the change. Under these circumstances, overcurrents, over-voltages, and rather fast pulses are not uncommon and these influence EMC behavior decisively.

A qualitative grasp of these phenomena may be achieved by recalling that a capacitor is a component that resists rapid voltage changes. Any attempt to force a high rate of change of voltage (dv/dt) results in a very high current. Similarly, an inductor is a component that resists rapid current changes. Any attempt to force a high rate of change of current (di/dt) results in a very high voltage. Improvements in EMC performance are very often achieved by controlling the rate of change of these quantities under all conceivable (normal and abnormal) conditions. Some examples of how circuits respond to change are given in the next two subsections.

5.3.5.2 Circuit Behavior during Switching Assuming an Idealized Switch

It is perhaps necessary before proceeding further, to state clearly what is meant by an "idealized switch." Let us assume, to start with, that when

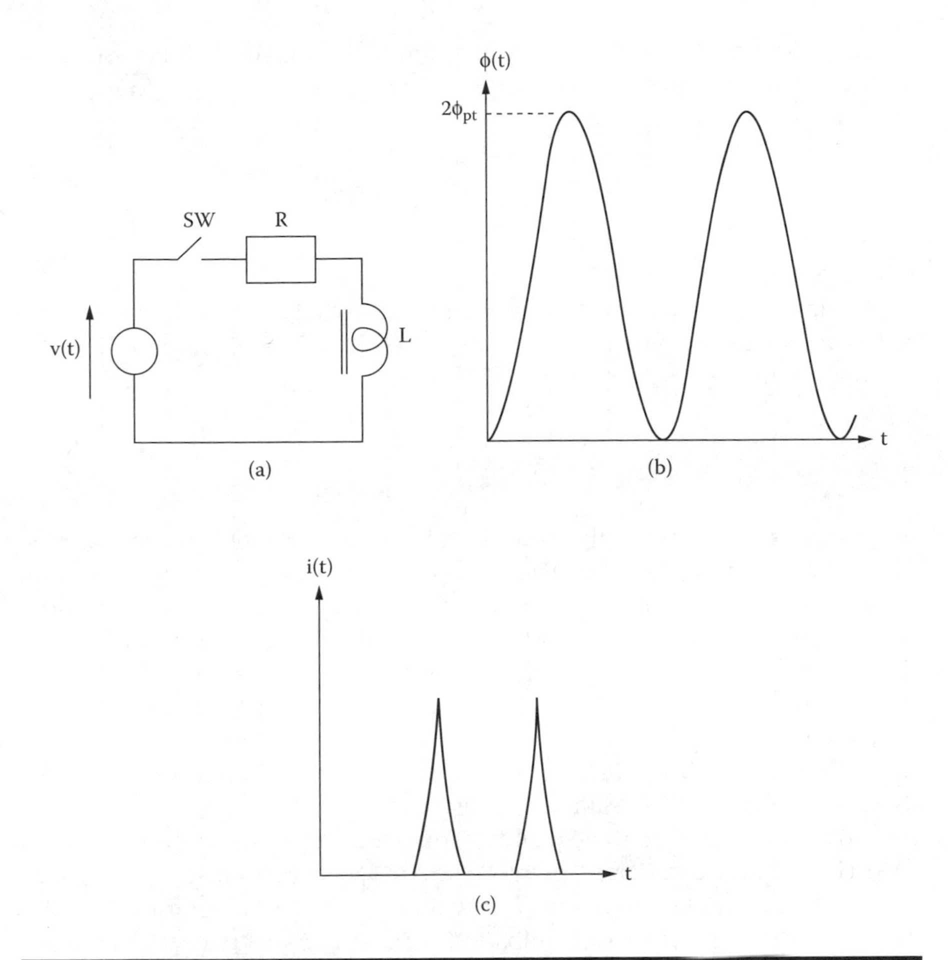

Figure 5.5 Closing transient on an inductor (a). Flux (b) and current (c) waveforms.

the switch is conducting (on) it presents zero impedance, whereas when it is not conducting (off) it presents infinite impedance. It is stressed that this behavior is neither available from practical switches, nor, in fact, desirable. This latter point will be demonstrated shortly after a few examples have been studied first. Let us examine what happens when an inductor is connected to an AC source, as shown in Figure 5.5a (closing transient in an inductor). It is assumed that the switch is closed at the instant in time when the source voltage has zero value and is rising. Neglecting the influence of losses ($R \rightarrow 0$) voltage balance in this circuit requires that

$$N \frac{d\phi}{dt} = V_{pk} \sin(\omega t) \tag{5.8}$$

where N and ϕ are the number of turns and the magnetic flux associated with the coil. Integrating this equation gives

$$\phi(t) = -\frac{V_{pk}}{\omega N}\cos(\omega t) + K \tag{5.9}$$

where K is a constant.

Assuming that at t = 0 (the moment the switch closes) there was no remanent magnetic flux linked with the coil ($\phi(0) = 0$) allows K to be calculated giving

$$\phi(t) = \phi_{pk}(1 - \cos(\omega t)) \tag{5.10}$$

where $\phi_{pk} = V_{pk}/(\omega N)$ is the peak steady-state magnetic flux in the coil. This expression is plotted in Figure 5.5b and it shows that the maximum excursion of the flux is to $2\phi_{pk}$, twice as large as the value expected under steady-state conditions. In most cases this flux corresponds to a value deep in the saturation region of the material used in the construction of the coil and hence it results in a current that is highly distorted and of a very large peak value, as shown schematically in Figure 5.5c. The waveform shown in this figure is rich in harmonics and therefore contributes to EMC problems. The situation is even worse if it cannot be assumed that the remanent flux is negligible. This calculation describes the worst case (losses and a different point-on-wave for switch closure result in a less severe, damped, transient). However, it is not uncommon for the peak magnetizing current of unloaded transformers to exceed, occasionally, twenty times normal values.

A transient of particular interest is that associated with *opening an inductive circuit* (Figure 5.5a). Even for this simple case no progress can be made in obtaining the response of the circuit without some explanation of the properties of the switch. Three possible scenarios are shown in Figure 5.6. In each case the switch is opened at time t_0 but the subsequent circuit behavior is different.

In case (a) the current I_0 is brought down to zero instantaneously. Notwithstanding the difficulties of engineering such a switch, it is interesting to speculate what will happen to the energy stored in the inductance ($\frac{1}{2}LI_0^2$) just prior to time t_0. Clearly, it cannot remain there since there can be no current through L. This is a classic case where the circuit shown in Figure 5.3a is not, even approximately, a reasonable model of the physical system. Neglecting losses again for clarity, the model can be improved considerably by adding the "stray" capacitance C shown in Figure 5.7. This capacitance could be that associated with the coil, bushing

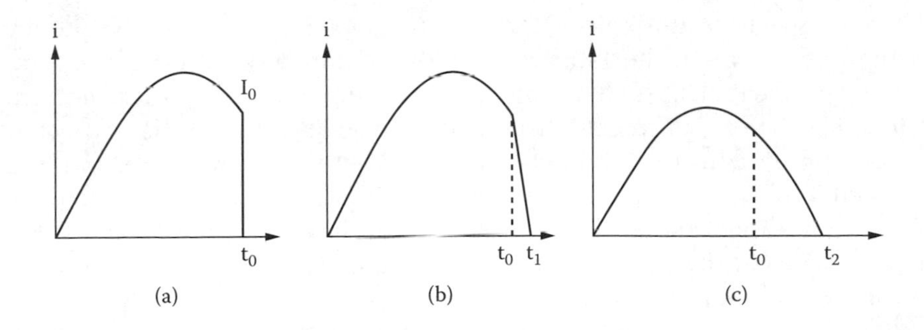

Figure 5.6 Current interruption, abrupt (a), gradual (b), and at the next current zero (c).

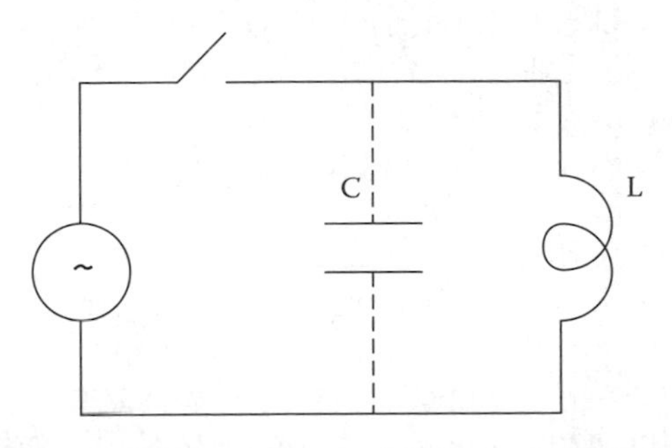

Figure 5.7 Opening transient in an inductive circuit.

(if the coil is of a substantial high-voltage construction), the switch open contacts, connecting wires, etc. Assuming that the voltage across C is small prior to the switch opening, then at t_0 the circuit to the right of the switch consists of an inductor with energy ($\frac{1}{2}$ LI_0^2) and a capacitor that is approximately uncharged. This circuit will undergo oscillations with a frequency equal to $1/2\pi\sqrt{LC}$. The peak value of the voltage V_{pk} across the coil and the capacitor can be estimated easily from energy conservation considerations. If losses are neglected, then when the current through the coil is zero, all the energy is stored as electrostatic energy, i.e.,

$$\frac{1}{2}\ CV_{pk}^2 = \frac{1}{2}\ LI_0^2 \tag{5.11}$$

Hence,

$$V_{pk} = I_0\sqrt{\frac{L}{C}} \tag{5.12}$$

Two aspects of this calculation warrant attention. First, the frequency of the oscillations in the L-C circuit is totally unrelated to normal operational specifications, such as the source frequency and, therefore, in principle, a very different oscillation frequency should be expected depending on construction methods and layout (value of C). Second, the magnitude of the peak voltage is totally unrelated to the source voltage. If the stray capacitance is very small, V_{pk} can have very large values that may cause flash-over across L and/or sparking across the switch contacts. In either case there are serious EMC implications.

Let us now examine case (b) in Figure 5.6 where the current is brought down more gradually from value I_0 at t_0 to zero at t_1. A linear variation is shown but in practice an exponential decay is normally obtained. Clearly, a switch so designed creates fewer problems for insulation and EMC. The rate of current decay is better defined and can be controlled so that any overvoltages can be kept within specified limits, i.e., $V = L\,\Delta i/\Delta t$ where $\Delta t = t_1 - t_0$. This case is described in more detail in the next subsection.

Finally, a switch may be designed so that in spite of contact separation (tripping) at t_0, the current is not effectively interrupted until time t_2 corresponding to the next available natural current zero (Figure 5.6c). This is clearly a case where overvoltages are minimized, since no energy is left stored in L to cause further oscillations. In practice it is difficult to achieve interruption exactly at t_2 and effectively the situation is not dissimilar to that shown in Figure 5.4a, but with I_0 being very small. The current at which the switching arc becomes unstable prior to the current zero is known as the chopping current and in a well-designed, high-power switch is equal to a few amperes. As an example, the interruption of the HV current in an unloaded 1000 kVA 11/3.3 kV power transformer is studied. The value of L is approximately 14H and C (mainly associated with the HV bushing) is 5000 pF. The "surge impedance" of this coil is therefore $\sqrt{L/C} \simeq 50\ \text{k}\Omega$. Assuming a chopping current of 1.5 A results in a peak voltage $1.5 \times 50 \times 10^3 = 75$ kV. This is considerably higher than the operating voltage of this transformer and illustrates the problem of interrupting inductive circuits. When losses are considered and when remanent flux is taken into account the problem appears less severe. For example, if only 50% of the coil energy is recoverable (the remainder is associated with the remanent flux), then Equation 5.12 is modified to give

$$V_{pk} = 0.63\, I_0 \sqrt{L/C}$$

The switch characteristic shown in Figure 5.6a is not practically achievable, that shown in Figure 5.6b is typical of switching at low voltages (e.g., fuses), and the characteristic shown in Figure 5.6c is typical of high-power circuit breakers.

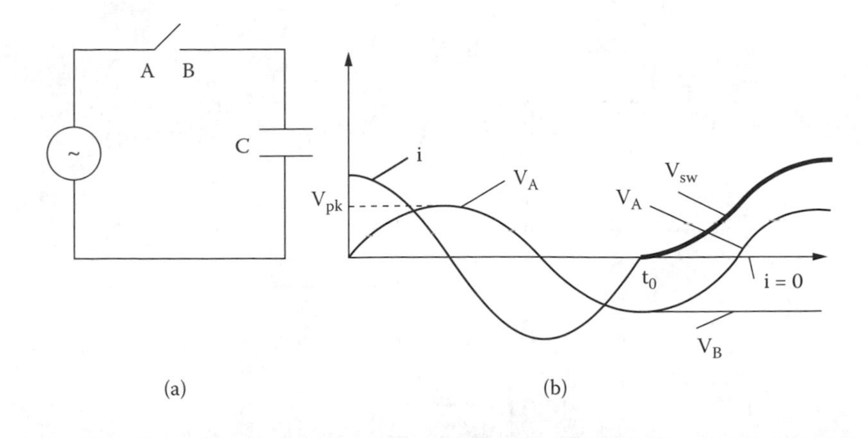

Figure 5.8 Opening transient in a capacitive circuit (a). The voltage across the switch $V_{sw} = V_A - V_B$ is shown in heavy outline in (b).

Let us now examine the case of the *opening transient in a capacitive circuit.* A typical circuit is shown in Figure 5.8a. It is assumed that the current is cleared at a natural current zero (t_0) as shown in Figure 5.8b. The source voltage V_A is also shown and it is clear that for $t > t_0$ the potential of point B remains constant at $-V_{pk}$. The potential difference across the switch is therefore $V_{sw} = V_A - V_B$ as shown in Figure 5.8b. The maximum excursion of V_{sw} is to $2V_{pk}$, thus causing substantial stress on the insulation of the switch. Should the switch fail under these circumstances and reignite, thus reestablishing conduction, fast oscillations will occur with any circuit inductance, as was the case for the inductive transient described earlier, but now with the initial energy store provided by the capacitor. Again, as before, there are important implications for EMC.

All the transients described above were studied using lumped circuit components. The situation is similar in principle, but considerably more complicated, when distributed parameter circuits are involved. For example, interrupting the charging current of an unloaded transmission line is similar to opening a capacitive circuit but with the added complication of the presence of traveling waves and reflections. Further information on transients and parameter values for transient calculations may be found in Reference 39.

5.3.5.3 Circuit Behavior during Switching Assuming a Realistic Model of the Switch

Switching elements, whether of the solid-state type or electromechanical, exhibit complex behavior. In most cases the user of such components has

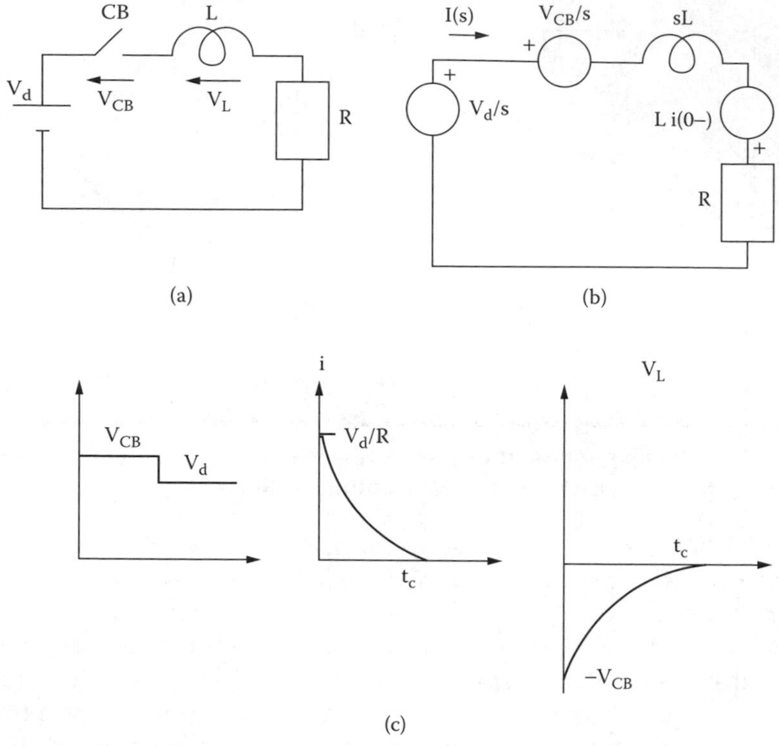

Figure 5.9 Influence of arcing voltage on current interruption (a). The breaker voltage, current, and voltage drop across L are shown in (c).

a limited grasp of their full operational features and even the specialist is hindered by the lack of well-documented, easily accessible information essential for modeling such components. Switch behavior is crucial in EMC studies. Three domains are important in the operation of a switch. First, switch behavior prior to actual current clearance; second, behavior around the point when the current is cleared; and third, its behavior after current clearance as it fully recovers its dielectric strength.

Let us first examine how current is cleared in a low-voltage circuit. The DC circuit shown in Figure 5.9a is studied. It is assumed that initially CB is closed and that steady-state conditions have been established. The electromechanical circuit breaker CB is tripped at t = 0 and it is assumed that the voltage drop across the arc is equal to V_{CB}, and remains constant until the current is brought to zero. The time taken to clear this circuit is required together with the voltage V_L across the inductor. The circuit may be solved using Laplace transforms with the operational equivalent shown in Figure 5.9b. The Laplace transform of the current is found to be

$$I(s) = \frac{V_d - V_{CB}}{L} \frac{1}{s(s + R/L)} + \frac{V_d}{R} \frac{1}{s + R/L}$$

Using tables, the current following the opening of the switch is found to be

$$i(t) = \frac{V_d}{R} - \frac{V_{CB}}{R}\left[1 - e^{-\frac{R}{L}t}\right] \tag{5.13}$$

The time it takes to force the current to zero is obtained from this expression and is

$$t_c = -\frac{L}{R} \ln \frac{V_{CB} - V_d}{V_{CB}}$$

The voltage across the inductor is

$$V_L = L\frac{di}{dt} = -V_{CB} e^{-\frac{R}{L}t}$$

The waveforms for V_{CB}, i, and V_L are sketched in Figure 5.9c. Clearly, the arcing voltage must be higher than V_d to effect current clearance. If $V_{CB} = 1.1\ V_d$ then $t_c \simeq 2.4\ (L/R)$. The magnitude and duration of the inductive voltage drop are crucially dependent on the arcing voltage. In practice the behavior of the switch during arcing is considerably more complex,[40,41] but the simple calculation above illustrates the essential features of switching.

Let us now examine in more detail how a practical electromechanical switch may behave following current interruption. A typical circuit that illustrates the essential parties to this phenomenon is shown in Figure 5.10. Following the parting of the contacts, the current I_{CB} through the switch consists of two parts. First, there is component I_1, which is due to the power source. In the short time after interruption of interest to this study, the source voltage may be considered to remain approximately constant, at its value at the moment of interruption $V_s(0)$. Hence I_1 is approximately equal to $V_s(0)t/(L_1 + L_2)$. Second, there will be a high-frequency component I_2, which is oscillatory in nature and is due to the redistribution of energy trapped in the energy storage circuit elements. The frequency of this component is very high compared with that of the source. Both components are sketched in Figure 5.10b. It is clear that there are moments, such as t_3, when the switch current I_{sw} is equal to

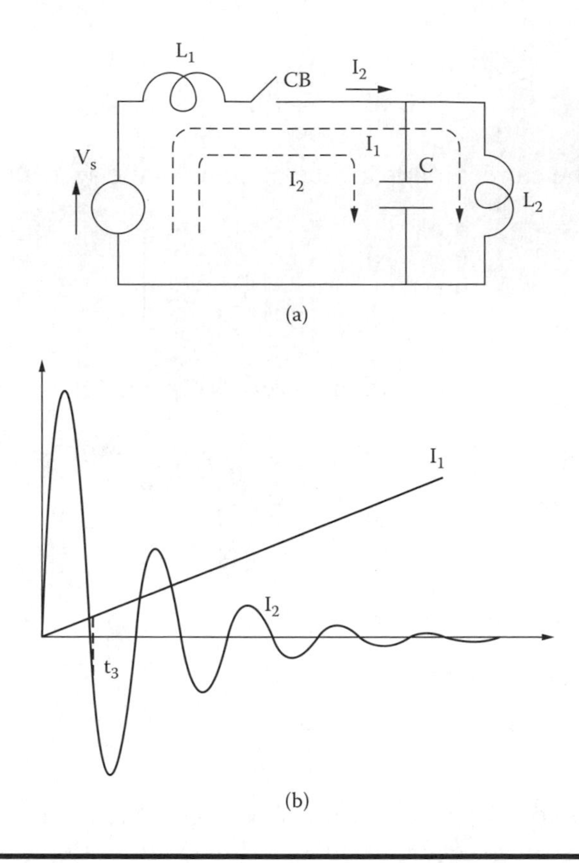

Figure 5.10 Circuit used to study arc reignition (a) and current components at power and high frequencies (b).

zero. The voltage V_{CB}, which appears across the switch, is similarly affected by high-frequency voltages. Following contact separation at t_0 and current clearance, say, at t_1, the circuit breaker recovers its dielectric strength relatively rapidly. The breaker recovery voltage shown by a broken line in Figure 5.11a illustrates this recovery. Following current interruption at t_1 a transient voltage appears across the breaker, as shown schematically by curve AB. At time t_2, the voltage applied to the breaker is about to exceed the level that can be sustained across the contacts. The breaker then reignites and conduction is reestablished. The breaker circuit is now made up of components I_1 and I_2 as explained earlier, and goes through zero at time t_3 as shown in Figure 5.11b. At this point the breaker may be able to clear the current again. The transient voltage difference across the breaker will then increase as shown by CD in Figure 5.11a. Another reignition is likely at time t_4 with a further current clearance at t_5. This process may be repeated several times until the breaker has recovered

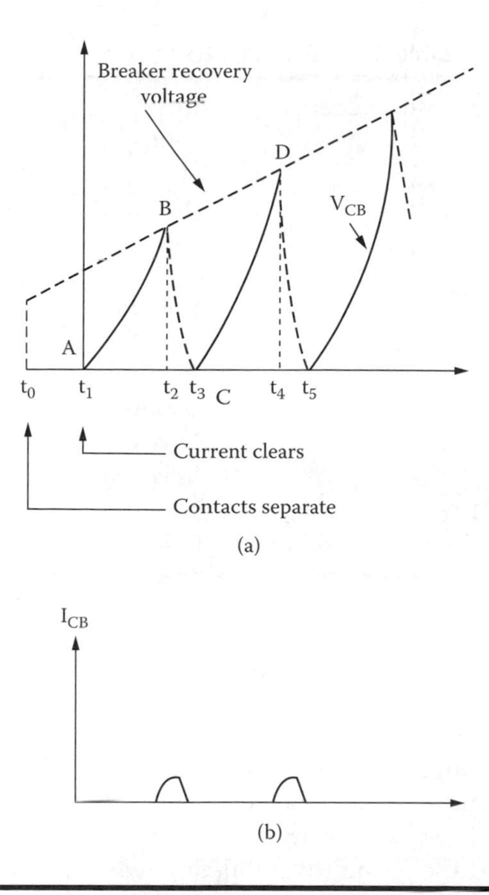

Figure 5.11 Breaker recovery and reignition (a) and breaker current (b).

sufficiently to sustain the maximum applied transient voltage. More details of calculations involving switching phenomena may be found in References 39, 42, and 43. The phenomenon just described is referred to in the literature as "showering arc" and it is capable of producing EMI in the megahertz range. Further details may be found in References 44, 45, and 46.

Another manifestation of the same phenomenon, but without the involvement of a circuit breaker, is the so-called "arcing ground." This is particularly important in ungrounded systems where a fault to ground combines with high-frequency transients involving capacitances to earth to produce a series of fault arc extinctions and reignitions.[39,47]

5.3.6 The Electrostatic Discharge (ESD)

Large material bodies are to a high degree electrically neutral. However, charge separation and charging can take place when two bodies, of which

Table 5.3 Triboelectric Series

Positive charging	Air
	Hands
	Glass
	Mica
	Nylon
	Wool
	Aluminum
	Paper
	Steel
	Acetate
	Polyester
	Polypropelene
	Pvc
Negative charging	Teflon

at least one is insulating, come into contact and subsequently separate. This is known as the triboelectric effect and it is one of the oldest known of electrical phenomena.[48] Some materials have a propensity to acquire electrons on contact, whereas others give away electrons easily. The former tend to charge negatively, whereas the latter tend to acquire positive charge. Materials may be placed in a triboelectric series according to their tendency to charge positively or negatively. An example is shown in Table 5.3.

Charging may also occur by induction whereby a charged body A causes charge separation in a neutral body B. Temporary grounding of B causes some of its charge to leak away, thus leaving B charged. The typical process involved is the charging of an insulator and the subsequent discharge (ESD) through a conductor to another conductor, which may be grounded. The most common problem is that of the charging of humans brought about by walking on low electrical conductivity materials (carpets). It has been found that the human body can charge to a potential often exceeding 10 kV. A phenomenological description of ESD in this case is as follows: Due to the triboelectric effect the shoes are left with negative charge. By induction, charge separation takes place in the body so that the lower extremities are positively charged and the upper body extremities are negatively charged. On approaching another body, an intense electric field is established that may result in an ESD. It has been found that the discharge has a shorter rise-time for faster approach and for lower voltage.

An approximate equivalent circuit is shown in Figure 5.12a. Component C_b represents the body capacitance to ground (typically 150 pF) and R_b the body resistance (typically 1.5 kΩ). Due to triboelectric effects C_b may

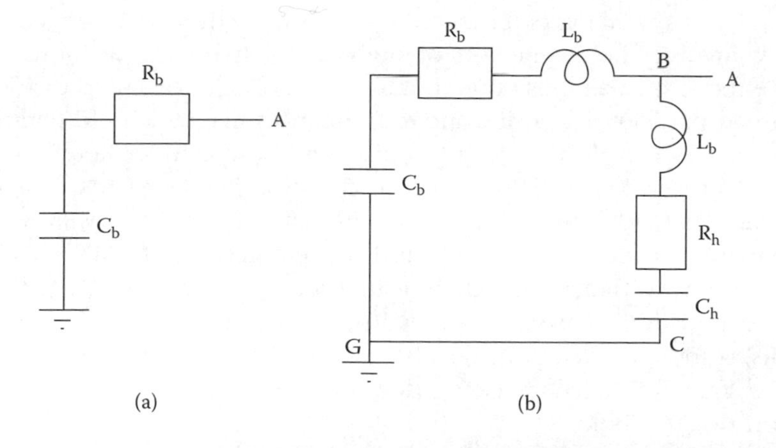

Figure 5.12 Simple (a) and more complete (b) equivalent circuit used to study ESD from the human body.

be charged to several kilovolts. When A is brought into contact with a grounded conducting body, the capacitor discharges, thus causing the flow of current, which results in conducted and radiated interference. An improved model is shown in Figure 5.12b where a small inductance L_b has been added and branch B–C is a model of the human hand. The discharge current has a fast rise-time component (less than 1 nsec) due to branch B–C and a slower component (100 nsec) due to branch B–G. Peak currents of several tens of amperes are typical. Measurements and analytical results of radiated fields due to ESD were reported in Reference 49. An electric field value of 75 V/m was measured at a distance of 1.5 m from a 2-kV ESD. Electrostatic charging problems are also observed in airborne systems, e.g., satellites.[50]

The fast rise-time pulses due to ESD may interfere with clock transitions in digital circuits and thus cause upsets and malfunctions.

5.3.7 The Nuclear Electromagnetic Pulse (NEMP) and High Power Electromagnetics (HPEM)

A high-power electromagnetic pulse (NEMP) is produced following a nuclear detonation. The basic phenomenology is as follows: Following a nuclear detonation a large number of photons (γ-rays) are produced and spread through space. These photons interact with the surrounding material to produce high-energy electrons (Compton effect). The Compton electrons travel at high speed and are thus the source current producing intense electromagnetic fields. These electrons produce further, secondary, electrons that increase the conductivity of the surrounding medium. The temporal and spatial development of the NEMP are therefore very complex as they involve the interaction of high-energy particles with materials. Other effects, such as those due to interaction of neutrons, and distortions in charged particle trajectories due to the earth's magnetic field contribute further to this complexity. More details may be found in References 51 and 52. Three types of NEMP may be distinguished. First, explosions at high altitudes (>100 km) described as high-altitude EMP (HEMP); second, explosions near the ground; and third, system-generated EMP (SGEMP).

The distinctive feature of HEMP is that it affects a very large geographical area and thus presents a simultaneous threat to a large number of systems. Although waveform details are classified, EMP studies are done using a standard NATO pulse to describe the electric field of a HEMP. The form of this field is

$$E(t) = 5.25\times 10^{4}\left[e^{-4\times 10^{-6}t} - e^{-4{\cdot}76\times 10^{8}t}\right] \quad (5.14)$$

This pulse has a rise-time of a few nanoseconds and a decay time of less than 1 μs. The peak electric field is of the order of 50 kV/m. The energy flux associated with this field is of the order of 1 J/m^2 and therefore represents a significant threat. A slower, less severe field lasting for hundreds of seconds follows this sharp pulse (magnetohydrodynamic EMP). Extensive exposed systems, such as power transmission networks, are particularly vulnerable to HEMP. Currents may be induced into these systems causing effects similar to those due to geomagnetic storms (protection malfunctions, transformer saturation, etc.).

Surface EMP refers to explosions near the ground and the associated EM effects have a much more limited geographical range than HEMP. Inevitably, studies of surface bursts involve ground effects and are almost always near the source (SREMP). Such studies are therefore particularly difficult and there are no guidelines regarding standard waveforms suitable for equipment hardening tests.

Finally, SGEMP involves the interaction of incident particles and photons with equipment casings, etc., thus producing very high fields and damage to electronic components.

The strategic objective of NEMP studies is to harden and thus maintain the integrity of critical paths and equipment, so that decision time is increased and enough capability survives a first strike to offer a credible deterrent.

Comparisons of the threat to systems posed by NEMP and by lightning (LEMP) may be found in References 10 and 11.

There is considerable interest in studying Intentional Electromagnetic Environments (IEME) that may result from hostile action or terrorism. A number of experimental systems were developed to evaluate impact. Narrow-band systems can deliver hundreds of megawatts of power in the frequency range 1 to 3 GHz. Moderate band systems delivering power in the range 100 to 700 MHz are also available. Broadband systems such as JOLT[53] can deliver a 100-ps pulse with a field-range product $rE_{pk} \simeq 5.3$ MV. The bandwidth of this system covers the range 40 MHz to 4 GHz. A survey of specialized HEMP systems may be found in Reference 54.

5.4 Surveys of the Electromagnetic Environment

Various organizations provide data characterizing the electromagnetic environment in specific areas or in general terms. A useful collection of survey results relating to mains transients may be found in Reference 33.

Large surveys of radio noise are prepared by the International Telecommunication Union (ITU). Data for external noise levels in the range 0.1 Hz to 100 GHz are reported in Reference 55. In broad terms, the external noise figure in decibels ranges from a value of 260 at 1 Hz to a value of 150 at 1 kHz. More details and values at higher frequencies may be found in the report quoted above. Important also is a survey by the ITU on man-made radio noise.[56] This report gives the mean value of noise power in decibels above thermal noise at 288 K, in the form

$$F = c - d \log f$$

where f is in megahertz and the coefficients c and d depend on the actual environment. As an example, in a residential environment $c = 72.5$ and $d = 27.7$. Further details may be found in the report quoted above. Similar information is provided in standards concerned with the establishment of EMC limits.[57]

In many cases, a systematic electromagnetic survey is necessary to establish the amplitude, frequencies, timing, and spatial details of electromagnetic signals on a particular site. This information is used to study the influence of EMI on equipment installed on this site. Recommendations as to how such surveys may be conducted can be found in Reference 58.

References

1. Kappenman, J G and Albertson, V D, "Bracing for the geomagnetic storms," IEEE Spectrum, Vol 27, pp 27–33, March 1990.
2. Rotkiewicz, W (editor), "Electromagnetic Compatibility in Radio Engineering," Elsevier, Warsaw, 1982.

3. Williams, E R, "The electrification of thunderstorms," Scientific American, pp 48–63, Nov 1988.
4. Chalmers, J A, "Atmospheric Electricity," Pergamon Press, Oxford, 1967.
5. Uman, M A, "The Lightning Discharge," Academic Press, NY, 1987.
6. Moreau, J P and Alliot, J C, "E and H field measurements on the Transall C160 aircraft during lightning flashes," 10th Int Aerospace and Ground Conf on Lightning and Static Electricity, Paris, pp 281–287, 1985.
7. Oh, L L, "Measured and calculated spectral amplitude distribution of lightning sferics," IEEE Trans, EMC-11, pp 125–130, Nov 1969.
8. Uman, M A and Krider, E P, "A review of natural lightning: experimental data and modelling" IEEE Trans on EMC, EMC-24, pp 79–112, 1982.
9. Uman, M A, "Natural and artificially-initiated lightning and lightning test standards," Proc IEEE, 76, pp 1548–1565, 1988.
10. Wik, M W, "Double exponential pulse models for comparison of lightning, nuclear and electrostatic discharge spectra," 6th Zurich Symp on EMC, pp 169–174, March 5–7, 1985.
11. Gardner, R L, Baker, L, Baum, C , and Andersh, D J, "Comparison of lightning with public domain HEMP waveforms on the surface of an aircraft," 6th Zurich Symp on EMC, pp 175–180, March 5–7, 1985.
12. Rubinstein, M and Uman, M A, "Methods for calculating the electromagnetic fields form a known source distribution: Application to lightning," IEEE Trans. on EMC, EMC-31, pp 183–189, 1989.
13. Safaeinili, A and Mina, M, "On the analytical equivalence of EM field solutions from a known source distribution," IEEE Trans. on EMC, EMC-33, pp 69–71, 1991.
14. Thottappillil, R and Rakov, V A, "On different approaches to calculating lightning electric fields," JGR, 106, pp 14191–14205, 2001.
15. Showers, R M, Schulz, R, and Lin, S-Y, "Fundamental limits on EMC," Proc IEE, Vol 69, pp 183–195, Feb 1981.
16. "United Kingdom Table of Radio Frequency Allocations," DTI, Her Majesty's Stationery Office, 1985.
17. Davenport, E M, Frank, P J, and Thomson, J M, "Prediction of field strengths near hf transmitters," The Radio and Electronic Engineer, 53, pp 75–80, 1983.
18. McLachlan, A S, "Radio frequency heating apparatus as a valuable tool of industry and a potential source of radio interference," 7th Int Zurich Symp on EMC, 3–54 March 1987 pp 261–266.
19. Sugiura, A and Okamura, M, "Evaluation of interference generated by microwave ovens," 7th Int Zurich Symp on EMC, pp 267–269, 3–5 March, 1987.
20. Koga, R, Wade, O, Hiraoka, T, Kosaka, M, and Sano, H, "Estimation of electromagnetic impulse noise radiated from a digital-circuit board," Proc Int Conf on EMC, pp 389–393, Nagoya, Japan, 1989.
21. Goldberg, G, "Low-frequency and high-frequency EMC in power systems," Proc 9th Int Zurich EMC Symp, Zurich, pp 635–642, 12–14 March, 1991.
22. Kloss, A, "Harmonics in power systems with converters," ABB Review, 491, pp 29–34, 1991.

23. Yacamini, R, "Harmonics caused by the various types of transformer saturation," Int Electrical Eng Educ, 19, pp 157–167, 1982.
24. Sundberg, Y, "The power circuit of arc furnaces," ASEA J, 45, pp 69–75, 1972.
25. Lai, L L and Johns, A T, "Harmonics and their effect on power system protection," 25th Universities Power Eng Conf, University of Aberdeen, pp 147–150, 1990.
26. Kopp, H and Kizilcay, M, "Statistical recordings of harmonics in low and medium voltage networks," 9th Int Zurich Symp on EMC, pp 649–654, 12–14 March, 1991.
27. Desquilbet, G, Corn, C, and Teisseire, L, "Measurement of distribution network voltage dips and short interruptions: measuring system, analysis method and result presentation," 9th Int Zurich Symp on EMC, pp 643–648, 12–14 March, 1991.
28. Cristina, S, D'Amore, M, and Feliziani, M, "Electromagnetic interference from digital signal transmission on power line carrier channels," IEEE Trans on Power Delivery, 4, pp 898–905, 1989.
29. Maddock, B J, "Overhead line design in relation to electric and magnetic field limits," Power Eng J, 6, pp 217–224, 1992.
30. Daley, M L, Benitez, H, Zajac, H, and Chartier, V, "Harmonic composition of low-frequency electromagnetic emissions associated with high-voltage transmission," IEEE Trans on EMC, EMC-27, pp 227–228, 1985.
31. Fraser-Smith, A C and Bower, M M, "The natural background levels of 50/60 Hz radio noise," IEEE Trans on EMC, EMC-34, pp 330–337, 1992.
32. Yoshino, T and Tomizaura, I, "Measurement of power line radiation over eastern Asia and northern Europe by the EXOS-A 'OHZORA' satellite," 7th Int Zurich Symp on EMC, pp 455–460, 1987.
33. Standler, R B, "Protection of Electronic Circuits from Overvoltages," John Wiley and Sons, NY, Chapter 3, 1989.
34. Don Russell, B, Harvey, S M, and Nilsson, S L, "Substation Electromagnetic Interference — Part 1," IEEE Trans on Power Apparatus and Systems, PAS-103, pp 1863–1870, 1984.
35. Fujimoto, N and Boggs, S A, "Characterisation of GIS disconnecter-induced short rise-time transients incident on externally connected power system components," IEEE Trans on Power Delivery, 3, pp 961–976, 1988.
36. Russell, B D et al., "Measurements and characterisation of substation electromagnetic transients," EPRI Report EL-2982, March 1983.
37. De Vore, R V, Kimball, D F, Kasten, D G, and Caldecott, R, "RF analysis of a 12-pulse HVDC converter," Proc IEE-C, 135, pp 210–218.
38. Working Group 36.04, "Guide on EMC in Power Plants and Substations," CIGRE Publ. 124, Dec. 1997, Paris.
39. Greenwood, A, "Electrical Transients in Power Systems," Second Edition, John Wiley and Sons, NY, 1991.
40. Erk, A and Schmelzle, M, "Grundlagen der Schaltgeratetechnik," Springer-Verlag, Berlin, p 18, 1974.
41. Lee, T H, "Physics and Engineering of High-Power Switching Devices," MIT Press, Cambridge, 1975.

42. Murano, M, Fujii, T, Mishikawa, S, and Okawa, M, "Voltage escalations in interrupting inductive current by vacuum switches," IEEE Trans, PAS-93, pp 264–280, 1974.
43. Greenwood, A and Glinkowski, M, "Voltage escalation in vacuum switching operations," IEEE Trans on Power Delivery, 3, pp 1698–1706, 1988.
44. Minegishi, S, Echigo, H, Ohmori, T, and Sato, R, "Frequency spectra of arc voltage due to electrical contacts opening with arc discharge," Int Symp on EMC, Tokyo, IEEE Rec No 84CH2097-4, pp 85–90, 1985.
45. Uchimura, K, Aida, T, and Takagi, T, "Showering arcs in breaking Au, Ag, Pd and W contacts and radio noise caused by these arcs," Int Symp on EMC, Tokyo, IEEE Rec No 84CH2097-4, pp 91–96, 1985.
46. Uchimura, K, Fujita, H, Ikesue, S, and Aida, T, "Noise induced by showering arc in switching relays and malfunction of digital circuits owing to its noise," Int Symp on EMC, Nagoya, Japan, IEEE Rec No 89TH0276-6, pp 400–405, Sept 8–10, 1989.
47. Nakagawa, S and Satakibara, T, "Harmonic overvoltage due to intermittent arcing fault in cv-cable distribution system," Int Symp on EMC, Nagoya, Japan, IEEE Rec No 89TH0276-6, pp 83–86, Sept 8–10, 1989.
48. Ott, H W, "Noise Reduction Techniques in Electronic Systems," Second Edition, John Wiley and Sons, NY, 1988.
49. Ma, M T, "How high is the level of EM fields radiated by an ESD?" Proc 8th Int Zurich EMC Symp, pp 361–365, 7–9 March 1989.
50. Purvis, C K, Garrett, H B, Whittlesey, A C, and Stevens, N I, "Design Guidelines for Assessing and Controlling Spacecraft Charging Effects," NASA Tech Paper 2361, 44 p, Sept 1964.
51. Longmire, C L, "On the electromagnetic pulse produced by nuclear explosions," IEEE Trans on EMC, Vol 29, pp 3–13, 1978.
52. Lee, K S H (editor), " Interaction: Principles, Techniques and Reference Data," Report AFWL-TR-80-402, Dec 1980.
53. Baum, C E et al., "JOLT: A highly directive, very intensive, impulse-like radiator," Proc. IEEE, Special Issue on Pulsed Power: Technology and Applications, pp 1096–1109, July 2004.
54. Giri, D V, Tesche, F M, and Baum, C E, "An overview of high-power electromagnetic (HPEM) radiating and conducting systems," Radio Science Bulletin, Number 318, pp. 6–12, Sept 2006.
55. ITU Report 670-1 "Worldwide minimum external noise levels, 0.1 Hz to 100 GHz," (1978–1990), Dusseldorf, Annex to Vol. I, 1990.
56. ITU Report 258-5 "Man-made radio noise," (1963–1990), Dusseldorf, Annex to Vol. VI, 1990.
57. American National Standard for "Electromagnetic Limits — Recommended Practice," ANSI C63.12-1987.
58. IEEE Recommended Practice for an Electromagnetic Site Survey (10 kHz to 10 GHz), IEEE Std 473-1985.

Chapter 6

Penetration through Shields and Apertures

In Chapter 5 the origin and nature of the sources of electromagnetic threats that contribute to electromagnetic interference (EMI) were described. EMC performance is dependent upon the presence of EMI sources but also on the existence and nature of coupling paths that permit EM energy from a source to appear as interference signals on sensitive electronic systems. In this chapter, coupling through shields and apertures is studied. It is followed by chapters dealing with propagation, conducting penetrations, and general multipath coupling.

6.1 Introduction

A shield is a layer of conducting material that partially or completely envelops an electronic circuit. It therefore affects the amount of EM radiation from the environment external to the circuit that can penetrate inside and, conversely, it influences how much of the EM energy generated by the circuit escapes to the external environment. Whether the source of EMI is inside or outside the shield, the shielding properties remain the same. A variety of materials are used for shielding with a wide range of electrical conductivity, magnetic permeability, and thickness. Shields invariably contain openings (apertures) used for access and cooling and a number of joints and seams through which EM radiation can penetrate.

Practical equipment also is attached to wires and/or pipes used for signaling and services. All these components constitute breaches of the shield integrity and play a decisive part in the overall performance of the shield. Shielding is important in EMC because, almost alone of all other EMC measures that the designer can adopt, it remains under his or her full control. It would be convenient to be able to specify reductions to EM radiation from sources, or to keep sources of EMI far away from equipment, or arrange the disposition of equipment in the most favorable manner for EMC. This is rarely possible, however, and the designer is left with few options, shielding being one of them, to engineer immunity to EM threats that are inadequately defined.

In this chapter, two specific aspects of shielding are considered. First, penetration via diffusion through conducting shield walls is examined. Second, the ingress of radiation through apertures (holes, windows) on the shield is studied. In both cases, the nature of the environment internal to the shield can strongly affect the distribution and spectral content of the field. This can be easily understood if it is recognized that the shield forms a cavity that resonates at specific frequencies. Such complications are not considered here as they require whole-system modeling and results are therefore problem dependent. Details of how such calculations can be done are given in Chapter 8. Similarly, practical issues of material selection, effect of seams and gasketing, and actual practical performance issues are discussed in Chapter 11.

Before embarking on attempts to quantify shielding effectiveness, it is useful to present a simple physical picture of the shielding mechanism. The generic shielding problem is shown in Figure 6.1. A wave E_i representing EMI is shown in Figure 6.1a incident on a shield of thickness t. The objective is to prevent this radiation from penetrating inside the shield, i.e., to minimize E_t. A similar situation is depicted in Figure 6.1b but now

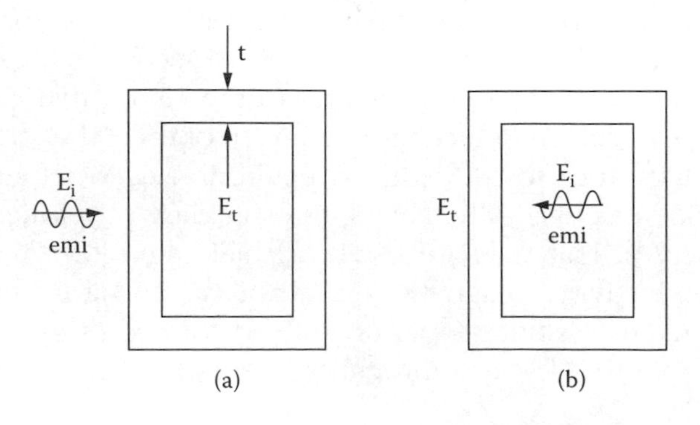

Figure 6.1 Penetration of EM radiation through a shield.

EMI originates from inside the shield. There are two physical principles that can be exploited to design shields. First, it is known that an impressed time-varying field induces currents (Faraday's Law) on adjacent conductors of such magnitude and phase as to oppose the incident field (Lentz's Law). A conducting layer would therefore provide shielding, provided sufficient currents were induced and they were allowed to flow without hindrance. Induction depends on the rate of change of the incident field and therefore becomes less effective at low frequencies. The effectiveness of such a shield is therefore diminished at low frequencies and this turns out to be a serious problem, particularly for magnetic field shielding. Similarly, any seams, joints, and openings on the shield disturb the flow of the induced currents and therefore reduce shielding effectiveness.

Second, it is known that electric and magnetic fields tend to concentrate in materials of high dielectric permittivity and magnetic permeability, respectively. This effect may be exploited to shield low-frequency magnetic fields by constructing a layer of high μ_r.

In practice, both the high electrical conductivity and magnetic permeability contribute to shielding to a greater or lesser extent, but it turns out that at low frequencies magnetic field shielding is μ_r dependent, whereas at higher frequencies, high electrical conductivity shields are more effective.

6.2 Shielding Theory

In this section, the penetration of fields via diffusion through a conductor of thickness t is considered. In order to obtain simple solutions and therefore enhance the understanding of the physics of the process and thus grasp the basic scaling laws and trends, several simplifying assumptions will be made. These assumptions, while making quantitative results of less practical use, will nevertheless help the designer to identify the correct approach to problems and have a grasp of what is possible under idealized conditions. Two approximate methods for calculating shielding effectiveness will be presented. Each method has its limitations and strengths but they both contribute to understanding shielding, which is invaluable in EMC problems. Some analytical solutions will also be referred to as they can be used as benchmarks in tests and calculations. The basic numerical simulation methods that can be used for shielding studies are also described. The shielding of cables is treated in a separate subsection.

6.2.1 Shielding Effectiveness

Shielding effectiveness (SE) is defined as the ratio in decibels of the incident and transmitted field

$$SE = 20 \log \left| \frac{E_i}{E_t} \right| \tag{6.1}$$

E_i normally represents a plane wave incident on one side of an infinite layer of the shielding material of thickness t and E_t represents the field at some point on the other side. As such, Equation 6.1 does not account for the actual shape of the shield. Shielding effectiveness may also be defined in terms of the magnetic field. It should be pointed out that the two definitions do not always result in the same value of shielding effectiveness. An alternative definition of the shielding effectiveness is to interpret E_i as the field at a point without the shield present and E_t the field at the same point when the shield is present. Under some circumstances, it may also be of interest to calculate the phase shift to the field introduced by the shield. Shielding effectiveness of the order of 20 dB is about the minimum worthwhile value, with 50 to 60 dB considered to be an average to cope with most problems. In some cases, such as transmitter equipment, at least 100 dB is desirable. Shielding above about 120 dB approaches the state of the art in this area.

6.2.2 Approximate Methods — The Circuit Approach

Before embarking on more detailed investigations, two mathematically simple problems will be tackled. The objective is to bring out some of the essential features of shielding and also illustrate some of the complexities inherent in this type of calculation. The two examples chosen are for the shielding provided by a thin-walled, very long conducting cylinder, subject to externally applied longitudinal and transverse magnetic fields.

Let us first start with the configuration shown in Figure 6.2a. It is clear that an impressed external field H_i along the z-direction will induce currents in the thin cylinder that tend to oppose the penetration of this field. For this to happen an azimuthal current density (K: amperes/meter) must flow as shown. The field inside the cylinder is that due to the current density K, i.e., the field inside a long solenoid of surface current density K amperes per unit length. It follows, therefore, that inside the cylinder H_t is uniform. In order to relate H_i to H_t, it is necessary to invoke Faraday's Law and apply it on the curve C shown in Figure 6.2a. Thus,

$$\int_C \mathbf{E}\, d\mathbf{l} = -\frac{d\Phi}{dt} \tag{6.2}$$

where Φ is the flux linked with C, i.e., $\mu_0 \pi a^2 H_t$. The electric field in the thin wall is related to the current density J by the expression $E = J/\sigma$

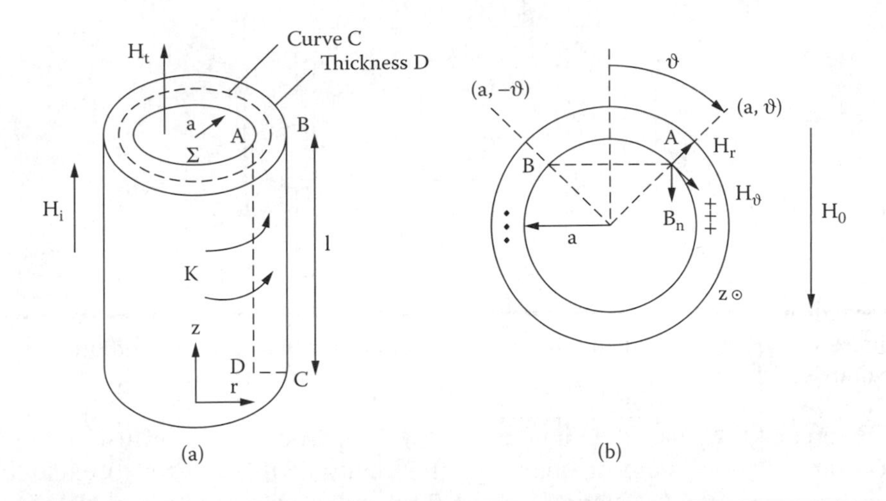

Figure 6.2 Penetration of a longitudinal (a) and transverse (b) magnetic field through a cylinder.

where σ is the electrical conductivity of the cylinder material. The current density J in A/m² is related to the surface current density K in A/m by the expression J D · 1 = K · 1 where D is the wall thickness. Hence J = K/D, and substituting this expression and the expression for the flux in Equation 6.2 gives

$$\frac{K(t)}{\sigma D}2\pi a = -\frac{d}{dt}(\mu_0 \pi a^2 \; H_t(t)) \tag{6.3}$$

where it has been recognized that K and H_t are, in general, time dependent. Similarly, from Ampere's Law applied on path ABCDA, $H_t l - H_i l = Kl$, hence $K = H_t - H_i$. Substituting for K in Equation 6.3 and rearranging gives[1]

$$\frac{dH_t(t)}{dt} + \frac{H_t(t)}{\tau_m} = \frac{H_i(t)}{\tau_m} \tag{6.4}$$

where $\tau_m = \mu_0 \; \sigma \; a \; D/2$. If the impressed field is a step function $H_i(t) = H_i$, then the solution of Equation 6.4 is

$$H_t(t) = H_i \; (1 - e^{-t/\tau_m}) \tag{6.5}$$

The surface current density is therefore

$$K(t) = H_t(t) - H_i(t) = -H_i \; e^{-t/\tau_m}$$

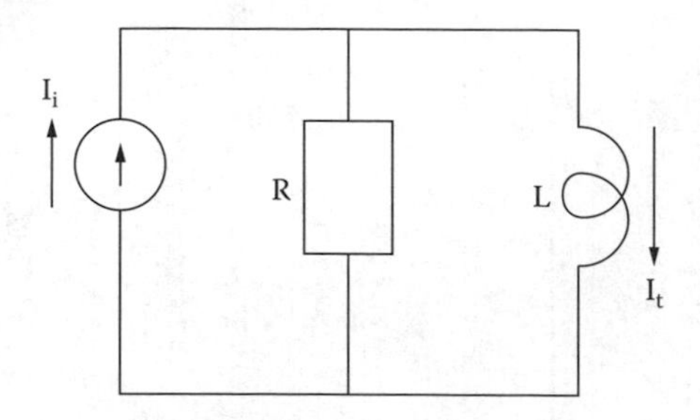

Figure 6.3 Circuit analogue for the penetration of a magnetic field though a cylinder.

The quantity τ_m has the dimensions of time and is the diffusion time constant of the system. It quantifies the ability of this system to exclude the impressed field. Clearly, a short field pulse of duration less than τ_m can be excluded relatively easily by the induced currents and shielding is effective. Conversely, a pulse lasting longer than τ_m will not be adequately attenuated and shielding is less effective. A perfect conductor ($\sigma \rightarrow \infty$) would be an ideal shield, since then τ_m would be infinitely large.

Looking at the structure of Equation 6.4 it is intuitively evident that the penetration of the field inside the cylinder may be understood and studied in terms of the circuit analogue shown in Figure 6.3. For this circuit, KCL gives

$$I_i = I_t + \frac{L(dI_t/dt)}{R}$$

or

$$\frac{dI_t}{dt} + \frac{I_t}{\tau} = \frac{I_i}{\tau} \tag{6.6}$$

where $\tau = L/R$, the time constant of the circuit. Clearly, an analogy may be established between the field and circuit quantities as follows

$$H_i \rightleftarrows I_i$$

$$H_t \rightleftarrows I_t$$

$$\frac{2\pi a}{\sigma D} \rightleftarrows R$$

$$\mu_0 \pi a^2 \rightleftarrows L$$

The circuit analogue approach offers another intuitively powerful tool for studying shielding. Further details of the circuit approach may be found in References 2 and 3.

The second example deals with the shielding properties of a thin-walled cylinder subject to a transverse magnetic field, as shown in Figure 6.2b. The physics of the problem remain the same as for the first example but the mathematics are more complex. Currents will be induced in the cylinder running along the z-direction as shown in (b), in a manner that tends to oppose the impressed field. The problem is best tackled by recognizing that the magnetic field in the inner and outer regions may be obtained as the gradient of a scalar, magnetic potential, i.e.,

$$H = -\nabla\Psi$$

The function Ψ is chosen to conform with the symmetry of the problem and the boundary conditions.[4,5] Thus, the following choices may be made:

$$\begin{aligned} &\text{inner region } \Psi_t = -H_t \, r \cos\theta \\ &\text{outer region } \Psi_i = -H_i \, r \cos\theta + \frac{C_1 \cos\theta}{r} \end{aligned} \tag{6.7}$$

where C_1 is a constant to be determined. The following three conditions must be met:

1. As $r \to \infty$ then $H \to H_i$, i.e., it becomes the uniform impressed field. For the potential chosen $\Psi_i \to H_i \, r \cos 0$ as $r \to \infty$. Hence

$$\begin{aligned} H_i &= \hat{r}\frac{\partial\Psi}{\partial r} + \hat{\theta}\frac{1}{r}\frac{\partial\Psi}{\partial\theta} + \hat{z}\frac{\partial\Psi}{\partial z} \\ &= -H_i \cos\theta \, \hat{r} + H_i \sin\theta \, \hat{\theta} \end{aligned}$$

   i.e., as expected, a uniform field in the transverse direction.

2. Continuity of normal component of B, i.e., at $r = a$

$$\frac{\partial\Psi_i}{\partial r} = \frac{\partial\Psi_t}{\partial r}$$

3. Ampere's Law, i.e., at $r = a$, $H_{t\theta} - H_{i\theta} = K$, where K is the induced current per meter length along the circumference.

The induced current may be related to the field inside the cylinder by applying Faraday's Law as for the first example. The law is applied on a curve running parallel to the axis at A and at B, as shown in Figure 6.2b. The chosen form of the potential Ψ_t in Equation 6.7 implies that the field inside the cylinder is constant of value H_t. The field component H_n normal to the surface bounded by the curve C (defined by the broken line AB and having a depth l into the paper) is

$$H_n = H_\theta \sin\theta - H_r \cos\theta$$
$$= -\frac{\partial \Psi_t}{\partial r}\cos\theta + \frac{1}{r}\frac{\partial \Psi_t}{\partial \theta}\sin\theta = H_t$$

Hence the flux linked with curve AB is

$$\phi = \mu_0\, H_t\, 2\, a \sin\theta\, l$$

From Faraday's Law

$$2\, E_z l = -\mu_0\, 2\, a\, l \sin\theta \frac{dH_t}{dt}$$

$$\frac{J_z}{\sigma} = -\,\mu_0\, a \sin\theta \frac{dH_t}{dt}$$

Hence the surface current density K in A/m is

$$K = -(\mu_0\; a\, D\, \sigma \sin\theta)\frac{dH_t}{dt} \tag{6.8}$$

Constant C_1 in Equation 6.7 may be obtained by applying the second boundary condition, namely

$$-H_t \cos\theta = -H_i \cos\theta - C_1 \frac{\cos\theta}{a^2}$$

Hence, $C_1 = (H_t - H_i)\, a^2$. Similarly, from boundary condition 3,

$$H_{t\theta} - H_{i\theta} = K \quad \text{at } r = a$$

or

$$H_t \sin\theta - H_i \sin\theta + (H_t - H_i)\sin\theta = K$$

Substituting K from Equation 6.8 in the equation above and rearranging gives an expression identical to that shown in Equation 6.4 where $\tau_m = \mu_0 a D \sigma/2$. Similar conclusions about shielding may be drawn as for the first example. If a harmonic field of frequency f is impressed, then the derivative term may be replaced by j $2\pi f\ \bar{H}_t$, where $\bar{H}_t$ indicates the magnetic field intensity phasor. Hence, in steady state

$$j\, 2\pi f\, \tau_m \bar{H}_t + \bar{H}_t = \bar{H}_i$$

and

$$\frac{\bar{H}_i}{\bar{H}_t} = 1 + j(2\pi f\, \tau_m) \tag{6.9}$$

The quantity in the brackets is $\alpha = \pi f\, \mu_0 a D \sigma$, and is a measure of the shielding afforded by the cylinder. Shielding is more effective the higher the frequency, radius, thickness, and electrical conductivity of the shield. As an example, for a cylinder made out of aluminum ($\sigma = 3.5 \times 10^7\ \Omega m^{-1}$) of radius a = 0.15 m and thickness D = 1 mm, $\alpha = 0.0207f$, hence

$$SE = 20 \log \left|\frac{\bar{H}_i}{\bar{H}_t}\right| = 20 \log \sqrt{1+\alpha^2} \tag{6.10}$$

For f = 50 Hz the shielding effectiveness is only just over 3 dB but at 1 kHz it rises to 26 dB. Much higher SE is possible at higher frequencies provided the assumptions made in this calculation hold. This is rarely the case in practice. In deriving Equation 6.9 it has been assumed that the shield thickness is very small. Thus, no account has been taken of the radial distribution of induced currents (significant if D is comparable to the skin depth), or of the magnetic properties of the shield.

Finite thickness effects may be included[6–8] and for a cylinder with an incident longitudinal field it is found that

$$\frac{\bar{H}_i}{\bar{H}_t} = \cosh(\gamma D) + \left(\frac{\gamma a}{2\mu_r}\right)\sinh(\gamma D) \tag{6.11}$$

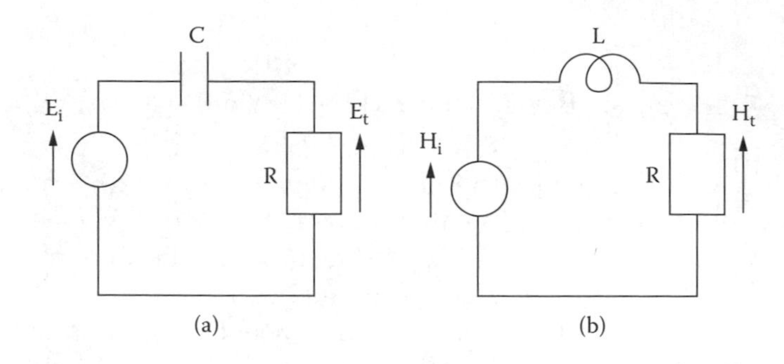

Figure 6.4 Circuit analogues for electric (a) and magnetic (b) field penetration.

where $\gamma = (1 + j)/\delta$ is the propagation constant in the conductor (Equation 2.52), δ is the skin depth (Equation 2.53), and μ_r is the relative magnetic permeability of the shield material. The result obtained above makes the quasistatic approximation, i.e., the displacement current is neglected. This is acceptable at low frequencies. Clearly, this also means that the wavelength of the incident field must be much larger than the cavity dimensions.

Let us now return to the circuit approach to shielding effectiveness. The circuit shown in Figure 6.3 may be modified to produce a series circuit, as shown in Figure 6.4b where now voltages model the incident and transmitted magnetic field strength. From Figure 6.3 or 6.4b, it is evident that low-frequency fields are not efficiently shielded. At low frequencies, most of the current flows throughout the low-impedance path (ωL) and hence H_t is high. Similarly, in the circuit shown in Figure 6.4a, there is significant voltage drop along the high-impedance element (ωC^{-1}) and hence the voltage across R is low. This implies a low transmitted field E_t. Any modification or imperfection in the shield that makes R larger inevitably decreases shielding effectiveness. This, for example, may come about by introducing slots or seams that prevent free flow of induced current and thus effectively increase R. This physical picture also suggests practical ways of introducing slits and apertures in such a way that shield integrity is least compromised. It is also evident from these figures that DC fields cannot be shielded at all if the thin shield is made of nonmagnetic material. A degree of magnetic field shielding may be afforded at DC and very low frequency if the shield is made out of relatively thick material of high magnetic permeability ($\mu_r >> 1$). Hence at low frequencies two approaches to magnetic field shielding may be employed:

1. Shielding by inducing eddy currents. This method is effective at low frequencies but not at DC or very low frequencies. This is obvious from Figures 6.3 and 6.4b and from basic physical principles.

2. Shielding by diverting the magnetic field into a high-permeability material (field ducting). This method exploits the fact that the magnetic field prefers to follow a low reluctance path and hence is mainly concentrated inside the shield walls. This naturally implies that high values of the magnetic field may be established inside the magnetic shield walls and the consequent risk of magnetic saturation. If this happens, then μ_r drops to very low values and shield performance then resembles that of a nonmagnetic shield, i.e., one based on the induced current method (1) described above.

Equation 6.11 may be used for studying the relative contribution of eddy currents and ducting at low frequencies. Some results to illustrate these effects are shown in Figure 6.5. Curve (a) shows the shielding efficiency for a cylindrical shield subject to a longitudinal magnetic field. The cylinder is made of Al and is 0.001 m thick. Curve (b) shows the

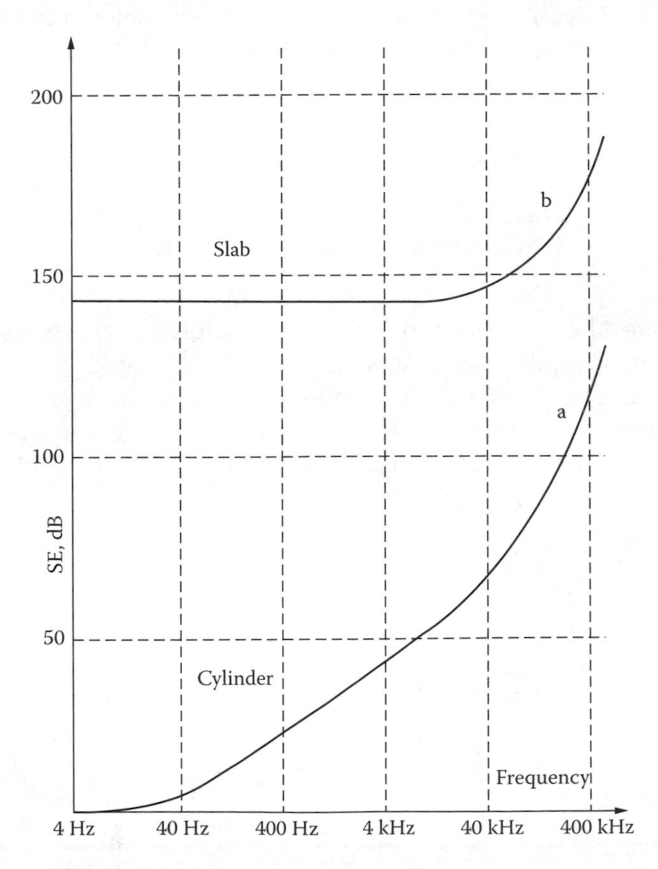

Figure 6.5 Shielding effectiveness of a cylindrical shield (a) and a slab (b).

shielding efficiency of an infinite slab of thickness 0.001 m made of Al. It is clear from these two curves that the shape of the shield is critical at low frequencies. Further details of magnetic field shielding at DC and low frequencies may be found in References 6, 8, and 9.

Let us now turn attention to the shielding of electric fields. A detailed development of the electric field circuit model may be found in Reference 3 and, in the quasistatic thin-shield approximation examined here, it results in the circuit shown in Figure 6.4a. The incident and transmitted electric field strength is described by the voltages shown, and it is apparent that at low frequencies, including DC, the shield is very effective. This situation should be contrasted with the magnetic field case where the opposite situation applies. Of more practical importance are the low-frequency shielding properties of a thin-walled spherical conducting shell.[3,10] For the case of a nonmagnetic ($\mu_r = 1$) shield of radius a and thickness D and at low frequencies $\lambda >> D$ it turns out that the circuits shown in Figure 6.4 hold with $C = 3\,\varepsilon_0\,a/2$, $L = \mu_0\,a/3$ and $R = 1/(\sigma D)$. The shielding effectiveness to electric and magnetic field can then be obtained directly and is

$$\begin{aligned} \text{SE electric} &\simeq 20\log\left|1+\frac{1}{j\omega RC}\right| \\ \text{SE magnetic} &\simeq 20\log\left|1+j\omega\, L/R\right| \end{aligned} \tag{6.12}$$

Following the approach described in Reference 11, some general remarks can be made regarding shielding. A conducting shield of largest dimension 2a and thickness D is shown schematically in Figure 6.6. An incident magnetic field pulse H_i of duration t_0 is also shown and the effectiveness of this shield in general terms is sought. Four characteristic times are important, namely:

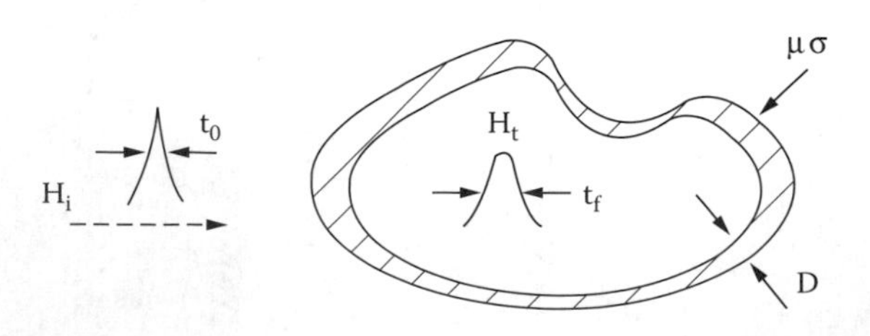

Figure 6.6 Configuration used to study penetration and shielding in general terms.

1. The duration t_0 of the incident pulse
2. The transit time across the entire shield $t_a \sim 2a/u$
3. The wall diffusion time constant $t_d \sim \mu_0 \sigma D^2$
4. The eddy current decay constant $t_f = L/R \sim \mu_0 \sigma Da$

Taking a typical shield as an example, made of Al ($\sigma = 3.5 \times 10^7$ S), two meters in diameter ($a \sim 1$ m) and a few millimeters thick ($D \sim 1$ mm) gives the following time constants for the shield:

$t_a \sim$ tens of nanoseconds (transit time)
$t_d \sim$ tens of microseconds (wall diffusion time)
$t_f \sim$ tens of milliseconds (eddy current decay time)

An incident magnetic field pulse $H_i(t)$ of short duration ($t_0 << t_d$) may be represented by an impulse of strength

$$H_0 = \int_{-\infty}^{\infty} H_i(t)\, dt$$

Following the application of this pulse the field diffuses through the wall and reaches its peak value inside the shield in a time of the order of t_d. Since $t_a << t_d$, it follows that the externally applied field has then passed by and hence at time $t > t_d$ the field is determined by the decay of eddy currents with a time constant t_f as indicated by the circuit in Figure 6.4b. The transmitted field during this phase is determined from the circuit shown in Figure 6.4b and is

$$H_t(t) = \frac{H_0}{t_f} e^{-t/t_r}$$

Hence the peak value of the field is H_0/t_f and the peak rate of change is estimated to be[12]

$$H_0/(t_f t_d)$$

The voltage induced on a single turn loop that fits inside a shield of radius a is

$$V_{ind} = \pi a^2 \mu_0 H_0/(t_f t_d) = \frac{H_0}{\mu_0 \sigma^2 D^3 a}$$

For a NEMP pulse with $E = 50$ kV/m and $t_0 = 250$ nsec, $H_0 = Et_0/Z = 3.3 \times 10^{-5}$ As/m and for an Al shield 3 m in radius and 1 mm thick, this expression gives an induced voltage of the order of a millivolt.

As far as time-harmonic incident fields are concerned, the important consideration is the magnitude of the skin-depth δ relative to the thickness of the wall D. A wall is regarded as electrically thick if $\delta < D$ and in such cases reflection and attenuation in the wall contribute substantially to shielding. In electrically thin walls $\delta > D$ the eddy current distribution and, therefore, the shape of the enclosure play a crucial role in shielding. An incident magnetic field with a period $T >> t_f$ cannot be adequately shielded. The only way to improve shielding under these circumstances is to use a thick wall of high magnetic permeability.

6.2.3 Approximate Methods — The Wave Approach

An alternative approach to calculating shielding effectiveness is to consider how waves incident on the shield wall suffer reflections and decay. A systematic approach may thus be devised that permits the calculation of the transmitted field E_t once the incident field E_i is known. A typical configuration is shown in Figure 6.7a, where the wall is regarded as an infinite slab of thickness D. This is a typical assumption in this type of calculation (planar wall approximation). A qualitative description of what happens to an incident field E_i as it impinges on the wall surface on the left-hand side as shown in Figure 6.7b, is as follows.

Upon the pulse E_i impinging on the wall, it encounters a discontinuity, some energy is reflected (E_1), and part of the energy penetrates into the conductor (E_2). Reflection at the left-hand side wall boundary represents, therefore, a first contribution to shielding. The pulse proceeds further to reach the wall surface on the right-hand side (E_3) and in this travel inevitably suffers some attenuation that thus contributes further to shielding.

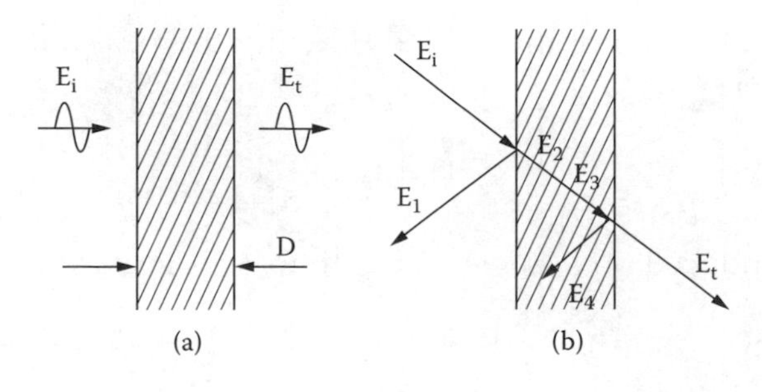

Figure 6.7 (a) Penetration through a shield of thickness D and (b) incident, reflected, and transmitted signals.

Subsequently, part of the signal is reflected back into the wall (E_4) and part proceeds further to form the transmitted signal E_t. The ratio E_i/E_t cannot, however, be regarded as the shielding effectiveness under all circumstances. In general, pulse E_4 will be reflected at the left-hand side wall boundary, thus causing further reflections inside the wall and making further contributions to the total transmitted signals. In a general case, there are therefore three aspects to be considered:

1. Reflection loss
2. Attenuation or absorption loss
3. Multiple reflection loss

An exact solution to the problem may be obtained by defining an incident plane wave in the medium to the left of the wall, which is assumed to be air (intrinsic impedance $Z_0 = \sqrt{\mu_0/\varepsilon_0}$ phase constant $\beta_0 = \omega\sqrt{\mu_0\varepsilon_0}$). It is further assumed that $H_i = E_i/Z_0$, which implies that the wall is in the far field of the source launching the plane wave. Similar conditions are assumed when considering the transmitted field in the air to the right of the wall. Inside the conducting wall of the shield a complex propagation constant γ and intrinsic impedance Z are assumed. These were obtained in Chapter 2, Equations 2.52 and 2.54. On the two interfaces between the three different media, boundary conditions are enforced for the tangential components of the electric and magnetic fields. The ratio of incident to transmitted field may then be obtained and is[13,14]

$$\frac{\bar{E}_i}{\bar{E}_t} = \frac{(Z_0 + Z)^2}{4\,Z_0 Z}\left[1 - \left(\frac{Z_0 - Z}{Z_0 + Z}\right)^2 e^{-2D/\delta}\, e^{\,j2\beta D}\right] e^{D/\delta}\, e^{j\beta D}\, e^{-j\beta_o D} \qquad (6.13)$$

This exact formula may be simplified for good conductors and for cases where $\delta \ll D$ to give the shielding effectiveness.[14]

$$SE = 20\log\left|\frac{Z_0}{4\,Z}\right| + 20\log e^{D/\delta} + 20\log\left|1 - e^{-2D(1+j)/\delta}\right| \qquad (6.14)$$

The first term in this expression represents the reflection losses in crossing the two conductor/air boundaries. The second term represents attenuation or absorption loss in the shield wall and the third term the multiple reflection loss. The same formulae may be obtained in an approximate manner by assuming that signal E_2 in Figure 6.7b may be obtained simply using the formula

$$\bar{E}_2 = \bar{E}_i \frac{2\,Z}{Z_0 + Z}$$

The reflection loss component of E_i/E_t is then

$$\left|\frac{\bar{E}_i}{\bar{E}_t}\right| = \left|\frac{Z_0 + Z}{2\,Z}\,\frac{Z_0 + Z}{2\,Z_0}\right| = \left|\frac{(Z_0 + Z)^2}{4\,Z_0\,Z}\right| \tag{6.15}$$

For good conductors $Z << Z_0$. Hence,

$$\left|\frac{\bar{E}_i}{\bar{E}_t}\right|_{\text{reflection}} \simeq \left|\frac{Z_0}{4\,Z}\right| \tag{6.16}$$

Since $Z << Z_0$, the largest contribution to the reduction of the electric field occurs at the left-hand side of the wall. In contrast, the major magnetic field decrease occurs at the right-hand side of the wall as it can be directly confirmed by substituting $E = H \times Z$ in the above expressions. This situation suggests that the shield thickness D is more important for magnetic field shielding. The attenuation in the wall (second term in Equation 6.14) can be calculated directly assuming exponential decay with a characteristic length equal to the skin depth. The multiple reflection loss may also be calculated by taking into account the successive reflections of pulse E_4 in Figure 6.7b between the two faces of the shield. Clearly, since the primary contribution to the reflection loss of the magnetic field is at the right-hand side of the wall, the largest contribution from multiple reflections will be on the magnetic field strength. It is relatively easy to confirm that multiple reflection contributes the third term in Equation 6.14. A systematic scheme for obtaining the shielding effectiveness is shown in Figure 6.8. The incident signal E_i is of unit amplitude as shown. Then the transmitted signal is the sum of the terms shown emerging from the right-hand side, i.e.,

$$E_t = (1 + S_{22})\,S_{12}\,\tau - S_{11}S_{12}S_{22},\,\tau^3(1 + S_{22})$$
$$[1 - \tau^2 S_{11}S_{22} + \tau^4 S_{11}^2 S_{22}^2 - \ldots]$$

where

$$S_{11} = (Z - Z_0)/(Z + Z_0),\; S_{12} = 2\,Z/(Z + Z_0)$$

$$S_{22} = (Z_0 - Z)/(Z_0 + Z) \text{ and } \tau = e^{-\gamma\delta}$$

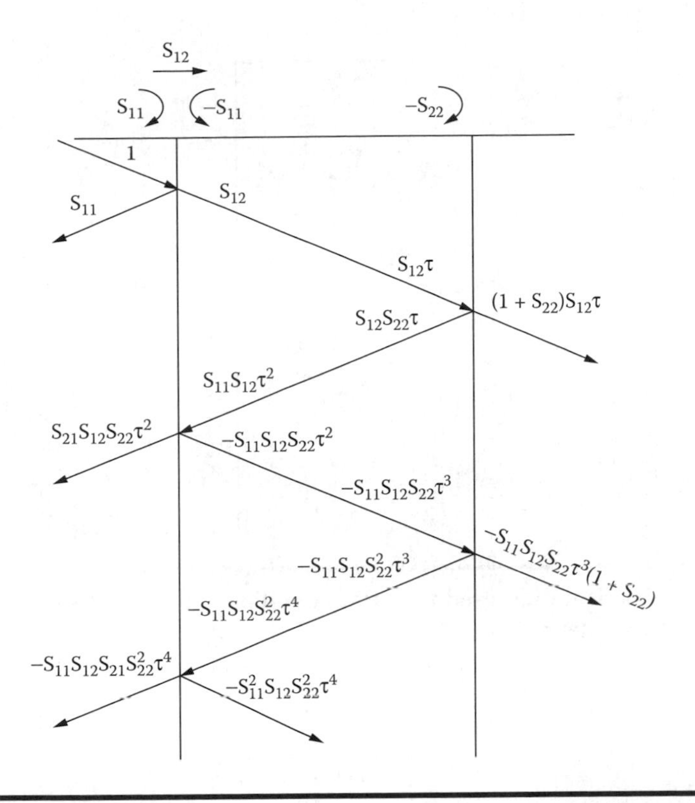

Figure 6.8 Scheme for calculating multiple reflections in slab.

Using the fact that $S_{11} = -S_{22}$ and that $S_{12} = 1 + S_{11} = 1 - S_{22}$, summing the infinite series above and rearranging gives

$$SE = 20 \log \left| \frac{E_i}{E_t} \right| = 20 \log \left| \frac{1 - \tau^2 S_{22}^2}{\tau(1 - S_{22}^2)} \right| \tag{6.17}$$

This formula includes all the contributions from reflections, attenuation, and multiple reflections.

The derivation above was based on the assumption that the source of EM waves is far from the shield. This allowed the use of the intrinsic impedance of the medium as the wave impedance. As discussed in Chapter 2, the wave impedance in the near field of an antenna differs from Z_0. The formulae for shielding effectiveness derived previously may still be used provided that the correct wave impedance is used. Assuming that a small coil (magnetic source) is placed at a distance z from the shield as shown in Figure 6.9, then the wave impedance at any point at the surface of the shield is

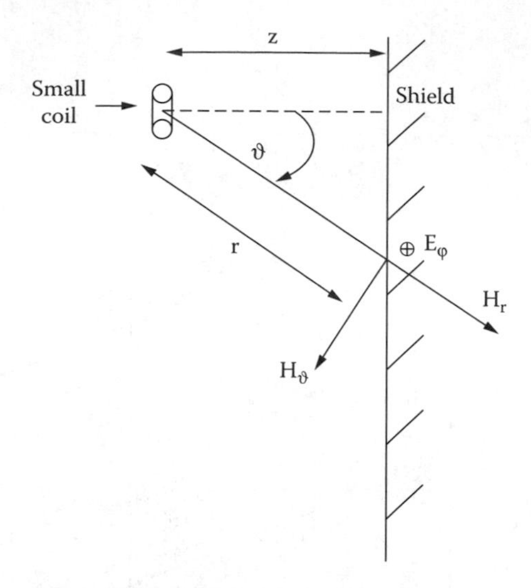

Figure 6.9 Configuration used to assess conditions at a shield surface in the near-field of an antenna.

$$Z_w = \frac{E_\rho}{H_r \sin\theta + H_\theta \cos\theta} \tag{6.18}$$

Substituting into Equation 6.18 for the fields due to a coil (Equation 2.74) and taking the limit for r << λ gives[15]

$$Z_w \cong j\,\omega\,\mu_0\,z/3 \tag{6.19}$$

This value must then be used for the wave impedance in the medium surrounding the source. It has been shown that the shielding effectiveness does not depend on the distance z. A full discussion of the difficulties associated with calculations in the near field may be found in References 15 and 16.

It is worth emphasizing that at low frequencies the shape of the shield is critical in calculating shielding effectiveness and that the planar wave theory described above does not make allowance for effects due to shape. Similarly, there is nothing in the theory presented here to account for further reflections that the transmitted signal E_t will suffer inside the cavity formed by the conducting shield. The formulae presented here must therefore be used with caution in practical situations.

6.2.4 Analytical Solutions to Shielding Problems

Very few shield configurations can be studied using analytical techniques. This is true whether a solution in the frequency or time domain is contemplated. Certainly no practical shields could be studied in this way; the best that can be expected is to obtain solutions for canonical shapes such as two infinitely long parallel plates, infinitely long cylinders, or spherical shells. Such solutions are very useful, however, in providing an insight into shield behavior and also accurate solutions against which approximate or numerical methods can be tested. The pioneering work in this area is due to Kaden.[17] Subsequently, further work appeared in the English literature on solutions to canonical problems of the type described above.

The attenuation of transient fields by an imperfectly conducting spherical shell was reported in Reference 18. Results for penetration through a slab, two parallel slabs, infinitely long cylinders, and a spherical shell presented in the frequency and time domains may be found in References 10 and 11. Results for the penetration of a NEMP through a slab are reported in Reference 19. The general functional form of the magnetic field shielding effectiveness of the canonical shapes mentioned above is given in Reference 10 and is

$$SE = 20 \log \left| \cosh(\gamma D) + \left(\alpha(\gamma D) + \frac{\beta}{(\gamma D)} \right) \sinh(\gamma D) \right| \tag{6.20}$$

where α, β have different values depending on the form and parameters of the shield. Equation 6.11 is a particular form of Equation 6.20 for the case of a cylindrical shield in an axial field and is plotted in Figure 6.5. For comparison, results are shown in the same figure for a slab of the same thickness and material properties as the cylinder. It can be seen that at low frequencies the performance of the two shields is very different, and hence that shape factors are important and must be considered in calculations.

6.2.5 General Remarks Regarding Shielding Effectiveness at Different Frequencies

In Sections 6.2.2 and 6.2.3, two approaches that may be used in the calculations of shielding effectiveness were presented. Both approaches have limitations and must therefore be used with caution. The circuit approach is generally favored at low frequencies, while the wave approach is normally used at high frequencies. What is more important, however, is to appreciate the general trends in shielding. Generally speaking, for conducting shields of a significant thickness of the order of a millimeter or more, shielding effectiveness exceeds 100 dB at frequencies exceeding a few tens of kilohertz (see Figure 6.5). In practice such high shielding effectiveness is possible, if at all, with great difficulty since apertures, seams, and wire penetrations dominate shielding. If very thin conducting coatings are used, or if poor conductors are employed as shielding materials (e.g., carbon fiber composites), further checks need to be made to ascertain the effectiveness of the shield. However, without doubt, penetration through the shield material becomes a problem at low frequencies and for the magnetic field in particular. Therefore, meeting magnetic field shielding requirements at low frequencies will in most cases also meet all other shielding requirements (subject to limitations due to wires and apertures). At very low frequencies and at DC it may be necessary to use high-magnetic-permeability materials to achieve a high degree of magnetic field shielding. Materials such as mumetal have values of μ_r exceeding 10,000 at low frequencies. However, this value decreases very considerably at frequencies above a few kilohertz and hence the material becomes less effective. Similarly, high values of magnetic field reduce the effective μ_r due to saturation effects. A detailed description of this type of shield may be found in Reference 9. As an illustration, the shielding effectiveness of a spherical shell of inner and outer diameters D_i and D_0, respectively, and with $\mu_r >> 1$, is, for static magnetic fields,

$$SE = 20 \log \frac{H_i}{H_t} = 20 \log \left[\frac{2}{9} \mu_r \left(1 - \frac{D^{i}_{3}}{D^{3}_{0}} \right) + 1 \right] \tag{6.21}$$

For a very thick shield $D_i/D_0 \rightarrow 0$ the shielding effectiveness tends to a limiting value of $2\mu_r/9$. Hence, if $\mu_r = 10{,}000$ it follows that shielding effectiveness above 67 dB cannot be attained even with a very thick wall. If higher shielding effectiveness is required, a double shield may be used. Formulae for calculating static magnetic field shielding effectiveness for double shields may be found in Reference 9. Naturally, multiple-layered shields may be used to increase shielding at all frequencies. Formulae for N-layered spherical shields may be found in Reference 10. As an example, the shielding effectiveness of a two-layered shield consisting of two shields of shielding effectiveness SE_1 and SE_2, respectively, is

$$SE = SE_1 + SE_2 - 20 \log \gamma_{12} \tag{6.22}$$

where $\gamma_{12} = (a_2/a_1)^3 \alpha_1 \alpha_2 Z_1 Z_2 \sinh Z_1 \sinh Z_2$ and a_1, a_2 are the radii of the outer and inner shields, $\alpha_i = a_i/(3D_i)$, $Z_i = j\omega \mu_0, \sigma_i D_i^2$ and D_1, D_2 are the thicknesses of the outer and inner shields.[10]

It may become possible in the future to use new high-temperature superconductors in the construction of shields.[19] The concept of skin depth applies in this case and the field strength at a depth z inside the material is $E = E_0 \exp(-z/\lambda)$ where E_0 is the field strength at the surface and λ is the London penetration length approximately constant up to microwave frequencies. For typical superconducting materials, λ is of the order of tens of nanometers and thus a very high degree of shielding could be achieved with very thin layers of such materials. Practical exploitation of these effects is dependent on progress in the development and manufacture of essentially room-temperature superconductors.

6.2.6 Surface Transfer Impedance and Cable Shields

The previous discussion of shielding was based on the "shielding effectiveness" as the parameter used to characterize shields. An alternative measure that can be advantageous under some circumstances, and which is used extensively in describing the quality and effectiveness of cable shields, is the transfer impedance Z_T. Cable shields are important in overall system performance as they are of considerable length and run through electromagnetically hostile environments. In this section, the concept of transfer impedance will be explained. The important topic of coupling through cable shields will be treated in more detail in future sections.

Let us first start by considering a slab of thickness D, as shown in Figure 6.10a. It is assumed that the shield carries in total a current I_s in A/m (the length measured into the paper). The transfer impedance Z_T may then be defined as the ratio of the electric field at the inner surface E(0) to the total current I_s.

$$Z_T = E(0)/I_s \tag{6.23}$$

Alternatively, Z_T may be defined as the ratio of E(0) to H(D), the magnetic field at the outer surface.

The transfer impedance may be calculated approximately ($|Z| << Z_0$) by obtaining the solution to the wave equation (Equation 2.49) inside the slab, i.e.,

$$\bar{E} = A\, e^{\gamma x} + B\, e^{-\gamma x}$$

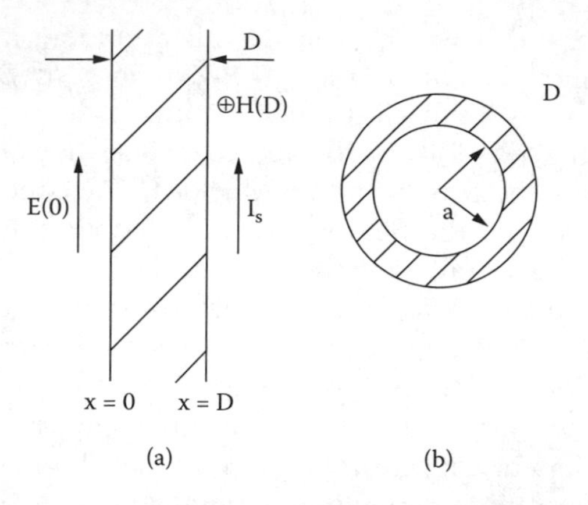

Figure 6.10 Configuration used to calculate the transfer impedance of (a) a slab and (b) a cylindrical shield.

The magnetic field is then obtained from Faraday's Law

$$\bar{H} = \frac{1}{j\omega\mu}\frac{dE}{dx} = \frac{\gamma}{j\omega\mu}(A\,e^{\gamma x} - B\,e^{-\gamma x})$$

Demanding that H(0) = 0 results in A = B and, similarly, demanding that H(D) = I_s (Ampere's Law) and exploiting the equality of A and B gives

$$A = \frac{j\omega\mu I_s}{2\,\gamma\,\sinh(\gamma D)}$$

Thus, the electric field at the inner surface is

$$E(0) = 2A = \frac{j\omega\mu I_s}{\gamma\,\sinh(\gamma D)}$$

and hence

$$Z_T = \frac{j\omega\mu}{\gamma\,\sinh(\gamma D)}$$

This expression may be modified further by using Equation 2.54 to give

$$Z_T = \frac{Z}{\sinh(\gamma D)} \tag{6.24}$$

At low frequencies, where the thickness D is much smaller than the skin depth δ, Equation 6.24 simplifies to $Z_T \simeq 1/\sigma D$, which is the DC resistance of the slab. At high frequencies when $\delta << D$ it turns out that $Z_T \simeq 2\, Z\, e^{-D/\delta}$.

The cylindrical shield (Figure 6.10b) commonly used in cables can be treated in the same way. Assuming that $D << a$ and neglecting r-variation in the shield gives

$$\frac{d^2E}{dr^2} \simeq \gamma^2 E$$

and hence

$$E = A\, e^{\gamma r} + B\, e^{-\gamma r}$$

$$H = \frac{\gamma}{j\omega\mu}(A\, e^{\gamma r} - B\, e^{-\gamma r})$$

From the boundary condition H(a) = 0 it follows that

$$A\, e^{\gamma a} = B\, e^{-\gamma a}$$

Similarly, from $H(a + D) = I_s/2\pi a$, where I_s is the shield current, after combining with the expression above it follows that

$$A\, e^{\gamma a} = B\, e^{-\gamma a} = \frac{j\omega\mu I_s}{4\pi\gamma a \sinh(\gamma D)}$$

Substituting in the expression for the electric field gives the following expression evaluated at the inner surface

$$E(a) = \frac{\gamma I_s}{2\pi a\sigma \sinh(\gamma D)}$$

The transfer impedance is

$$Z_T = \frac{E(a)}{I_s} = \left(\frac{1}{2\pi a\sigma D}\right)\frac{(\gamma D)}{\sinh(\gamma D)} \qquad (6.25)$$

The quantity in the brackets is the DC resistance of the shield, hence

$$Z_T = R_{DC} \frac{(\gamma D)}{\sinh(\gamma D)} \tag{6.26}$$

At low frequencies ($D << \delta$)

$$\frac{Z_T}{R_{DC}} = 1 - \frac{(\gamma D)^2}{6}$$

but

$$\gamma D = \frac{1+j}{\delta} D \simeq 0$$

Hence

$$|Z_T|/R_{DC} \simeq 1 \text{ (low frequencies)}$$

Similarly, at high frequencies ($D >> \delta$)

$$\frac{Z_T}{R_{DC}} = 2(\gamma D)\, e^{-\gamma D}$$

and

$$|Z_T|/R_{DC} \simeq 2\sqrt{2}\frac{D}{\delta} e^{-D/\delta} \text{ (high frequencies)}$$

These expression show that as the frequency increases, the electric field and hence the voltage drop $E(a)l$ along a length l of the shield decreases. This voltage appears in the signal path of the cable and therefore constitutes EMI. The transfer impedance calculated above accounts only for diffusion through the shield and only applies to solid shields. In many practical situations, notably braided shields, there are holes between the partially overlapping strands, incomplete electrical contact between strands, and slight differences in the spacing between strands and the shielded conductor.[21–23] All these factors make varying contributions to shielding effectiveness and may be balanced against each

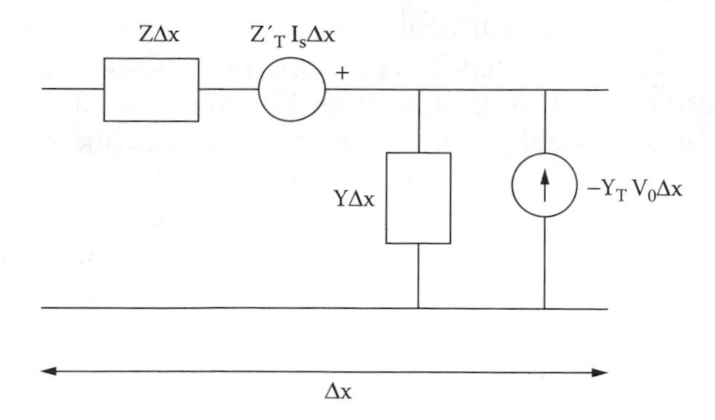

Figure 6.11 Circuit equivalent of a shield taking account of transfer impedance and admittance.

other for optimum design.[24] The overall result is that an extra term appears in the expression for the transfer impedance shown in Equation 6.26 for a solid shield. This expression is modified to account for the different DC resistance of the braided shield and an additional term $j\omega M_{12}$ is introduced that accounts for magnetic field penetration through holes and other similar effects. The addition of this term means that the overall transfer impedance does not fall continuously as predicted by Equation 6.26, but after a certain frequency (typically ~ 1 MHz for most cables) it starts to increase. Similarly, electric field penetration may be accounted for by a transfer admittance $Y_T = j\omega\ C_{12}$.[24] A circuit model of the coaxial line may thus be constructed as shown in Figure 6.11. Terms $Z\Delta x$ and $Y\Delta x$ represent the normal line parameters. The voltage source represents diffusion and inductive coupling

$$Z'_T = Z_T + j\omega\ M_{12}$$

and the current source represents capacitive coupling through the holes to the external circuit where $Y_T = -j\omega\ C_{12}$ and V_0 is the potential difference between the inner conductor and shield common path. A more detailed discussion of cable shielding is presented in the chapter on propagation.

6.3 Aperture Theory

It has already been mentioned that the overall shielding effectiveness of an enclosure is more often than not determined by the presence of holes, slits, and other openings with or without wire penetrations, and that diffusive penetration through the material is, at least for highly conducting shields, only a serious issue at low frequencies and for magnetic fields.

An approximate intuitive calculation will suffice to establish the severity of coupling through apertures.[12] Let us consider a shield with an aperture of area A subject to a powerful external EM pulse. Let the component of the electric field normal to the aperture be E and the magnetic field component tangent to the aperture be H. Both these fields have been calculated assuming that the aperture has been shorted out by a perfectly conducting sheet. The maximum current that can be induced on an internal conductor may be obtained assuming all the electric flux is coupled to the conductor, i.e.,

$$i \cong A \frac{dD}{dt} = \varepsilon A \frac{dE}{dt}$$

Assuming a severe EMP threat (E = 50 kV/m in 10 nsec) and an aperture of diameter 0.1 m gives a current equal to 0.35 A. Similarly, the voltage induced on a conductor loop is

$$v \cong \mu a^2 dH/dt$$

where a is the diameter of the aperture. For a magnetic field rising to 133 A/m in 10 nsec this formula gives an induced voltage equal to 167. This value is considerably larger compared to that obtained for diffusion in Section 6.22, emphasizing the importance of coupling through apertures.

It is therefore of considerable importance to examine the general principles that determine the penetration of fields through apertures. The treatment will be limited to electrically small apertures, that is, apertures for which the largest linear dimension is much smaller than the wavelength. A rigorous examination of practical configurations is very complex and to date few canonical problems have been solved analytically. To illustrate the principles of such a calculation and to provide the basis for the treatment of more complex problems, the simple case of a very thin, highly conducting infinite sheet with a very long narrow slot of width 2a, as shown in Figure 6.12a, will be studied in detail. To simplify the problem further, it is assumed that a DC electric field E_0, which is uniform and points upward far from the slot, is established in the external region ($y > 0$). The problem posed is to obtain the electric field that penetrates through this aperture into the internal region ($y < 0$). It may come as no surprise to the reader that even this highly idealized simple problem presents some considerable difficulty in trying to obtain an analytical solution. The general approach is to transform the conductor configuration shown in Figure 6.12a into a simpler one, namely, that shown in Figure 6.12b, for which a solution for the potential can be easily found. This is done by using a

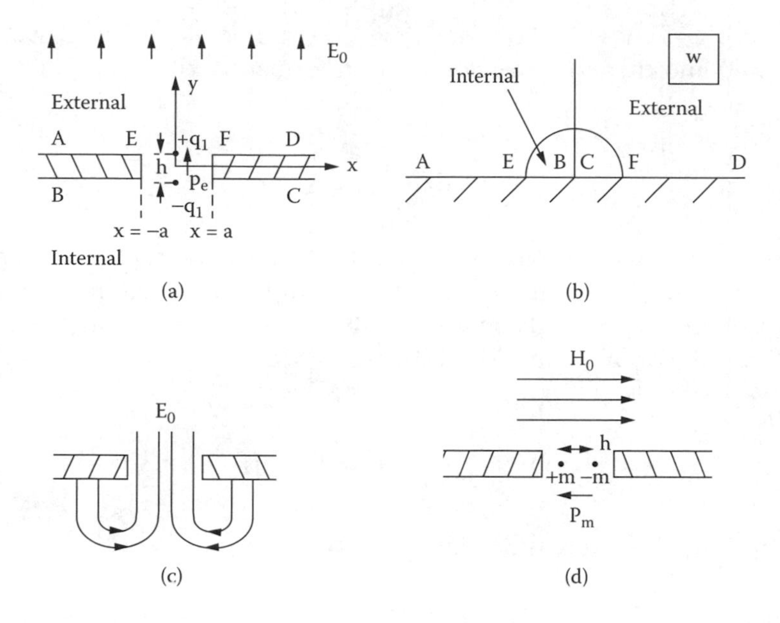

Figure 6.12 Penetration of a static electric field through a slot (a), w-plane configuration (b), and schematic of the electric field distribution (c); a description of the dual problem of magnetic field penetration (d).

conformal transformation from the z-plane (Figure 6.12a) to the w-plane (Figure 6.12b).[17,25] A general transform for polygonal shapes from the z- to the w-plane, known as the Schwarz-Christoffel transform, may be used, and for this example is

$$\frac{dz}{dw} = A(w-b)^{\frac{\beta}{\pi}-1}(w-c)^{\frac{\gamma}{\pi}-1}(w-d)^{\frac{\delta}{\pi}-1}$$

From Figure 6.12a, $b = -a/2$, $c = 0$, $d = +a/2$, $\beta = 2\pi$, $\gamma = -\pi$, and $\delta = 2\pi$. Substituting in the formula above gives

$$\frac{dz}{dw} = A\left(1 - \frac{a^2}{4w^2}\right)$$

or

$$z = Aw + A\frac{a^2}{4w} + B$$

where A and B are constants to be determined. At $z = -a$, w must be $-a/2$, hence $-a = -aA + B$. Similarly, at $z = a$, w must be equal to $a/2$, and hence a = aA + B. From these two constraints it follows that A = 1 and B = 0 and therefore the desired transformation is

$$z = w + \frac{a^2}{4w}$$

It is easy to show using this expression that the exterior space is mapped into the region $r_w > a/2$ in the w-plane. A complex potential function may then be chosen in the w-plane, namely $\Phi = -E_0\,w$, where w may be found by solving the above equation and is

$$w = \frac{1}{2}\left(z \pm z\sqrt{1 - a^2/z^2}\right)$$

For $|z| >> a$ this expression simplifies to

$$w = \frac{1}{2}\left[z \pm z\left(1 - \frac{1}{2}\frac{a^2}{z^2}\right)\right]$$

For the exterior region, we may choose the solution with the positive sign

$$\Phi_{ext} = -\frac{E_0}{2}\left(2z + \frac{1}{2}\frac{a^2}{z}\right) \simeq -E_0 z, \text{ for } |z| >> a$$

The potential function is then the imaginary part of Φ_{ext}, i.e.,

$$-E_0 r \sin\phi, \text{ where } z = r\,e^{j\phi}$$

It can be confirmed by calculating E_r and E_ϕ from this potential, that E_x = 0 and E_y = E_0 away from the aperture as desired. Of more interest is the potential solution for the interior (solution with negative sign), i.e.,

$$\Phi_{int} = -E_0\frac{a^2}{4z}$$

with imaginary part

$$\frac{E_0 a^2}{4r}\sin\phi$$

This potential is the same as that due to a linear dipole distributed along the slit length, namely,

$$\phi_{dip} = \frac{p \sin\phi}{2\pi \varepsilon r}$$

where p is the dipole moment $p = q_l \cdot h$, q_l is the charge and h is the distance between the positive and negative dipole charges as shown in Figure 6.12a. All that is needed to establish the equivalence is to choose as $p_e = \pi \varepsilon a^2 E_0/2$. The field penetrating through this slit can thus be obtained from this equivalent dipole and is shown schematically in Figure 6.12c.

This derivation suggests a general method for dealing with the penetration of an electric field through an aperture. In applying this method the electric field at the position of the aperture is first obtained assuming that the aperture is replaced by a perfectly conducting sheet (E_0 in this particular example). This field is referred to in the literature as the short-circuit field E_{sc}. Penetration then is calculated as if it is due to an equivalent dipole of dipole moment given by

$$p_e = 2\,\varepsilon\,\alpha_e\,E_{sc} \tag{6.27}$$

The quantity α_e is known as the aperture electric polarizability and has been calculated for a number of shapes.[26,27] As an example, the polarizability of a narrow slot of length l and width 2a (2a << 1) is $\pi l\ a^2/4$. Substituting into Equation 6.27 the polarizability per unit length from this formula gives the same value of p as calculated for the infinite slot. In practice the calculation of the electric field in the internal region is done by replacing the aperture by a perfectly conducting sheet and placing a small dipole just inside the shield at the position where the aperture would be. To take account of the reinforcing effect of the image dipole, the value of p chosen is then half that calculated in Equation 6.27. For a circular hole of diameter d the electric polarizability is $\alpha_e = d^3/12$. Values for other shapes may be found in the literature.

The penetration of a magnetic field may be treated in a similar way and results in an identical formulation, where now the electric dipole is replaced by a magnetic dipole as shown in Figure 6.12d. The magnetic dipole moment is now given by

$$p_m = \frac{\pi}{2} a^2\,H_0 \tag{6.28}$$

In general, the equivalent magnetic aperture dipole moment is given by the expression

$$p_m = -2\,\alpha_m\,H_{sc} \tag{6.29}$$

where α_m is the aperture magnetic polarizability and H_{sc} is the short-circuit field, i.e., the magnetic field at the position of the aperture when the aperture is replaced by a perfectly conducting sheet. The value of α_m for a narrow slit (2a << 1) and for a field running at right angles to the narrow side is $\pi l\, a^2/4$. Substituting the dipole moment per unit length into Equation 6.29 gives the value obtained in Equation 6.28. For a circular hole of diameter d, $\alpha_m = d^3/6$.

In Reference 17 a formula for the total power radiated into the interior space through a hole of radius r has been obtained:

$$P = \frac{2}{27\pi}\frac{(2\pi/\lambda)^4 E_0^2 r^6}{Z_0}$$

This formula shows that the power that leaks through the aperture scales as the fourth power of frequency and the sixth power of the radius. It is clear, therefore, that large holes have very deleterious effects on shielding. It is more advantageous to use instead many smaller holes if that is an option permitted in the design. This conclusion is also confirmed by the circuit approach described in the last section. Large holes disturb the path of eddy current flow and thus reduce shielding more, compared to a configuration that consists of several but smaller holes.

Another issue that must be addressed regards penetration through apertures in thick walls. In the calculations described above, the thickness of the shield walls was neglected. A derivation of the equivalent aperture dipole moment where the finite thickness of the screen is taken into account may be found in Reference 17. However, as the thickness of the screen becomes considerable, more elaborate models for penetration become necessary. A general formulation for this type of problem, based on the application of the equivalence principle, may be found in Reference 28. The case of a narrow slit in a thick conducting screen is treated in Reference 29. For screen thicknesses equal to multiples of the half wavelength, considerable penetration through the slot may occur. Under these resonant conditions and for narrow slit widths $2a \ll \lambda$ it is found that[29]

$$P_{trans}/P \simeq \frac{\lambda}{\pi}$$

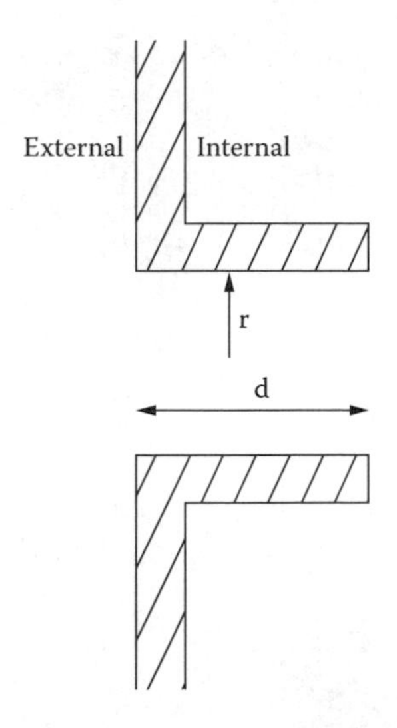

Figure 6.13 Cut-off properties of an aperture of finite length d.

where P_{trans} is the total power transmitted per unit length of the slit and P is the incident power density at the slit. Hence, under these conditions, the effective width of the slit is l/π times the wavelength, irrespective of the actual value of the slit width 2a. A survey of work in this area and typical results may be found in Reference 30.

The depth of the slot may be artificially increased above the thickness of the shield material, as shown in Figure 6.13. The structure thus formed resembles a waveguide and thus frequencies well below its cut-off frequency suffer additional attenuation. For a circular air-filled waveguide and a TE_{11} mode, the cut-off wavelength is 3.41r and hence the cut-off frequency is $8.8 \times 10^7/r$.[31] For frequencies below cut-off, the attenuation constant for propagation along the waveguide is

$$\alpha \simeq \frac{54.6}{\text{cut-off wavelength}} = \frac{16}{r} \text{ in dB/m}$$

The length d can thus be selected to achieve the required additional attenuation.

It can be shown that for any antenna, the effective aperture A_e and the gain G are related by the formula

$$A_e = G\frac{\lambda^2}{4\pi}$$

Assuming that for a small loop of radius r, $A_e \simeq \pi r^2$ and substituting to the formula above gives

$$G = \left(\frac{2\pi r}{\lambda} \right)^2$$

The shielding performance of circular opening may be obtained by using the formula above (duality)

$$SE = 10 \log \frac{\text{Power without opening}}{\text{Power with opening}} = 10 \log \left(\frac{\lambda}{2\pi r} \right)^2$$

$$= 20 \log (\lambda/(2\pi r))$$

More practical aspects of aperture penetration and design will be discussed in Chapter 10.

6.4 Rigorous Calculation of the Shielding Effectiveness (SE) of a Conducting Box with an Aperture

In the previous section we have dealt with penetration through an aperture using the concepts of polarizability and equivalent magnetic/electric dipoles. These are essentially low-frequency techniques and are useful in getting a basic assessment of SE. The general case of high-frequency penetration through an aperture that is backed by a conducting cavity is of the most practical significance as it represents realistic engineering configurations (e.g., an equipment cabinet with an opening for access, ventilation, etc.).

A typical arrangement is shown in Figure 6.14. The problem may be posed as follows: An external EM field, normally a plane wave, is incident on the cabinet wall with the aperture. This wave suffers a partial reflection and also partially penetrates through the aperture to establish a field inside the cabinet. It is desired to calculate the relative shielding afforded by the cabinet with the aperture, i.e., the SE. The difficulty in performing this calculation is that the tangential fields at the aperture are not known a priori and the field inside the cabinet exhibits strong resonant behavior. It is by no means obvious that the cabinet will provide shielding at all frequencies, although we may be tempted to believe that it might do so. Such problems may be treated in several ways using analytical techniques, intermediate models, or full-field models as is explained in this and the following sections.

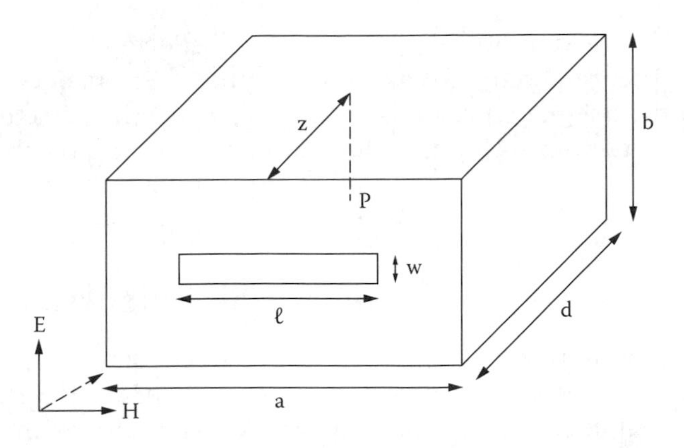

Figure 6.14 Rectangular cabinet with a rectangular aperture. The observation point is P and the wall thickness is t.

Analytical approaches when they are possible are accurate and can be used to benchmark other solution techniques. However, they become intractable when shapes are not canonical. The essence of these calculations is described at the end of this section.

Intermediate models are a compromise whereby semi-analytical results are used to establish models that, although not rigorous, have known and controllable accuracy. They are most useful in rapid conceptual and "what if" studies early at the design stage. They are described in Section 6.5.

Full-field numerical models are rigorous within the limitations of the accuracy of numerical methods, but they are computer intensive and require access to sophisticated CAD tools. A brief survey of available numerical techniques is presented in Section 6.6.

Ways of speeding up computation of SE problems using full-field numerical models in the case of apertures with complex perforations are presented in Section 6.7. These techniques are best employed when a basic design has evolved in order to check assumptions and detailed behavior at points of interest.

An effective way of dealing with SE calculations using analytical techniques is presented in References 32 and 33. The field inside the cavity is expressed as a combination of cavity or waveguide modes and outside the cavity by a combination of free-space modes. The transverse magnetic field inside the cavity is expressed in terms of the total aperture electric field using the waveguide admittance operator. Similarly, the incident and reflected transverse magnetic field outside the cavity is expressed in terms of the incident and reflected electric field at the aperture. Enforcing field continuity at the aperture results in an integral equation that can be solved for the aperture fields. Once these have been calculated, the field inside the cabinet and reflected from it may be calculated. The required admittance operators are obtained from the modes and the integral equation is solved using Galerkin's method. These calculations, although well established and accurate, are computationally intensive.

6.5 Intermediate Level Tools for SE Calculations

As already indicated, rapid calculation of the SE effectiveness early in the design stage is essential as it permits a basic conceptual design to evolve where potential interference problems are identified, optimum placement of sensitive equipment is decided, and the basic sizing is established. It is obvious that at this stage only basic details of the design are available and the designer is primarily interested in exploring rapidly the design parameter space and thus deciding on the basic configuration. A suitable simulation tool for this type of study is one where large-scale behavior can be rapidly established (full details are not yet available!). Full-field numerical models are too slow and simulate a level of detail that is not required for this stage in the design. Empirical models, on the other hand, are inaccurate when used outside their limited range of applicability and cannot therefore be used with confidence for creative design. The type of tool that is best suited to this task is what we describe as an "intermediate level tool." It is a compromise between the rigor of a full-field numerical model and the simplicity of an empirical model. Its nature will become clearer as we examine in more depth its structure for the problem depicted in Figure 6.14.

Before we start on model development we consider first the worst case of field polarization with regard to penetration through the slot in Figure 6.14. To develop the argument we consider conditions with regard to the reflection and penetration of a wave from a conducting wall with a slot as shown in Figure 6.15 for an electric field polarized parallel to the long dimension of the slot (a) and perpendicular to it (b). Reflection from a conducting wall takes place because currents **j** are induced such that they generate a reflected (scattered) electric field to cancel out the incident field and thus satisfy the boundary conditions on the wall ($\mathbf{E}_t = 0$). We see from Figure 6.15a that the slot does not impede the flow of currents in the direction shown and thus the conducting wall is able to generate a full reflection without any appreciable field penetration through the slot. Another intuitive way to reach the same conclusion is to reason that the electric field component parallel to the long side of the slot must be zero in order to satisfy the boundary conditions on the slot long sides. Thus

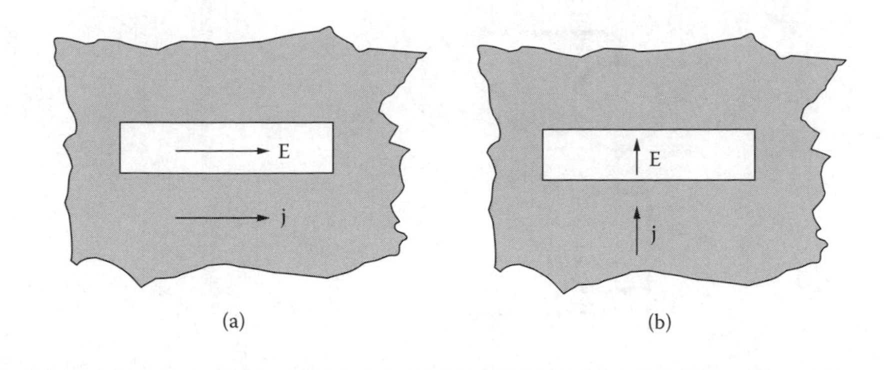

Figure 6.15 Penetration through an aperture when the field is polarized parallel to the longest dimension (a), and perpendicular to it (b).

fields find it difficult to penetrate. In contrast, for this electric field polarization in Figure 6.15b, currents must flow as shown and clearly the presence of the slot has a significant impact. Reflection is significantly impeded and therefore the wave penetrates. For this reason we focus on the case depicted in Figure 6.15b as it represents the worst case.

Before embarking on building a mathematical model of the configuration in Figure 6.14 for the purposes of calculating SE, we first consider the essential component of such a model. First, we need a representation of the incident field impinging on the cabinet with the slot. Second, the impact of the slot on the field as it negotiates its passage through it must be represented. Third, the way that EM energy distributes inside the cabinet must be accounted for (box resonances). We develop a suitable model for each of the three components before we finally put them together.[34–36]

We assume a plane wave incident onto the cabinet and we represent it simply by the Thevenin equivalent shown in Figure 6.16a. The voltage source represents the strength of the incident field and η_0 is the intrinsic impedance of free space (= 377). The exact value of V_0 is of no significance as the SE is expressed as the ratio of two fields and V_0 should cancel out.

The model of the aperture is more difficult to envisage. This part of the configuration is depicted in Figure 6.16b and it has the appearance of a co-planar stripline shorted at both ends. This is shown more clearly in Figure 6.17 where

$$Z_{ocs} = 120\pi^2 \left[\ln\left(2\frac{1+\sqrt[4]{1-(w_e/b)^2}}{1-\sqrt[4]{1-(w_e/b)^2}} \right) \right]^{-1} \tag{6.30}$$

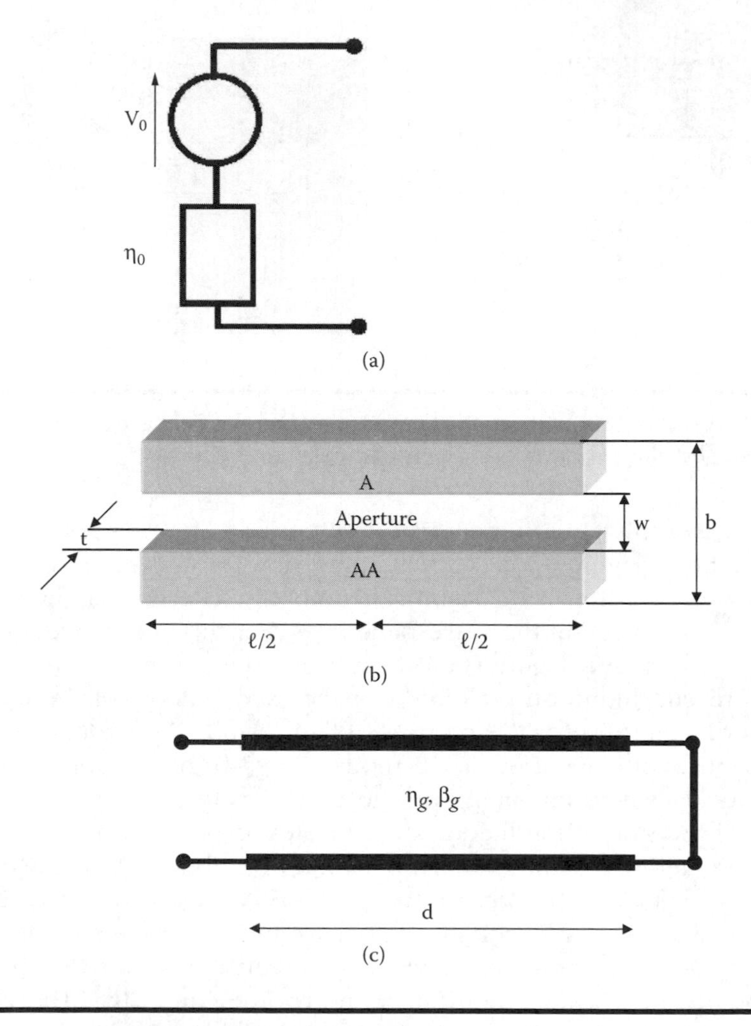

Figure 6.16 The three elements of the shielding problem: (a) source of EM threat, (b) aperture, and (c) cabinet.

where

$$w_e = w - \frac{5t}{4\pi}\left(1 + \ln\frac{4\pi w}{t}\right)$$

is the characteristic impedance of the co-planar stripline.[37] The input impedance of each of the two shorted lines looking to the left and to the right of the slot center is

$$Z_{IN} = jZ_{ocs}\tan\left(\beta_0 \ell/2\right)$$

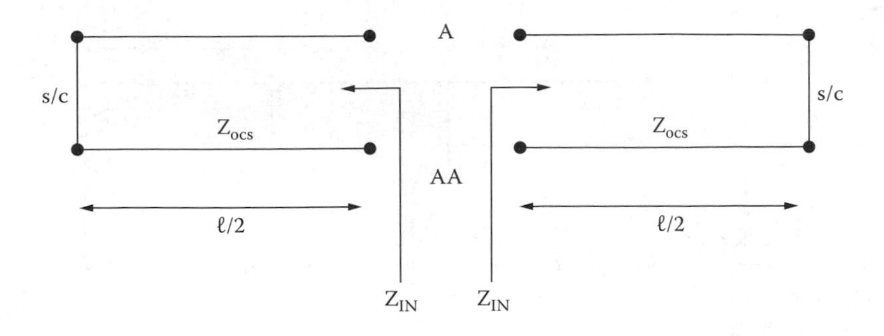

Figure 6.17 Equivalent circuit of the slotline.

where $\beta_o = 2\pi/\lambda$. Hence, the slot may be simply represented by a shunt impedance across its center given by the parallel combination of Z_{IN} to the left and to the right, i.e.,

$$Z_{slot} = \frac{1}{2}\frac{\ell}{a} jZ_{ocs} \tan\left(\beta_0 \ell/2\right) \tag{6.31}$$

where the scaling factor (ℓ/a) accounts for the different length of the slot and the cabinet side.

The final element of the model is the cabinet, which is excited by the slot fields at its front walls and is shorted at the far end as shown in Figure 6.16c. We are therefore inclined to represent the cabinet as a rectangular shorted waveguide where propagation is along the z-direction. From standard waveguide theory[38] the cutoff wavelength is $\lambda_c = 2a/n$ and therefore the waveguide wavelength and characteristic impedance are

$$\lambda_g = \frac{\lambda}{\sqrt{1-\left(\frac{\lambda}{\lambda_c}\right)}}$$

$$\eta_g = \frac{\sqrt{\mu/\varepsilon}}{\sqrt{1-\left(\frac{\lambda}{2a}\right)^2}} \tag{6.32}$$

Retaining only the lowest (n = 1) mode and expressing all quantities in terms of the dimensions we get the guide characteristic impedance and propagation constant

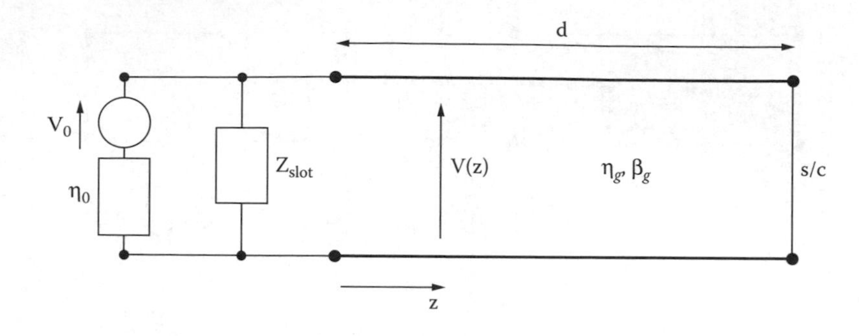

Figure 6.18 Equivalent circuit of the complete system.

$$\eta_g = \frac{\eta_0}{\sqrt{1-\left(\lambda/2a\right)^2}}$$

$$\beta_g = \beta_0\sqrt{1-\left(\lambda/2a\right)^2} \tag{6.33}$$

where, $\eta_0 = 377\ \Omega$, $\beta_0 = 2\pi/\lambda$. In this reasonable approximation, the cabinet is represented as a shorted transmission line with the parameters calculated above. Putting the three models together gives the simplified model of the entire configuration as shown in Figure 6.18. Calculation of the electric field inside the cabinet at a distance z from the front face with the aperture is equivalent to the calculation of the voltage V(z) in the transmission line shown in Figure 6.18. This can be done simply by reducing this circuit to the configuration shown in Figure 6.19 where V_s, Z_s are the components of the Thevenin equivalent representing the incident field and slot in Figure 6.18. A further reduction to the Thevenin equivalents looking left and right from the position z inside the cabinet gives the circuit shown in Figure 6.20 where

$$V_1 = \frac{V_s}{\cos(\beta_g z) + j\frac{Z_s}{\eta_g}\sin(\beta_g z)}$$

$$Z_1 = \frac{Z_s + j\eta_g \tan(\beta_g z)}{1 + j\frac{Z_s}{\eta_g}\tan(\beta_g z)} \tag{6.34}$$

$$Z_2 = j\eta_g \tan\left[\beta_g(d-z)\right]$$

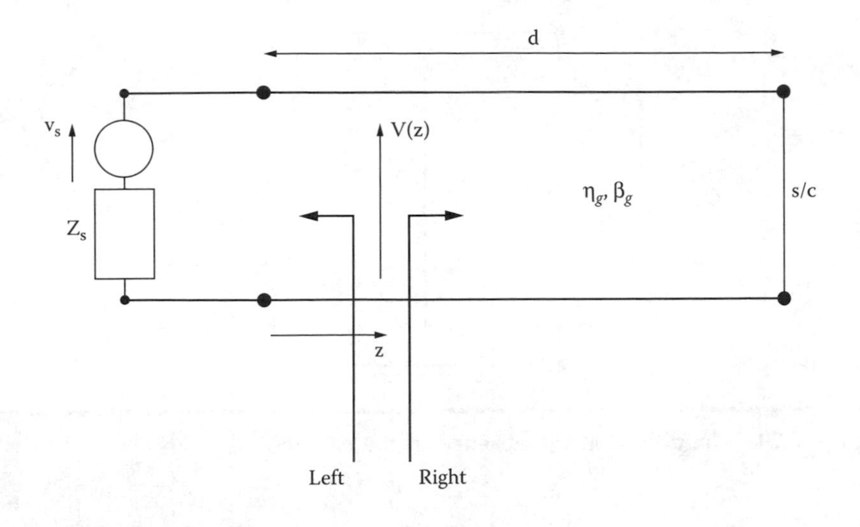

Figure 6.19 Reduction of the circuit in Figure 6.18.

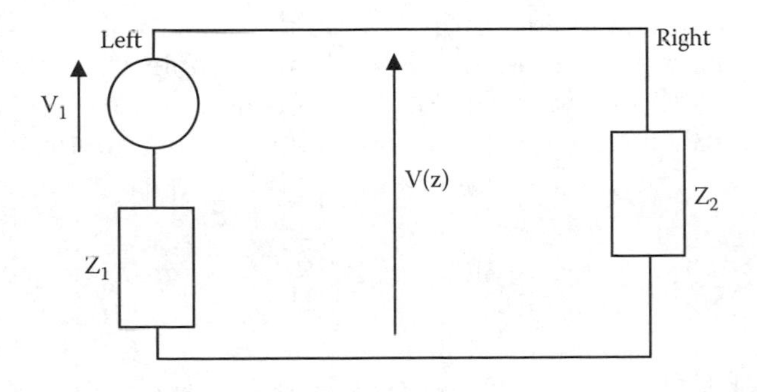

Figure 6.20 Reduction of the circuit in Figure 6.19 to a Thevenin equivalent.

Hence, the voltage at distance z is

$$V(z) = V_1 \frac{Z_2}{Z_1 + Z_2}$$

In the absence of the cabinet the electric field at the same position is equivalent to the voltage V_{nc} obtained for the equivalent circuit shown in Figure 6.21, $V_{nc} = V_0/2$. The electric field shielding effectiveness is therefore given by the expression

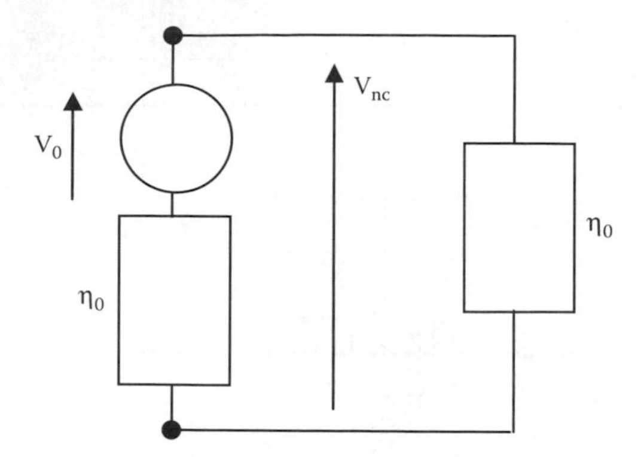

Figure 6.21 Conditions in the absence of the cabinet.

$$SE_{electric} = 20\log\left(\frac{E_{nc}}{E_c}\right) = 20\log\left[\frac{V_0}{2V(z)}\right] \qquad (6.35)$$

The magnetic shielding effectiveness may be similarly obtained by calculating the current at position z. A numerical example of electric shielding effectiveness is shown in Figure 6.22. In this figure the solid curve is the result of the calculation based on Equation 6.35, and the broken curve is obtained from

$$SE = 20\log\left(\frac{\lambda}{2\ell}\right)$$

for a slot of length ℓ.[39] It should be noted that such a simple formula, which does not account in any way for the presence of the cabinet, fails naturally to predict the cabinet resonance at around 700 MHz but also over the entire spectrum of frequencies far removed from the resonance. Another feature worthy of note is that at resonance the SE effectiveness is negative. This implies that at this frequency the cabinet is not providing any shielding but on the contrary enhances the field above its value at that point when the cabinet is absent. It is therefore necessary to consider a model of the entire system before any remotely reliable estimates of SE can be made. Statements commonly found in trade journals that a wall material has a particular value of SE offer little guidance for the value of SE effectiveness of a finished product made using this material. Figure 6.23 shows further results and comparisons with measurements confirming the good performance of this model. It is important to realize the limitations of the model presented here. Most notable among them is accounting for

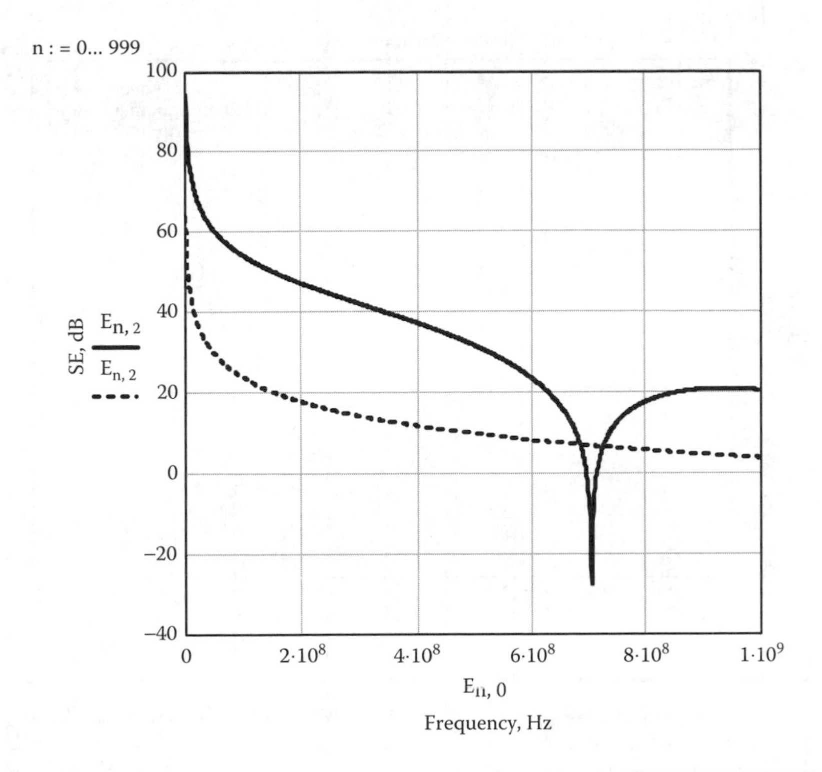

Figure 6.22 Shielding effectiveness of a box as shown in Figure 6.14. Cabinet dimensions 0.3 × 0.12 × 0.3 m³, slot 0.1 × 0.005 m², wall thickness t = 1.5 mm, observation point z = 0.15 m. Solid curve is this model and the broken curve Ott's formula.

only one waveguide mode, neglecting the contents (loading) of the cabinet, and multiple and off-center apertures.

The treatment of SE described above is limited by the inclusion of only the first cavity mode. This restriction may be removed at the expense of additional model complexity. Oblique incidence and off-center apertures may be similarly treated. The case of a loaded cabinet (e.g., containing PCBs) may be approximately treated in one of two ways. First, the propagation characteristic impedance and propagation constant in Equation 6.33 may be made complex to introduce losses through a correction factor that is of the order of the inverse of the Q-factor of the cabinet

$$\eta_g' = \left(1 - \zeta - j\zeta\right)\eta_g$$

$$\beta_g' = \left(1 - \zeta - j\zeta\right)\beta_g$$

where $\zeta \simeq Q^{-1} << 1$.

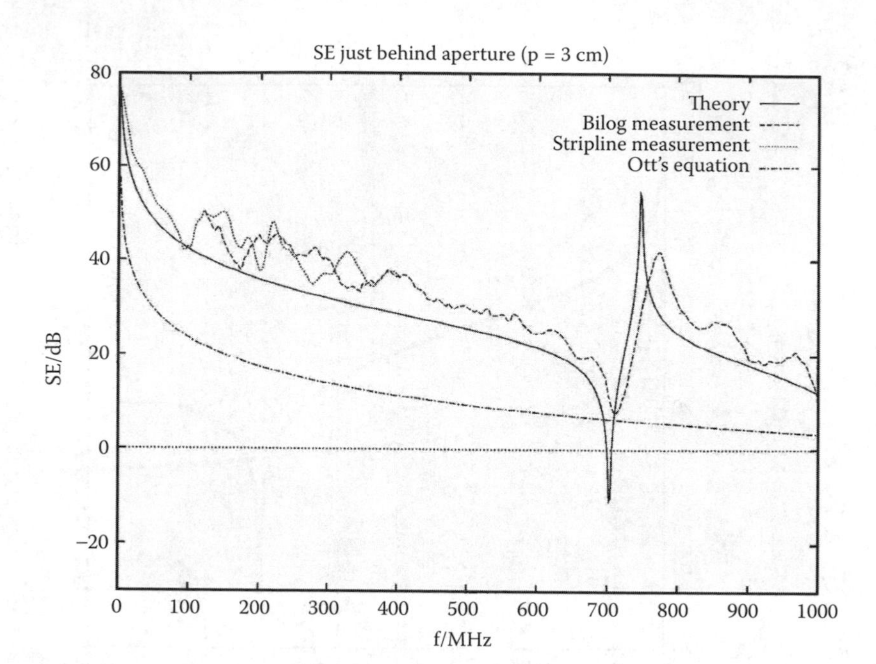

Figure 6.23 Typical results for the SE of a cabinet. The solid curve is this model compared with measurements.

In this approach the load is distributed uniformly throughout the cabinet. In the second approach loads representing PCBs are located at specific point in the cabinet by introducing shunt impedances at corresponding points along the transmission line representing the cabinet.[40,41] The impact of losses is generally to reduce the severity resonances and therefore to partially or totally eliminate the negative part of the SE. This may be interpreted as an improvement in SE; however, if the damping was provided by the PCBs it means that they have absorbed EM energy and hence may suffer the consequences of interference. The presence of damping material specifically placed inside the cabinet to absorb EM energy (so to speak "sacrificial" material) will naturally reduce the impact of the energy distributed inside the cabinet.

The model described above can also be used to account approximately for several slots by simply modifying Equation 6.31 to include the number of slots n:

$$Z_{slot} = n\frac{1}{2}\frac{\ell}{a}jZ_{cs}\tan\left(\beta_0\ell/2\right)$$

A circular aperture of diameter d_h can be described to a good approximation by setting the slot dimensions to

$$\ell = w = \frac{\sqrt{\pi}}{2} d_h$$

A fairly useful estimate of the SE of practical cabinets (e.g., a eurocabinet), which contain a multitude of holes and slots, may be obtained using techniques demonstrated in Reference 42.

6.6 Numerical Simulation Methods for Penetration through Shields and Apertures

The analytical treatment of electromagnetic penetration in previous sections covered a rather limited range of canonical highly simplified geometries. Several approximations and simplifications were necessary in order to obtain solutions. More complex configurations may be treated using analytical or approximate techniques but solution difficulties increase rapidly, thus putting an effective limit on what can be done even by the specialists. The engineer interested in EMC is concerned, however, with practical systems, which at a first glance exceed in complexity anything that at present can be tackled analytically. The ability of the engineer to study trends in EMC design and to understand the behavior of systems is considerably enhanced by the development and use of numerical computer-based models. These models, while at present are incapable of dealing with every conceivable practical system, offer, nevertheless, a significant extension of the range of problems that can be tackled. This section is therefore devoted to describing some of the main numerical simulation methods and their capabilities in the area of EMC. It is not possible in this text to cover in detail the various methods available. However, it is hoped that sufficient detail will be presented to help the reader grasp what can or cannot be done with numerical methods, and which methods are best suited to a particular problem. It is perhaps worth pointing out at this stage that analytical methods and solutions to problems are still important as they provide insight into the significance of various parameters and also benchmarks against which numerical methods can be checked and evaluated.

6.6.1 Classification of Numerical Methods

Problems in EMC are, in their simplest form, posed in the following manner. An excitation or input representing a source of EMI is applied to the system and the response or output from the system is required. This is shown schematically in Figure 6.24. Solution of the problem by

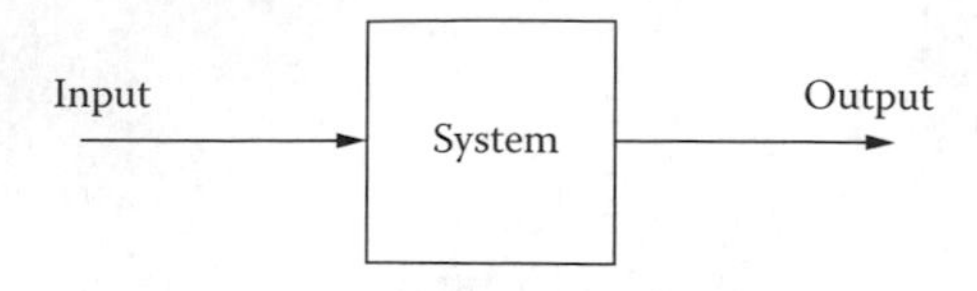

Figure 6.24 Stimulus and response of a system.

numerical means consists of obtaining a transfer function relating input and output. This function is not in a closed analytical form. Instead, it may be given as a set of numbers or graphs. If the input is a sinusoidal signal, then the transfer function H(ω) relates to this specific frequency and the problem is said to be solved in the frequency domain. If the response to another frequency is required, the simulation must be repeated. In this way, the numerical function H(ω) may be obtained over a wide frequency range. Alternatively, the input may be in the form of an impulse, or in practice a pulse of very short duration, in which case the impulse response h(t) of the system is obtained. The problem is then said to be solved in the time domain. Naturally, other time signals may be applied, but in most cases an impulse response is sought. Numerical methods normally fit into one of these two classes. Knowing the frequency response H(ω) allows the calculation of the impulse response h(t) and vice versa, since these two functions form a Fourier transform pair, i.e.,

$$h(t) \underset{FT^{-1}}{\overset{FT}{\rightleftarrows}} H(\omega) \tag{6.37}$$

In a problem where the steady-state response at a particular frequency is required, a frequency-domain method may be advantageous. On the other hand, in transient problems or where the response over a wide band of frequencies is required, a time-domain method could be the best choice. Nonlinearities such as saturation are normally best dealt with in the time domain.

Another criterion that may be used to classify methods is that of the type of formulation used in the description of the system. Let us consider as an example the solution of an electrostatic problem. This may be formulated by using either Gauss's Law in integral form (Equation 2.5) or its equivalent in differential form (Equation 2.6). In a formulation based on the integral form, conditions are imposed on surfaces in the system. This requires discretization of these surfaces into a number of segments, imposition of the appropriate boundary conditions, and solution of the resulting equations. In contrast, in a differential formulation the entire volume of the problem must be discretized and the relevant laws applied at each point or cell. It should be evident that in an integral formulation, there are fewer points to deal with compared to a differential formulation. However, in the former case the resulting equations are more difficult to solve. Integral methods deal quite rigorously with open boundaries, whereas differential methods require special, approximate, techniques to deal with such problems. In homogeneous, nonlinear, problems with complex boundaries and apertures are normally best dealt with by differential methods. Finally, there are methods that are based on principles borrowed from optics that are applicable at high frequencies. It should be evident from this brief discussion that no method excels in all cases and that, therefore, a mixture of methods or a hybrid method may offer the best general solution technique.

In the following two sections, the application of time- and frequency-domain methods to penetration problems will be discussed. Many generic numerical techniques may be applied either in the frequency or time domain. However, in most cases one particular formulation is the most common and efficient and the subsequent discussion is limited to this, the most popular implementation.

6.6.2 The Application of Frequency-Domain Methods

Two methods are well established in this area, namely the finite element (FE) method, and the method of moments (MM). Each method is briefly described below.

The finite-element (FE) method[43] — This is one of the oldest numerical simulation methods, first developed in connection with structural problems. Its use in electromagnetic problems is relatively recent. Most of the EM applications are at low frequencies in eddy current problems encountered in machine design, but higher-frequency formulations are gradually being developed. A simple example from electrostatics will serve to illustrate the method. Let us consider a two-dimensional problem where the potential is required. In applying the FE method, space is divided into triangular-shaped elements as shown in Figure 6.25. The potential at any point inside the element is $\phi(x, y) = A + Bx + Cy$ where A, B, and C can be obtained from the potential at the vertices of the triangle. A solution for the potential at the vertices of all triangles in the problem is obtained by demanding that a suitable energy functional is minimized. For an electrostatic problem a suitable functional is the stored electrostatic energy. Minimization of this energy, together with the boundary conditions, yields a sufficient number of equations to determine the potentials at the vertices.

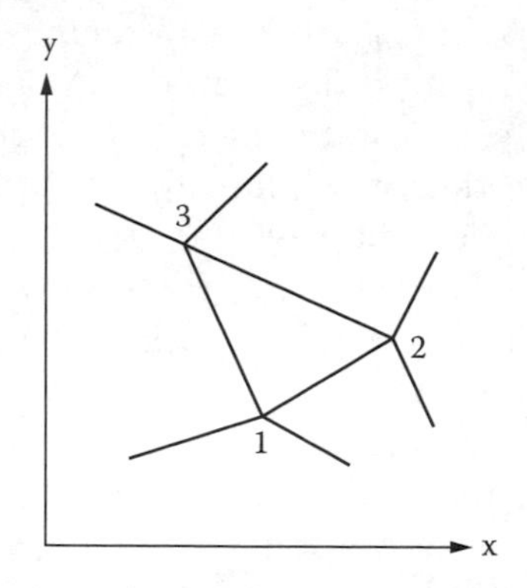

Figure 6.25 A basic two-dimensional element used in FE calculations.

However, solution of these equations requires matrix inversion and thus the number of elements that can be treated is limited to a few thousand. The use of triangles as basic elements gives a lot of flexibility in describing complex shapes. The method described is of a differential type, thus there are difficulties when dealing with open boundary problems. One approach is to specify the potential at the artificial numerical boundaries, but this results in numerical artifacts in the solution. The improvement of these boundaries is the subject of continuing research. In one development of the method, the wave equation in three dimensions is converted into a surface integral equation. These integrals are evaluated on each element on the boundary, rather than the whole problem space, thus resulting in what is known as the boundary element method (BEM) or boundary integral method (BIM).[44] The resulting equations are fewer in number but more difficult to handle computationally. A hybrid formulation whereby the conventional FE method is used for interior problems and the BIM for exterior, open-boundary regions and the two solutions are matched at the common boundary is also possible.[45] An example of a simulation of penetration through apertures using the FE method is reported in Reference 46.

The method of moments (MM)[47] — One of the difficulties when applying the FE method is the identification of a suitable energy functional before a minimization procedure may begin. An alternative approach, which overcomes this difficulty, is known as the generalized Galerkin's method, or the method of weighted residuals, or the method of moments. The basis of the method is best illustrated by a simple example. Let us consider the solution of an equation of the type

$$L\{\Phi\} = \rho \tag{6.38}$$

where L is an operator (e.g., ∇^2), Φ is the unknown function (e.g., potential), and ρ is a known term (e.g., electric charge density representing a source). The unknown function may be expanded in terms of N known basis function Φ_n so that

$$\Phi = \sum_{n=1}^{N} C_n \Phi_n \tag{6.39}$$

This expansion is analogous to the description of an unknown space vector in terms of its projections to three coordinate directions. In the case of function Φ, the larger N is, the better the accuracy, but a larger number of coefficient C_n must then be determined. Naturally, the estimate of Φ represented by Equation 6.39 when substituted into Equation 6.38 will yield a residual which is

$$\sum_{n=1}^{N} C_n L\{\Phi_n\} - \rho$$

The next step in the calculation consists of demanding that this residual is zero. This can best be done by projecting the residual into a set of trial functions X_m and demanding that this projection (inner product of residual and X_m) is zero. This condition yields a set of N equations with N unknowns, the set of coefficient C_n, of the form

$$\sum_{n=1}^{N} K_{mn} C_n = b_m \tag{6.40}$$

Equation 6.40 may then be solved using standard techniques for the coefficient C_n and thus the unknown function Φ in Equation 6.39 may be found. This basic scheme may be implemented in a number of ways and, depending on the choice of the basis Φ_n and test X_m functions, is known by different names. If the basis and test function are chosen to be the same, then the scheme is known as Galerkin's method. A simple implementation whereby Equation 6.40 is satisfied at discrete points is known as the point, matching, or colocation method. In problems found in electromagnetics, Equation 6.38 is normally an integral equation. Two formulations are common. The first formulation is expressed in terms of the incident magnetic field and results in the magnetic field integral equation (MFIE). An alternative formulation based on the incident electric field results in the electric field integral equation (EFIE). The former formulation is particularly suited to smooth-closed bodies and large structures, whereas the latter is most often used to study thin wires and electrically thin structures. The EFIE may also be used to represent surfaces by wire-grid models. Hybrid formulations are also possible to study more general problems. Particular care must be taken to avoid artificial coupling between solutions for the internal and external regions of a problem. Examples of simulation of EM wave penetration through apertures using moment methods may be found in References 30, 46, and 48. A collection of papers on the method of moments and its applications may be found in Reference 49.

6.6.3 The Application of Time-Domain Methods

Two methods dominate time-domain simulation, namely the finite-difference time-domain (FD-TD) method, and the transmission-line modeling (TLM) method. The principles and applications of each method are described briefly below.

The finite-difference time-domain (FD-TD) method[50] — The finite-difference method is one of the oldest and has been used extensively to solve scalar problems such as those found in electrostatics, magnetostatics, fluid flow, etc. A particular implementation, familiar to most readers from electrostatics, is one whereby the potential at a point, part of a large grid of points describing the problem, is obtained by averaging the potential of its nearest neighbors. This process is repeated several times until the solution converges. A large number of problems have been solved in this way. FD-TD is a direct solution of the full set of Maxwell's differential equations in the time-domain. Space is discretized using what is known as the Yee-mesh,[51] a basic element of which is shown in Figure 6.26. The electric and magnetic field components are calculated at different points in the space lattice as shown. A large number of such lattice elements are used to fill the problem space and to model various features in it. It is clear therefore that FD-TD is a differential method. The largest total number of elements that can be dealt with in a single problem depends on the computational resources available. For a typical modern workstation, a million elements is a useful guideline. In order to get a reasonable modeling accuracy, it is recommended that about ten elements per wavelength are used. This means that the largest three-dimensional problem that can be tackled without excessive computational demands cannot be larger than a few wavelengths. In open-boundary problems, where effectively the boundary is free space, an artificial numerical boundary must be introduced into the simulation. The boundary conditions appropriate

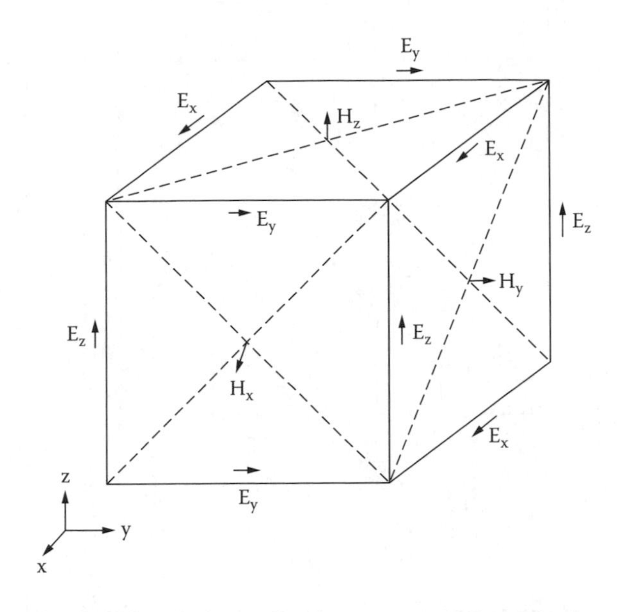

Figure 6.26 Schematic of an element of a three-dimensional Yee-mesh.

for free-space boundaries are difficult to establish accurately. This is a common problem to all differential methods. However, techniques have been developed to deal with such problems and these are continuously improved.

Computation proceeds as follows: All derivative terms in Maxwell's equations are approximated using the central-difference scheme, and each field component is calculated at successive time-steps. The resulting algorithm describes, therefore, a marching-in-time procedure that simulates the manner in which an initial electromagnetic disturbance propagates through space and interacts with structures until a steady state is reached. Steady state is approached after a number of periods of the excitation signal used, the exact number of periods depending on conditions. The maximum time-step that can be used is limited by stability considerations. A single computational run in the time domain is sufficient to obtain the impulse response. Results in the frequency domain can then be obtained at any frequency, within the bandwidth of the mesh, from the Fourier transform of the impulse response. The bandwidth of the mesh is conventionally determined by an upper frequency limit corresponding to a wavelength equal to ten space discretization lengths.

A large number of problems of some considerable complexity have been studied using the FD-TD method. It is particularly suited to studying penetration through slots and apertures in enclosures of a general shape.

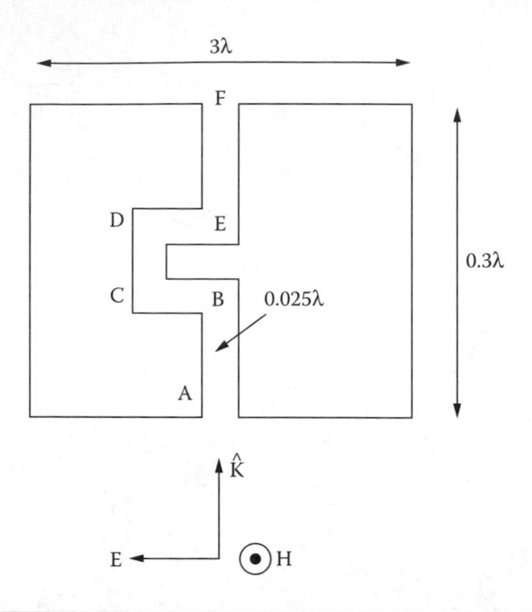

Figure 6.27 Example of a lapped joint treated by finite-difference techniques. (From Taflove et al., IEEE Trans, AP-36, 247, 1988. With permission.)

Fairly complicated structures may be described. Of particular interest are thin slits of complicated shape. In such cases, it is unreasonable to use a space discretization length that is dictated by the finest feature of the slit. Techniques have been developed to deal with this situation by defining equivalent ε and μ to describe a narrow slit in a problem described by a coarse mesh[52] or by a multigrid mesh,[53] or, alternatively, by a contour path model based on Faraday's Law.[54] These techniques are described as thin-slot formulations. A typical problem of a lapped U-shaped joint in a screen 0.3 wavelengths long is shown in Figure 6.27. The path length through the lapped joint ABCDEF is chosen to be equal to 0.45, thus giving highly resonant conditions. Results for the field along path ABCDEF are presented in References 50 and 54. They show good agreement between results obtained using a fine mesh to describe the joint ($\Delta l = 0.025\lambda$), and coarse mesh ($\Delta l = 0.09\lambda$) coupled with a thin-slot formulation. A number of other similar problems have also been solved using FD-TD.[55,56]

The transmission-line modeling (TLM) method[57] — A time-domain differential method that has many similarities with the FD-TD method is the transmission-line modeling (TLM) method. Many issues concerning problem definition, selection of space discretization length, and open-space boundaries, are the same as for the FD-TD method. However, the modeling philosophy is different. While the FD-TD scheme is an algorithm

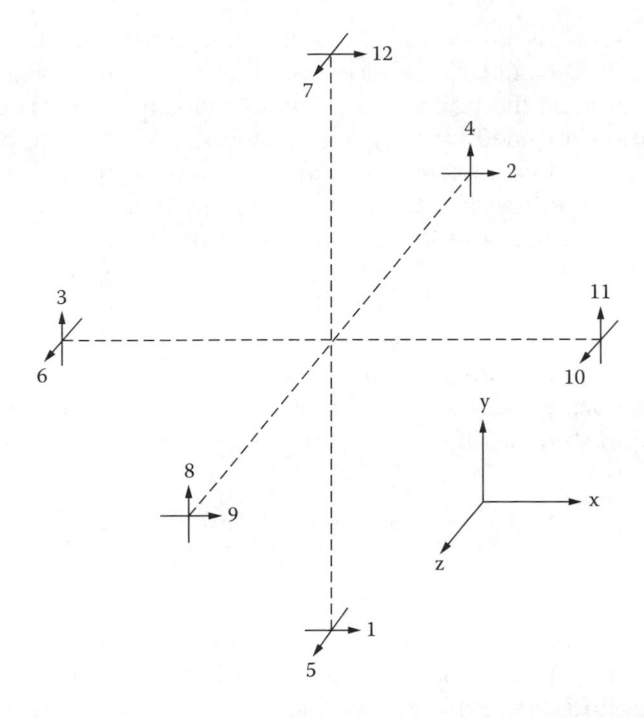

Figure 6.28 Schematic of a three-dimensional symmetrical condensed node of a TLM mesh.

for solving the differential equations describing the EM field, TLM approaches the problem differently by constructing in effect a network of interconnected transmission lines on which voltage and current pulses are exact analogues of electric and magnetic fields. It can be shown that the resulting transmission-line equations and Maxwell's equations have the same form and thus the network can be used as a model to describe the EM field. A block in space of volume $(\Delta l)^3$ is represented by twelve transmission-line segments (two per coordinate direction) interconnected at a node. The node most commonly used for three-dimensional problems is the symmetrical condensed node described in Reference 58 and shown in Figure 6.28. The space length l, time-step t, and the wave propagation velocity u in the modeled medium are related by the expression $\Delta t = \Delta l/(2u)$. Computation in TLM proceeds as follows: at each time-step, voltage pulses V^i are incident at each of the twelve transmission-line ports. These pulses are scattered at the node according to transmission-line theory, to produce twelve reflected pulses V^r where

$$\mathbf{V}^r = [S]\,\mathbf{V}^i \tag{6.41}$$

The matrix [S] is known as the scattering matrix and it has a very simple form.[58] The initial conditions provide the values of all incident voltage pulses at all the nodes at the start of the simulation. The scattered pulses are then obtained by applying Equation 6.41 at all the nodes. The incident voltages at the next time-step are obtained easily from the position of each node in the mesh. For example, the incident voltage at port 1 of node (x, y, z) at time-step k + 1, is equal to the voltage scattered from port 12 of node (x, y – 1, z) at the previous time-step k. This process, which is described as "connection," is repeated for all nodes and ports in the problem. Thus, "scattering" and "connection" follow each other for as long as desired. At any time, field components may be obtained from the incident voltage pulses. For example, the electric field component in the x-direction at node (x, y, z) and time-step k is

$$E_x = -\frac{V_1^i + V_2^i + V_9^i + V_{12}^i}{2\,\Delta l}$$

where the incident voltages are obtained from the relevant ports of node (x, y, z) at time-step k. Similar expressions apply for the other field components. Losses and different values of ε and μ may be incorporated in TLM by introducing stubs.[58–60] Irregularly shaped nodes may be modeled efficiently using the hybrid node.[61,62] A multigrid formulation has also been developed to model fine features in more detail.[63] Results of simulations for EM wave penetration through imperfectly conducting walls (such as carbon fiber composite panels) may be obtained using a multisection transmission-line model of the wall interposed between the symmetrical condensed nodes placed at either side of the wall, as described in Reference 64. A narrow slot formulation is also available to deal with penetration through apertures.[65] Further details on the fundamentals and application of this method may be found in Reference 66.

6.7 Treatment of Multiple Apertures through a Digital Filter Interface

In the previous section we have seen ways by which SE may be estimated using a range of analytical and numerical tools. Broadly speaking, accuracy has to be traded against model complexity. Full-field models are accurate and permit us to accommodate a range of features in simulations. There are, however, computer intensive and thus for many practical configurations their use is not realistic. A particular example, which is common in practical situations, is the calculation of SE in a cabinet with a large number

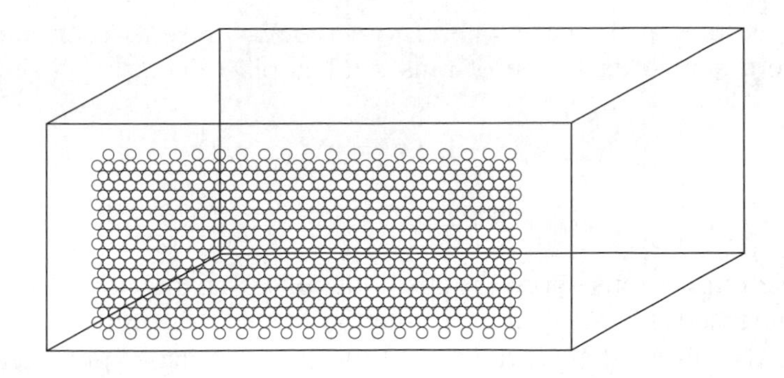

Figure 6.29 A conducting cabinet with perforations for cooling.

of apertures, e.g., ventilation holes as shown in Figure 6.29. Neither the cabinet nor the apertures need have a canonical shape (e.g., round holes and rectangular cabinet) and the cabinet may contain particular features (wires, PCBs, etc.) that we need to model. The intermediate modes cannot accommodate such general configurations and the full-field ones become seriously inefficient. One has to contemplate modeling perhaps one hundred holes in a full-field code where the outlines of each have to be mapped on a very fine mesh. This task, apart from leading to an enormous computation, is extremely tedious, time consuming, and prone to errors at the data entry stage. It is therefore important to seek ways where such features may be incorporated into full-field codes in a more efficient manner. The purpose of this section is to introduce a technique for doing just this.

We illustrate the approach by reference to the TLM full-field modeling technique. In Figure 6.30 two adjacent TLM nodes (computational cells) are shown. We will treat the case where at the interface between the cells we insert a perforated conducting screen as shown in Figure 6.29. But

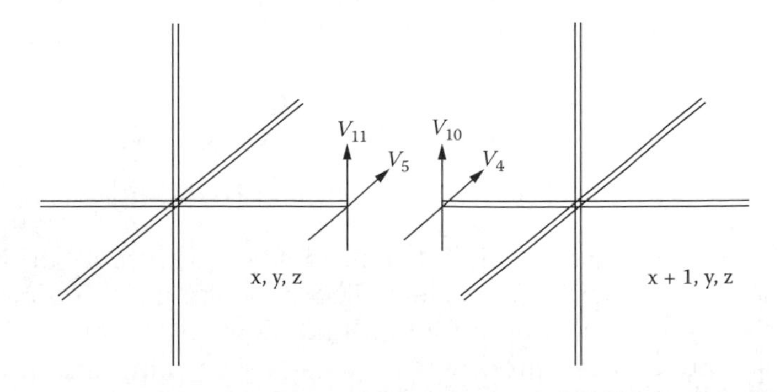

Figure 6.30 Computation across a boundary where a thin perforated panel may be placed.

first let us look at the case where no screen is present. Then across the interface a simple exchange of pulses takes place, i.e.,

$$\begin{aligned} {}_{k+1}V_4^i(x+1,y,z) &= {}_kV_5^r(x,y,z) \\ {}_{k+1}V_{10}^i(x+1,y,z) &= {}_kV_{11}^r(x,y,z) \end{aligned} \tag{6.42}$$

These expressions show the incident pulses at the node to the right at the next (k + 1) time-step as a function of the pulse reflected from the node to the left at the previous (k) time-step. Similar expressions apply for the pulses incident at the node to the left. This process is extremely simple — a straightforward exchange. How would this simple exchange be modified if a conducting screen were introduced? Let us first consider the extreme case of a perfectly conducting screen (no perforations) inserted between the two nodes. In such a case any pulses reflected from a node encounter the perfect electric conductor (PEC) and are returned to the same node with opposite sign. The expressions are therefore modified to

$$\begin{aligned} {}_{k+1}V_4^i(x+1,y,z) &= -{}_kV_4^r(x+1,y,z) \\ {}_{k+1}V_{10}^i(x+1,y,z) &= -{}_kV_{10}^r(x+1,y,z) \end{aligned} \tag{6.43}$$

It is now time to consider the case where the conducting screen is perforated. It is obvious that this is an intermediate case between no screen and a perfectly conducting screen — part of the reflected signal will pass through as in Equation 6.42 and part will be reflected as in Equation 6.43. It stands to reason that the amount of transmission through and reflection from the screen will be dependent on the relative size of the wavelength and of the perforations. The exchange of pulses between the two nodes is a scattering event that is frequency dependent. The exchange of pulses is described through the scattering process in the equation

$$\begin{bmatrix} {}_{k+1}V_{10}^i(x+1,y,z) \\ {}_{k+1}V_{11}^i(x,y,z) \end{bmatrix} = \begin{bmatrix} R & T \\ T & R \end{bmatrix} \begin{bmatrix} {}_kV_{10}^r(x+1,y,z) \\ {}_kV_{11}^r(x,y,z) \end{bmatrix} \tag{6.44}$$

where R and T are reflection and transmission coefficients through the screen that are frequency dependent. The difficulty in incorporating this frequency-dependent procedure into a time-domain model can be overcome by adopting formal procedures extensively used in signal processing to obtain a digital filter (a procedure in the time domain) that is equivalent to an analogue filter (specified in the frequency domain). An outline of the steps taken is shown in Figure 6.31.

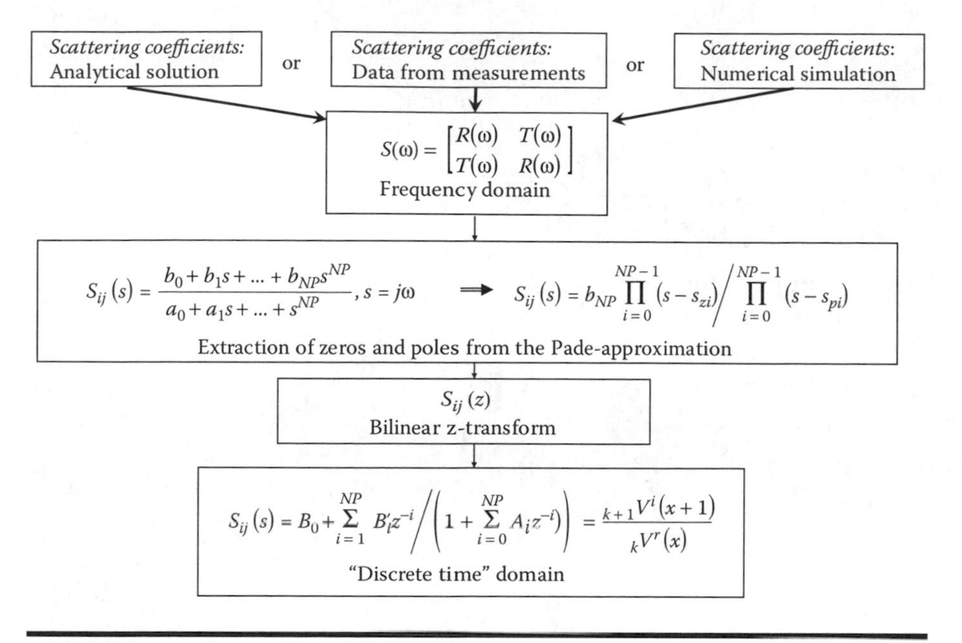

Figure 6.31 Schematic of the digital filter interface (DFI).

The first step is to specify the scattering matrix S(ω). This can be done either analytically, from measurements, or by numerical simulations on a single cell of a periodic cluster of perforations. In this way the frequency-dependent elements $S_{ij}(\omega)$ of the scattering matrix may be obtained. It is a distinct advantage that this can be done in three different ways as indicated, thus giving a compact and flexible specification of the screen properties. Each scattering element may be described by a number of poles and zeros and expressed in the form

$$S_{ij} = \frac{b_{NP}(s - s_{z0})(s - s_{z1})\ldots(s - s_{z(NP-1)})}{(s - s_{p0})(s - s_{p1})\ldots(s - s_{p(NP-1)})} \tag{6.45}$$

where NP is the number of poles and s = jω. The extraction of the poles and zeros is done using the frequency-domain Prony method.[67] The required number of poles NP depends on the complexity of the screen response and also on the required accuracy. We will not dwell on this technique as it is a well rehearsed mathematical procedure. We will focus on the implementation of Equation 6.45 in the time domain. Multiplying out in Equation 6.46 gives us the coefficients in Padé form

$$S_{ij}(s) = \frac{b_0 + b_1 s + b_2 s^2 + \ldots + b_{NP} s^{NP}}{a_0 + a_1 s + a_2 s^2 + \ldots + s^{NP}} \tag{6.46}$$

Applying the pulse invariant transformation[68]

$$s - s_{zi} \leftrightarrow \frac{-s_{zi}}{1-\beta_{zi}}(1 - z^{-1}\beta_{zi}), \quad \beta_{zi} = e^{s_{zi}\Delta t} \tag{6.47}$$

Substituting into Equation 6.46 gives

$$S_{ij}(z) = B_0 \prod_{i=0}^{NP-1} \frac{1 - z^{-1}\beta_{zi}}{1 - z^{-1}\beta_{pi}} \tag{6.48}$$

where B_0 is a z-independent coefficient. Multiplying out in Equation 6.48 reduces this expression to the form

$$S(z) = \frac{B_0 + \sum_{i=1}^{NP} B_i z^{-i}}{1 + \sum_{i=1}^{NP} A_i z^{-i}} \tag{6.49}$$

This expression can be further manipulated and put in the form

$$S(z) = B_0 + \frac{\sum_{i=1}^{NP} B_i' z^{-i}}{1 + \sum_{i=1}^{NP} A_i z^{-i}} \tag{6.50}$$

Typically, Equation 6.50 relates incident and reflected voltages interacting with the screen, i.e.,

$$V^r = B_0 V^i + \frac{V^i \sum_{i=1}^{NP} B_i' z^{-i}}{1 + \sum_{i=1}^{NP} A_i z^{-i}} \tag{6.51}$$

We define state variables X_i such that

$$V^r = B_0 + \sum_{i=1}^{NP} B_i' X_i$$

or, in more compact form,

$$V^r = B_0 V^i + \mathbf{B}'^T \mathbf{X} \tag{6.52}$$

In the state-space representation of the system in Equation 6.52 is the output equation. In a similar manner a state equation may be obtained in the form

$$\boldsymbol{X} = z^{-1}\mathbf{A}'\mathbf{X} + z^{-1}1'V^i \tag{6.53}$$

A full definition of these coefficients may be found in Reference 69. Equations 6.52 and 6.53 describe a procedure in the time-domain that is embodied in the diagram shown in Figure 6.32. It effectively replaces Equation 6.42 for the case without a screen and Equation 6.43 for the case for a perfectly conducting screen without perforations. It describes the scattering by the screen in the time domain and it has the form of an equivalent digital filter. We may describe it as a digital filter interface (DFI). Using the DFI, any type of perforations, materials, and composites may be represented efficiently without the need to mesh directly the configuration

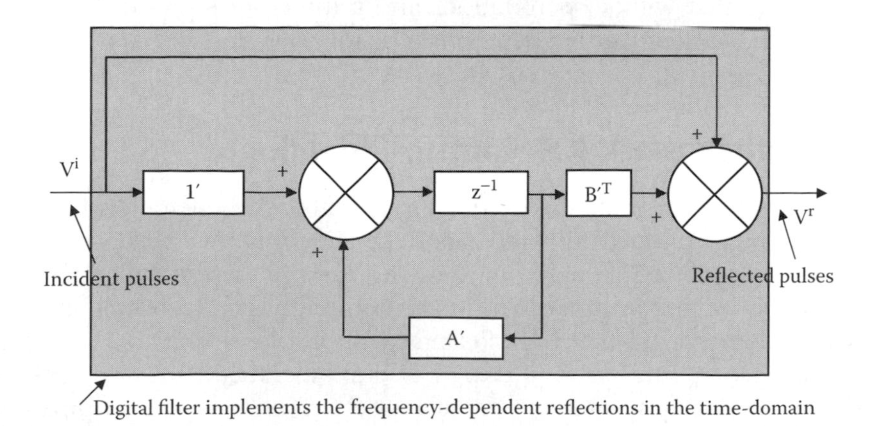

Figure 6.32 Digital filter to implement the frequency-dependent scattering in the time domain.

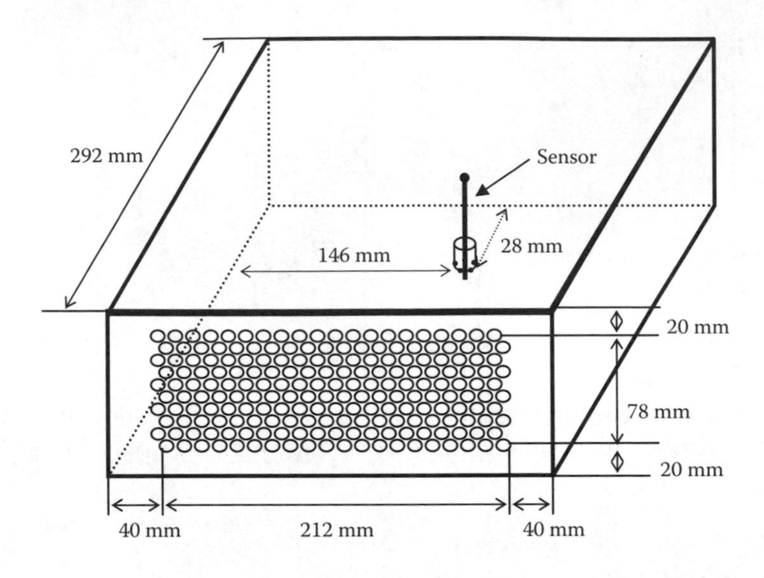

Figure 6.33 Sketch of a cabinet with a perforated wall used for SE studies (hole diameter = 4 mm, hole spacing = 10 mm).

of the screen. This offers great economy both in data preparation and during the simulation run. As an illustration, we show results for the configuration shown in Figure 6.33. For this type of perforations. analytical results for the scattering coefficient in the frequency-domain are available.[70] The electric SE is shown in Figure 6.34 for two model discretization lengths and is compared to measurements. This simulation is comparable in efficiency to that without perforations and it therefore represents a major model improvement. Further details may be found in References 67 and 71.

6.8 Further Work Relevant to Shielding

In this chapter, the basic theory relating to diffusive penetration through finite electrical conductivity screens and through apertures was presented. Emphasis was placed on understanding the basis of these calculations and the inherent approximations. The main numerical simulation techniques that may be used to study complex practical problems were described in the last sections. The practical design of equipment cabinets and screens and the electromagnetic environment inside and around such equipment is also dependent on wire penetrations, propagation, and crosstalk between transmission systems, etc. These topics are examined in the next two chapters. More practical aspects and design rules for shielding are discussed in Chapter 10.

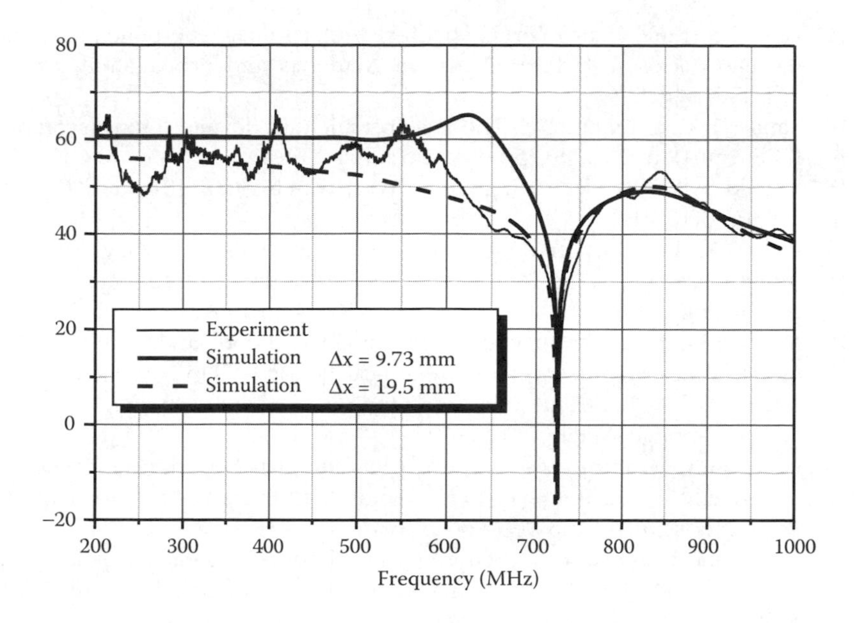

Figure 6.34 Comparisons of shielding effectiveness in decibels from experiments and simulation.

References

1. Haus, H A and Melcher, J R, "Electromagnetic Fields and Energy," Prentice-Hall, NJ, 1989.
2. Franceschetti, G, "Fundamentals of steady-state and transient electromagnetic fields in shielding enclosures," IEEE Trans on EMC, 21, pp 335–348, Nov 1979.
3. Bridges, J E, "An update on the circuit approach to calculate shielding effectiveness," IEEE Trans on EMC, 30, pp 211–221, Aug 1988.
4. Wilson, W R and Mankoff, L L, "Short-circuit forces in isolated-phase busses," Trans AIEE, 73, Pt III, pp 382–391, April 1954.
5. Greenwood, A, "Electrical Transient in Power Systems," 2nd Edition, Wiley, NY, 1991.
6. Cooley, W W, "Low-frequency shielding effectiveness of nonuniform enclosures," IEEE Trans on EMC, 10, No 1, pp 34–43, Mar 1968.
7. King, L V, "Electromagnetic shielding at radio frequencies," Philos Mag, 15, No 97, pp 201–223, Feb 1933.
8. Thomas, A K, "Magnetic shielded enclosure design in the DC and VLF region," IEEE Trans on EMC, 10, pp 142–152, 1968.
9. Mager, A J, "Magnetic shields," IEEE Trans on Magnetics, 6, pp 67–75, 1970.
10. Lee, K S H, "Electromagnetic shielding," in Recent Advances in Electromagnetic Theory, H N Kritikos and D L Jaggard (editors), Springer-Verlag, NY, 1990.

11. Lee, K S H and Bedrosian, G, "Diffusive electromagnetic penetration into metallic enclosures," IEEE Trans on Antennas and Propagation, 37, pp 194–198, March 1979.
12. Vance, E F and Graf, W, "The role of shielding in interference control," IEEE Trans on EMC, 30, pp 294–297, Aug 1988.
13. Schelkunoff, S A, "Electromagnetic Waves," Van Nostrand, Toronto, Canada, pp 303–306, 1943.
14. Paul, C R, "Introduction to Electromagnetic Compatibility," Wiley, NY, 1992.
15. Whitehouse, A C D, "Screening: new wave impedance for the transmission-line analogy," Proc IEE, 116, pp 1159–1164, July 1969.
16. Miedzinski, J, "Electromagnetic screening — theory and practice," The Electrical Research Association, Technical Report M/T135, 65 pages, 1959.
17. Kaden, H, "Wirbelstrome und Schirmung in der Nachrichtentechnik," 2nd Edition, Springer-Verlag, NY, 1959.
18. Harrison, C W and Papas, C H, "On the attenuation of transient fields by imperfectly conducting spherical shields," IEEE Trans on Antennas and Propagation, AP-13, pp 960–966, 1965.
19. Ari, N and Hansen, D, "Durchdringung schnellveranderlicher elektromagnetischer impulsfelder durch eine leitende platte," etz Archiv, 7, pp 177–181, 1985.
20. Excell, P S and Belshaw, J C, "Electromagnetic compatibility and the superconducting revolution," Sixth Int Conf on EMC, IERE Conf Publ No 81, pp 279–285, York, England, 12–15 Sept 1988.
21. Vance, E F, "Shielding effectiveness of braided-wire shields," IEEE Trans on EMC, 17, pp 71–77, 1975.
22. Fowler, E P, "Superscreened cables," The Radio and Electronic Eng, 49, pp 38–44, 1979.
23. Casey, K F, "EMP coupling through cable shields," IEEE Trans on EMC, 20, pp 100–106, 1978.
24. Hoeft, L O, Hofstra, J S, and Peel, R J, "Experimental evidence for purposing coupling and optimisation in braided cables," 8th Int Zurich Symposium on EMC, Zurich, 7–9, pp 505–509, March 1989.
25. Smythe, W R, "Static and Dynamic Electricity," McGraw-Hill, NY, 1950.
26. Cohn, S , "The electric polarizability of apertures of arbitrary shape," Proc IRE, pp 1069–1071, 1952.
27. Lee, K S H (editor), ' Handbook — Principles, Techniques and Reference Data," Report AFWL-TR-80-402, Albuquerque, NM, 1980.
28. Harrington, R F and Mautz, J R, "A generalised network formulation for aperture problems," IEEE Trans, AP-24, pp 870–873, 1976.
29. Harrington, R F and Auckland, D T, "Electromagnetic transmission through narrow slots in thick conducting screens," IEEE Trans, AP-28, pp 616–622, 1980.
30. Butler, C M, Rahmat-Samii, Y, and Mittra, R, "Electromagnetic penetration through apertures in conducting surfaces," IEEE Trans, AP-26, pp 82–92, 1978.
31. Kraus, J D, "Electromagnetics," 4th Edition, McGraw-Hill, NY, 1992.

32. Sewell, P, Turner, J D, Robinson, M P, Thomas, D W P, Benson, T M, Christopoulos, C, Dawson, J F, Ganley, M D, Marvin, A C, and Porter, S J, "Comparison of analytic, numerical and approximate models for shielding effectiveness with measurements," IEE Proc.-Sci. Meas. Technol., 145(2), pp 61–66, 1998.
33. Cerri, G, De Leo, R, and Primiani, V M, "Theoretical and experimental evaluation of the electromagnetic radiation from apertures in shielded enclosures," IEEE Trans. on Electromagnetic Compatibility, 34(4), pp 423–432, 1992.
34. Robinson, M P, Turner, J D, Thomas, D W P, Dawson, J F, Ganley, M D, Marvin, A C, Porter, S J, Benson, T M, and Christopoulos, C, "Shielding effectiveness of a rectangular enclosure with a rectangular aperture," Electronics Letters, 32, pp 1559–1560, 1996.
35. Robinson, M P, Benson, T M, Christopoulos, C, Dawson, J F, Ganley, M D, Marvin, A C, Porter, S J, and Thomas, D W P, "Analytical formulation of the shielding effectiveness of enclosures with apertures," IEEE Trans. on Electromagnetic Compatibility, 40(3), pp 240–248, 1998.
36. Konefal, T, Dawson, J F, Marvin, A C, Robinson, M P, and Porter, S J, "A fast multiple mode intermediate level circuit model for the prediction of shielding effectiveness of a rectangular box containing a rectangular aperture," IEEE Trans. on Electromagnetic Compatibility, 47(4), pp 678–691, 2005.
37. Gupta, K C, Garg, R, and Bahl, I J, "Microstrip Lines and Slotlines," Artech House, 1979.
38. Ramo, S, Whinnery, J R, and Van Duzer, T, "Fields and Waves in Communication Electronics," 2nd edition, Wiley, 1984.
39. Ott, H W, "Noise Reduction Techniques in Electronic Systems," 2nd edition, Wiley, 1988.
40. Denton, A, Thomas, D W P, Konefal, T, Benson, T M, Christopoulos, C, Dawson, J F, Marvin, A C, and Porter, S J, "A simple method for representing PCBs in equipment enclosures," EMC Europe 2000, Sept. 11–13, July 2000, pp 509–513.
41. Thomas, D W P, Denton, A, Konefal, T, Benson, T M, Christopoulos, C, Dawson, J F, Marvin, A C, and Porter, S J, "Characterization of the shielding effectiveness of loaded equipment enclosures," EMC York 1999, July 12–13 1999, pp 89–94.
42. De Smedt, R, De Moerloose, J, Criel, S, De Zutter, D, Olyslager, F, Laermans, E, Wallyn, W, and Lietaert, N, "Approximate simulation of the shielding effectiveness of a rectangular enclosure with a grid wall," Proc. Int. Conf. EMC Denver, 1998, pp 1030–1034.
43. Silvester, P P and Ferrari, R L, "Finite-elements for Electrical Engineers," 2nd Edition, Cambridge University Press, Cambridge, 1990.
44. Kagami, S and Fukai, I, "Application of boundary element method to electromagnetic field problems," IEEE Trans, MTT-32, pp 455–461, 1984.
45. Salon, S J, "The hybrid finite element-boundary method in electromagnetics," IEEE Trans, MAG-21, pp 1829–1834, 1985.

46. Sibbald, C L, Stuchly, S S, and Costache, G I, "Numerical analysis of waveguide apertures radiating into lossy media," Int J Num Modeling, 5, pp 259–274, 1992.
47. Harrington, R F, "Field Computation by Moment Methods," Macmillan, NY, 1962.
48. Gobin, V and Alliot, J C, "Modelling of electromagnetic wave penetration through loaded apertures," Int J Num Modelling, 4, pp 163–174, 1991.
49. Miller, E K, Medgyesi-Mitschang, L, and Newman, E H, "Computational Electromagnetics: Frequency-Domain Method of Moments," IEEE Press, NY, 1992.
50. Taflove, A and Umanshankar, K R, "The finite-difference time-domain method for numerical modelling of electromagnetic wave interactions with arbitrary structures," in Progress in Electromagnetic Research, J Amsterdam, A Kong (editor), pp 287–373, Elsevier, 1990.
51. Yee, K S, "Numerical solution of initial value problems involving Maxwell's equations in isotropic media," IEEE Trans, AP-14, pp 302–307, 1966.
52. Gilbert, J and Holland, R, "Implementation of the thin-slot formalism in the finite-difference EMP code THREDII," IEEE Trans, NS-28, pp 4269–4274, 1981.
53. Yee, K S, "A numerical method of solving Maxwell's equations with a coarse grid bordering a fine grid," Doc D-DV-86-0008, D Division, Lawrence Livermore National Laboratory, California, 1986.
54. Taflove, A et al., "Detailed FD-TD analysis of electromagnetic fields penetrating narrow slots and lapped joints in thick conducting screens," IEEE Trans, AP-36, pp 247–257, 1988.
55. Holland, R, "THREDE: A free-field EMP coupling and scattering code," IEEE Trans, NS-24, pp 2416–2421, 1977.
56. Kunz, K S and Lee, K M, "A three-dimensional finite-difference solution of the external response of an aircraft to a complex transient EM environment," IEEE Trans, EMC-20, pp 328–333, 1978.
57. Christopoulos, C, "Field analysis software based on the transmission-line modelling method" in Advances in Electrical Engineering Software, P. P. Silvester (ed.), Springer-Verlag, NY, pp 135–148, 1990.
58. Johns, P B, "A symmetrical condensed node for the TLM method," IEEE Trans, MTT-35, pp 370–377, 1987.
59. Naylor, P and Desai, R A, "New three-dimensional condensed node for solution of EM wave problems by TLM," Electronic Lett, 26, pp 492–494, 1990.
60. German, F J, Gothard, G, and Riggs, L S, "Modelling of materials with electric and magnetic losses with the symmetrical condensed TLM method," Electronic Lett, 26, pp 1307–1308, 1990.
61. Scarammuza, R A and Lowery, A J, "Hybrid symmetrical condensed node for the TLM method," Electronic Lett, 26, pp 1947–1949, 1990.
62. Scaramuzza, R A and Christopoulos, C, "Developments in transmission-line modelling and its applications in EM field simulation," COMPEL, 11, pp 49–52, 1992.

63. Christopoulos, C and Herring, J L, "Developments in the transmission-line modelling (TLM) method," Proc 8th Annual Rev of Progress in Applied Comp Electromagnetics, pp 523–530, March 16–20, NFS, Monterey, CA, 1992.
64. Johns, D P, Wlodarczyk, J, and Mallik, A, "New TLM models for thin structures," Proc Int Conf on Computation in Electromagnetics, pp 335–338, 25–27, London, IEE Conf Publ No 350, Nov 1991.
65. Mallik, A, Johns, D P, and Wlodarczyk, A J, "TLM Modelling of wires and slots," Proc Zurich EMC Symposium, Mar 9–11, 1993.
66. Christopoulos, C, "An introduction to the Transmission-Line Modelling (TLM) Method," IEEE Press, 1995.
67. Paul, J D, Podlozny, V, and Christopoulos, C, "The use of digital filtering techniques for the simulation of fine features in EMC problems solved in the time-domain," IEEE Trans. on Electromagnetic Compatibility, 45(2), pp 238–244, 2003.
68. Rabiner, L R and Gold, B, "Theory and Application of Dgital Signal Processing," Prentice Hall, 1975.
69. Paul, J D, Christopoulos, C, and Thomas, J D, "Equivalent circuit models for the time-domain simulation of ferrite electromagnetic wave absorbers," Proc. 13th Int. Zurich EMC Symp., 1999, pp 345–350.
70. Chen, C C, "Transmission of microwaves through perforated flat plates of finite thickness," IEEE Trans. on Microwave Theory and Techniques, 21, pp 1–6, 1973.
71. Paul, J D, Podlozny, V, Thomas, D W P, and Christopoulos, C, "Time-domain simulation of thin material boundaries and thin panels using digital filters in TLM," Turkish Journal of Electrical Eng. and Computer Sci., 10(2), pp 185–198, 2002.

Chapter 7

Propagation and Crosstalk

In Chapter 6 the penetration of electromagnetic energy through shields and apertures was studied. Irrespective of the exact manner of penetration, this energy most often couples into wires that are part of transmission lines, such as ribbon cables and printed circuit boards. Energy thus propagates, guided by the wires, to spread throughout the system and cause further interference. The purpose of this chapter is to address the problem of propagation and coupling to the field and between wires.

7.1 Introduction

Electrical signals may appear on a transmission line due to a variety of influences. One obvious possibility is the case where the transmission line shares a common path with another line. In such cases current flow on one line will produce an interference signal on the other line, in what is called "conducted" interference. This type of interference is relatively easy to identify as it involves a visible physical connection between the circuits. A different coupling path, which does not involve a physical connection between circuits, is the origin of what is described as "radiated" interference. This mechanism in turn involves several processes that contribute in a different measure in each practical problem. It is useful to divide "radiated" interference into several regimes, namely, capacitive, inductive, and radiated interference. Although a sharp dividing line cannot be drawn between these different regimes, each appears dominant under particular circumstances. Each coupling regime is described briefly below.

Capacitive coupling is particularly evident when two high-impedance circuits are in close proximity to each other and carry low-frequency signals. A schematic of such a circuit is shown in Figure 7.1a. Any potential difference v_1 appearing on conductor 1 will cause an interference signal to appear on conductor 2 equal to

$$v_2 = \frac{C_{12}}{C_{12} + C_2} v_1 \tag{7.1}$$

If the resistance to ground of conductor 2 is comparable to the capacitive impedance at the signal frequency, then it can be added in parallel to C_2, thus modifying Equation 7.1 at high frequencies.

Capacitive coupling may be modified by the addition of a shield around conductor 2, as shown in Figure 7.1b. If the shield has no DC connection to ground its potential will rise to

$$v_s = \frac{C_{1s}}{C_{1s} + C_s} v_1 \tag{7.2}$$

and since there is no current flow through C_{2S} conductor 2 will rise to the same potential. If, however, the shield is connected to ground, then $v_s = 0$ and so is the induced voltage on conductor 2. The situation whereby the shield over conductor 2 is not complete is an intermediate situation between the two cases just considered.

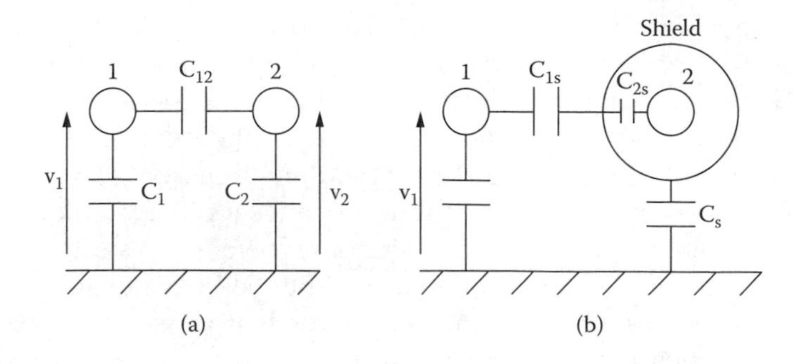

Figure 7.1 Configuration used to study coupling between circuits 1 and 2 (a) and modifications in the presence of a shield (b).

Inductive coupling appears dominant in low-impedance circuits in close proximity to each other carrying low-frequency signals. Referring to the circuit shown in Figure 7.1, the voltage induced on conductor 2 when a current i_1 flows in conductor 1 is

$$v_2 = M_{12}\, di_1/dt \tag{7.3}$$

where M_{12} is the mutual impedance between the two circuits. If a non-magnetic shield is added to conductor 2 as shown in Figure 7.1b, the voltage induced on the shield is

$$v_s = M_{1s}\, di_1/dt \tag{7.4}$$

If the shield is floating or connected to earth at one point only, no shield current flows and hence the voltage induced in conductor 2 remains unchanged, as given by Equation 7.3. If, however, the shield is grounded at both ends then the induced voltage given by Equation 7.4 will drive a shield current i_s and an additional voltage will be induced on conductor 2 due to this current. This additional voltage is equal to $M_{2S}(di_s/dt)$ where M_{2S} is the mutual inductance between the shield and conductor 2. If the geometry of the shield and the distribution of current on its circumference is such that no magnetic field is established within the shield then all the magnetic flux Φ established by i_s is also linked with conductor 2, hence $M_{2S} = \Phi/i_s = L_s i_s/i_s = L_s$ where L_s is the self-inductance of the shield.

The total induced voltage, assuming harmonic variation, is thus

$$\begin{aligned}\overline{V}_2 &= j\omega\, M_{12}\, \overline{I}_1 - j\omega\, L_s\, \overline{I}_s \\ &= j\omega\, M_{12}\, \overline{I}_1 \left[1 - \frac{L_s}{M_{12}} \frac{\overline{I}_s}{\overline{I}_1}\right]\end{aligned}$$

But

$$\overline{I}_s = \frac{j\omega\, M_{1s}\, \overline{I}_1}{R_s + j\omega\, L_s} \simeq \frac{j\omega\, M_{12}\, \overline{I}_1}{R_s + j\omega\, L_s}$$

where R_s and L_s are the resistance and self-inductance of the shield, and the induction due to i_2 has been neglected. Eliminating the shield current between the two expressions gives

$$\overline{V}_2 = j\omega\, M_{12}\, I_1 \left[\frac{R_s / L_s}{R_s / L_s + j\omega} \right] \tag{7.5}$$

As the frequency increases, and for $\omega >> R_s/L_s$, the induced voltage tends to

$$\overline{V}_2 = M_{12}\, \overline{I}_1\, R_s/L_s \quad \text{(high-frequency)}$$

At low frequencies, the induced voltage is unaffected by the shield and tends to

$$\overline{V}_2 = j\omega\, M_{12}\, \overline{I}_1$$

More detailed calculations for a variety of configurations may be found in References 1 and 2.

The methodology used above for capacitive and inductive coupling calculations breaks down as the frequency increases and lines become electrically long, i.e., their physical length becomes comparable to the wavelength. Coupling between lines separated by larger distances cannot be regarded as either capacitive or inductive. Instead, energy from the sources of interference propagates in the form of electromagnetic waves that impinge on other lines, thus causing EMI. This mode of coupling is referred to as radiated interference. Depending on the distance between the source and the victim of interference, a further subdivision to far-field radiated interference and near-field radiated interference can be made. If the victim circuit is sufficiently far from the source of EMI so that fields decay with distance as $1/r$ (typically for distances larger than $\lambda/2\pi$), then far-field conditions prevail. Otherwise, coupling takes place under near-field radiative conditions with a mixture of field components decaying as $1/r$, $1/r^2$, and $1/r^3$. This can be confirmed by studying the structure of the field near a dipole given by Equations 2.62 and 2.63.

Coupling under these conditions, and propagation in electrically long lines, is therefore a very complex matter and cannot be easily understood by resorting to simple lumped circuit concepts alone. The purpose of this chapter is to address these issues and present appropriate tools and methodologies for predicting propagation and field-to-wire coupling.

7.2 Basic Principles

The basic principles of propagation in multiconductor systems are presented in this section. A simple two-wire above-ground system is studied

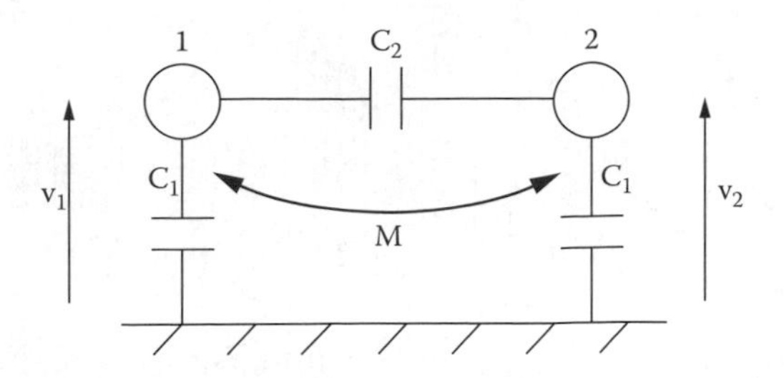

Figure 7.2 A two-wire line above ground.

in detail in order to establish the main features of propagation in multi-conductor systems. This treatment is followed by a generalization to more complex configurations.

Let us consider the system shown in Figure 7.2 consisting of two wires, each with capacitance C_1 to ground, self-inductance L_s, and with mutual capacitance and inductance C_2 and M, respectively. Applying KVL gives the following equations:

$$
\begin{aligned}
-\frac{\partial v_1}{\partial x} &= L_s \frac{\partial i_1}{\partial t} + M \frac{\partial i_2}{\partial t} \\
-\frac{\partial v_2}{\partial x} &= M \frac{\partial i_1}{\partial t} + L_s \frac{\partial i_2}{\partial t}
\end{aligned}
\tag{7.6}
$$

Similarly, from KCL

$$
\begin{aligned}
-\frac{\partial i_1}{\partial x} &= C_1 \frac{\partial v_1}{\partial t} + C_2 \frac{\partial}{\partial t}(v_1 - v_2) \\
-\frac{\partial i_2}{\partial x} &= C_1 \frac{\partial v_2}{\partial t} + C_2 \frac{\partial}{\partial t}(v_2 - v_1)
\end{aligned}
\tag{7.7}
$$

Rearranging Equations 7.6 and 7.7 and expressing them in matrix form gives

$$
-\left[\frac{\partial v}{\partial x}\right] = [L]\left[\frac{\partial i}{\partial t}\right] \tag{7.8}
$$

$$
-\left[\frac{\partial i}{\partial x}\right] = [C]\left[\frac{\partial v}{\partial t}\right] \tag{7.9}
$$

where

$$[L] = \begin{bmatrix} L & M \\ M & L \end{bmatrix}, \quad [C] = \begin{bmatrix} C_1 + C_2 & -C_2 \\ -C_2 & C_1 + C_2 \end{bmatrix}$$

Taking the derivatives of Equations 7.8 and 7.9 with respect to x and t, respectively, and combining gives

$$\left[\frac{\partial^2 v}{\partial x^2}\right] = [L][C]\left[\frac{\partial^2 v}{\partial t^2}\right] \tag{7.10}$$

where

$$[L][C] = \begin{bmatrix} a & b \\ b & a \end{bmatrix}$$

and

$$a = L_s(C_1 + C_2) - C_2 M, \quad b = M(C_1 + C_2) - C_2 L_s$$

A similar equation may be obtained for the current. Equation 7.10 indicates that voltages v_1 and v_2 are coupled through the off-diagonal elements b. It is convenient to seek another set of voltages, say v_{m1} and v_{m2}, referred to as modal voltages, which are decoupled. The modal voltages are related to v_1 and v_2 through a transformation matrix [S] yet to be determined, i.e.,

$$[v] = \begin{bmatrix} v_1 \\ v_2 \end{bmatrix} = [S]\begin{bmatrix} v_{m1} \\ v_{m2} \end{bmatrix} = [S][v_m] \tag{7.11}$$

Substituting Equation 7.11 into Equation 7.10 gives

$$[S]\frac{\partial^2}{\partial x^2}[v_m] = \begin{bmatrix} a & b \\ b & a \end{bmatrix}[S]\frac{\partial^2}{\partial t^2}[v_m]$$

Multiplying both sides of this equation by $[S]^{-1}$ from the left gives

$$\frac{\partial^2}{\partial x^2}\left[v_m\right]=\left[S\right]^{-1}\begin{bmatrix} a & b \\ b & a \end{bmatrix}\left[S\right]\frac{\partial^2}{\partial t^2}\left[v_m\right] \tag{7.12}$$

Decoupling is achieved if [S] is chosen in a way that makes the product of the three matrices above a diagonal matrix, i.e.,

$$\left[S\right]^{-1}\begin{bmatrix} a & b \\ b & a \end{bmatrix}\left[S\right]=\begin{bmatrix} \lambda_1^2 & 0 \\ 0 & \lambda_2^2 \end{bmatrix} \tag{7.13}$$

The elements of the diagonal matrix are referred to in mathematics as the eigenvalues. Multiplying both sides of Equation 7.13 by [S] from the left and rearranging gives

$$\begin{bmatrix} S_{11}\,\lambda_1^2 - a\,S_{11} - b\,S_{21} & S_{12}\,\lambda_2^2 - a\,S_{12} - b\,S_{22} \\ S_{21}\,\lambda_1^2 - b\,S_{11} - a\,S_{21} & S_{22}\,\lambda_2^2 - b\,S_{12} - a\,S_{22} \end{bmatrix} = 0$$

For this to be the zero matrix, all elements must be equal to zero. Equating the elements of the first column to zero gives

$$\left\{\begin{bmatrix} a & b \\ b & a \end{bmatrix} - \begin{bmatrix} \lambda_1^2 & 0 \\ 0 & \lambda_1^2 \end{bmatrix}\right\}\begin{bmatrix} S_{11} \\ S_{21} \end{bmatrix} = 0 \tag{7.14}$$

Similarly, from the second column

$$\left\{\begin{bmatrix} a & b \\ b & a \end{bmatrix} - \begin{bmatrix} \lambda_2^2 & 0 \\ 0 & \lambda_2^2 \end{bmatrix}\right\}\begin{bmatrix} S_{12} \\ S_{22} \end{bmatrix} = 0 \tag{7.15}$$

For nontrivial solutions, the determinants of the expressions in the angled brackets must be zero. This requirement leads to the following eigenvalues

$$\lambda_1^2 = a + b$$

$$\lambda_2^2 = a - b$$

The variation of the two modal voltages is obtained from Equation 7.12 and is thus

$$\frac{\partial^2 v_{m1}}{\partial x^2} = (a+b)\frac{\partial^2 v_{m1}}{\partial t^2}$$
$$\frac{\partial^2 v_{m2}}{\partial x^2} = (a-b)\frac{\partial^2 v_{m2}}{\partial t^2} \tag{7.16}$$

These equations describe two independent modes of propagation. Mode 1 propagates with a velocity $u_1 = 1/\sqrt{a+b}$, whereas mode 2 has a velocity $u_2 = 1/\sqrt{a-b}$.

The modal voltages are related to the conductor voltages, through matrix [S], Equation 7.11. The elements of matrix [S] may be found from Equations 7.14 and 7.15. Choosing $S_{11} = S_{12} = 1$ and substituting into these equations gives the remaining elements, i.e.,

$$[S] = \begin{bmatrix} 1 & 1 \\ 1 & -1 \end{bmatrix} \tag{7.17}$$

Thus, from Equation 7.11

$$v_{m1} = (v_1 + v_2)/2$$
$$v_{m2} = (v_1 - v_2)/2 \tag{7.18}$$

It is evident that the modal voltages in Equation 7.18 are none other than the common and differential mode signals. Equation 7.16 suggests that the modal components are waves traveling with velocities u_1 and u_2 and that the forward-traveling components are

$$v_1 = v_{m1}(x - u_1 t) + v_{m2}(x - u_2 t)$$
$$v_2 = v_{m1}(x - u_1 t) - v_{m2}(x - u_2 t) \tag{7.19}$$

These expressions may be substituted into Equation 7.9, which after integration gives

$$i_1 = u_1 C_1 v_{m1} + u_2 (C_1 + 2C_2) v_{m2}$$
$$i_2 = u_1 C_1 v_{m1} - u_2 (C_1 + 2C_2) v_{m2} \tag{7.20}$$

The modal current components are thus

$$\begin{bmatrix} i_{m1} \\ i_{m2} \end{bmatrix} = \begin{bmatrix} u_1 C_1 & 0 \\ 0 & u_2\left(C_1 + 2C_2\right) \end{bmatrix} \begin{bmatrix} v_{m1} \\ v_{m2} \end{bmatrix} \tag{7.21}$$

The 2 × 2 matrix on the right-hand side is the modal impedance matrix $[Z_m]$. The modal propagation velocities u_1 and u_2 corresponding to mode 1 (common mode) and mode 2 (differential mode) are in general not equal to each other. Only when b = 0 are the two velocities the same. It is interesting to examine further under what circumstances b = 0. Substituting for b and equating to zero gives

$$\frac{M}{L} = \frac{C_2}{C_1 + C_2} \tag{7.22}$$

In physical terms this expression is equivalent to demanding that if voltage v_1 and current i_1 are established on conductor 1, the voltage inductively coupled to 2 is equal to the capacitively coupled voltage. Under these circumstances, the two modes propagate at the same velocity. In a lossless system, with a uniform dielectric, Equation 7.22 holds. However, losses and/or a nonuniform dielectric, such as the presence of insulator sleeves on the wires, make the modal velocities different. A signal induced or generated at a point on a line will couple to adjacent lines in amounts determined by the parameters of the line. Energy then propagates along the lines toward their load terminations. If the modal velocities are the same, the modal signals arrive at the loads simultaneously and combine to produce the total signal on each of the lines. If, however, the modal velocities are different, modal components arrive at the loads at different times and thus combine to impress total signals at the loads, which are different in magnitude and shape from those at the point of origin of the signal. Thus, different modal velocities result in dispersion and different severity of crosstalk depending on the line length.

In a multiconductor system with n conductors plus ground, there are, in general, n modes of propagation.

The basic model described above for the two-wire plus ground system may be generalized to deal with any number of wires. The treatment of the two-wire line above ground presented so far is clearly in the time domain. Voltages and currents on the lines are time-dependent functions. The general multiconductor equations in the time domain are

$$\frac{\partial^2}{\partial x^2}[v]=[L][C]\frac{\partial^2}{\partial t^2}[v] \tag{7.23}$$

$$\frac{\partial^2}{\partial x^2}[i]=[C][L]\frac{\partial^2}{\partial t^2}[i] \tag{7.24}$$

where [v] and [i] are column vectors with elements the voltage to ground and the current for each conductor. The square matrices [L] and [C] represent inductive and capacitive coupling between the lines and are discussed further in Section 7.2. Losses may be included in this formulation in a straightforward manner.[3]

Equations 7.23 and 7.24 describe propagation and mutual coupling between lines (crosstalk) in a multiconductor system. If the lines are subject to an external incident electromagnetic field, additional source terms must be incorporated into the model. Thus, Equations 7.8 and 7.9, generalized to include coupling to the external field, are

$$\frac{\partial}{\partial x}[v]+[L]\frac{\partial}{\partial t}[i]=[v_s] \tag{7.25}$$

$$\frac{\partial}{\partial x}[i]+[C]\frac{\partial}{\partial t}[v]=[i_s] \tag{7.26}$$

where $[v_s]$ and $[i_s]$ are suitable source term matrices defined further in Section 7.3.

The equations given above may be easily expressed in the frequency domain. In this formulation $\bar{V}(x)$ and $\bar{I}(x)$ represent the voltage and current phasors of a harmonic signal of frequency ω. Thus, the equivalents of Equations 7.25 and 7.26 in the frequency-domain are[4,5]

$$\frac{d}{dx}[\bar{V}]=-[Z][\bar{I}]+[\bar{V}_s] \tag{7.27}$$

$$\frac{d}{dx}[\bar{I}]=-[Y][\bar{I}]+[\bar{I}_s] \tag{7.28}$$

where for lossless lines [Z] = jω [L] and [Y] = jω [C]. Losses may be incorporated by adding resistance and conductance matrices to [Z] and [Y], respectively. Column matrices $[\bar{V}_s]$ and $[\bar{I}_s]$ are source terms representing coupling to external fields.

The general procedures for solving these equations, either in the time or frequency domains, is to decompose into modal components and solve the resulting simpler, uncoupled equations. In a lossless line with uniform dielectric, such decomposition is possible in both time and frequency domains. For asymmetrical lines with losses there is no unique decomposition matrix valid at all frequencies. In such cases, strict modal decomposition is not possible in the time domain. Approximate decomposition may then be achieved by obtaining the [S] matrix in the frequency domain at a particular frequency, corresponding to the peak of the wave spectrum, and using the real part of these matrices for time-domain decomposition. This approach normally gives small errors but must be used with caution. Further details of the formulation and solution of multiconductor equations are given in Section 7.4.

The previous derivation illustrates the complexity of coupling in interconnect configurations. Such a practical configuration is shown in Figure 7.3. The line generating interference (indicated by the letter G for generator of interference) is fed at one end by a voltage source in series with a resistance (the "near end" — NE) and terminated at the other end by a resistive load (the "far end" — FE). The victim line (indicated by the letter R for the receiver of interference) is terminated at both ends by resistive loads. This is a classic crosstalk problem whereby signals propagating on line G couple through to line R. We are interested on calculating the voltages V_{NE} (near-end coupling) and V_{FE} (far-end coupling). The problem can be solved using the principles already outlined subject to the appropriate boundary conditions. Further details may be found in the work of Paul.[6]

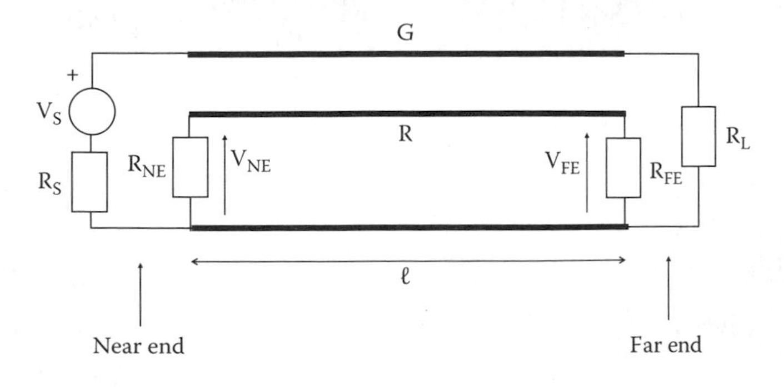

Figure 7.3 Schematic showing the wire generating interference (G) and the victim wire (R).

We summarize here the main results. The starting point is the [L] and [C] matrices of the configuration that are obtained from the geometrical details and material properties of the lines. This is discussed further in Section 7.3 and some formulae are provided in Appendix B.

$$
\begin{aligned}
&[L] = \begin{bmatrix} L_G & M \\ M & L_R \end{bmatrix} \\
&[C] = \begin{bmatrix} c_G + c_M & -c_M \\ -c_M & c_R + c_M \end{bmatrix} = \begin{bmatrix} C_G & -C_M \\ -C_M & C_R \end{bmatrix}
\end{aligned}
\tag{7.29}
$$

The near- and far-field coupling is then obtained after some considerable algebra from

$$
V_{NE} = \frac{S}{Den} \left\{ \begin{aligned} &\frac{R_{NE}}{R_{NE}+R_{FE}} j\omega M\ell \left[C + \frac{j2\pi\ell/\lambda}{\sqrt{1-k^2}} \alpha_{LG} S \right] I_{G_{DC}} \\ &+ \frac{R_{NE}R_{FE}}{R_{NE}+R_{FE}} j\omega C_M \ell \left[C + \frac{j2\pi\ell/\lambda}{\sqrt{1-k^2}} \frac{1}{\alpha_{LG}} S \right] V_{G_{DC}} \end{aligned} \right\}
\tag{7.30}
$$

$$
V_{FE} = \frac{S}{Den} \left\{ -\frac{R_{FE}}{R_{NE}+R_{FE}} j\omega M\ell I_{G_{DC}} + \frac{R_{NE}R_{FE}}{R_{NE}+R_{FE}} j\omega C_M \ell V_{G_{DC}} \right\}
$$

where the following auxiliary parameters have been defined

$$
Den = C^2 - S^2\omega^2\tau_G\tau_R \left[1 - k^2 \frac{(1-\alpha_{SG}\alpha_{LR})(1-\alpha_{LG}\alpha_{SR})}{(1+\alpha_{SR}\alpha_{LR})(1+\alpha_{SG}\alpha_{LG})} \right] + j\omega CS(\tau_G + \tau_R)
$$

$$
C = \cos(\beta\ell)
$$

$$
S = \frac{\sin(\beta\ell)}{\beta\ell}
$$

$$
k = \frac{M}{\sqrt{L_G L_R}} = \frac{C_M}{\sqrt{C_G C_R}} \le 1
$$

$$
\tau_G = \frac{L_G\ell}{R_S + R_L} + C_G\ell \frac{R_S R_L}{R_S + R_L}
$$

$$\tau_R = \frac{L_R \ell}{R_{NE} + R_{FE}} + C_R \ell \frac{R_{NE} R_{FE}}{R_{NE} + R_{FE}}$$

$$V_{G_{DC}} = \frac{R_L}{R_S + R_L} V_S$$

$$I_{G_{DC}} = \frac{V_S}{R_S + R_L}$$

$$Z_{CG} = \sqrt{\frac{L_G}{C_G}} \quad , \quad Z_{CR} = \sqrt{\frac{L_R}{C_R}}$$

$$\alpha_{SG} = \frac{R_S}{Z_{CG}}, \alpha_{SG} = \frac{R_L}{Z_{CG}}, \alpha_{SG} = \frac{R_{NE}}{Z_{CR}}, \alpha_{SG} = \frac{R_{FE}}{Z_{CR}}$$

In these equations k represents the strength of coupling between the lines and the other symbols have their usual meaning.

These are exact equations within the limitations of TL theory, i.e., it is assumed that we have TEM conditions. Under TEM conditions line parameters can be calculated as shown in the next section. TEM conditions means that the electric and magnetic field components are perpendicular to each other and transverse to the direction of propagation. Consequences of this are that waves propagate along the axis of the TL; the E and H field are solutions of the two-dimensional electrostatic and magnetostatic equations respectively; the line integral of the electric field from one conductor to another (the potential difference V) has a unique value and thus a unique value of capacitance may be defined ($C = Q/V$); the line integral of the magnetic field around each conductor has a unique value (conductor current I) and thus a unique value of inductance may be defined ($L = \Phi/I$); and the velocity of propagation is independent of the cross-sectional dimensions of the line and depends only on the properties of the surrounding medium. These conditions hold for many practical configurations but there are cases where TEM conditions do not apply and hence the transmission line equations are not strictly accurate. Examples of violations of the TEM conditions are that as the skin effect becomes prominent at high frequencies, R increases and L decreases with frequency resulting in longitudinal field components. Normally R variations are accounted for and variations in L are neglected (the latter violates causality and results in signal precursors). Another violation is due to the presence of inhomogeneous materials that results in small longitudinal field components. It is naturally implied in the use of these equations that there is no significant radiation loss from the lines. Provided TEM holds approximately (quasi-TEM approximation), reasonable estimates can be made using the equation with the line parameters calculated as shown in Section 7.4.

We can simplify Equation 7.30 under certain conditions. For electrically short lines ($\ell << \lambda$) and for weak coupling ($k << 1$), we obtain

$$
\begin{aligned}
V_{NE} &= \frac{1}{Den}\left[\frac{R_{NE}}{R_{NE}+R_{FE}} j\omega M\ell I_{G_{DC}} + \frac{R_{NE}R_{FE}}{R_{NE}+R_{FE}} j\omega C_M \ell V_{G_{DC}}\right] \\
V_{NE} &= \frac{1}{Den}\left[-\frac{R_{FE}}{R_{NE}+R_{FE}} j\omega M\ell I_{G_{DC}} + \frac{R_{NE}R_{FE}}{R_{NE}+R_{FE}} j\omega C_M \ell V_{G_{DC}}\right] \qquad (7.31) \\
Den &\simeq \left(1 + j\omega\tau_G\right)\left(1 + j\omega\tau_R\right)
\end{aligned}
$$

We see already in these equations two terms — one inductive and the other capacitive — in the near- and far-end coupling. If in addition we approximate further to assume low frequencies ($\omega\tau << 1$), then we find that $Den \approx 1$ and the coupled equations simplify to

$$
\begin{aligned}
V_{NE} &= \frac{R_{NE}}{R_{NE}+R_{FE}} j\omega M\ell I_{G_{DC}} + \frac{R_{NE}R_{FE}}{R_{NE}+R_{FE}} j\omega C_M \ell V_{G_{DC}} \\
V_{FE} &= -\frac{R_{FE}}{R_{NE}+R_{FE}} j\omega M\ell I_{G_{DC}} + \frac{R_{NE}R_{FE}}{R_{NE}+R_{FE}} j\omega C_M \ell V_{G_{DC}}
\end{aligned}
\qquad (7.32)
$$

These equations are simple enough to make it possible to comment in simple physical terms about the nature of coupling. The following observations are relevant:

- The first term on the RHS represents inductive coupling
- The second term represents capacitive coupling
- Coupling is proportional to frequency (faster signal transitions result in more crosstalk)
- Capacitive coupling at the near and far ends is of the same sign
- Inductive coupling is of opposite sign at the near and far ends; in principle, it is therefore possible to select terminations to reduce or cancel out coupling at the far end

The simplified equations in Equation 7.32 may be represented by the simple coupling circuit shown in Figure 7.4 where the impact of the proximity of the generator line is accounted for by the voltage and current sources shown.

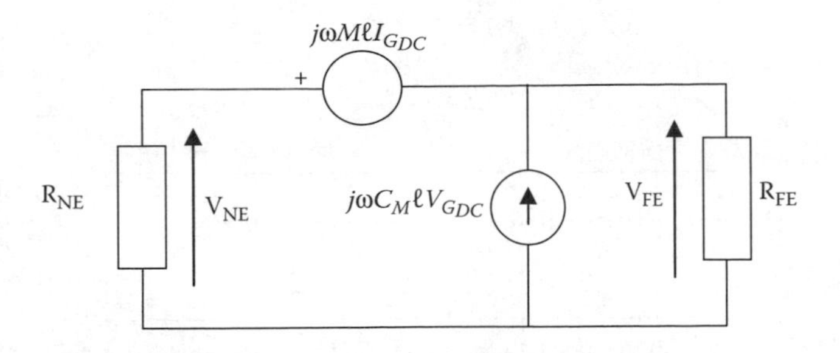

Figure 7.4 Schematic of the coupling model at low frequencies.

It is repeated here that the expressions in Equation 7.30 are applicable in the TEM approximation and the following expressions are subject to further low-frequency conditions.

Also, as already pointed out in connection with Equation 7.22, propagation takes place at different modes that have a different velocity if Equation 7.22 is not satisfied. It is easy to see that this will be the case in lines that are surrounded by inhomogeneous dielectric. In Figure 7.5a two conductors are placed inside dielectric ε_1, which is then surrounded by dielectric ε_2. The equivalent circuit is shown in Figure 7.5b. If $\varepsilon_1 = \varepsilon_2$, Equation 7.22 is satisfied and propagation proceeds at a single velocity (degenerate case). If, however, $\varepsilon_1 > \varepsilon_2$, then C_2 is larger than for the same dielectric case and C_1 smaller. Therefore the RHS of Equation 7.22 is larger than before, while the LHS has the same value since inductance does not depend on the values of dielectric permittivity. Equation 7.22 is therefore not satisfied and each mode propagates at different velocity (nondegenerate case). This can have significant impact on crosstalk and signal

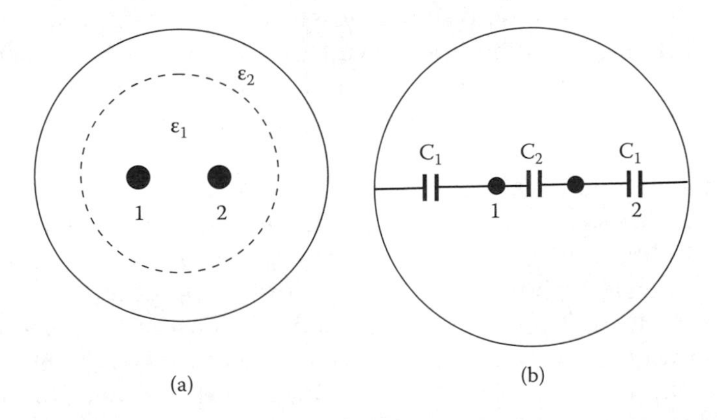

Figure 7.5 Schematic showing two wires inside two dielectrics (a) and the corresponding circuit equivalent (b).

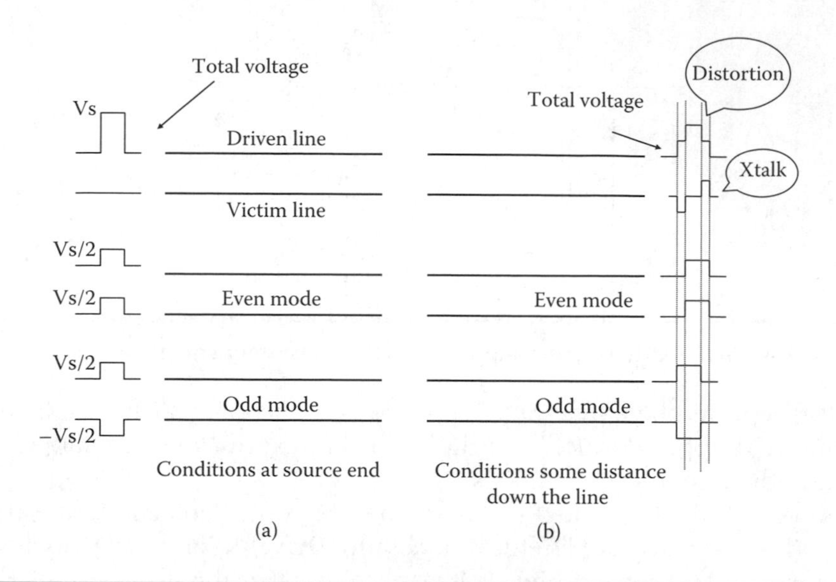

Figure 7.6 Explanation of propagation with even (common) and odd (differential) modes (a) and conditions further down the line (b).

distortion (EMC and Signal Integrity [SI] implications). This is demonstrated in Figure 7.6a where the two lines in a three-line configuration are shown schematically. At the source end the driven (generator) line has voltage V_s impressed on it and the victim (receiver) line has zero volts. The excitation conditions may be decomposed into even (common) and odd (differential) mode components using Equation 7.18 as shown schematically. Assuming for illustration that the even mode propagates faster, then the modal conditions some distance down the lines will be as shown in Figure 7.6b. Using Equation 7.18 solved for the total voltages and using the modal voltage component indicated in the figure gives the total voltages on each line shown at the top of the illustration in Figure 7.6b. We observe that the total voltage on the driven line is distorted (SI issue) and that due to crosstalk there are very short-duration pulses on the victim line (EMC issue). When one considers the large number of interconnects with mixed dielectrics in real systems, the potential of these very sharp spikes to generate high-frequency noise can be appreciated.

At very high frequencies where the line dimensions become comparable to the wavelength a full-field solution is required based on Maxwell's equations. Normally, numerical solutions are sought to take full account of line behavior.

A particular example, where adjustments must be made to the transmission line equations to extend their validity, is in cases where the line has a bend. It is known that a sharp bend on a PCB track violates TEM conditions as at the bend point significant radiation takes place. Naturally, this phenomenon may be studied fully using full-field numerical solutions of Maxwell's equations. Nevertheless, TL models may still be used to get approximate solutions provided special sections are inserted into the model to represent the line near the bend. Traditionally, reactive components are inserted to account for the discontinuity caused by the bend. No account can thus be taken of radiation from the bend. More sophisticated techniques were developed where these sections (spanning a distance of approximately one wavelength around the bend) contain frequency and position-dependent reactive and dissipative elements accounting for radiation loss and signal distortion near the bend. This approach is based on the theoretical framework described in Reference 7.

Details of the development of this scheme and its implementation for EMC and SI applications may be found in References 8 and 9.

7.3 Line Parameter Calculation

Formulae for calculating the inductance and capacitance of typical conductor configurations are given in Appendix B. The complexity of practical multiconductor configurations is such that it is not possible to give a comprehensive coverage of this topic. In most cases, analytical solutions are not available and the designer has to resort to numerical methods in order to obtain numerical values. Thus, at first sight, it might seem unnecessary even to attempt to obtain analytical solutions. There are, however, two strong reasons to spend some time looking at analytical solutions. First, a systematic examination of the problem, which is inherent in the process of obtaining analytical solutions, makes assumptions, simplifications, and major features of the problem explicit and thus helps the designer to ask the right questions when confronted with a more intractable problem. Second, although analytical solutions are only possible in a few idealized configurations, they may still be used with minor adaptations as a first approximation to more complex configurations. The basic electrical parameters, inductance and capacitance, are normally logarithmic functions of the geometrical dimensions and shape factors, and thus vary slowly with changes in these parameters. In the next two subsections, analytical and numerical methods for obtaining line parameters are presented.

7.3.1 Analytical Methods

The background to electromagnetic calculations was described in Chapters 2 and 3. The ideas developed there are applied to a few typical configurations shown schematically in Figure 7.7.

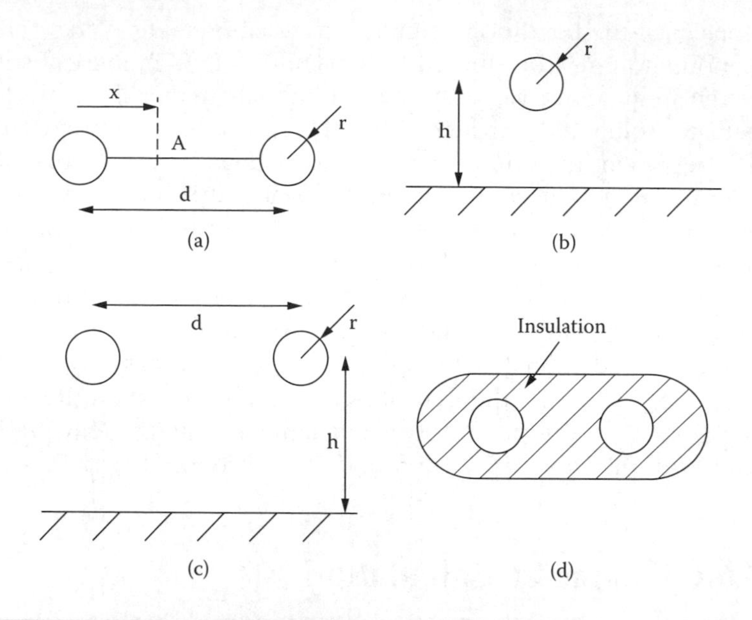

Figure 7.7 Various line configurations: (a) two-wire, (b) wire above ground, (c) two wires above ground, and (d) plastic insulated two-wire line.

The parameters of the parallel conductor line shown in Figure 7.7a may be easily obtained if the assumption is made that the charge or current distribution on each conductor is uniform, unaffected by the proximity of the other conductor. This is clearly approximately true when d >> r. In such cases the electric field at point A a distance x away from one of the conductors carrying a charge q coulombs per meter length is

$$E(x) = \frac{q}{2\pi\varepsilon x} + \frac{q}{2\pi\varepsilon(d-x)}$$

where it has been assumed that a charge –q resides on the other conductor. The potential difference between the two conductors is then

$$V = \frac{q}{2\pi\varepsilon}\int_{x=r}^{d-r}\left(\frac{1}{x} + \frac{1}{d-x}\right)dx = \frac{q}{\pi\varepsilon}\ln\left(\frac{d-r}{r}\right)$$

Hence, the capacitance per unit length between the two conductors is

$$C \equiv \frac{q}{V} = \frac{\pi\varepsilon}{\ln\left(\frac{d-r}{r}\right)}, \text{ in F/m} \tag{7.33}$$

If proximity effects cannot be ignored, an analytical solution may still be obtained, using the method of images, and is

$$C = \frac{\pi\varepsilon}{\cosh^{-1}(d/2r)} \tag{7.34}$$

Similarly, the inductance may be obtained by first calculating the flux density at point A

$$B(x) = \frac{\mu I}{2\pi}\left(\frac{1}{x} + \frac{1}{d-x}\right)$$

Hence, the flux linked per meter length is

$$\phi = \frac{\mu I}{2\pi}\int_{x=r}^{d-r}\left(\frac{1}{x} + \frac{1}{d-x}\right)dx = \frac{\mu I}{\pi}\ln\left(\frac{d-r}{r}\right)$$

The inductance per unit length is then

$$L \equiv \frac{\phi}{I} = \frac{\mu}{\pi}\ln\left(\frac{d-r}{r}\right), \text{ in H/m} \tag{7.35}$$

An alternative approach that is often used is to recognize that the velocity of propagation on the line shown in Figure 7.7a is

$$\frac{1}{\sqrt{\mu\varepsilon}} = \frac{1}{\sqrt{LC}} \tag{7.36}$$

where C and L are given by Equations 7.29 and 7.31. Thus, the capacitance only need be calculated from the conductor configuration and the inductance can then be obtained directly from Equation 7.36.

For the conductor above perfect ground configuration shown in Figure 7.7b, the method of images may be employed whereby an equivalent, image conductor is assumed. This results in two parallel conductors a distance 2h apart. The capacitance between conductor and earth is then obtained by calculating the potential difference V in the equivalent problem between one of the conductors and the midplane, and by dividing q by V, i.e.,

$$C = \frac{2\pi\varepsilon}{\ln\left(\frac{2h - r}{r}\right)}, \text{ in F/m} \tag{7.37}$$

A similar approach may be used to calculate inductance. In cases where a perfectly conducting plane cannot be assumed, such as when a conductor runs above rocky soil, simple image theory cannot be used and a more sophisticated approach is necessary. In a formulation due to Carson,[11] the depth d_g of the image conductor used for calculating inductance is not taken equal to h but is given by the formula $d_g = 660\sqrt{\rho/f}$ meters, where ρ and f are the ground resistivity and signal frequency, respectively. A more complete description of propagation in such a configuration may be found in Referenced 12.

Conductor configurations such as the one shown in Figure 7.7c are easily dealt with by applying the method of images. This results in a conductor configuration in free space consisting of the two original conductors plus the two images. The charge per unit length q_1 and q_2 on each conductor and the respective voltages v_1 and v_2 with respect to ground are related by the expressions

$$q_1 = c_{11} V_1 + c_{12} V_2$$

$$q_2 = c_{21} V_1 + c_{22} V_2$$

where the capacitance coefficients are

$$c_{11} = -\frac{p_{22}}{\Delta}, c_{12} = \frac{p_{12}}{\Delta}, c_{21} = \frac{p_{21}}{\Delta}, c_{22} = -\frac{p_{11}}{\Delta}$$

$$p_{11} = p_{22} = \frac{1}{2\pi\varepsilon_0}\ln\frac{2h}{r}, p_{12} = p_{21} = \frac{1}{2\pi\varepsilon_0}\ln\frac{D}{d} \tag{7.38}$$

$$\Delta = p_{12}p_{21} - p_{11}p_{22}$$

D is the distance between one charge and the image of the other charge. Taking the case of d = 8r, h = 3r, which corresponds, approximately, to two tracks above a ground plane on a printed circuit board, and substituting in Equation 7.38 gives

$$c_{11} = c_{22} = 63 \text{ pF/m and } c_{12} = c_{21} = -7.8 \text{ pF/m}$$

The capacitance between each conductor and ground is

$$C_1 = C_2 = c_{11} + c_{12} = 55 \text{ pF/m}$$

and the mutual capacitance between the two conductors is

$$C_{12} = -c_{12} = 7.8 \text{ pF/m}$$

Configurations where the conductor cross section is not circular may be treated approximately in a similar way by using equivalent radii and the geometric mean distance between conductors.

Difficulties sometimes arise when considering the inductance of a wire segment part of a loop. An efficient approach is to obtain the loop inductance through the vector potential A

$$L = \frac{\int_c \mathbf{A}\, d\mathbf{l}}{I} \tag{7.39}$$

where I is the current in the loop and the integration path c coincides with the loop. Thus, the inductance of a loop segment i may be interpreted as the contribution from the integral in Equation 7.35 along that segment, and it consists of a self-inductance term L_{ii} (due to magnetic flux produced by current in the segment) and mutual-inductance terms L_{ij} due to fluxes produced by current in other segments j and linked to i. The total inductance of the segment is the sum of all partial inductances L_{ii} and L_{ij}. In geometrical terms, the partial self-inductance L_{ii} may be interpreted as the ratio of the magnetic flux produced by I_i crossing an area extending between the segment and infinity, over the current I_i. Similarly, L_{ij} is the ratio of the magnetic flux produced by I_j crossing an area extending from segment i and infinity, over the current I_j. These ideas are illustrated in Figure 7.8 where we have used symbols L_p and M_p for the self and mutual partial inductances of a segment, respectively. We illustrate these concepts by examples for a two-wire and three-wire line. Two segments of a line are shown in Figure 7.9a where L_p and M_p are self and mutual partial inductances as indicated earlier. $V_1$and V_2 are the voltage drops along each segment and are given by

$$\begin{aligned} V_1 &= L_{p1}\frac{dI_1}{dt} + M_p\frac{dI_2}{dt} \\ V_2 &= M_p\frac{dI_1}{dt} + L_{p2}\frac{dI_2}{dt} \end{aligned} \tag{7.40}$$

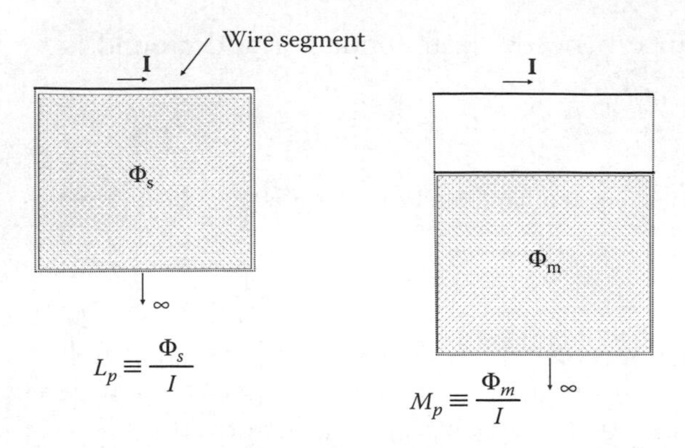

Figure 7.8 Schematic showing quantities required to calculate self and mutual partial inductances.

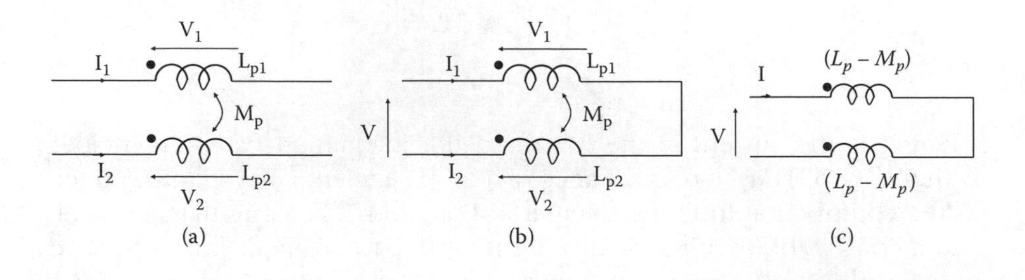

Figure 7.9 Schematic of two coupled lines (a) and differential mode current flow (b) and (c).

Assuming that the two segments are carrying differential mode currents (I_1 = I, I_2 = –I) as shown in Figure 7.9b, substituting in Equation 7.40 we obtain

$$V_1 = \left(L_{p1} - M_p\right)\frac{dI}{dt}, \quad V_2 = -\left(L_{p2} - M_p\right)\frac{dI}{dt}$$

Assuming that the two segments have the same radius and hence the same partial self inductance L_p, we calculate the voltage V

$$V = V_1 - V_2 = \left(L_{p1} + L_{p2} - 2M_p\right)\frac{dI}{dt}$$

$$= 2\left(L_p - M_p\right)\frac{dI}{dt}$$

The equivalent circuit to differential mode current is therefore as shown in Figure 7.9c.

We repeat this calculation for the three-wire line shown in Figure 7.10a. The voltage drops along each segment is given by

$$\begin{aligned}
V_1 &= L_{p11}\frac{dI_1}{dt} + L_{p12}\frac{dI_2}{dt} + L_{p13}\frac{dI_3}{dt} \\
V_2 &= L_{p21}\frac{dI_1}{dt} + L_{p22}\frac{dI_2}{dt} + L_{p23}\frac{dI_3}{dt} \\
V_3 &= L_{p31}\frac{dI_1}{dt} + L_{p32}\frac{dI_2}{dt} + L_{p33}\frac{dI_3}{dt}
\end{aligned} \tag{7.41}$$

Assuming for simplicity that the lines are symmetrical we obtain

$$\begin{aligned}
V_1 &= L_p\frac{dI_1}{dt} + M\frac{dI_2}{dt} + M\frac{dI_3}{dt} \\
V_2 &= M\frac{dI_1}{dt} + L_p\frac{dI_2}{dt} + M\frac{dI_3}{dt} \\
V_3 &= M\frac{dI_1}{dt} + M\frac{dI_2}{dt} + L_p\frac{dI_3}{dt}
\end{aligned} \tag{7.42}$$

We will now consider the inductance presented by this arrangement to common-mode currents, i.e., when the three segments are as shown in

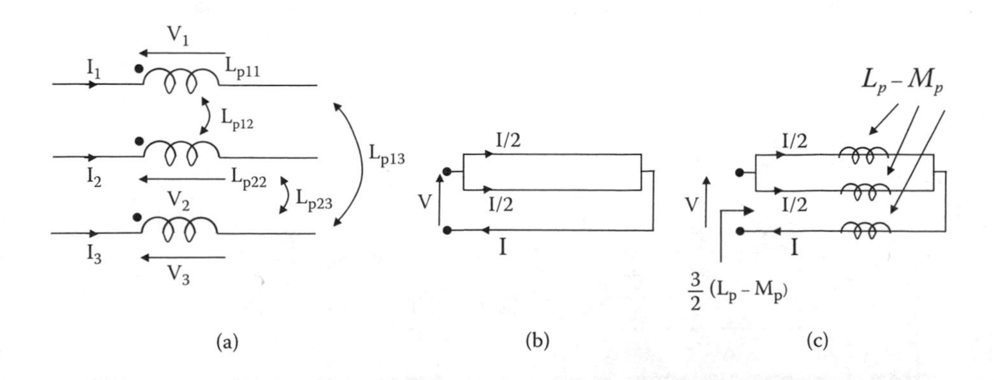

Figure 7.10 Schematic of a three-wire system (a), common-mode current flow (b), and equivalent CM inductance (c).

Figure 7.10b. Here we have $I_1 = I_2 = I/2$ and $I_3 = -I$. Substituting in Equation 7.42

$$V_1 = V_2 = \left(L_p + M_p\right)\frac{d}{dt}(I/2) - M_p\frac{dI}{dt}$$

$$V_3 = 2M_p\frac{d}{dt}(I/2) - L_p\frac{dI}{dt}$$

Hence the voltage V is

$$V = V_1 - V_3 = \ldots = \frac{3}{2}\left(L_p - M_p\right)\frac{dI}{dt}$$

The equivalent circuit for this arrangement is shown in Figure 7.10c. We see, therefore, that depending on the nature of the currents (common or differential form) the inductance presented will be different. We give for reference in Figure 7.11 the partial self inductance of a segment and the mutual partial inductance of two segments. A more thorough discussion of these matters may be found in Reference 13. Formulae for the calculation of inductance may be found in Appendix B and in References 14 and 15.

It should be emphasized that inductance and capacitance are quasistatic concepts and that the values calculated above apply for quasi-TEM wave propagation. At high frequencies, when the line spacing becomes comparable to the wavelength, and a modal field structure is established between the lines, complete field solutions must be sought.

Another factor that sometimes must be taken into account is the contribution due to end effects. A parallel wire line such as the one shown in Figure 7.7a has a capacitance per unit length adequately described by

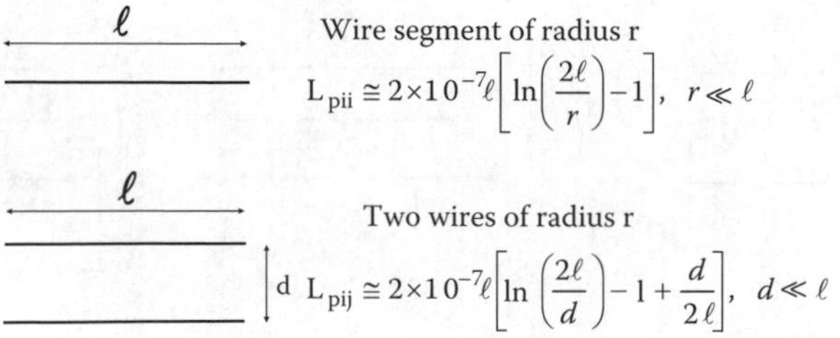

ℓ

Wire segment of radius r

$$L_{pii} \cong 2\times10^{-7}\ell\left[\ln\left(\frac{2\ell}{r}\right)-1\right],\quad r \ll \ell$$

ℓ

Two wires of radius r

d

$$L_{pij} \cong 2\times10^{-7}\ell\left[\ln\left(\frac{2\ell}{d}\right)-1+\frac{d}{2\ell}\right],\quad d \ll \ell$$

Figure 7.11 Formulae for the calculation of self and mutual partial inductances.

Equations 7.33 or 7.34 away from its ends. At an open-circuit end there will be a fringing electric field that adds extra capacitance C_f to the line. This may be added at the open-circuit termination to describe better the behavior of the line. Alternatively, the fringing capacitance may be viewed as increasing the electrical length of the line by an amount $\Delta l = C_f/C$ where C is the line capacitance per meter length away from the ends. An empirical formula used to quantify these effects is given below.[16]

$$\frac{\Delta l}{d} = \frac{1}{-3.954 + \left\{\left[2.564 \cosh^{-1}\left(d/2r\right)\right]^2 + 3.954^2\right\}^{1/2}} \tag{7.43}$$

7.3.2 Numerical Methods

The analytical approach may be used in many practical configurations but, inevitably, there are many problems that are not amenable to this type of treatment. Irregularly shaped conductors or configurations, such as the one shown in Figure 7.7d, where the two conductors are partially covered by plastic insulating material (ribbon cable) must be analyzed using numerical techniques. Rough approximations can sometimes be made by assuming that the conductors are immersed in a uniform dielectric of an "equivalent" permittivity.

Numerical solutions involve considerable complexity and are often based on the method of moments (MM). There are several ways of applying the MM to such problems.[17–19] A popular approach is to expand the unknown surface charge density in terms of known harmonic bases functions. Thus, for a dielectric-coated conductor i there will be two expansions. First, the surface charge density on the conductor (free plus bound charge) at any angular position θ_i is expressed in the form

$$\sigma_i\left(\theta_i\right) = a_{io} + \Sigma\, a_{im} \cos\left(m\,\theta_i\right) + \Sigma\, b_{im} \sin\left(m\,\theta_i\right)$$

Coefficients a_i and b_i are unknowns to be determined.

Second, the surface-bound charge density at the outer surface of the dielectric coating is expressed in a similar way with a different set of unknown coefficients a_i'. This process is repeated for each conductor in the system. The equations necessary for calculating the unknown coefficients are obtained by enforcing the boundary conditions on a number of points sufficient to yield the correct number of equations. Solving these equations gives the expansion coefficients and thus all charge densities.

The capacitance coefficients are then obtained from the matrix equation [q] = [c] [V]. The boundary conditions are that the surface of each conductor is an equipotential and that the normal component of the electric flux density is continuous across the boundary between two ideal dielectrics.

This technique, and its adaptations, can be used in a variety of problems. A PC-based computer package for obtaining the matrix parameters of multiconductor lines is described in Reference 20.

It should be noted that once the capacitance coefficients [c] have been obtained, the inductance coefficients [L] may be calculated from $[c_o]$ obtained by replacing the coating by the surrounding medium (normally air). Since this change does not affect the magnetic properties, it follows that

$$[L] = [1/u_0^2][c_o]^{-1}$$

where u_o is the propagation velocity in the surrounding medium.

7.4 Representation of EM Coupling from External Fields

A fundamental problem in EMC analysis is the determination of the coupling of energy transported by an electromagnetic field onto a transmission line. The source of the EM wave may be other adjacent circuits. In this section the basic approaches to tackling such problems will be described. These allow the determination of the amplitude and spectral content of induced signals for a range of electromagnetic threats and thus assist in assessing the susceptibility of equipment to such threats.

It is assumed, for simplicity throughout, that quasi-TEM conditions apply and that the parallel wire line carries equal and opposite currents (differential mode). The starting point in establishing a coupling model is to express Faraday's Law on the loop shown by a broken line in Figure 7.12.[17,22] Moving in the anticlockwise direction gives

$$\int_c \mathbf{E}\, d\mathbf{l} = -\frac{\partial}{\partial t}\int_s \mathbf{B}\, d\mathbf{s} \tag{7.44}$$

where the term on the right-hand side is calculated on a surface bounded by the loop. In the expressions that follow, the time dependence of fields, voltages, and currents is not shown in the interests of clarity. Evaluating each term gives

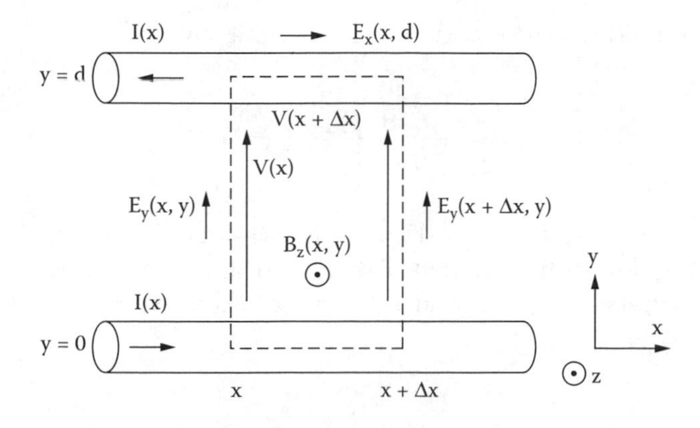

Figure 7.12 Configuration used to study field-to-wire coupling.

$$\int \mathbf{E}\,d\mathbf{l} = \int_x^{x+\Delta x} E_x(x,0)\,dx + \int_o^d E_y(x+\Delta x, y)\,dy$$

$$- \int_x^{x+\Delta x} E_x(x,d)\,dx - \int_o^d E_y(x,y)\,dy$$

$$= \int_o^d E_y(x+\Delta x,\ y)\,dy - \int_o^d E_y(x,y)\,dy = V(x) - V(x+\Delta x)$$

where it has been assumed that the total value of the electric field along each wire is zero (perfect conductors). Similarly,

$$-\mu\frac{\partial}{\partial t}\int \mathbf{H}\,d\mathbf{s} = -\mu\frac{\partial}{\partial t}\int_x^{x+\Delta x}\int_{y=0}^{d} H_z(x,y)\,dy dx$$

$$\simeq -\mu\frac{\partial}{\partial t}\int_o^d H_z(x,y)\,dy\ \Delta x$$

Substituting in Equation 7.44 gives

$$V(x) - V(x+\Delta x) = -\mu\frac{\partial}{\partial t}\int_o^d H_z(x,y)\,dy\ \Delta x$$

Dividing both sides by Δx and taking the limit gives

$$\frac{\partial V(x)}{\partial x} = \mu \frac{\partial}{\partial t} \int_o^d H_z(x, y)\, dy \tag{7.45}$$

The magnetic field component H_z may be separated into two parts. The first part H_z^i is due to the incident EM field and the second part H_z^s is due to the field generated by the induced current (scattered component). Thus, the total field is

$$H_z = H_z^i + H_z^s \tag{7.46}$$

Substituting Equation 7.46 into Equation 7.38 and recognizing that the flux linkage due to the scattered component is equal, by definition, to $-L_d I(x)$ where L_d is the line inductance per unit length, gives

$$\frac{\partial V(x)}{\partial x} + L_d \frac{\partial I(x)}{\partial t} = \mu \frac{\partial}{\partial t} \int_o^d H_z^i(x, y)\, dy \tag{7.47}$$

The term on the right-hand side may be regarded as a source term $V_s(x)$ representing the coupling of the field into the line

$$\frac{\partial V(x)}{\partial x} + L_d \frac{\partial I(x)}{\partial t} = V_s(x) \tag{7.48}$$

Another expression may be obtained from the continuity equation applied on a surface S enclosing a segment from x to x + Δx of one conductor

$$I(x + \Delta x) - I(x) = -\frac{\partial}{\partial t} \int_v \rho\, dv \tag{7.49}$$

where the term on the right-hand side is the rate of change of charge in the volume inside S. From Gauss's Law

$$\int_v \rho\, dv = \int_s \mathbf{D}\, d\mathbf{s} = \int_s \mathbf{D}^s\, d\mathbf{s} = q$$

since only the "scattered" electric flux density is associated with the free charge q per meter length of the wire. By definition, this charge is equal to the capacitance of the wires (length Δx) times the potential difference associated with the scattered field, i.e.,

$$\begin{aligned} q = C_d\, \Delta x\, V^s_{(x)} &= -C_d \int_o^d E^s_y(x,y)\,dy\,\Delta x \\ &= -C_d \int_o^d E_y(x,y)\,dy\,\Delta x + C_d \int_o^d E^i_y(x,y)\,dy\,\Delta x \end{aligned} \tag{7.50}$$

Combining Equations 7.50 and 7.49 and taking the limit as $\Delta x \to 0$ gives

$$\frac{\partial I(x)}{\partial x} + C_d \frac{\partial V(x)}{\partial t} = -C_d \frac{\partial}{\partial t} \int_o^d E^i_z(x,y)\,dy \tag{7.51}$$

The term on the right-hand side of Equation 7.51 may be regarded as a source term representing the coupling of the field into the line

$$\frac{\partial I(x)}{\partial x} + C_d \frac{\partial I(x)}{\partial t} = I_s(x) \tag{7.52}$$

Equations 7.48 and 7.52 suggest that the equivalent circuit for a segment Δx of a line subject to an incident EM field is as shown in Figure 7.13a. In this circuit V(x) and I(x) are the total voltage and current on the line and V_s, I_s represent coupling with the field through the transverse component of the total magnetic field (H_z) and the total component of the electric field perpendicular to the direction of propagation (E_y). An alternative, equivalent formulation is possible, where the total current and scattered voltage on the line are determined from the longitudinal component of the incident electric field E^i_x.[23]

Equations 7.41 and 7.45 and the associated circuit model in Figure 7.13 may be solved in a variety of ways. A common approach is to seek a solution in the frequency domain.[5,24,25] In this approach, the time derivative operator is replaced by jω and the system equations thus reduce to

$$\begin{aligned} \frac{d\bar{V}(x)}{dx} &= -j\omega L_d \bar{I}(x) + \bar{V}_s(x) \\ \frac{d\bar{I}(x)}{dx} &= -j\omega C_d \bar{V}(x) + \bar{I}_s(x) \end{aligned} \tag{7.53}$$

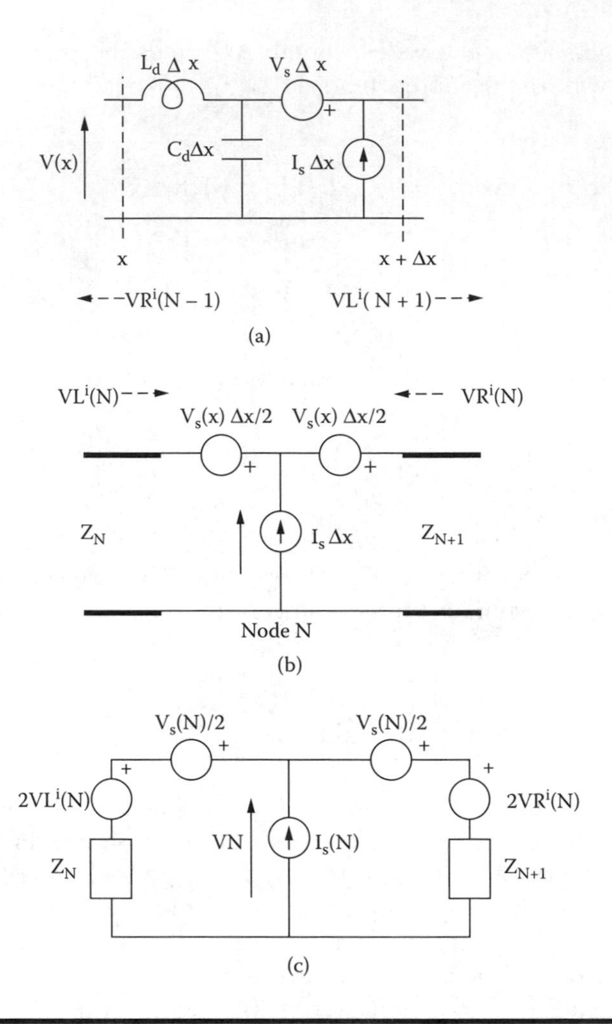

Figure 7.13 A wire segment with sources representing field coupling (a), TLM (b), and Thevenin (c) equivalents.

where

$$\overline{V}_s(x) = j\omega\mu \int_o^d \overline{H}_z^i(x, y)\,dy$$
$$\overline{I}_s = -j\omega\, C_d \int_o^d \overline{E}_y^i(x, y)\,dy \tag{7.54}$$

The solution to these equations is obtained using state variable methods and is

$$\bar{V}(x) = \cos(\beta x)\,\bar{V}(0) - juL_d \sin(\beta x)\,\bar{I}(0) + \bar{V}_s'(x)$$
$$\bar{I}(x) = -juC_d \sin(\beta x)\,\bar{V}(0) + \cos(\beta x)\,\bar{I}(0) + \bar{I}_s'(x) \tag{7.55}$$

where $\beta = 2\pi/\lambda$, u = velocity of propagation, and

$$\bar{V}_s'(x) = \int_0^x \left[\cos\left[\beta(x-\tau)\right]V_s(\tau) - juL_d \sin\left[\beta(x-\tau)\right]I_s(\tau)\right]d\tau$$
$$\bar{I}_s'(x) = \int_0^x \left[-juC_d \sin\left[\beta(x-\tau)\right]V_s(\tau) + \cos\left[\beta(x-\tau)\right]I_s(\tau)\right]d\tau \tag{7.56}$$

Equations 7.55 and 7.56, together with the conditions at the terminations V(0) = –I(0) Z_0 and V(l) = I(1) Z_l, provide the solution at any desired frequency.

A typical problem is the determination of the signal coupled to a line subject to incident EM plane waves. Three configurations are the most common, sidefire, broadside, and endfire as shown in Figure 7.14. In this figure the electric field vector and direction of propagation are shown.[26]

We summarize results here for a number of cases of increasing complexity. We can distinguish three cases: lines that are electrically short (low frequency case), lines of medium electrical length, and lines that are electrically long.

For the simplest case of *electrically short lines* we neglect the distributed L, C, lump excitation sources together, neglect reradiation from the line, and assuming line terminations of the same order as the line characteristic impedance, we obtain for sidefire excitation

$$V_{NE} = -\frac{Z_s}{Z_s + Z_\ell}\left[-j(d\ell)\beta E_0 \frac{\sin(\beta d/2)}{(\beta d/2)} e^{-j(\beta d/2)}\right] \tag{7.57}$$

$$V_{FE} = \frac{Z_\ell}{Z_s + Z_\ell}\left[-j(d\ell)\beta E_0 \frac{\sin(\beta d/2)}{(\beta d/2)} e^{-j(\beta d/2)}\right]$$

where the near- and far-end induced voltages are phasors, $j = \sqrt{-1}$, $\beta = 2\pi/\lambda$ and $E^i = E_0 + j0$.

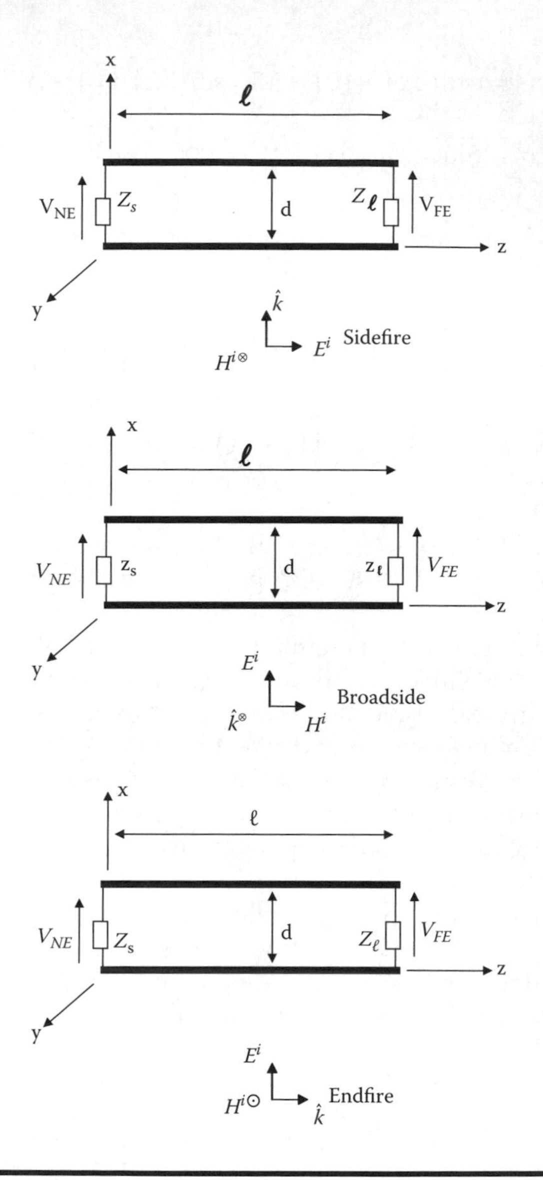

Figure 7.14 Sidefire, broadside, and endfire excitation configurations.

For broadside excitation

$$V_{NE} = V_{FE} = \frac{Z_s Z_\ell}{Z_s + Z_\ell}\left[-j\omega C_d (d\ell) E_0\right] \tag{7.58}$$

where C_d is the line capacitance per unit length.

For endfire excitation

$$V_{NE} = -\frac{Z_s}{Z_s + Z_l} V_{src} + \frac{Z_s Z_\ell}{Z_s + Z_\ell} I_{src} \tag{7.59}$$

$$V_{FE} = \frac{Z_\ell}{Z_s + Z_\ell} V_{src} + \frac{Z_s Z_\ell}{Z_s + Z_\ell} I_{src}$$

where

$$V_{src} = j(d\ell)\beta E_0 e^{-j\beta\ell/2} \text{ and } I_{src} = -j\omega C_d (d\ell) E_0 e^{-j\beta\ell/2}$$

At *medium frequencies* we can use the TL equations (TEM approximation conditions must be met) to study the coupling. Results are summarized below for the three configurations.

For sidefire excitation

$$V_{NE} = Z_s \frac{E_0}{D} de^{-j\beta d/2} \frac{\sin(\beta d/2)}{(\beta d/2)} \left[\frac{Z_\ell}{Z_c} \left[\cos(\beta\ell) - 1 \right] + j\sin(\beta\ell) \right] \tag{7.60}$$

$$V_{FE} = -Z_\ell \frac{E_0}{D} de^{-j\beta d/2} \frac{\sin(\beta d/2)}{(\beta d/2)} \left[\frac{Z_s}{Z_c} \left[\cos(\beta\ell) - 1 \right] + j\sin(\beta\ell) \right]$$

where

$$D = \cos(\beta\ell)(Z_s + Z_\ell) + j\sin(\beta\ell) \left(Z_c + \frac{Z_s Z_\ell}{Z_c} \right)$$

Similarly for broadside excitation

$$V_{NE} = -Z_s \frac{dE_0}{D} \left[\cos(\beta\ell) - 1 + j\frac{Z_\ell}{Z_c} \sin(\beta\ell) \right] \tag{7.61}$$

$$V_{FE} = Z_\ell \frac{dE_0}{D} \left[1 - \cos(\beta\ell) - j\frac{Z_s}{Z_c} \sin(\beta\ell) \right]$$

where

$$D = \cos(\beta\ell)(Z_s + Z_\ell) + j\sin(\beta\ell)\left(Z_c + \frac{Z_s Z_\ell}{Z_c}\right).$$

Finally, for endfire excitation

$$V_{NE} = -jZ_s \frac{dE_0}{D} \sin(\beta\ell)\left[1 + \frac{Z_\ell}{Z_c}\right] \tag{7.62}$$

$$V_{FE} = Z_\ell \frac{dE_0}{2D}\left[1 - \frac{Z_s}{Z_c}\right]\left[1 - \cos(2\beta\ell) + j\sin(2\beta\ell)\right]$$

where

$$D = \cos(\beta\ell)(Z_s + Z_\ell) + j\sin(\beta\ell)\left(Z_c + \frac{Z_s Z_\ell}{Z_c}\right).$$

It remains to discuss the *high-frequency* case where radiation losses cannot be neglected. In this case a numerical solution is normally the best way to proceed. Further details may be found in References 26 and 27.

The case of a wire above ground (e.g., a PCB track above a conducting backplane) can be similarly treated by transforming it into a parallel wire case using the method of images.

An alternative approach is to seek a solution in the time domain. An efficient formulation is to use the transmission-line modeling (TLM) method, whereby the line is described by a cascade of sections similar to that shown in Figure 7.13a.[28] The lossless section may be replaced by a transmission line segment of characteristic impedance $Z = \sqrt{L_d/C_d}$ as shown in Figure 7.13b and for symmetry the voltage source term is shared equally either side of the current source term. For generality, the characteristic impedance is taken to be different on adjacent sections to represent a nonuniform line. The propagation time along each transmission line segment is chosen to be the same Δt along the line to simplify computation. The voltage VN at node M a distance $x = N \cdot \Delta x$ along the line can be obtained at any time, multiple of Δt, from the value of the incident voltages $VL^i(N)$ and $VR^i(N)$ and the voltage and current source. Replacing the lines to the left and right of node N by their Thevenin equivalents gives the circuit shown in Figure 7.13c. The voltage at node N may then be directly obtained:

$$VN = I_s(N)\frac{Z_N Z_{N+1}}{Z_N + Z_{N+1}} + \frac{\dfrac{2VL^i(N) + V_s(N)/2}{Z_N} + \dfrac{2VR^i(N) - V_s(N)/2}{Z_{N+1}}}{\dfrac{1}{Z_N} + \dfrac{1}{Z_{N+1}}} \tag{7.63}$$

Other quantities such as the current may be similarly obtained. The voltages reflected away from node N to travel to the left $VL^r(N)$ and to the right $VR^r(N)$ are given by the formulae

$$\begin{aligned} VL^r(N) &= VN - \frac{V_s(N)}{2} - VL^i(N) \\ VR^i(N) &= VN + \frac{V_s(N)}{2} - VR^i(N) \end{aligned} \tag{7.64}$$

Similar expressions may be derived at the source and load nodes. Computation proceeds as follows:

1. The value of all incident voltages is determined at the start of the calculation from the initial conditions
2. Voltages and current at all nodes are obtained from expressions such as Equation 7.63 and its equivalent for source and load nodes
3. The voltages reflected from each node are obtained from Equation 7.64 and its equivalent for source and load nodes
4. The incident voltages at the next time step n + 1 are obtained from the reflected voltages at the previous time step n and depend on the topology of the network, i.e.,

$$\begin{aligned} {}_{n+1}VL^i(N) &= {}_{n}VR^r(N-1) \\ {}_{n+1}VR^i(N) &= {}_{n}VL^r(N+1) \end{aligned} \tag{7.65}$$

5. Voltages at time step n + 1 are obtained as in step (2)
6. This process is repeated for as long as desired

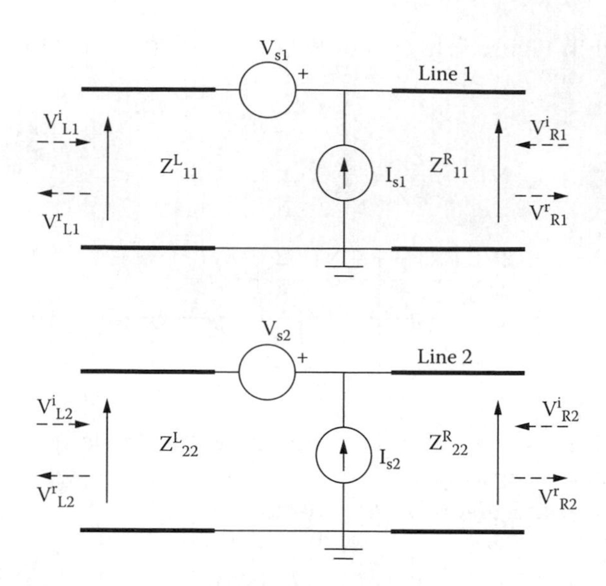

Figure 7.15 Model of a two-wire line above ground and its coupling to an external field.

The frequency- and time-domain solution techniques may be generalized to deal with multiconductor multimode problems such as those depicted in Figure 7.7d. A description of the frequency-domain method applied to such a problem may be found in Reference 29. The time-domain solution method is presented here in some detail because of its applicability to a wide range of EMC problems.

Lossless multiconductor problems where all modes propagate at the same velocity, may be tackled by the methods described in Reference 30. The approach adopted may be described by reference to a two-conductor system above ground. A schematic of each line showing incident and reflected voltages is shown in Figure 7.15, where solid arrows indicate polarity and broken arrows indicate direction of propagation. The total voltage on any line segment is the sum of incident and reflected voltages, i.e.,

$$V_{L1} = V_{L1}^{i} + V_{L1}^{r}$$

In matrix form

$$\left[V_L\right] = \left[V_L^i\right] + \left[V_L^r\right] \tag{7.66}$$

and, similarly, for the current

$$\left[I_L\right]=\left[I_L^i\right]+\left[I_L^r\right] \tag{7.67}$$

where [ ] indicates a column vector, e.g., $\left[V_L\right]=\left[V_{L1},V_{L2}\right]^T$. Incident voltages and currents are related through the impedance matrix

$$\left[V_L^i\right]=\left[Z_L\right]\left[I_L^i\right] \tag{7.68}$$

where

$$\left[Z_L\right]=\begin{bmatrix} Z_{11}^L & Z_{12}^L \\ Z_{21}^L & Z_{22}^L \end{bmatrix}$$

Similarly, for the reflected quantities

$$\left[V_L^r\right]=-\left[Z_L\right]\left[I_L^r\right] \tag{7.69}$$

Equivalent expressions apply for the section to the right of the sources

$$\left[V_R^i\right]=\left[Z_R\right]\left[I_R^i\right] \tag{7.70}$$

$$\left[V_R^r\right]=-\left[Z_R\right]\left[I_R^r\right] \tag{7.71}$$

The total voltage and current and the field coupling sources are related by the expressions

$$\left[V_S\right]=\left[V_R\right]-\left[V_L\right] \tag{7.72}$$

$$\left[I_S\right]=\left[I_R\right]-\left[I_L\right] \tag{7.73}$$

Substituting Equation 7.66 and the equivalent expression for $[V_R]$ into Equation 7.71 gives

$$\left[V_S\right]+\left[V_L^i\right]+\left[V_L^r\right]=\left[V_R^i\right]+\left[V_R^r\right] \tag{7.74}$$

Similarly, Equation 7.73 may be expressed in the form

$$\left[I_S\right]+\left[I_L^i\right]+\left[I_L^r\right]=\left[I_R^i\right]+\left[I_R^r\right]$$

and, using Equations 7.68 and 7.69 and their equivalent for $[V_R^i]$ and $[V_R^r]$ into the expression above, gives

$$\left[I_S\right]+\left[Y_L\right]\left[V_L^i\right]-\left[Y_L\right]\left[V_L^r\right]=-\left[Y_R\right]\left[V_R^i\right]+\left[Y_R\right]\left[V_R^r\right] \tag{7.75}$$

where $[Y_L] = [Z_L]^{-1}$ and $[Y_R] = [Z_R]^{-1}$.

Multiplying Equation 7.74 by $[Y_L]$ and combining with Equation 7.75 gives

$$\begin{aligned}\left[V_R^r\right]=&\left\{\left[Y_R\right]+\left[Y_L\right]\right\}^{-1}\left\{\left[I_S\right]+\left[Y_L\right]\left[V_S\right]\right.\\&\left.+2\left[Y_L\right]\left[V_L^i\right]+\left(\left[Y_R\right]-\left[Y_L\right]\right)\left[V_R^i\right]\right\}\end{aligned} \tag{7.76}$$

Similarly, multiplying Equation 7.74 by $[Y_R]$ and combining with Equation 7.75 gives

$$\begin{aligned}\left[V_L^r\right]=&\left\{\left[Y_R\right]+\left[Y_L\right]\right\}^{-1}\left\{\left[I_S\right]-\left[Y_R\right]\left[V_S\right]\right.\\&\left.+2\left[Y_R\right]\left[V_R^i\right]+\left(\left[Y_L\right]-\left[Y_R\right]\right)\left[V_L^i\right]\right\}\end{aligned} \tag{7.77}$$

Equations 7.76 and 7.77 express the reflected voltages in terms of the known source terms and the incident voltages. They are considerably simplified for a uniform line where $[Z_R] = [Z_L] = [Z]$ to the following expressions

$$\begin{aligned}\left[V_L^r\right]&=0.5\left[Z\right]\left[I_S\right]-0.5\left[V_S\right]+\left[V_R^i\right]\\\left[V_R^r\right]&=0.5\left[Z\right]\left[I_S\right]+0.5\left[V_S\right]+\left[V_L^i\right]\end{aligned} \tag{7.78}$$

Scattering of pulses at the start (source end) and the end of the lines (load end) may be similarly treated.[30] Computation proceeds by following the same five steps as described in connection with propagation along a single line. Losses may be introduced by adding suitable resistive elements between lossless sections.

The case of multiconductor propagation where modes have different propagation velocities has been treated in Reference 31. The first step in the computation is to obtain the modal components and the corresponding propagation velocities for each mode. The problem is then separated into external coupling regions, mode conversion, and mode propagation regions, as shown in Figure 7.16 for a three-conductor line above ground. At each section Δx, modal voltage and currents are converted to line components and are combined with source terms to account for external coupling with an EM field. The resulting line components are converted into modal components, which then propagate on the three lines represented by the appropriate modal impedances. Upon traversing a distance Δx, the modal components are converted into line components to be combined with the external coupling sources. This process is repeated for the total number of sections in the model. One complication that arises is due to the different velocity of propagation and hence to the different transit time in each section for each of the three modes. To simplify computation a single time step is desirable. This is achieved by defining as the time step the transit time of the fastest mode and using the same parameters for all three modal lines. Additional capacitance is introduced

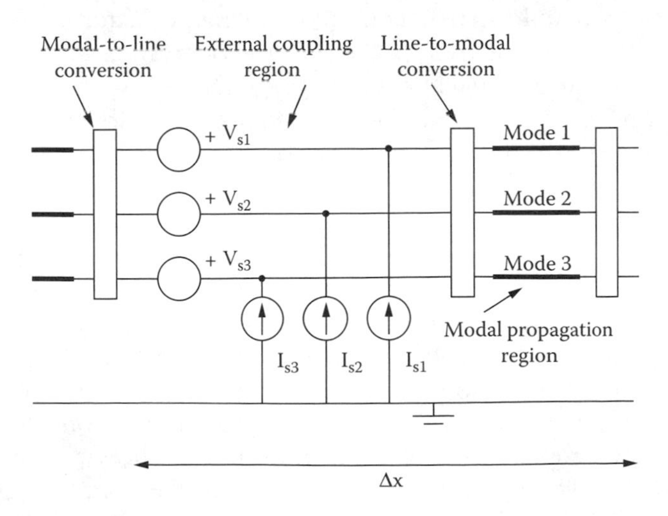

Figure 7.16 Model of multimode propagation on lines coupled to an external field.

for the slower modes in the form of stubs to obtain the correct propagation velocity.[31] The technique described applies to any number of conductors as is described fully in the reference given.

It remains to mention briefly the determination of the source terms $[V_s]$ and $[I_s]$. In simple configurations where the incident field is easily defined, the source terms may be obtained analytically from the integral expressions in Equations 7.47 and 7.51. Often, however, a numerical technique must be employed. This topic is discussed further in Chapter 8. The models presented account fully for crosstalk and coupling from an external field.

The techniques described for dealing with multiconductor propagation may be applied to a variety of configurations such as shielded wires[32] and twisted wires.[33] In the latter case, because of the twist, adjacent segments couple to the external field through a voltage source of opposite polarity, thus reducing overall magnetic coupling. In wire bundles where the number of individual wires is large, special techniques are necessary to reduce computational complexity. A discussion of such problems may be found in Reference 34.

7.5 Determination of the EM Field Generated by Transmission Lines

In the previous sections, techniques for calculating the current flowing on conductors due to crosstalk and coupling to external fields were described. From the EMC point of view, it is often necessary to calculate the strength of the electromagnetic fields generated by these currents so that emissions may be quantified and compared with permitted levels described in EMC standards. These emissions are commonly referred to as "radiated." It is important to emphasize that this term is best interpreted as meaning air-borne emissions, their precise character being capacitive/inductive, near or far field, depending on the distance between the observation point and the radiating object. Details of the "radiated" fields may be found in Chapter 2. Another aspect of emission from transmission lines of all kinds, such as cables and printed circuit boards, is the nature of the currents flowing on these lines. In the previous sections the possibility of decomposing line currents to modal components was described. The simplest case of a two-wire line is shown in Figure 7.17. The impedances shown represent schematically the coupling of the line to adjacent objects and are normally capacitive in nature. Other influences may also be present that are not shown explicitly in Figure 7.17. Line currents I_A and I_B are not necessarily equal in magnitude and in antiphase. However, they can be decomposed as follows:

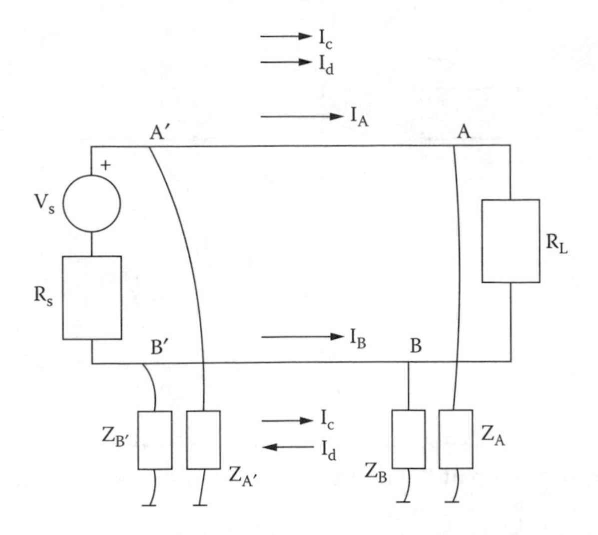

Figure 7.17 Configuration used to study differential- and common-mode currents.

$$I_A = I_c + I_d \qquad I_B = I_c - I_d \tag{7.79}$$

Component $I_d = (I_A - I_B)/2$ is the normal differential current and the only one present if coupling to other structures is negligible. It is the current component intended to flow in this circuit to meet operational requirements. In contrast, component $I_c = (I_A + I_B)/2$ described as the common mode current returns through coupling paths and adjacent structures and it is not essential to circuit operation. More specifically, it does not flow through the load. The situation is in general quite complex, especially at high frequencies, but the origin and significance of these currents does not change. It is unlikely that the circuit designer would have paid significant attention to common-mode currents and these are, therefore, normally unknown. Standard transmission-line analysis, which neglects coupling to other structures, accounts only for the differential currents. Only a comprehensive model that includes all coupling paths is capable of accurately predicting common mode currents. Even in cases where $I_c << I_d$ and therefore impact on normal operation is minimal, the common-mode current may make a major contribution to emissions.[35] The reason for this is that the contributions to emission due to I_d from segments A′A and B′B tend to cancel each other, especially at large distances from the line. In contrast, contributions from I_c assist each other.

At the low-frequency limit and in the far field it is possible to obtain approximate formulae for the electric field due to two parallel tracks

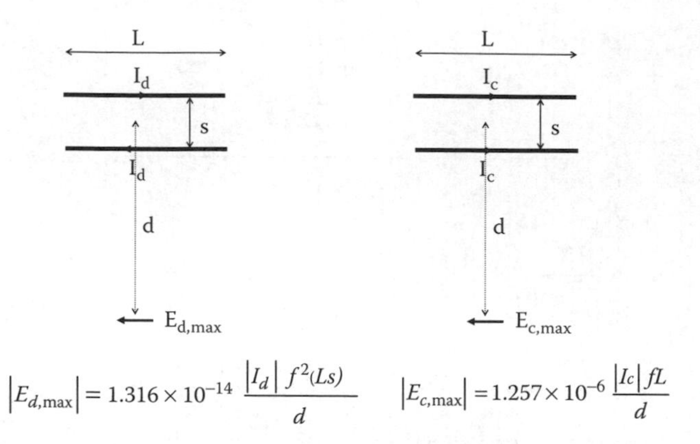

Figure 7.18 Calculation of the maximum radiated electric field due to DM and CM currents.

carrying differential and common mode current at distance d and on the same plane as the tracks as shown in Figure 7.18.[6]

Example: Emission from current loops — In a PCB design the clock frequency is 20 MHz and it is estimated that the currents flowing in a loop (area 15 cm²) at the 3rd, 18th, and 37th harmonic are 70, 45, and 20 dBμA respectively. The loop is above a ground plane. Make an estimate of the maximum field at each frequency and compare with CISPR 22 Class B limits. (These are given in Chapter 13.)

Solution: We use the formula for DM emissions in Figure 7.18 and multiply it by 2 since the image of the tracks on the ground plane may interfere constructively (worst case). This limit is set at 10 m:

$$\left|E_{d,max}\right| = 2 \times 1.316 \times 10^{-14} \frac{\left|I_d\right| f^2 (Ls)}{d} = 2.632 \times 10^{-14} \frac{\left|I_d\right| f^2 (Ls)}{d}, \quad Vm^{-1}$$

$$= 2.632 \times 10^{-14} f^2(MHz) \times 10^{12} I \frac{14 \times 10^{-4}}{10} = 3.948 \times f^2(MHz) I, \ \mu Vm^{-1}$$

Now we calculate the current at the three harmonics:

$$I_3 = \log^{-1}(70/20) = 3162 \ \mu A$$

$$I_{18} = \log^{-1}(45/20) = 178 \ \mu A$$

$$I_{37} = \log^{-1}(20/20) = 10 \ \mu A$$

Hence the electric field for comparison with the Class B limits is

$$E_3 = 3.948 \times 60^2 \times 3162 \times 10^{-6} = 45\,\mu\text{Vm}^{-1} \Rightarrow 33\,\text{dB}\mu\text{Vm}^{-1} \quad \text{Limit } 30\,\text{dB}\mu\text{Vm}^{-1}$$

$$E_{18} = 3.948 \times 360^2 \times 178 \times 10^{-6} = 91\,\mu\text{Vm}^{-1} \Rightarrow 39\,\text{dB}\mu\text{Vm}^{-1} \quad \text{Limit } 37\,\text{dB}\mu\text{Vm}^{-1}$$

$$E_{37} = 3.948 \times 740^2 \times 10 \times 10^{-6} = 22\,\mu\text{Vm}^{-1} \Rightarrow 27\,\text{dB}\mu\text{Vm}^{-1} \quad \text{Limit } 37\,\text{dB}\mu\text{Vm}^{-1}$$

where we have indicated the appropriate limit on the RHS. We see that the electric field exceeds the limit value at the 3rd and 18th harmonics and therefore the products fails.

More accurate calculations can be made provided the DM and CM current distribution is known along the line.

The generated EM field can then be calculated at any distance from the line using antenna theory. The most common approach is to separate each line segment into small enough sections of length l and treat the radiation from each section as that due to a short electric dipole as described in Section 2.3.2. This is acceptable provided that l is much smaller than the wavelength. Useful formulae for such calculations may be found in Reference 36. Several approximations commonly made in such calculations introduce errors that may be unacceptable. For example, describing radiation from segment A′A as the superposition of contributions from a number of sections of length l leaves uncompensated charges at A′ and A, and thus introduces errors in the electrostatic term. These are pronounced at low frequencies and near the line. These errors may be avoided by taking account of segments A′A and AB, but this requires high resolution to account properly for phase differences. The so-called near-field equations account for line spacing only in the phase term and not in the amplitude. This approximation introduces errors when calculating fields near the line. These difficulties and ways to overcome them are discussed more fully in Reference 37.

It is emphasized here that in EMC work identifying the modes of electromagnetic emission from circuits needs particular attention and they are not always the most obvious ones. We give below some simple examples of the type of work needed to predict emissions and hopefully identify remedies.[38] The basic configuration is shown in Figure 7.19a consisting of an interconnect (two parallel tracks) supplied by a source and terminated by a load with two long cable attachments connected to the ground trace as shown. These typically could be the braids of coaxial cables connected to this circuit. The question asked is "How does this section radiate?" There are three possible mechanism and we will examine one at a time.

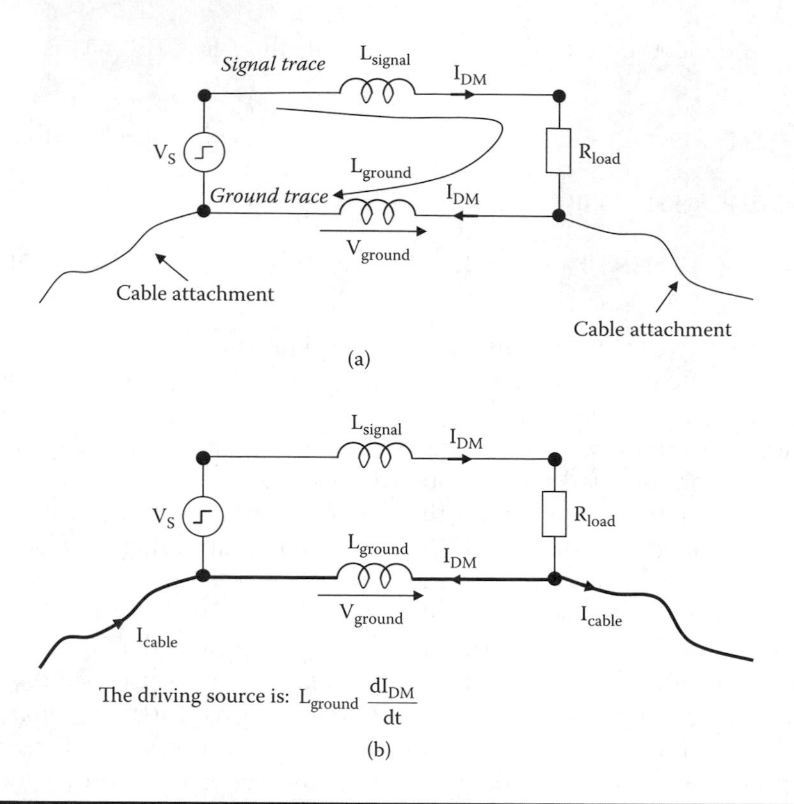

Figure 7.19 Emissions from DM and CM currents. (a) DM currents, (b) emission from the antenna mode driven by the ground noise source, (c) emissions due to asymmetrically fed line, (d) decomposition to DM and CM currents, (e) circuits showing DM and CM current flows.

First, there is an obvious emission mechanism due to the flow of the DM current in the loop shown. As this current is the normal current flow responsible for the functionality of the circuit it is normally well understood and it is unlikely that the designer will be unaware of its impact. If appropriate, the two tracks could be brought closer together to reduce loop area and hence reduce emissions. However, there are two other, less obvious ways that this circuit may emit.

The second mechanism involves the sections shown highlighted in Figure 7.19b. During the operation of this circuit fast signal transitions generate current with a very high rate of change (di/dt) and hence along the ground trace a voltage is generated equal to

$$V_s = L_{gnd} \frac{dI_{DM}}{dt} \tag{7.80}$$

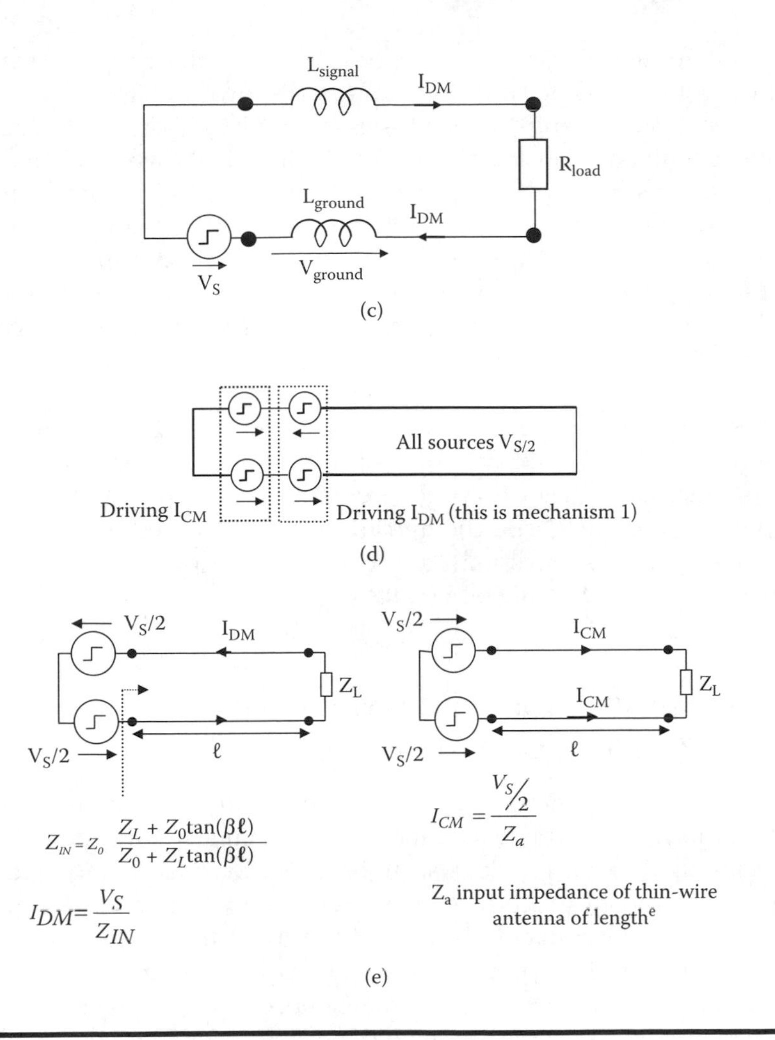

Figure 7.19 (continued).

where L_{gnd} is the partial inductance of the ground trace. This voltage acts like a source driving a dipole antenna with the two arms the two cable attachments. This mode of emission may be described as an antenna mode driven by the ground noise source in Equation 7.80. We see, therefore, that significant voltages may be generated along the ground trace that drive antenna-like structures in the circuit. These unintentional antennas emit in just the same manner as intentional antennas. If we wish to reduce these emissions we need to reduce the rate of change of current (slower transitions) and/or reduce the ground trace partial inductance. This means wide ground tracks widths and preferably a large ground plane.

The third mechanism of emissions has something to do with the symmetry of the circuit. The configuration is shown in Figure 7.19c where the source is not feeding the circuit symmetrically. Following Reference 39 we may obtain an equivalent to this configuration shown in Figure 7.19d. The first set of sources clearly drives an antenna mode (a monopole driven by a common-mode current) while the second set drives differential current similar to that described by the first mechanism described above. The CM and DM current for this case can be obtained from the configurations in Figure 7.19e where Z_a is the input impedance of a thin wire antenna of length ℓ. It can be calculated numerically or in simple cases obtained analytically from antenna theory.

These examples are meant to illustrate the complexity of the issues involved when EMC is considered. Normal circuit operation is concerned with the first mechanism only (differential mode currents). All the other complications are EMC specific and must be considered as they may be dominant as far as emissions are concerned, especially at very high frequencies and in fast digital circuits.

7.6 Numerical Simulation Methods for Propagation Studies

The basic electromagnetic interactions present in multiconductor propagation problems, including crosstalk and coupling to the environment, were described in the previous sections. It is clear that the complexity of such calculations in practical systems is great and, therefore, inevitably, numerical simulation packages are used to assist the designer in making EMC assessments. The purpose of this section is not to present a comprehensive picture of available commercial packages in this area; this task, in a rapidly changing market, would be impossible. Instead, the basic generic methods and some examples of packages and capabilities will be presented to familiarize the designer with currently available computational tools.

The most common propagation and crosstalk codes are based on a frequency-domain formulation of the transmission line equations. Software based on this technique is described in Reference 25. For electrically short lines, very simple models may be implemented using T or Π equivalents and solved using standard packages such as SPICE.

Time-domain solutions are described in Reference 3. Computation is facilitated if a decomposition into modes is made. An example of available software based on this technique may be found in Reference 40. An alternative formulation based on SPICE is described in Reference 41. Models based on the transmission-line modeling (TLM) method are described in References 30, 31, and 42. An example of the application of the finite-difference time-domain (FD-TD) method to crosstalk problems is described in Reference 43. Another code, which is widely used for calculating transients in power systems, is the Electromagnetic Transients Program (EMTP).[44] This program may be adapted to study general propagation problems. Time-domain formulations are ideal for studying transients and nonlinear circuits but not so flexible in allowing the modeling of frequency-dependent components. An approach commonly used to overcome this difficulty is to devise a combination of lumped components that has the desired frequency response and to solve the resulting network in the time domain.[45] Examples of codes suitable for propagation and emission studies are described in Reference 46 based on the finite-element method (FE), and in Reference 47 based on the method of moments (MM). A literature review of EMC effects in printed circuit boards may be found in Reference 48.

References

1. Mohr, R J, "Coupling between open and shielded wire lines over a ground plane," IEEE Trans, EMC-9, pp 34–45, 1967.
2. Ott, H W, "Noise Reduction Techniques in Electronic Systems," 2nd Edition, John Wiley, NY, 1988.
3. Djordjevic, A R, Sarkar, T K, and Harrington, R F, "Time-domain response of multiconductor transmission lines," Proc IEE, 75, No 6, pp 743–764, June 1987.
4. Paul, C R, "Frequency response of multiconductor transmission lines illuminated by an electromagnetic field," IEEE Trans, EMC-18, pp 183–190, Nov 1976.
5. Smith, A A, "Coupling of External Electromagnetic Fields to Transmission Lines," John Wiley, NY, 1977.
6. Paul, C R, "An Introduction to Electromagnetic Compatibility," Wiley, 1992.
7. Nakamura, T and Hayashi, N, "Radiation form a transmission line with an cute bend," IEEE Trans. on EMC, EMC-37, 3, pp 317–325, 1995.
8. Liu, X, Christopoulos, C, and Thomas, D W P, "Prediction of radiation losses and emission from a bent wire by a network model," IEEE Trans. on EMC, EMC-48, 3, pp 476–484, Aug 2006.
9. Liu, X, Christopoulos, C, and Thomas, D W P, " Distortion of voltage pulses propagating on a bent interconnect," 9th IEEE Workshop on Signal Propagation on Interconnects, May 10–13, 2005, Garmisch Partenkirchen, pp 177–180.
10. Zahn, M, "Electromagnetic Field Theory: A Problem Solving Approach," R E Krieger, Malabar, FL, 1987.
11. Wagner, C F and Evans, R D, "Symmetrical Components," McGraw-Hill, NY, 1933.

12. Wait, J R, 'Tutorial note on the general transmission line theory for a thin wire above the ground," IEEE Trans, EMC-33, pp 65–67, 1991.
13. Paul, C R, "Modelling electromagnetic interference properties of printed circuit boards," IBM J Res Develop, 33, No 1, pp 33–50, 1989.
14. Grover, F, "Inductance Calculations: Working Formulas and Tables," Dover, NY, 1962.
15. Ruehli, A E, "Inductance calculations in a complex integrated circuit environment," IBM J Res Develop, 16, pp 470–481, Sept 1972.
16. Green, H E and Cashman, J D, "End effect in open-circuited two-wire transmission lines," IEEE Trans, T-34, pp 180–182, 1986.
17. Clements, J C, Paul, C R, and Adams, A T, "Computation of the capacitance matrix for systems of dielectric-coated cylindrical conductors," IEEE Trans, EMC-17, pp 238–248, Nov 1975.
18. Wei, C, Harrington, R F, Mautz, J R, and Sarkar, T K, "Multiconductor transmission lines in multilayered dielectric media," IEEE Trans, MTT-32, pp 439–450, 1984.
19. Sali, S, Benson, F A, and Sitch, J E, "Coupling in a dielectric-coated multicoaxial-cable system — An analysis based on a quasi-TEM model," Proc IEE-A, 131, pp 67–73, 1984.
20. Djordjevic, A, Harrington, R F, Sarkar, T, and Bazdar, M, "Matrix Parameters for Multiconductor Transmission Lines: Software and User's Manual," Artech House, Norwood, MA, 1989.
21. Taylor, C D, Satterwhite, R S, and Harrison, C W, "The response of a terminated two-wire transmission line excited by a non-uniform electromagnetic field," IEEE Trans, AP-13, pp 987–999, 1965.
22. Nucci, C A and Rachidi, F, " On the contribution of electromagnetic field components in field-to-transmission line interaction," IEEE Trans. on Electromagnetic Compatibility, 37, pp 606–508, 1995.
23. Smith, A A, "Representation of source terms in field-to-wire coupling solutions," Int Conf on EMC, pp T17.16–T17.18, Washington, DC, June 1986.
24. Paul, C R, "Frequency response of multiconductor transmission lines illuminated by an EM field," IEEE Trans, EMC-18, pp 183–190, 1976.
25. Abraham, R T and Paul, C R, "Basic EMC technology advancement for C^3 systems — Coupling of EM fields onto transmission lines," RADC-TR-82-286, Vol IVA, Rome Air Development Center, Griffiss AFB, NY, Nov 1982.
26. Paul, C R, "Analysis of Multiconductor Transmission Lines," Wiley, New York, 1994.
27. Tesche, M, Ianoz, M V, and Karlsson, T, "EMC Analysis Methods and Computational Models," Wiley, New York, 1997.
28. Naylor, P, Christopoulos, C, and Johns, P B, "Analysis of the coupling of EM radiation into wires using TLM," 5th Int Conf on EMC, 1–3 Oct 1986, York, IERE Publ No 71, pp 129–135.
29. Beech, W E and Paul, C R, "Basic EMC technology advancement for C^3 systems — Prediction of crosstalk in flatpack coaxial cables," RADC-TR-82-286, Vol IVf, Rome Air Development Center, Griffiss AFB, NY, Dec 1984.

30. Christopoulos, C and Naylor, P, "Coupling between EM fields and multiconductor transmission systems using TLM," Int Num J Mod, 1, pp 31–43, 1988.
31. Naylor, P and Christopoulos, C, "Coupling between EM fields and multimode transmission systems using TLM," Int Num J Mod, 2, pp 227–240, 1989.
32. Paul, C R, 'Transmission-line modelling of shielded wires for crosstalk prediction," IEEE Trans, EMC-23, pp 345–351, 1981.
33. Paul, C R and McKnight, J A, "Prediction of crosstalk involving twisted pairs of wires. Part I: A transmission line model for twisted wire pairs," IEEE Trans, EMC-21, pp 92–105, 1979.
34. Paxton, A H and Gardner, R L, "Application of transmission line theory to networks with a large number of component wires," 7th Zurich EMC Symp, pp 307–312, 3–5 Mar, 1987.
35. Paul, C R, "A comparison of the contributions of common-mode and differential-mode currents in radiated emissions," IEEE Trans, EMC-31, pp 189–193, 1989.
36. Rubinstein, M and Uman, M A, "Methods for calculating electromagnetic fields from a known source distribution: Application to lightning," IEEE Trans, EMC-31, pp 183–189, 1989.
37. Thomas, D W P, Christopoulos, C, and Pereira, E T, "Calculation of radiated electromagnetic fields from cables using time-domain simulation," IEEE Trans on EMC, 36, pp 201–205, 1994.
38. German, R F, Ott, H W, and Paul, C R, "Effect of an image plane on printed circuit board radiation," IEEE Int. Symp. On EMC, Aug. 21–23, 1990, Washington, DC, pp 284–291.
39. Weeks, W L, "Antenna Engineering," McGraw Hill, 1968.
40. Djordjevic, A, Sarkar, T, Harrington, R F, and Bazdar, M, "Time-Domain Response of Multiconductor Transmission Lines: Software and User's Manual," Artech House, 1989.
41. Paul, C R, "A simple SPICE model for coupled transmission lines," Proc IEEE Int Symp on EMC, pp 327–333, Seattle, WA, Sept 1988.
42. Christopoulos, C, "Propagation of surges above the corona threshold on a line with a lossy earth return," Int Comp Math in Electrical and Electronic Eng (COMPEL), 4, pp 91–102, 1985.
43. Pothecary, N M and Railton, C J, "Analysis of crosstalk on high speed digital circuits using the FDTD method," Int J Num Mod, 4, pp 225–240, Sept 1991.
44. The Leuven EMTP Centre, "ATP — Alternative Transients Program," July 1985.
45. Yen, C S, Fazarinc, Z, and Wheeler, R L, "Time-domain skin-effect model for transient analysis of lossy transmission lines," Proc IEEE, 70, pp 750–757, 1982.
46. Khan, R L and Costache, G I, "Finite element method applied to modelling crosstalk problems on printed circuit boards," IEEE Trans, EMC-31, pp 5–15, 1989.

47. Oing, S, John, W, and Mrozynski, G, "Calculation of radiated electromagnetic fields from electronic circuits," Int J Num Mod, 4, pp 241–258, 1991.
48. Gravelle, L B and Wilson, P F, "EMI/EMC in printed circuit boards — a literature review," IEEE Trans, EMC-34, pp 109–116, 1992.

Chapter 8

Simulation of the Electromagnetic Coupling between Systems

8.1 Overview

In previous chapters, various aspects of electromagnetic coupling relevant to EMC have been considered. The physical basis of the interactions involved was outlined and analytical or numerical models suitable for describing them were introduced. It is inevitable that in a complex system several of these mechanisms are present simultaneously and that many of the configurations of interest are rarely subject to simple treatment. In many cases it is not possible to isolate and treat different aspects of the problem separately, e.g., penetration and propagation. The resulting complexity can only be handled by using sophisticated numerical computer-based techniques. The purpose of this chapter is to present an introduction to the main simulation methods and the manner in which they can best be employed to solve complex EMC problems. A useful framework for classifying and tackling EMC problems is as follows.[1,2]

External/source region — In susceptibility problems, the source region may be other equipment, radio transmitters, etc. Certain severe forms of EM threat such as EMP and lightning may need special characterization, especially if the victim of the interference is very near the source. In emission problems, the external environment is not normally specified in detail, except in various EMC standards, e.g., emitting equipment in an open site with measurements taken a distance of 10 m away.

Penetration and coupling — Emphasis was placed in previous chapters on assessing penetration through apertures and non-perfectly conducting walls. In practice all kinds of conducting penetrations from signal leads to water pipes breach the relative isolation between regions internal and external to a barrier shield.

Propagation and crosstalk — The penetrating EM radiation couples onto various components and networks and propagates to interfere with other systems through crosstalk.

Device susceptibility and emission — In cases of susceptibility, a portion of the coupled EM energy becomes incident on individual components or subsystems causing damage or malfunction. Conversely, EM emissions from such subsystems may become the source of EMI added to the external environment.

In many practical problems it is not easy to identify clearly and treat separately these various aspects of EMC. It is helpful, however, to be able to identify important factors and solution methods suitable for tackling all these aspects of the problem so that accurate models may be developed and suitable numerical simulation methods may be selected.

8.2 Source/External Environment

In most cases, the external environment for emission studies is either taken to be free space or, alternatively, to conform to recommended test environments. Of the latter, the open-site environment rarely presents problems for simulation. However, emission assessment by simulation inside screened rooms presents several difficulties. In unlined screened rooms it is acceptable to assume perfectly conducting walls and then construct models that predict the interaction of the device-under-test (DUT) with the screened room and associated instrumentation. Rooms lined with radiation absorbing material (RAM) may, depending on frequency and the quantity of RAM, be regarded as partially or fully anechoic. The latter condition is difficult to achieve, especially at low frequencies. For accurate prediction, sophisticated models of the entire environment, including the RAM, are necessary. Examples of such simulations may be found in References 3 to 5.

In susceptibility studies, the greatest difficulty occurs when simulating coupling to systems placed in the source region of EMP and lightning. In both cases the assumption of an incident plane wave is not valid as the entire interaction takes place under near-field conditions. In addition, severe nonlinearities are present. In the source region of EMP, the time-dependent ionization of air and the presence of rapidly varying intense fields requires a self-consistent simulation of all aspects of the problem. Such simulations have been done but they are important for certain military systems only and unlikely to be of interest to ordinary commercial EMC. General guidance as to likely levels of the EM threat may be found in Chapter 5.

Lightning affects a large number of systems and it is of particular interest to mobile systems such as aircraft and missiles. The presence of a craft in the path of a lightning channel can be studied by simulation to predict current levels and resonances. In these studies, the lightning channel may be modeled in various degrees of detail. A fully three-dimensional self-consistent solution presents formidable difficulties. These are caused by the very different physical scales of the problem (~ km for the separation of electric charge centers, ~ m for aircraft, and ~ mm for the highly ionized lightning channel), and the complexity and inadequate understanding of ionization in a lightning channel. Numerical models of the channel have been described by several authors.[6–8] Studies of lightning-aircraft interactions have been reported using a stick model of an aircraft and the method of moments (MM),[9] the finite-difference time-domain (FD-TD) method,[10] and the transmission-line modeling (TLM) method.[11] The last two methods can deal with a variety of situations but at a considerable computational effort. A decision to investigate source-region coupling effects in detail should not be taken lightly as it invariably involves work at the limits of computational and modeling capabilities.

8.3 Penetration and Coupling

Penetration and coupling through shields and apertures were studied in Chapter 6. Radiated interference coupling by diffusion through thin walls and directly through electrically small and large apertures has been studied in some detail. Simulation packages suitable for diffusion studies are normally based on frequency-domain formulations using the finite-element (FE) method. Time-domain formulations are less popular since, in general, diffusion times are long and excessive computer run times are required. In most practical situations, penetration through apertures dominates. The most general simulation codes suitable for any size apertures, modeling simultaneously regions internal and external to shields, are based on differential time-domain formulations such as FD-TD and TLM.[12,13] Alternative formulations based on the MM are possible but must be applied with care to avoid solution artifacts due to "numerical" coupling between inner and outer regions.

Conducted interference makes a major contribution to the overall interference budget and has hitherto not been considered in detail. The term "conducted interference" refers to EM energy guided by conductors

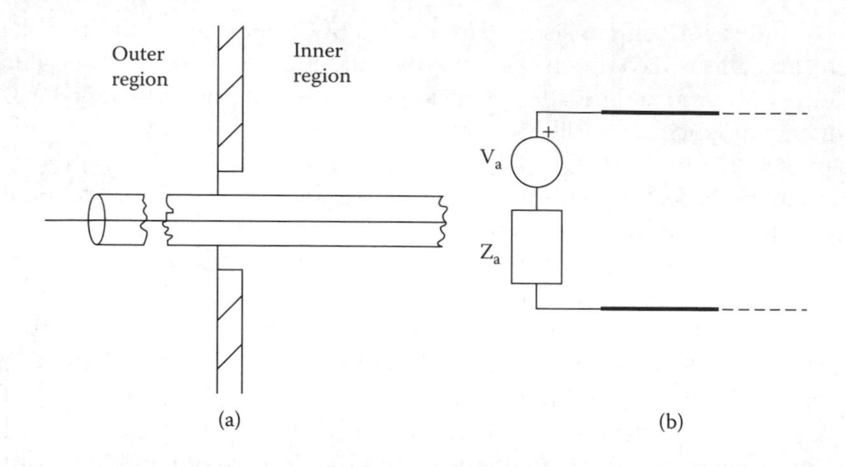

Figure 8.1 A shielded wire penetration (a) and its circuit equivalent (b).

such as signal cables and their shields, pipes used to bring various services inside enclosures (e.g., water, compressed air), and other structural parts. The easiest penetration to deal with is that due to a shielded wire, shown schematically in Figure 8.1a. Coupling from an external source onto exposed parts may be represented by an appropriate Thevenin equivalent circuit, as shown in Figure 8.1b, or, alternatively, in more complex situations, by a complete field coupling model such as that shown in Figure 7.13. Assuming that the wire shield is joined solidly by a 360° connection to the barrier shield, separating inner and outer regions results in the equivalent circuit shown in Figure 8.1b, which can be used to determine the penetrating signal. Conversely, if emission from the inner region is considered, quantities V_a and Z_a represent the Thevenin equivalent of the radiating parts considered as sections of antennas. These quantities may be obtained efficiently from antenna codes based on MM techniques.[14,15] An unshielded conducting penetration may be brought through the shield via feed-through capacitor, via an electrical filter, or may be directly connected to it. Coupling of the conducting penetration with EM fields in the inner and outer regions may be represented as in Figures 8.1b and 7.13. Of particular interest is to ascertain the modifications to the signal traveling along the conducting penetration as it interacts with the barrier shield and any associated filtering components deliberately or unavoidably placed there.

Let us consider first the case of a conductor going through the barrier shield without a DC connection to it. The capacitance C_b between the conductor and the barrier represents either stray capacitance or a feed-through capacitor specifically inserted for the purpose. Assuming that an assessment of EM penetration needs to be made and that coupling from

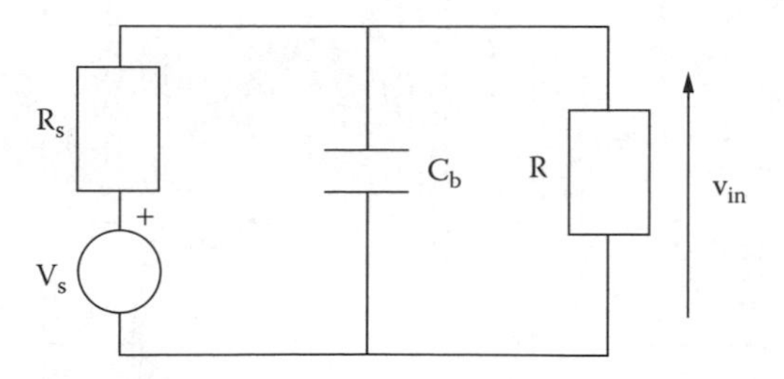

Figure 8.2 Circuit used to study the penetration of a conductor through a shield without a DC connection to it.

the external environment may be presented as in Figure 8.1b gives the equivalent circuit shown in Figure 8.2, where R is the input impedance seen by the conductor in the inner region. In obtaining this circuit, it has been implicitly assumed that a simple lumped component representation of the feed-through and the inner circuit is possible (low-frequency approximation). The presence of capacitor C_b affords a certain amount of shielding from the interfering signal V_s as the ratio of the voltage seen across R when C_b is absent and when it is present, is equal to

$$1+\frac{j\omega RC_b}{1+R/R_S} \tag{8.1}$$

From this expression it is clear that C_b is effective in introducing some attenuation in high-impedance circuits.

At very high frequencies transmission-line effects begin to dominate the behavior of feed-through capacitors. A typical cylindrical arrangement is shown in Figure 8.3.[16] The cylindrical feed-through capacitor is in fact an open-circuit transmission line of length 1. A useful measure of effectiveness is the ratio between the voltage V_2 induced at the capacitor end inside the barrier shield, and the noise current I_1. This ratio can be referred to as the transfer impedance $Z_C = \overline{V}_2/\overline{I}_1$. Using the transmission line Equations 3.25 and 3.27 for $I_2 = 0$ gives

$$Z_c = \frac{Z_0}{j\sin(2\pi 1/\lambda)} \tag{8.2}$$

where Z_0 is the characteristic impedance of the coaxial line (cylindrical capacitor) and λ is the wavelength. At low frequencies ($\lambda >> 1$), Equation 8.2 may be simplified by recognizing that $\sin(2\pi l/\lambda) \simeq 2\pi 1/\lambda$ to give

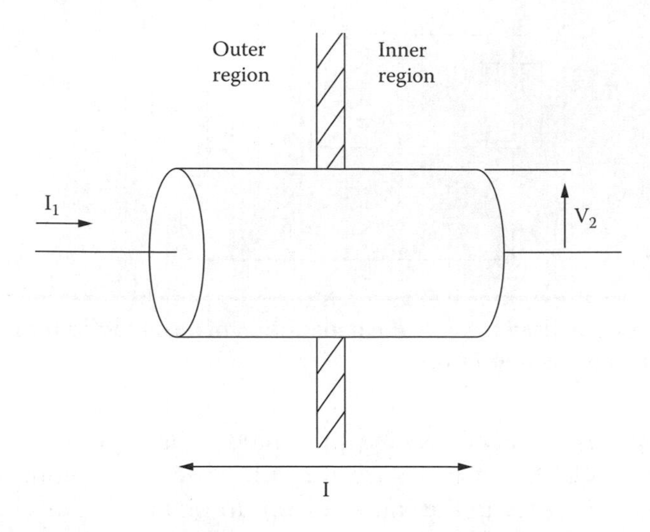

Figure 8.3 Penetration via a feed-through capacitor.

$$Z_c = \frac{\sqrt{L_d / C_d}}{j\,2\pi 1/\lambda} = \frac{1}{u\,C_d j 2\pi 1/\lambda} = \frac{1}{j\frac{2\pi}{\lambda}u\left(c_d l\right)} = \frac{1}{j\omega C_b}$$

This confirms that at the low-frequency limit a lumped representation of the feed-through capacitor is acceptable. However, as the wavelength becomes comparable to the length of the line, the behavior of the feed-through component becomes very complex. As an example, when $l/\lambda = 0.5$, Equation 8.2 gives $Z_c \to \infty$.

It may be imagined that a DC connection to the barrier shield will completely prevent interfering current from inducing voltages inside the shielded region. Such an arrangement is shown schematically in Figure 8.4a and it is suitable for non-signal-carrying conducting penetrations. Assuming that the contact resistance between the barrier shield and the conductor is R_c, the induced voltage is $V = IR_c$. This may be further reduced by the double contact arrangement shown in Figure 8.4b. Using the approach described in Reference 16, the equivalent circuit for this arrangement may be obtained where L is the inductance of the coaxial line (inner and outer radii r_1, r_2, and length 1) between the double contacts. Assuming that $\omega L >> R_{c1}$, R_{c2} gives $\bar{V}_1 \sim \bar{I}R_{c1}$ and $\bar{V}_2 = R_{c2}\bar{V}_1/(R_{c2} + j\omega L) \simeq \bar{I}R_{c1}R_{c2}/j\omega L$.

For contact resistances a fraction of a milliohm and L of the order of a nanohenry, the voltage V_2 becomes, at high frequencies, a small fraction of V_1, thus improving shielding. Similar benefits are obtained by using a

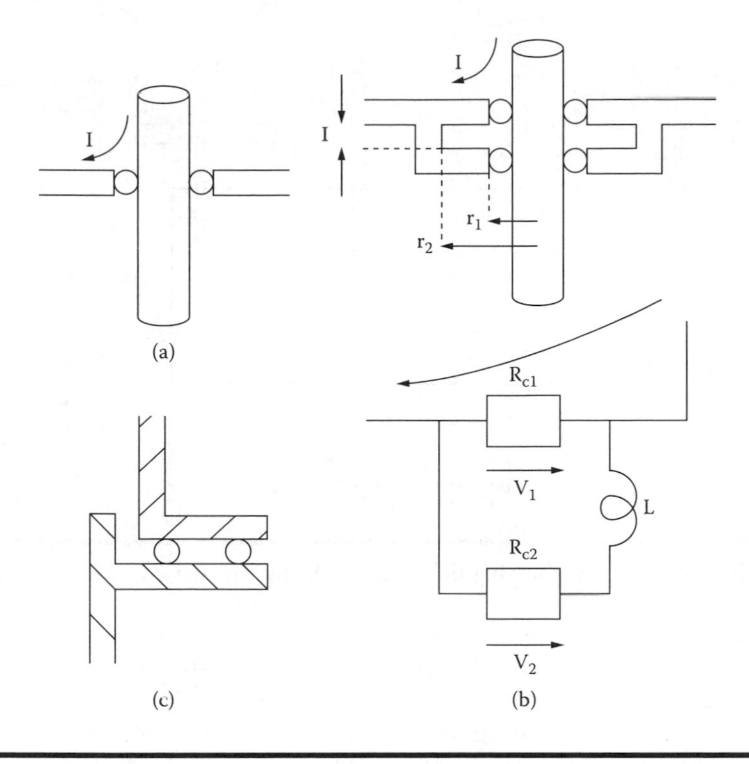

Figure 8.4 Various gasketing arrangements (a) and (c) and an equivalent circuit (b).

double contact arrangement to effect a temporary joint between two panels, as shown in Figure 8.4c. In some cases specially designed filters are installed between the barrier shield and the conducting penetration. In these circumstances, a complete analysis can be made, taking into account all factors, to determine the amount of interference.

Simple calculations, such as those presented here, can be made to establish the value of each component and the severity of penetration. Alternatively, numerical simulation may be used to obtain a better measure of each type of penetration. Examples of such calculations for antenna penetrations and the effects of "pigtail" connections may be found in References 17 to 19.

Every attempt is made to characterize penetrations in terms of equivalent circuits and induced voltages. These are then used as source terms in propagation studies aiming at establishing the spreading of interference away from the point of entry by conduction and radiation. Although description of such penetrations as an integral part in a three-dimensional problem is possible, the small size and complexity of these features make simulations quite difficult.

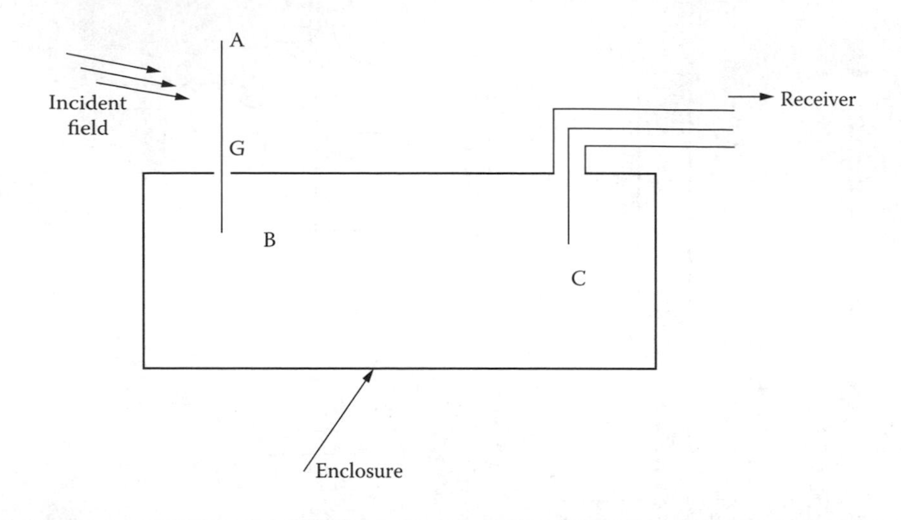

Figure 8.5 Schematic showing the coupling between two wires inside a cabinet.

A generic case of penetration and coupling is shown in Figure 8.5. This illustrates several of the problems arising in the electromagnetic characterization of cabinets and shows intermediate level models (models not based on a full-field description) that may be employed to characterize quantitatively the level of coupling. Let us first begin with a qualitative description of the coupling mechanisms relevant to the configuration shown in Figure 8.5. The cabinet has a wire penetration AGB, typically part of a communications or power connection. The segment AG is external to the cabinet and therefore subject to interference from external incident fields. Thus the induced current penetrates into the segment GB internal to the cabinet after negotiating the barrier at G. Segment GB acts now as an antenna exciting EM fields inside the cabinet, which distribute according to the demands of the resonant structure that the cabinet is. Energy is thus coupled to the wire segment C, which appears at the receiver as interference. The EMC engineer wishes to establish a quantitative relationship between the incident field and the interference voltage measured at the receiver. This is a typical EMC problem and as already indicated requires an understanding of several stages in a complex interaction. Naturally, this problem could be tackled by constructing a full-field numerical model of the entire configuration to establish the required transfer function between the external field and the receiver voltage. This, however, leads to a large computation and we will therefore focus instead on constructing an intermediate level model where we simplify interactions as far as possible.[20]

At the input side we recognize that we have, effectively, two monopole antennas, AG and BG. Each may be represented by an equivalent circuit

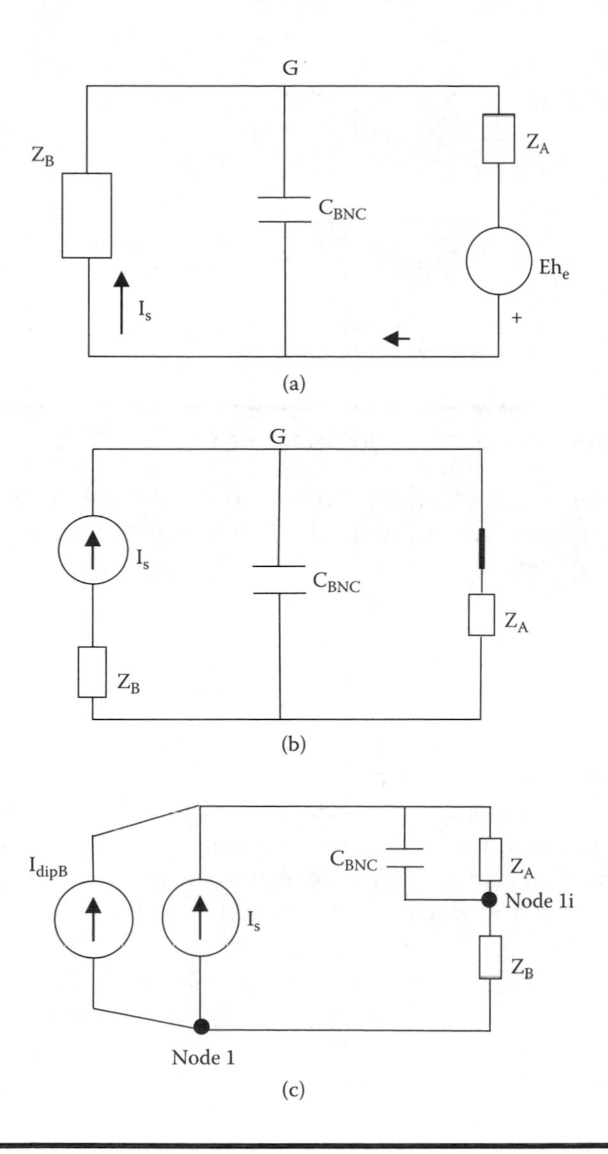

Figure 8.6 Equivalent circuit of two monopoles AG, BG (a); replacing the voltage source in (a) by an equivalent current source (b); introducing I_{dipB} to represent the coupling of the cavity fields to GB (c).

consisting of its self impedance in series with a voltage source. The equivalent circuit is shown in Figure 8.6a where, since the incident field only couples directly to monopole GA, the voltage source in GB is set to zero. The barrier capacitance C_{BNC} is also included, representing in the simplest case the connector capacitance (e.g., a BNC type connector). In this figure Z_A and Z_B represent the self impedance of each monopole and

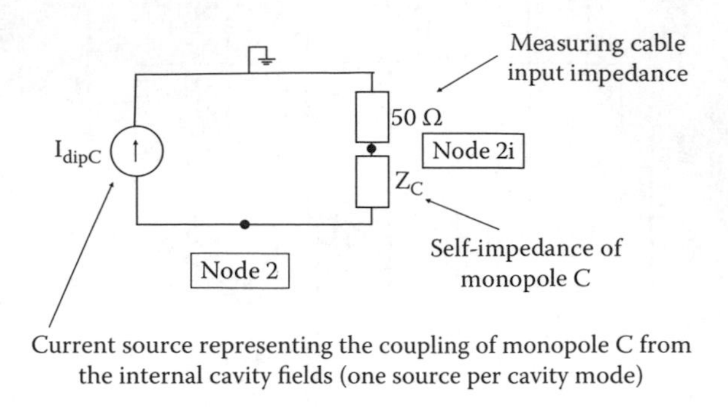

Figure 8.7 Equivalent circuit of the monopole GC.

h_e the effective length of monopole GA. These are frequency-dependent parameters and may be obtained from References 20 and 21. The current flowing in the internal monopole is then

$$I_s = \frac{Eh_e(j\omega)}{Z_B + Z_A(1 + j\omega C_{BNC} Z_B)} \tag{8.3}$$

Using the reciprocity circuit theorem, we may short out the voltage source in Figure 8.6a and insert a current source in branch Z_B and the current flow remain unchanged. This is shown in Figure 8.6b. The circuit is further modified in Figure 8.6c where we have introduced an additional current source I_{dipB} that represents coupling to the GB due to the cavity fields. Several such current sources may be added to account for the modes of the cavity.

We now turn our attention to the output circuit at C. Here the model is far simpler to visualize as shown in Figure 8.7 where Z_C is the self impedance of GC and the 50-Ω resistor represents the input impedance of the receiver. The current source I_{dipC} represents the coupling of the cavity fields to GC.

We have dealt with the model building for the monopoles GA/GB and GC. The third and final element of the complete model is a representation of the coupling of the cavity fields and their coupling with a dipole or monopole antenna. The cavity may in a first approximation be represented by a short-circuited transmission line representing a waveguide mode, where at the antenna (wire) position $z = -z_C$ a current source I_{WG} is added to represent the current induced on the line due to the antenna voltage. The configuration is shown in Figure 8.8. The coupling from the line to the antenna is also shown in the same figure by a current source I_{DIP}. The antenna self impedance is Z_d. The two current sources are given by the expressions

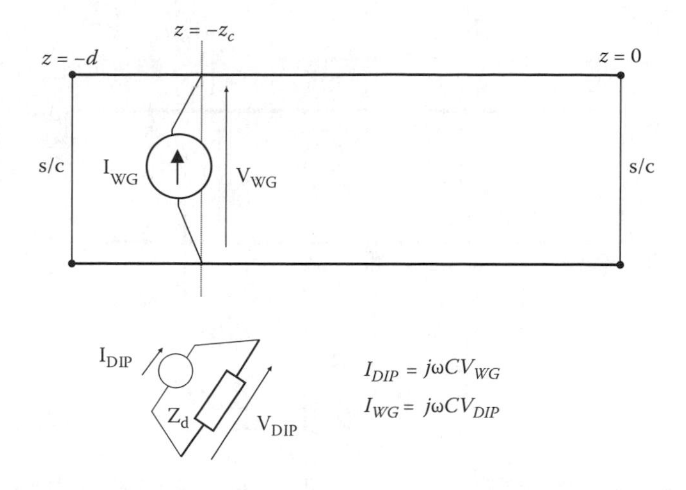

Figure 8.8 Coupling between a dipole and a line representing a waveguide mode.

$$\begin{aligned} I_{DIP} &= j\omega C V_{WG} \\ I_{WG} &= j\omega C V_{DIP} \end{aligned} \tag{8.4}$$

where C is the mutual capacitance between the dipole and the line. In our application, we have a source antenna A1 (GB) and a victim antenna A2(GC) positioned at locations 1 and 2 as shown in Figure 8.9. We need to develop this model for each waveguide mode n. We first consider the middle section (between antennas A1 and A2) shown in Figure 8.10a with the appropriate notation for mode n. The Norton equivalent circuit for this section is shown in Figure 8.10b where the two current sources are given by the expressions

$$\begin{aligned} P_n &= T_{2n} V_{A1}^{(n)} + T_{1n} V_{A2}^{(n)} \\ H_n &= T_{1n} V_{A1}^{(n)} + T_{2n} V_{A2}^{(n)} \end{aligned} \tag{8.5}$$

The coefficients T_{1n} and T_{2n} are obtained from the TL equations as follows

$$\begin{aligned} V_{A1}^{(n)} &= \cosh(\gamma_{n0} h) V_{A2}^{(n)} + Z_{n0} \sinh(\gamma_{n0} h) I_{A2}^{(n)} \\ I_{A1}^{(n)} &= \frac{1}{Z_{n0}} \sinh(\gamma_{n0} h) V_{A2}^{(n)} + \cosh(\gamma_{n0} h) I_{A2}^{(n)} \end{aligned} \tag{8.6}$$

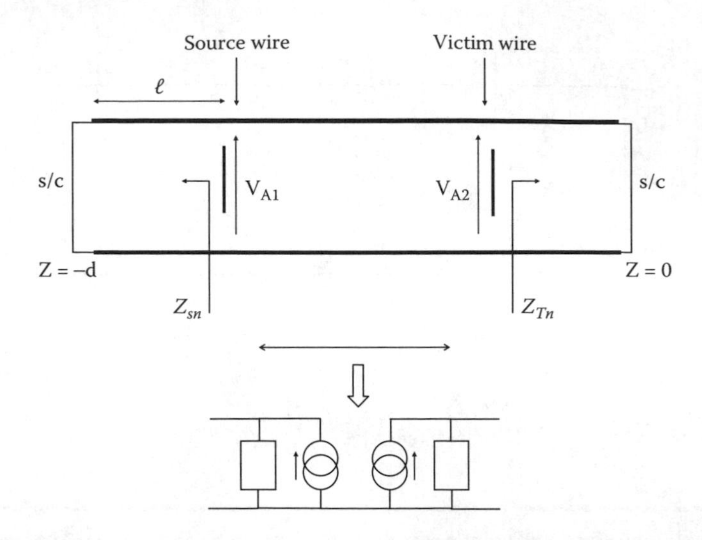

Figure 8.9 Schematic of coupling between two wires (source and victim) inside the cavity. The equivalent circuit of the middle section is shown schematically directly below.

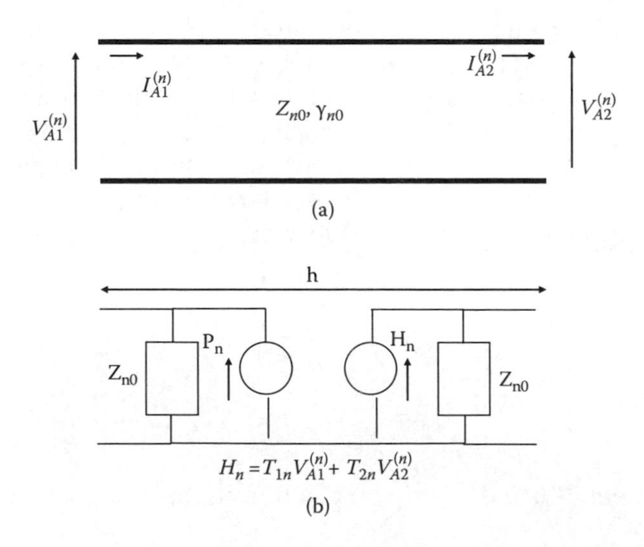

Figure 8.10 The middle segment between the coupled wires (a) and its equivalent circuit (b).

$$= \frac{1}{Z_{n0}} \sinh(\gamma_{n0}h) V_{A2}^{(n)}$$

$$+ \cosh(\gamma_{n0}h) \frac{V_{A1}^{(n)} - \cosh(\gamma_{n0}h) V_{A2}^{(n)}}{Z_{n0} \sinh(\gamma_{n0}h)}$$

$$= V_{A1}^{(n)} \frac{\cosh(\gamma_{n0}h)}{Z_{n0} \sinh(\gamma_{n0}h)} + V_{A2}^{(n)} \left\{ \frac{\sinh(\gamma_{n0}h)}{Z_{n0}} - \frac{\cosh^2(\gamma_{n0}h)}{Z_{n0} \sinh(\gamma_{n0}h)} \right\}$$

$$= V_{A1}^{(n)} \frac{\coth(\gamma_{n0}h)}{Z_{n0}} + V_{A2}^{(n)} \left\{ \frac{\sinh(\gamma_{n0}h)}{Z_{n0}} - \frac{1 + \sinh^2(\gamma_{n0}h)}{Z_{n0} \sinh(\gamma_{n0}h)} \right\}$$

$$= V_{A1}^{(n)} \frac{\coth(\gamma_{n0}h)}{Z_{n0}} - V_{A2}^{(n)} \frac{1}{Z_{n0} \sinh(\gamma_{n0}h)}$$

From Figure 8.10b we have that

$$I_{A1}^{(n)} = \frac{V_{A1}^{(n)}}{Z_{n0}} - P_n = V_{A1}^{(n)} \left[\frac{1}{Z_{n0}} - T_{2n} \right] - V_{A2}^{(n)} T_{1n} \tag{8.7}$$

Comparing this equation with the RHS of Equation 8.6 wc obtain

$$\begin{aligned} T_{1n} &= \frac{1}{Z_{n0} \sinh(\gamma_{n0}h)} \\ T_{2n} &= \frac{1}{Z_{n0}} \left[1 - \coth\left(\gamma_{n0}h\right) \right] \end{aligned} \tag{8.8}$$

Looking from the source wire toward the left we replace the line by its input impedance, which is that of a short-circuited section of length

$$Z_{sn} = Z_{n0} \tanh\left(\gamma_{n0}\ell\right)$$

Similarly, from the victim wire looking to the right, we obtain

$$Z_{Tn} = Z_{n0} \tanh\left(\gamma_{n0}\ell'\right)$$

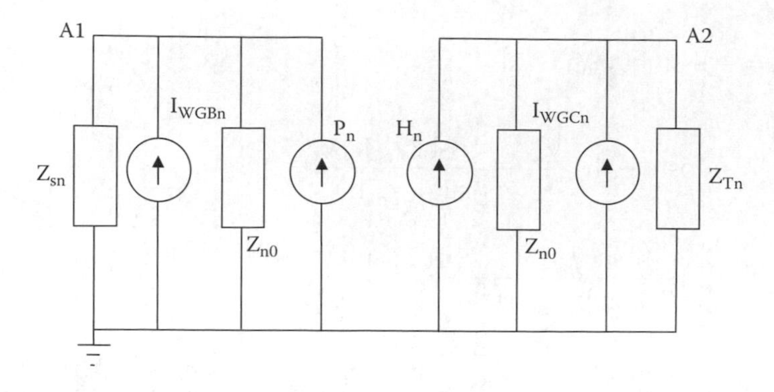

Figure 8.11 Line model for mode n.

where ℓ' is the length of the line from z = 0 to the position of the victim wire. Therefore the model for the line representing mode n is as shown in Figure 8.11. The entire coupling model for the problem where we have included two modes is shown in Figure 8.12. Figure 8.12a depicts the source wire (GB) model where current sources I_{dipB1} and I_{dipB2} represent induced current due to modes TE_{10} and TE_{20}, respectively. Figure 8.12b represents the victim wire (GC) where I_{dipC1} and I_{dipC2} represent induced current by the two modes. Figures 8.12c and 8.12d are the transmission line equivalents of the two waveguide modes included in this computation. The notation is self-explanatory, e.g., I_{WGB1} is the current due to source wire GB induced on the TL representing mode n = 1 (TE_{10}) etc. More modes can be included if necessary. Further details of the computation and results may be found in Reference 20.

8.4 Propagation and Crosstalk

Simulations of propagation and crosstalk, possibly involving coupling to an incident field, are possible using frequency- or time-domain techniques. The general approach to solving such problems was described in Chapter 7. Difficulties in such studies are associated with the complexity of actual circuits, the presence of nonuniformities, fast signals, and generally the number of parameters to be described. Progress in this area hinges on seeking to describe increasingly complex systems in terms of simpler equivalents, which are then used in simulations. A powerful technique, which can be used to describe electrically small radiators, is to devise a system of three electric and three magnetic equivalent dipoles (one for each coordinate direction).[22] An entire subsystem can be described

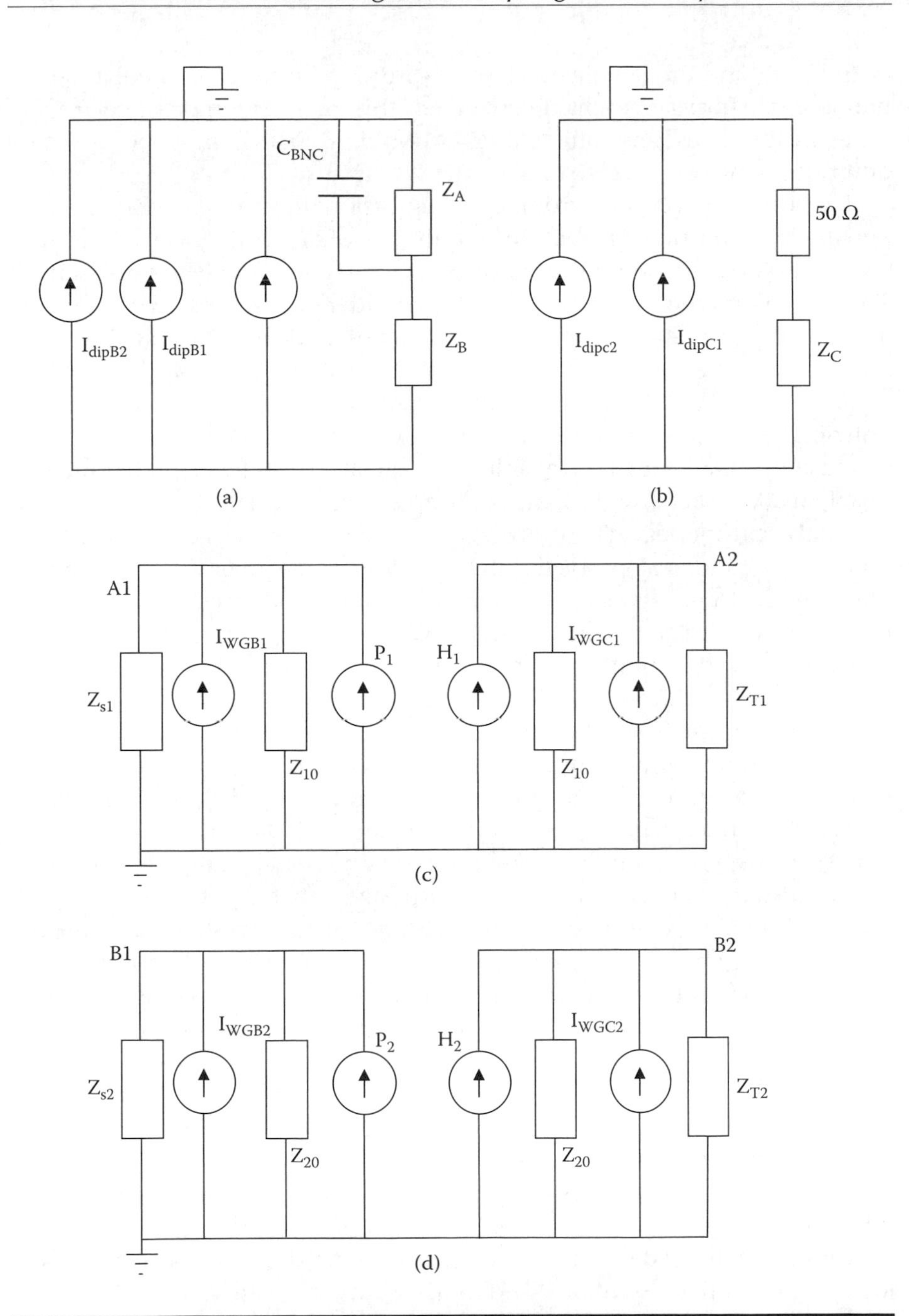

Figure 8.12 Coupling model for two modes, source wire model (a), victim wire model (b), TL model for mode TE_{10} (c), and TL model for mode TE_{20} (d).

in this way and the simplified description may be used in simulations. A number of subsystems characterized in this manner, interconnected by transmission lines and unintentional coupling paths, may be used to describe more complex systems.

It should be obvious from this discussion that, in general, source, penetration, and propagation are not always easily distinguishable or separately solvable and that in some cases whole-system modeling is essential. To solve such complex problems with confidence requires advancements in modeling techniques, computational power, and validation against careful experiments, and is an ongoing task for research in EMC.

A permanent difficulty in such studies is the extreme complexity of practical systems. It is therefore essential that a *hierarchy of models* of increasing complexity is available to deal at some level with practical problems. We sketch below the nature of these models.

Analytical models: These are accurate solutions in closed form for a number of canonical problems that possess some degree of symmetry. They are generally too simple as representations of practical systems but nevertheless are very useful as benchmark solutions for model evaluation and for illustrating trends in design in an easily accessible manner that maximizes physical understanding.[23–25]

Expert systems/rule checkers: These are computer programs that in some way emulate human thinking.[26] Unlike ordinary programs, which handle a large number of data through precisely defined algorithms, expert systems use rules that are based on the "good judgement" of someone who is an expert in the field. They may warn users against bad practices and suggest remedies. It must be emphasized that the validity of these tools is limited to a relatively narrow design space and that caution must be exercised when using them in new or unfamiliar applications.

Lumped circuit models: Insofar as a network can be considered electrically small in all directions, standard circuit techniques and solvers such as SPICE may be employed. The test of electrical size must be done at the highest frequency of interest (smallest wavelength) for EMC. Parameters L, C, etc. may be obtained analytically or through numerical simulations.

Distributed parameter models: These are appropriate to describe interconnects when the TEM approximation holds as explained elsewhere in this book.

Intermediate models: These are a compromise between semi-empirical ideas and full-field solutions and borrow, as necessary, concepts from lumped distributed and full-field models. The basic idea is to reduce model complexity but keep accuracy under control. Such models are mentioned throughout this book.

Full-field models: These invariably are solutions to Maxwell's equations and are sophisticated tools requiring years of development. They are accurate within the limitations of numerical accuracy. Providing full input data for these models can be a problem and they also make heavy use of computational resources. Many practical problems cannot be solved using such models because of the huge computational cost. Thus techniques are developed to embed more efficient numerical structures in the form of macronodes that describe particularly complex or fine features. This speeds up computational accuracy significantly. A review of such multiscale models may be found in Reference 27.

Hybrid models: Human ingenuity enables innovative approaches to problem solving. A particular example is the combination of different models in the same problem to maximize model advantages and minimize their disadvantages. Such hybrid models can be very powerful but care must be take to ensure that they remain versatile and easy to use. Examples are combinations of techniques such as FDTD and GTD or TLM with Integral Equation methods. The development of such tools is for the specialist but users for EMC applications should be aware of their existence and also of the prospect of more such tools becoming available as they mature beyond the research stage.

8.5 Device Susceptibility and Emission

The complexity and detail found in modern electronic equipment is such that in any attempt to predict behavior by simulation, boundaries must be set to limit the extent of computation to a manageable level. For example, if emissions originating from a clock circuit are considered, the clock chip is characterized by a simple Thevenin equivalent circuit and, normally, no account is taken of the internal circuit structure. Similarly, in a susceptibility study the clock chip is normally replaced by a suitable input impedance. A more detailed description of the chip is possible by using macromodels such as those available in SPICE. Progress is being made in interfacing general three-dimensional field solvers with lumped circuit models of actual devices in order to give an integrated self-consistent model. The complexity of such calculations should not be underestimated. It should also be borne in mind that actual accurate models of devices are not always available and that the variability between nominally similar components from the same or different manufacturers can be very significant. This topic is addressed in more detail in the next chapter. The presence of interference at the input to a device, in the form of fast pulses of magnitude outside the normal range, requires detailed knowledge of the input circuit of the device including stray components and nonlinearities. Only then can accurate numerical models be constructed. Such information is rarely available in sufficient detail in device specifications.

8.6 Numerical Simulation Methods

There are many techniques suitable for solving electromagnetic and circuit problems. It is not the purpose of this section to give a comprehensive survey of all available methods. There are techniques that, by exploiting symmetry or some other simplifying feature of the problem, can produce accurate solutions very rapidly for specific classes of problems. A survey of numerical methods may be found in References 28 and 29. Attention is focused here only on methods that are general and versatile enough to provide solutions to the wide range of problems encountered in EMC studies.

A classification of numerical methods and a discussion of their special features may be found in Reference 30. Broadly speaking, methods are classified as time or frequency domain, depending on the type of formulation employed. Fast transients, nonlinearities, and time-dependent parameters are generally best treated in the time domain. However, problems where the steady-state response at a few frequencies is desired, or when frequency-dependent components are present, are best dealt with in the frequency domain. Whichever domain is adopted, results in the other domain may be found by recognizing that the impulse and frequency responses of a system are Fourier transform pairs.

An alternative classification scheme is into differential and integral methods. In the former case the physical laws are enforced on a number of points occupying the entire volume of the problem space, whereas in the latter, enforcement of laws is limited to points on certain surfaces of the problem.

Differential methods are better suited to solving inhomogeneous nonlinear problems but suffer from difficulties when modeling open-boundary problems. In contrast, integral methods cope well with open-boundary problems but they are not ideally suited to solving inhomogeneous problems. There are some methods suitable for dealing with very high frequency problems that do not fit any of these categories. They are generally referred to as ray methods.[31] Ray methods are suitable for tackling problems where the wavelength is much smaller than the physical size of the objects under consideration. Differential and integral methods require a space discretization that typically is smaller than a tenth of a wavelength. As a result, the size of the problem that can be treated with modest computational resources cannot exceed a few wavelengths in each coordinate direction.

Before embarking on a brief description of each method, it is worth identifying what the main requirements are for simulation codes suitable for EMC studies.

1. Geometrical features — the code should be capable of modeling a variety of objects and shapes, including semi-enclosed regions commonly found in EMC problems
2. Material properties — EMC problems are characterized by a wide range of materials and the code must be capable of dealing with nonuniform, nonlinear materials
3. Physical scale — it is not unusual in EMC to have a problem that is infinite in extent and where fine features (e.g., thin wires or narrow slots) must also be modeled; the code must have adequate facilities to deal with fine features and open boundaries
4. Time scale — interest in EMC is over a wide range of time signals or, equivalently, a wide range of frequencies; adequate and efficient coverage over a broad band of frequencies is an essential requirement from any code used in EMC

The main field solvers suitable for whole-system modeling are based on one or more of the techniques described in the next four subsections.

8.6.1 The Finite-Difference Time-Domain (FD-TD) Method

This method is a differential, time-domain method based on a time and space discretization of Maxwell's equations. Derivatives are evaluated using central differences on a spatial grid referred to as the Yee-mesh.[32] Formulations are available to deal with fine features, thin wires, and with open-boundary problems through specially developed radiation-absorbing conditions. The method is a powerful and versatile tool and has been used successfully to study a range of problems in EM. A more detailed description of the method may be found in References 33 to 35. FD-TD provides a complete simulation of the build-up of the EMC field from the start of the excitation to the time when steady-state conditions have been reached. Typically, the excitation is a Gaussian pulse. The response to any other input may then be obtained by convolution.

8.6.2 The Transmission-Line Modeling (TLM) Method

This is a differential time-domain method similar in many respects to the FD-TD method. The main difference is that field solutions are obtained through equivalence to voltages and currents flowing in a network of transmission lines, clustered into nodes, occupying the problem space. The basic node, which permits the determination of all field components at each point in space, is the symmetrical condensed node described in

Reference 36. Further details of the method may be found in References 37 to 39. The method is implemented in a similar way as the FD-TD method. Adaptations to deal with fine features, thin wires, and panels are described in the references given.

8.6.3 The Method of Moments (MM)

The method is normally implemented in the frequency domain and consists of the imposition of suitable conditions on the boundary surfaces of the problem. For example, in a scattering problem the appropriate condition is that on the surface of a perfect conductor the incident and reflected tangential components of the electric field must sum to zero. This type of formulation is referred to as the electric field integral equation (EFIE), and it is suitable for modeling thin wires. An alternative formulation based on the magnetic field (MFIE) is more suitable for smooth large bodies. In implementing the method, the unknown function on each surface element is expressed as a combination of known basis functions. The unknown coefficients of this expansion are obtained by solving a system of equations expressing an optimality criterion (e.g., minimum error). A collection of papers describing the method and its applications may be found in Reference 40. Method of moment techniques are particularly suitable for modeling wire structures and open-boundary problems.

8.6.4 The Finite-Element (FE) Method

Normally, the FE method is formulated in the frequency domain. The problem space is divided into elements (e.g., triangles in two dimensions) on which the potential is defined as a combination of the potentials at the vertices. A suitable functional is found and the unknown potentials at the vertices are calculated by minimizing this functional. For example, in an electrostatic problem a suitable functional is the electrostatic energy. The difficulties in terminating open-boundary problems accurately introduce spurious components in the response. In a development of the method, known as the boundary element method (BEM), the number of unknowns is reduced by restricting the definition of elements on suitable boundaries in the problem. The resulting equations, although fewer in number, are more difficult to solve. Details of the FE method and its applications may be found in References 41 and 42. The FE method is very well developed as a tool for predicting eddy current distributions at low frequencies. One of its advantages is its flexibility in defining boundaries by using differently shaped elements.

The techniques described above are continuously developed to extend their capabilities and minimize their disadvantages. For example, time-domain techniques may be reformulated and applied in the frequency-domain. The information given above refers to the most common implementation.

No method excels in every conceivable application. It appears inevitable that, in a sophisticated design environment, a number of codes should be available to deal with the variety of problems likely to be encountered. All codes, however friendly, are based on assumptions and simplifications that may be unwarranted in certain applications. It is important that the user is familiar with such limitations. The complexity of these codes is a demotivating factor for those who wish to understand more deeply the nature and significance of implicit or explicit code limitations. It appears to be good advice to suggest that it is better to use a number of simpler tools with which one is familiar, rather than a complex tool poorly understood. Using two methods in a single solver to enhance the advantages and minimize the disadvantages, i.e., to hybridize two methods, appears attractive and such codes are continuously being developed.[43] An example of a general code using MM to model features smaller than a wavelength, the geometrical theory of diffraction (GTD) to model structures larger than a wavelength and finite-differences for internal (cavity) problems, is known as the general EM model for the analysis of complex systems (GEMACS).[44]

Under practical circumstances, it is difficult to assess all important interactions to the level of accuracy and detail implied by these analysis methods. Although increasingly more sophisticated and comprehensive simulations are possible, it is nevertheless useful to be able to determine EMC compliance without resort to full-scale EM simulation. Such a capability is useful when the complexity of a full-field simulation is prohibitive, or when the system is not yet sufficiently well defined to embark on such a calculation. An EMC budget can be established for the entire system by identifying the main emitters and receptors of EM radiation, and the coupling paths between them. Examples of such paths are antenna to antenna, wire to wire, field or antenna to wire, or enclosure to enclosure. The frequency response may then be established for each coupling path individually, and eventually the total interference signal may be calculated. Design modifications may then be necessary so that interference levels are brought within desired limits. Such calculations are inevitably based on several simplifications and they must not be regarded as very accurate. They are nevertheless useful in identifying problem areas and in directing the major effort on these specific problems. An example of such system-level EMC code is IEMCAP (Intrasystem EMC Analysis Program).[45]

Further details on the application of numerical simulation methods to EMC problems may be found in References 46 to 48.

References

1. Lee, K S H (editor), "E Interaction: Principles, Techniques and Reference Data," Report AFWL-TR-80-402, Dec. 1980.
2. Tesche, F , "E interaction with aircraft and missiles: a description of available coupling, penetration and propagation analysis tools," AGARD Lecture Series LS-144, 10.1–10.21, 1986.
3. Herring, J L and Christopoulos, C, "Numerical simulation for better calibration and measurements," 5th British Electromagnetic Measurements Conference, Malvern, UK, pp 37/1–37/4, 11–14 Nov 1991.
4. Herring, J L, Naylor, P, and Christopoulos, C, "Transmission-Line Modelling in electromagnetic compatibility studies," Int J Num Mod, 4, pp 143–152, 1991.
5. Corona, P, Latrmiral, G, and Paolini, E, "Performance and analysis of a reverberating room with variable geometry," IEEE Trans, EMC-22, pp 2–5, 1980.
6. Uman, M A and Krider, E P, "A review of natural lightning: experimental data and modelling," IEEE Trans, EMC-24, pp 79–112, 1982.
7. Baker, L, "Lightning return-stroke transmission line model," Proc 11th Int Aero and Ground Conf on Lightning and Static Electricity, Dayton, pp 35.1–35.11, June 1986.
8. Mattos, da F M A and Christopoulos, C, "A model of the lightning channel including corona and prediction of the generated electromagnetic fields," J Phys D — Applied Physics, 23, pp 40–46, 1990.
9. Deng, J, Balanis, C A, and Barber, G C, "NEC and ESP codes: Guidelines, limitations and EMC applications," IEEE Trans, EMC-35, pp 124–133, 1993.
10. Kunz, K S, "Finite element/finite difference modelling in electromagnetic compatibility," 10th Int Zurich EMC Symp, pp 487–492, 9–11 Mar 1993.
11. Johns, P B and Mallik, A, "EMP response of aircraft structures using TLM," 6th Zurich EMC Symp, pp 387–389, Mar 5–7, 1985.
12. Taflove, A, Umashankar, K R, Baker, B, Harfoush, F, and Yee, K S, "Detailed FD-TD analysis of EM fields penetrating narrow slots and lapped joints in thick conducting screens," IEEE Trans, AP-36, pp 247–257, 1988.
13. Scaramuzza, R, Naylor, P, and Christopoulos, C, "Numerical simulation of field-to-wire coupling problems using transmission line modelling," Proc Int Conf on Computation in Electromagnetics, IEE Conf Publ 350, London, pp 63–66, 25–27 Nov, 1991.
14. Rockway, J W, Logan, J C, Tam, D W S, and Li, S T, "The MININEC System: Microcomputer Analysis of Wire Antennas," Artech Ho., Norwood, MA, 1988.
15. Popovic, B D, "CAD of Wire Antennas," Research Studies Press, England, 1991.

16. Kaden, H, "Wirbelstrone and Schirmung in der Nachrichtentechnik," Springer-Verlag, NY, 1959.
17. Dudley, D G and Casey, K F, "A measure of coupling efficiency for antenna penetrations," IEEE Trans, EMC-33, pp 1–9, 1991.
18. Hejase, H A N, Adams, A T, Harrington, R F, and Sarkar, T K, "Shielding effectiveness of 'pigtail' connections," IEEE Trans, EMC-31, pp 63–68, 1989.
19. Hejase, H A N, Adams, A T, and Harrington, R F, "A quasi-static technique for evaluation of pigtail connections," IEEE Trans, EMC-31, pp 180–183, 1989.
20. Thomas, D W P, Denton, A, Benson, T M, Christopoulos, C, Paul, J D, Konefal, T, Dawson, J F, Marvin, A C, and Porter, S J, "Electromagnetic coupling to an enclosure via a wire penetration," IEEE Int. EMC Symp., Montreal, Aug. 13–17, 2002, pp 183–188.
21. Tang, T G, Tieng, Q M, and Gunn, M W, "Equivalent circuit of a dipole antenna using frequency-independent lumped elements," IEEE Trans. on Antennas and Propagation, 41(1), pp 100–103, 1993.
22. Ma, M T and Koepke, G H, "A method to quantify the radiation characteristics of an unknown interference source," NBS Technical Note 1059, 1982.
23. Kaden, H, "Wirbelstrome und Schirmung in der Nachtrichtentechnik," 2nd edition, Springer, 1959.
24. Collin, R E, "Field Theory of Guided Waves," 2nd edition, IEEE Press, 1991.
25. Tesche, M, Ianoz, M, and Karlsson, T, "EMC Analysis Methods and Computational Models," Wiley, New York, 1997.
26. Nebendahl, D, "Expert Systems: Introduction to the Technology and Applications," Siemens A G and Wiley, 1988.
27. Christopoulos, C, "Multi-scale modelling in time-domain electromagnetics," Int. J. Electron. Commun. (AEÜ), 57(2), pp 100–110, 2003.
28. Itoh, T (editor), "Numerical techniques for microwave and millimeter-wave passive structures," John Wiley, NY, 1989.
29. Booton, R C, "Computational Methods for Electromagnetics and Microwaves," John Wiley, NY, 1992.
30. Miller, E K, "A selective survey of computational electromagnetics," IEEE Trans, AP-36, pp 1281–1305, 1988.
31. Hansen, R C (editor), "Geometrical Theory of Diffraction," IEEE Press, NY, 1981.
32. Yee, K S, "Numerical solution of initial boundary-value problems involving Maxwell's equations in anisotropic media," IEEE Trans, AP-74, pp 302–307, 1966.
33. Taflove, A and Umanshankar, K R, "The finite-difference time-domain method for numerical modelling of EM wave interactions with arbitrary structures," in PIERS Progress in Electromagnetics Research, J A Kong (editor), pp 287–373, Elsevier, Amsterdam, 1990.
34. Kunz, K S and Luebbers, R J, "The Finite Difference Time Domain Method for Electromagnetics," CRC Press, Boca Raton, FL, 1993.
35. Sadiku, M N O, "Numerical Techniques in Electromagnetics," CRC Press, Boca Raton, FL, 1992.

36. Johns, P B, "A symmetrical condensed node for the TLM method," IEEE Trans, MTT-35, pp 370–377, 1987.
37. Hoefer, W J R, "The transmission-line matrix method — theory and applications," IEEE Trans, MTT-33, pp 882–893, 1985.
38. Christopoulos, C and Herring, J L, "The application of transmission-line modelling (TLM) to electromagnetic compatibility problems," IEEE Trans, EMC-35, pp 185–191, 1993.
39. Christopoulos, C, "An Introduction to the Transmission-Line Modelling (TLM) Method," IEEE Press, NY, 1995.
40. Miller, E K, Midgyesi-Mitschang, L, and Newman, E H (editors), "Computational Electromagnetics — Frequency Domain Method of Moments," IEEE Press, NY, 1992.
41. Silvester, P P and Ferrari, R L, "Finite Elements for Electrical Engineers," 2nd Edition, Cambridge University Press, Cambridge, 1990.
42. Hoole, S R H, "Computer-Aided Analysis and Design of Electromagnetic Devices," Elsevier, Amsterdam, 1989.
43. Medgyesi-Mitshang, L N and Wang, D-S, "Hybrid methods in computational electromagnetics: a review," Comp Phys Commun, 68, pp 76–94, 1991.
44. Siarkewicz, K R, "GEMACS — An executive summary," IEEE Symp on EMC, Wakefield, Mass, Aug 20–22, 1985.
45. Lee, G and Ellersick, S D, "Methodology for determination of circuit safety margins for MIL-E-6051 EMC system test," IEEE Symp on EMC, pp 502–510, Wakefield, Mass, 20–22 Aug, 1985.
46. Kubina, S J, "Numerical analysis and modelling techniques," in AGARD-LS177, Electromagnetic Interference and Electromagnetic Compatibility, pp 3.1–3.18, Neuilly sur Seine, France, 1991.
47. Perez, R, "Solution of EMC problems via analysis methods: a review," IEEE Regional Symp on EMC, pp 2.3.1/1–8, Tel Aviv, Israel, 2–5 Nov, 1992.
48. Special Issue on EMC Applications of Numerical Modelling Techniques, T Hubing (Guest Editor), IEEE Trans, MTT-35, No 2, May 1993.

Chapter 9

Effects of Electromagnetic Interference on Devices and Systems

Electromagnetic interference affects systems in various ways. Depending upon the severity of the EM threat and the sensitivity of the system, it may lead to an upset or even damage. The behavior of a system may be altered by various degrees and still be acceptable to the user. A burst of interference may cause a momentary malfunction, with the system returning to normal operation within a few milliseconds. However, in safety-critical or in mission-critical systems, such as those found in many military or mobile applications, such a momentary malfunction may have catastrophic results and cannot be tolerated. Another type of upset is that causing a change in the system (such as overwriting parts of a computer memory), which is not immediately detrimental but can cause problems in the future if the problem is not detected and rectified. Finally, interference can cause an irreversible change to the system, such as semiconductor damage, which renders the system incapable of functioning. The degree of upset and the amount of damage that can be tolerated is system dependent and no general advice can be given. However, it is of value to give some guidelines as to how devices fail so that the designer has some grasp of the issues involved.

Electromechanical components, such as motors and transformers, normally fail due to breakdown of insulation. This may be due to overvoltages applied suddenly, or more often to incipient faults that gradually lead to complete failure. Insulation that is regularly stressed and overheats eventually fails from thermal ageing (a ten-degree rise in operating temperature halves the lifetime of insulation). However, this type of device has a high level of immunity to overvoltages compared to semiconductor devices. Trends in the design of digital systems are toward smaller and faster devices. This leads to smaller cross-sectional areas and spacings and hence to higher current densities (several kA/mm^2) and a risk of voltage breakdown. Semiconductor devices are therefore the weak link as far as immunity to interference is concerned and their threshold to failure is of critical importance. This is a specialized and complex subject and it is not the intention to offer complete coverage in this chapter. Instead, a brief introduction and references to literature on this subject will be given, so that the designer is aware of damage thresholds, their statistical variability, and therefore the likely safety margin between the level of interference present at the device pins and its damage threshold.

Most semiconductors fail due to overheating in the junction area and interconnections. Local melting and resolidification establish a low-impedance path across the junction. Similar thermal effects are thought to be responsible whether the junction is pulsed in the forward or reverse direction.[1] However, since the voltage across a forward-biased junction is lower, higher currents are necessary to deposit the required amount of energy at the junction. Hence, the threshold in this case is higher compared to the reverse-biased case. The maximum power that can be absorbed by a reverse-biased device without damage depends critically on the duration of the interfering pulse. It is commonly assumed that the relationship between the maximum power P and the duration t is of the form

$$P = k_1 \, t^{k_2} \tag{9.1}$$

where k_1 and k_2 are constants.

It has been shown, based on some assumptions, that $k_2 = -0.5$,[2] and that Equation 9.1 is an adequate model provided the duration t is in the microsecond region. It is important to emphasize the statistical nature of Equation 9.1 and its limitations. The constants k_1 and k_2 are dependent on the particular device and the shape of the EMI pulse, and vary widely. Studies have shown that even for the same device type and number, the variability in maximum power damage threshold can be very large, both between devices produced by the same manufacturer or by different manufacturers.[3–5] This threshold depends critically on the details of manufacture, which vary considerably even for the same device type. The whole statistical variation of this parameter has led some authors to suggest that thresholds should be conservatively determined from the maximum steady-state ratings provided by manufacturers (e.g., twice the rated junction voltage).[3] Undoubtedly, more needs to be known about the failure modes and characteristics of semiconductor components. Additional material on this important topic may be found in Reference 6.

The requirements for military systems exposed to intentionally generated EMI and to radiation effects require special measures and hardening procedures that are beyond the scope of this book. Useful material and references for this type of application may be found in Reference 7.

It is extremely difficult to conduct comprehensive studies of damage due to EMI on complete systems such as VLSI chips. The designer is normally content to determine interference signals on input pins to complex devices and ascertain on a semi-empirical basis the likelihood of malfunction or damage. Studies that involve more detailed models of SSI chips, which are then incorporated into general circuit solvers such as SPICE, have been reported. An example of such a study, using a macromodel for the popular 741 op-amp, is described in Reference 8.

9.1 Immunity of Analogue Circuits

There are differences in the way analogue and digital circuits respond to interference. An everyday example is the whistles one occasionally hears in analogue radio due to interference. This is not the case for digital radio. Similarly, EMI in analogue TV manifests itself by the appearance of "snow" on the screen. In digital TV the likely effect is a frozen screen. In analogue circuits noise coupled to the circuit propagates throughout modified according to circuit impedances. In contrast, in a digital circuit if noise at the input to the gate is rejected (within the noise margin) the output of the gate is free from interference. It goes without saying that proper RF design to minimize interference is paramount for both analogue and digital circuits. These common rules are summarized below and are expanded further in Part 3:

- Understand and control possible stray circuit components that affect current flows
- Understand and control common and differential mode currents and employ balancing and isolation as required
- Assess the impact of radiation and crosstalk especially from circuits carrying fast pulses or high-frequency currents
- Minimize as far as possible logic transition times (use the slowest logic possible compatible with operational requirements) and system bandwidth

- Pay particular attention to shielding and grounding so that current return paths are fully understood by the designer and are kept as short as possible
- Develop strategies for segregating circuits that are likely to generate interference and circuits that are sensitive to it

Although transients affect all types of circuit, they are particularly threatening to digital circuits. Analogue circuits normally carry low-level signals and thus are particularly prone to RF interference. If this interference is in-band (within the normal frequency spectrum of operation), it can only be kept out by proper screening of the circuit and connections to it. Excessive levels of in-band interference may be controlled by nonlinear devices (e.g., diodes connected as clamps). Out-of-band interference can be similarly dealt with and in addition by filtering. The difficulty with this type of interference is that it invariably encounters nonlinear components and thus may be rectified (demodulation) and appear as a DC offset. Typically, the envelope of the high-frequency signal is obtained, which may contain lower frequency components that may be in-band signals. The addition of a high-frequency (by-pass) capacitor across the input of amplifiers can help minimize demodulation products. Another nonlinear effect is the generation of other frequencies that, again, may coincide in frequency with operational components. Further details may be found in Reference 9.

9.2 The Immunity of Digital Circuits

We summarize here the operation of logic gates and their associated noise margins.[10] The transfer function of an ideal gate is shown by the thin line in Figure 9.1a; there is a sharp transition at the mid-point between the "low" and "high" logic levels. However, real gates show behavior that varies around the ideal (Figure 9.1b). Output high is not necessarily V_{CC} but may be as low as V_{OH}. Similarly, an output low is not necessarily 0V but can be as high as V_{OL}. A gate will recognize as an input high a voltage that is as low as V_{IH} and as an input low a voltage that is as high as V_{IL}. These different levels are shown in Figure 9.2. It can be seen that if noise were superimposed on the output V_{OH} of the driving gate, this would not affect the input of the driven gate as long as the voltage did not fall below V_{IH}. Hence there is a margin of safety equal to $V_{OH} - V_{IH}$. This is designated as the "high noise margin" NM_H. Similarly a "low noise margin" can be defined as $NM_L = V_{IL} - V_{OL}$. This is essentially a static picture. If we apply a pulse that is above a threshold, the logic will switch provided the pulse duration is longer than the response time of the gate. Therefore, a dynamic

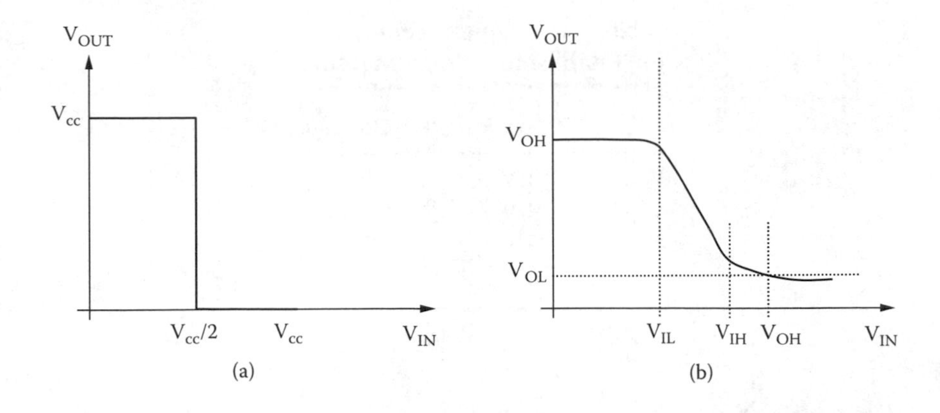

Figure 9.1 Ideal inverter transfer characteristic (a) if the input is smaller than $V_{cc}/2$ then the output goes high, and typical inverter transfer characteristic (b).

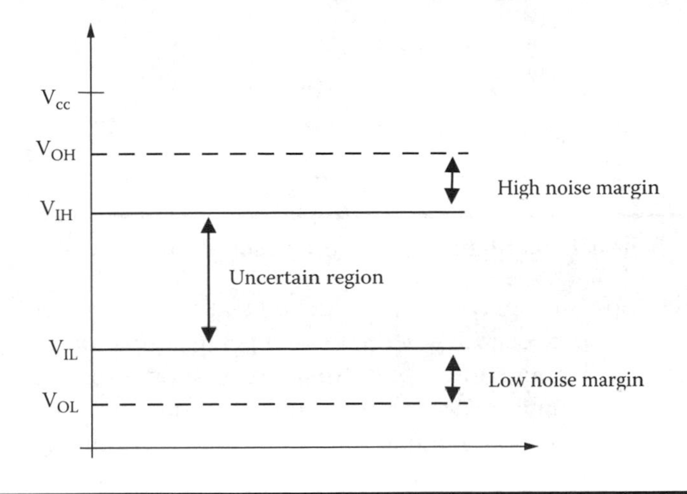

Figure 9.2 Typical noise margins.

noise margin recognizes that the shorter the pulse the higher its peak can be before it causes an upset.[9] Typical static values for two common logic families are given in Table 9.1.

Another factor affecting gate performance refers to delays. Figure 9.3 depicts a typical situation indicating how the driving gate pulse is related to the driven gate pulse. In synchronously clocked circuits it is important to control delays in gates and also interconnects. Gate propagation delays are several nanoseconds depending on logic family. Timing depends on the various thresholds, interconnect delays, and any additional delays that may be caused by reflections, ringing, etc. Referring to the ringing on the

Table 9.1 Typical Parameters for Establishing Noise Margins

	TTL Logic	*CMOS Logic*
V_{OH}	2.4 V	4.9 V
V_{OL}	0.4 V	0.1 V
V_{IH}	2 V	3.5 V
V_{IL}	0.8 V	1.5 V

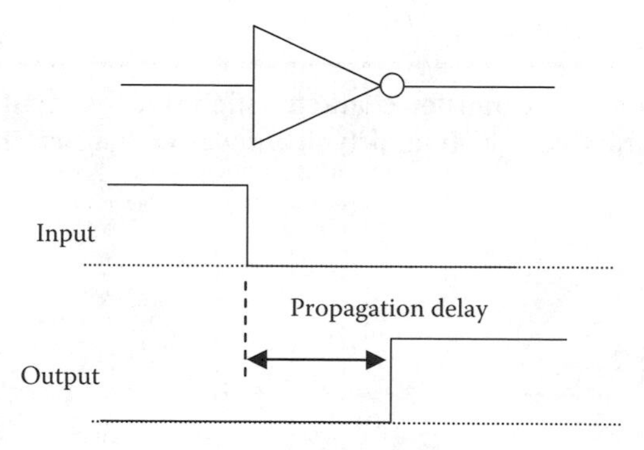

Figure 9.3 Schematic illustrating gate propagation delay.

output waveform in Figure 9.4, we see that at the points indicated by thick arrows the logic state is uncertain and glitches may occur. In Figure 9.5 a schematic of input and output transitions is shown for a hypothetical gate depicting the timing relationships, the gate propagation delay, and area where the signal and logic states are uncertain. Any pulse components due to interference that occur during signal transitions may modify timing relationships and logic states to the detriment of circuit operation.

The same design principles already mentioned in Section 9.1 apply with obvious additional considerations, i.e., the use of logic families with high noise margins and the use of defensive programming to achieve EMC by software design (see Section 12.3.1). High levels of interference likely to exceed the noise margins of the device are described as static failures. More modest levels of interference are not sufficient to breach the noise margin but nevertheless lead to failures through their effect on time delays (dynamic failures). Software error correction codes can to some extend address some of these issues. Testing digital circuits for immunity is very challenging due to the many parameters affecting failure. As an example, an interfering pulse may or may not cause interference depending on the

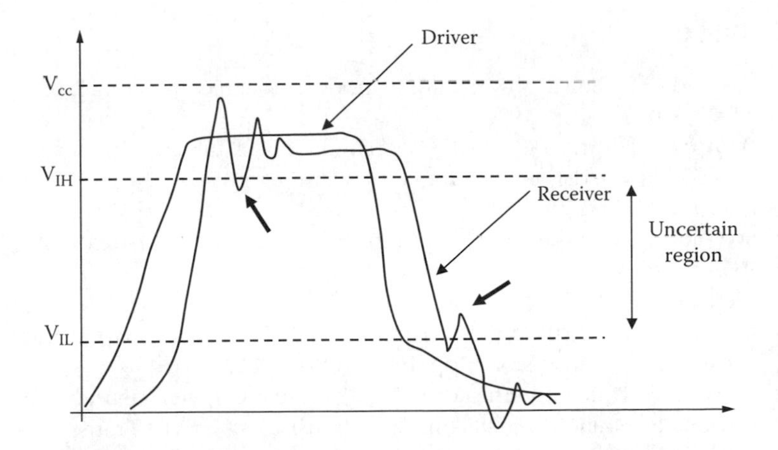

Figure 9.4 Driver and receiver signal relative to various noise thresholds.

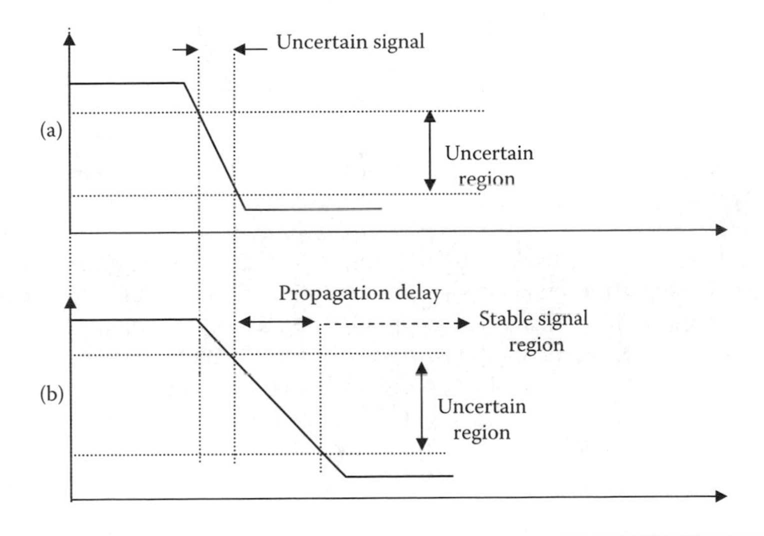

Figure 9.5 Timing diagram of a hypothetical gate showing the input (a) and output (b) states.

instant in time during the cycle that it is applied. Testing at a single frequency is not very helpful as nonlinear effects can produce a multitude of other frequency components. Digital systems appear to be more susceptible to broadband noise. Current standards and test procedures are rooted to the traditional need to protect radio systems and hence less suitable for digital technology. Some of these issues are aired in References 11 and 12.

References

1. Tasca, D M, "Pulse power failure modes in semiconductors," IEEE Trans, NS-17, pp 364–372, 1970.
2. Wunsch, D C and Bell, R R, "Determination of threshold failure levels of semiconductor diodes and transistors due to pulse voltages," IEEE Trans, NS-15, pp 244–259, 1968.
3. Standler, R B, "Protection of Electronic Circuits from Overvoltages," John Wiley, NY, 1989.
4. Alexander, D R, Enlow, E W, and Karaskiewicz, R J, "Statistical variations in failure thresholds of silicon npn transistors subjects to electrical overstress," IEEE Trans, NS-27, pp 1680–1687, 1980.
5. Alexander, D R and Enlow, E W, "Predicting lower bounds on failure power distribution of silicon npn transistors," IEEE Trans, NS-28, pp 4305–4310, 1981.
6. Scuka, V and Demoulln, B, "Effects of transients on equipment," Proc 10th EMC Zurich Symp, Supplement pp 93–100, 9–11 March 1993.
7. Pierre, J R, "Evaluation of methodologies for estimating vulnerability to EM pulse effects," Report published by the National Academic Press, Washington DC, 1984.
8. Graffi, S, Masetti, G, and Golzio D, "New macromodels and measurements for the analysis of EMI effects in 741 op-amp circuits," IEEE Trans, EMC-33, pp 25–34, 1991.
9. Williams, T, "EMC for Product Engineers," 3rd Edition, Newnes, 2001.
10. Sedra, A S and Smith, K C, "Microelectronic Circuits," 3rd Edition, Harcourt Brace Jovanovich College Publishers, 1982.
11. Townsend, D A, Pavlasek, T J F, and Segal, B N, "Breaking all the rules: Challenging the engineering and regulatory precepts of EMC," IEEE Int. Symp. on EMC, Atlanta, USA, pp 194–199, 1995.
12. Flintoft, I D, "Preliminary investigation into the methodology of assessing the direct RF susceptibility of digital hardware," Report prepared for the Radiocommunications Agency, UK, contract no AY3326, 1999, http://www.radio.gov.uk/topics/research/topics/emc/r99042/r99042.pdf.

INTERFERENCE CONTROL TECHNIQUES

In Parts I and II the concepts underlying electromagnetic compatibility and the main analytical and numerical predictive models were described. The intention has been to provide the reader with a feel for the issues relevant to EMC and with the necessary tools to predict the behavior of systems. It was repeatedly pointed out that typical systems are complex and that simplifications must be made to reduce problems to tractable proportions. This reduction should always be done on the basis of scientific understanding and engineering intuition and not simply in order to address the problem by using ready-made semi-empirical recipes. Notwithstanding the lofty nature of this statement, it has to be realized that continuing progress will be required to develop more sophisticated tools before a rigorous, self-consistent analysis of all EMC problems is possible using a modest investment in time and facilities.

An engineer confronted with designing a system to meet an EMC specification needs to have access to a number of remedies and control techniques that may be used to reduce interference or increase immunity. The analyses presented in the first two parts would have defined the electromagnetic environment and, hopefully, have identified areas or systems that must be specifically addressed from the EMC point of view. The aim of the next three chapters is to offer general guidance on practical measures that may be applied during EMC design, and suggestions as to what constitutes good EMC practice. The author is aware of the danger of trying to oversimplify matters so that the advice given sounds clear and authoritative. Confidence based on false premises can be counterproductive and such an attitude would be counter to the tone of this book. However, every attempt will be made to give reference, whenever possible, to more practical texts and publications, so that the reader is able to draw selectively on the experience of others who have faced similar problems.

The first design of a system that conforms to good practice can be used for further optimization by using an experimental or a simulation approach. A good first design will allow a straightforward application of these tools. Any changes that are required at this stage may then be simply implemented without a lengthy and expensive redesign of the entire system.

Chapter 10

Shielding and Grounding

Shielding is employed to reduce the coupling of undesired fields with equipment. Its proper design to meet this objective has to be considered in association with other issues, such as weight, appearance, and ease of access. The need for shielding must be examined after all other means for reducing coupling have been investigated. Total shielding is impossible without imposing unacceptable restrictions in the design of equipment. At low frequencies unacceptably thick shields may be required and the presence of wire penetrations, ventilation slots, and access points will compromise even the most carefully designed shield. Hence, as is common in design, a compromise must be sought to balance all the relevant technical, economic, and esthetic issues. The theoretical background to shielding was examined in Chapter 6. The objective of the sections that follow is to offer a more practical view to the technology of shielding and the associated problem of grounding.

10.1 Equipment Screening

The construction of equipment cabinets that afford a certain amount of shielding (or screening) from electromagnetic fields is considered in this section. The problem of screening cables is addressed in Section 10.2.

10.1.1 Practical Levels of Attenuation

The amount of attenuation required from a shield is strongly dependent on the type of equipment under consideration. In some systems, such as aircraft, a certain amount of shielding is afforded by the structure of the craft. Shielding in such cases may be in the range of a few tens of decibels. In most applications attenuation of about 50 dB is adequate, except in cases of sensitive measurement instruments, high-frequency transmitting equipment, and tracking and guidance systems, where attenuation in excess of 100 dB may be required. Clearly, the materials and construction practice used in each case reflects the type of specification that must be met. A simple wire-mesh made of galvanized steel with wire spacing of 10 mm has a shielding effectiveness ranging, approximately, from 80 dB at 1 MHz to 20 dB at 1 GHz. More often than not, the overall specification, which includes the effects of all other factors, such as cooling slots, small gaps in modular construction designs, instrument panels, wire penetrations, connectors, etc., must be considered. Further details of theoretical shielding effectiveness of shields made of various materials and thicknesses may be found in Reference 1.

10.1.2 Screening Materials

In many applications, equipment screens may be constructed using conducting materials such as steel, aluminum, and copper. With proper control of manufacture, in particular of joints between panels, a very high degree of shielding may be achieved. For example, an aluminum sheet with a thickness of 0.125 mm can offer a screening effectiveness to plane waves exceeding 100 dB at 1 MHz. There are, however, applications where for esthetic, cost, or weight reasons, fully metallic enclosures cannot be used. Widespread use is now being made of various plastics that by themselves offer no electromagnetic shielding. An approach to making the plastic partially conducting is to use a filler such as carbon fiber. The electrical conductivity of the resulting composite is higher the larger the ratio of the filler to the plastic material. In practice, shielding effectiveness of a few tens of decibels is possible. An alternative approach is to use a conductive coating on the plastic base material. This can be applied by painting, spraying, or any other suitable process. In practice only very thin conducting layers can be applied in this way and the low-frequency performance of such shields has to be considered carefully by comparing the skin depth for the conducting layer to its thickness. Materials used for coating depend on the manufacturer and the coating process and are typically nickel, copper, zinc, etc.

In every case, it is of paramount importance to ascertain the electrical effectiveness of joints made between different panels of the plastic material and of other attachments to them. This assessment must cover the entire lifetime of the equipment. Manufacturers quote the shielding effectiveness of a panel but this quantity is not a sure guide to the shielding effectiveness of a cabinet made by using such panels. In addition to the problem of joints already referred to, the presence of cavity resonances inside the partially conducting cabinet walls has now to be considered. At resonance frequencies, the electromagnetic environment inside the cabinet cannot be ascertained from the shielding effectiveness of panels alone and a full study along the lines indicated in Chapters 6 and 8 is necessary.

In certain applications, e.g., displays, there is a requirement for electromagnetic shielding, while at the same time visual transmission must not be significantly impaired so that the display remains readable. Various techniques are available to achieve this. One option is to enclose a thin wire mesh between two layers of plastic or glass. Another option is to deposit a transparent conductive coating. Whichever technology is used, a partial loss of light transmission (approximately 50%) can be expected, with shielding effectiveness at high frequencies claimed to be in excess of 50 dB. Data on the attenuation achieved by different materials used in construction may be found in Reference 2.

Some of the more recent material developments are likely to have a major impact on shielding. For this reason, a brief introduction to this exciting area is given below.

Developments in nanotechnology are enhancing material choices in EMC for better shielding. It is possible to construct metalodielectric multilayered structures that have the same thickness as traditional thin metal screens but with a much higher transmittance at optical frequencies. These so-called "transparent metals" are opaque at microwave frequencies but transparent at optical frequencies. The physics behind their optical transmission properties are rooted to photonic bandgap theory and beyond the scope of this book.[3–5] It is expected that in the future nanomaterials will find further application in high-performance applications.

Recent developments have led to a systematic study of bandgap structures not simply for photonic applications but also at much lower microwave frequencies. These are generally referred to as *electromagnetic bandgap (EBG)* structures.[6] The area most commonly exploited in EMC is the design of surfaces that present a high impedance over a band of frequencies. Thus, current flows may be controlled on such surfaces selectively at specific frequency bands. Current flow is paramount in shielding and coupling and hence the importance of these developments. It is worth clarifying how a conducting surface shields against incident fields. For a perfect smooth electrical conductor (PEC) the electric field tangential to the surface must be zero; hence, currents are induced that generate a reflected electric field that is out of phase (π radians) to the incident field. Thus, in total, the tangential electric field at the surface is zero. The field away from the PEC is the combination of the original field and the reflected component, which is phase shifted to meet the boundary conditions at the conductor surface. Conditions are analogous to those encountered by a wave on a transmission line encountering a short-circuit termination. An antenna placed parallel and very near a PEC does not radiate well due to the destructive interference between the incident and reflected waves. This is the reason why a ground plane is, in general, beneficial in reducing emissions provided it is very close to the radiating structure. A finite PEC introduces some complications due to radiation from its edges. Let us now examine how a high-impedance surface (HIS) behaves. Conditions are analogous to those encountered when a wave traveling on a TL encounters an open circuit — it is reflected, but the reflected component is in phase with the incident wave component. Thus, for a radiating element near an HIS, incident and reflected components are likely to interfere constructively. This is primarily of interest to antenna designers. We are primarily interested here on the suppression of current flows brought about by an HIS and how such a surface may be constructed. Consider the basic element shown in Figure 10.1 consisting of the top plate, bottom plate, and the post joining the two plates. All the components are perfectly conducting. To illustrate the principle of how an HIS works, imagine that a number of these cells are arranged in one dimension to form the structure shown schematically in Figure 10.2a. Compared to a flat smooth metal surface this arrangement sustains TM waves propagating at a much lower velocity and also TE modes. An equivalent circuit is

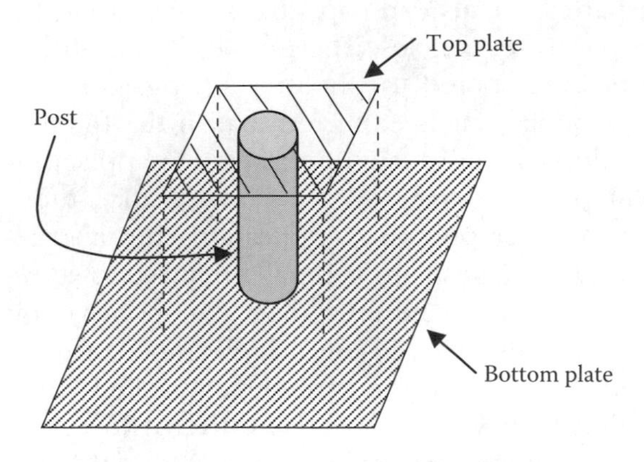

Figure 10.1 Basic element of a high-impedance surface (HIS).

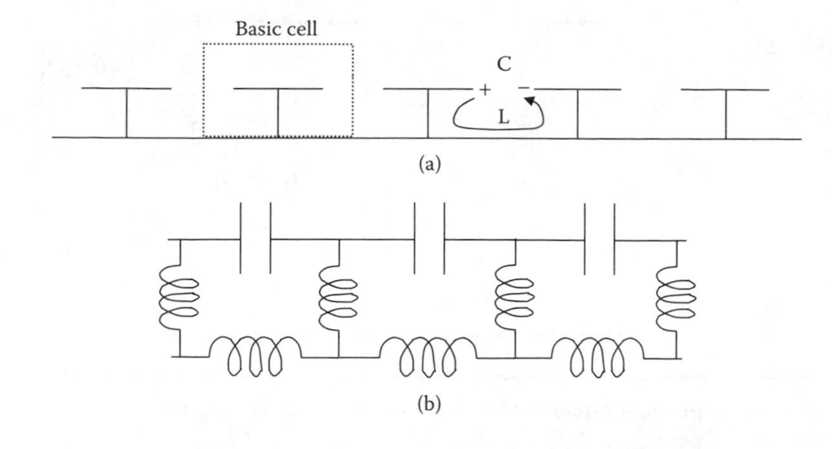

Figure 10.2 Schematic of a one-dimensional HIS (a), and its electrical equivalent (b).

shown in Figure 10.2b where the L and C account for the electric and magnetic field storage (potential differences and current flows in the HIS). Clearly, the parallel LC resonance gives a higher impedance at selected frequency. A practical way in which such a scheme can be implemented is shown in Figure 10.3.[7–10] An HIS is implemented in part (a) of this figure. A chip with a via connection is also shown in (b). High-frequency

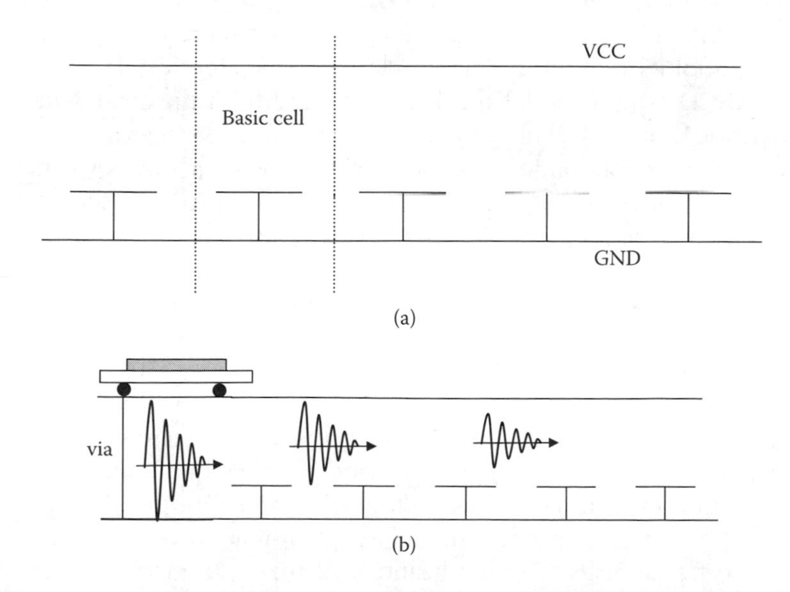

Figure 10.3 A parallel plate line with an HIS as reference (a), and with a chip and via added to illustrate the propagation of interference (b).

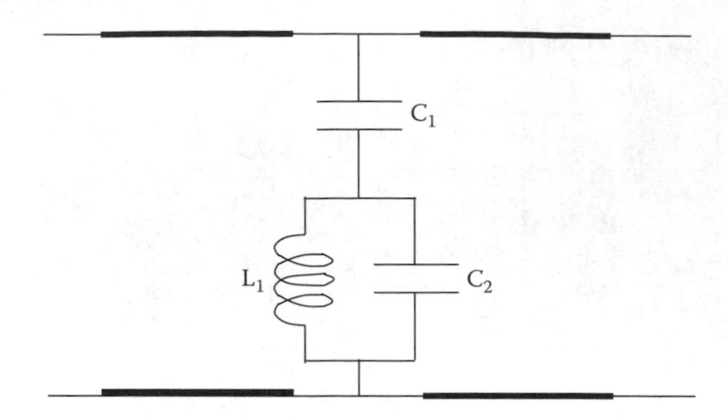

Figure 10.4 Electrical equivalent of the basic cell in Figure 10.3.

currents flowing through the via generate radiation that propagates between the two conducting planes. The presence of the HIS will suppress to some extent this radiation that otherwise would spread and couple to several other circuits through vias, common ground paths, etc. The dimension of the features of the HIS can be varied to achieve attenuation over the desired band of frequencies.

The basic cell in the configuration shown in Figure 10.3 is shown in electrical circuit equivalent form in Figure 10.4, where the parameters of the shunt branch may be obtained from the dimensions and material properties. Solution in the frequency domain is straightforward by obtaining the ABCD parameters of the basic cell and then the total parameters for a cascade of cells. Details may be found in the references already given.

We also mention ongoing general developments on new materials collectively referred to as "metamaterials," the word "meta" deriving from the Greek for "coming after." Conventional materials are used as found in nature, and as far as electrical properties are concerned, the available range of ε, μ is a limited one. The most obvious limitation is that both are constrained to be positive. Plotting the available materials on a ε-μ diagram will populate partially only one quadrant (μ, $\varepsilon > 0$). It is possible by the judicious use of loaded loops and dipoles incorporated into a background material to generate artificial media — metamaterials — with a far wider range of parameters covering all four quadrants of the ε-μ diagram. Although the beginnings of this work were at optical frequencies, substantial progress has been made in extending these ideas down to microwave frequencies. The implications of these developments for EMC and SI are likely to be substantial and new approaches to shielding and controlling coupling are likely to emerge in the years to come. The topic is an extensive one and cannot be covered here. The interested reader can find an introduction to the techniques involved in References 11 and 12.

10.1.3 Conducting Penetrations

It is often necessary to pass conductors through EM screens. These may be used to connect services (e.g., gas, water, compressed air), connect to the mains, or exchange signals.

Piping used to bring services can sometimes be electrically decoupled by placing insulating inserts on either side of the penetration. Earthed penetrations must be connected to the shield around the entire perimeter (360° joint) to minimize coupling between the EM environment external and internal to the shield. The electrical conductivity of the shield and the conducting path between the point of entry and the earthing connection to the shield must be checked for integrity and, if possible, a path through panel joints must be avoided. Mains cables must not penetrate the shield without adequate filtering, as explained in Chapter 11. The same attention to earthing must be paid as explained earlier. In ascertaining the effectiveness of earthing arrangements, it must be borne in mind that it is not only the DC resistance that matters. At high frequencies it is the impedance to earth that counts and, unfortunately, the magnitude of this quantity is difficult to establish. Similarly, signal cables must pass through the shield without, as far as possible, breaching its integrity. The screening of cables is discussed in Section 10.2. Whether filtering is a realistic option depends on the particular circumstances, such as the frequency spectrum of the EMI in relation to that of the signal. In many situations standard connectors are used to connect cables to equipment. It should be emphasized that the connector must be regarded as an important part of the cable shield system and must conform to the same EMC specification as other parts of the system.

10.1.4 Slits, Seams, and Gasketing

Any gaps in the shield allow the penetration of fields, as explained in Chapter 6. Accurate calculation of the shielding effectiveness under these circumstances is not an easy matter and resort must be made to numerical techniques as explained in Chapter 8. However, during an initial conceptual design of a system, it is necessary to estimate the likely deterioration in shielding due to various openings so that likely problem areas are identified and addressed in greater detail. The shielding effectiveness in the presence of a small hole of radius r has been estimated in Section 6.3.

$$SE = 20\log\left(\lambda/2\pi r\right), \text{ in dB}$$

For holes of other shapes the factor $2\pi r$ may be replaced by 2l where l is the largest dimension of the hole,[13] hence

$$SE = 20 \log(\lambda/2l) \tag{10.1}$$

This formula indicates that a slot with greatest dimension equal to 1/20 of the wavelength has a shielding effectiveness of 20 dB. At a frequency of 1 GHz this level of performance implies that 1 = 1.5 cm. The smaller the opening (hole, slot, gap, etc.), the better the shielding effectiveness. The shielding effectiveness is reduced as the number of openings N is increased and it scales as the square root of N. This can be easily confirmed by substituting in the formula for the effective aperture in Section 6.3 $A_e \simeq N \pi r^2$. Long thin apertures can be made less likely to contribute to poor shielding by replacing them with several smaller apertures. A rule of thumb often used in designs is to keep the largest dimension of all openings smaller than 1/20 of the smallest wavelength of interest. This can be done, depending on the physical requirements, by introducing conducting fingers, bolts, metal grills, conducting mesh, etc.

A seam is in many ways similar to a very narrow slot but where imperfect electrical contact is being made at various points along its length. Perfect contact can only be ensured by welding along the seam, but this is impractical in many cases. Riveting and bolting are reasonable alternatives provided their spacing is smaller than 1/20 of the wavelength. A plain seam exhibits shielding behavior that depends on the electrical quality of the contact between the two panels. On contact points a resistance is present that depends primarily on the condition of the surfaces. This resistance can be very high when the surfaces are painted, corroded, etc. Where there is no contact, a small capacitance is present between the panels. Hence, in electrical terms, the seam presents an impedance that is effectively the parallel combination of resistance and capacitance. Shielding properties will therefore be frequency dependent. A simple expedient that can reduce penetration due to seams is to increase the area of overlap between panels.

The same difficulties arise around doors, hatches, and other removable panels, where effectively a seam is formed. Improvements can be made by using finger strips made of beryllium copper to establish a more reliable electrical connection between panels. Another alternative is to use conductive gasket material. This can be made out of knitted wire mesh or conducting rubber. This material is made in various shapes and is compressed by appropriate means between adjacent panels. Details of mounting methods and properties of various gasketing materials may be found in Reference 14.

In all the techniques used to improve the quality of joints and to minimize penetration through slots, the guiding principle must be the minimization of any disturbance to the normal flow of current in the shield.

If normal current flow can be ascertained, the design and placement of slots, seams, etc., may be selected to adhere to this principle. It must always be borne in mind that whatever design choices are made, these must be appropriate not only during the initial use of the equipment as new, but also throughout its lifetime. This is particularly important for demountable panels, doors, etc., where corrosion and normal wear and tear must be taken into account at the design and maintenance stage.

10.1.5 Damping of Resonances

It has already been mentioned that equipment cabinets are effectively resonant cavities with, normally, a low Q-factor. Hence, everything that has been said so far must be modified at specific frequencies where the cabinet becomes resonant. Formulae are available to calculate the resonances of simple cavities, but for a loaded cavity resort must be made to numerical methods. It is possible to seek the damping of resonances by introducing damping material, so that emission is reduced at these critical frequencies. Radiation-absorbing material (RAM) is routinely used to damp resonances in screened rooms and newer materials such as ferrite tiles can also be used. It is possible to place a small amount of absorbing material in certain positions inside a screened room to achieve maximum dampings.[15] Similar methods may be used for damping selected resonances in equipment cabinets.

10.1.6 Measurement of Screening Effectiveness

Often, especially with coated materials, the "resistance per square" is given by manufacturers. However, by far the most useful quantity for EMC studies is the shielding effectiveness (SE). Several methods have been developed to test the SE of materials. An example is the ASTM D 4935 test used to characterize materials illuminated by a plane wave. It consists of an expanded coaxial line with the inner and outer conductors interrupted to insert the test specimen. Other tests have been developed to test the performance of the specimen to E- and H-fields. These are referred to as near-field measurements and are variations of a configuration where the specimen is used to partition the test cell into two volumes. A transmitting and receiving antenna are placed at a specified distance on either side of the specimen and measurements are taken with and without the specimen to establish the SE. Linear dipoles or loops are used depending on whether E- or H-field measurements are required. In all these tests care must be taken to check the repeatability of the measurements, especially for various conducting plastics where the establishment of good electrical contact with the test cell is difficult to ensure. Further details maybe found in References 16 and 17.

10.2 Cable Screening

Three aspects of cable screening are considered in this section, namely the transfer impedance, connectors, and earthing. These are all important in establishing EMC in a cable system. There are cases where cable screening is not required, either because emission is negligible or when the equipment has a high immunity threshold. Twisted-pair and ribbon cable are widely used in many applications without screening. Twisted-pair cable is quite effective in minimizing interference at low frequencies and may also be shielded to provide additional isolation. Ribbon cable is used extensively in information technology equipment. Screening may be used, if required, but in most cases some effort on improving ground returns is worthwhile. Placing the ribbon cable near a ground plane, providing an integral ground plane, or providing several ground returns in the ribbon cable to minimize ground loops are all matters to be considered in design.[18] In the most demanding applications, however, fully screened cables are used.

10.2.1 Cable Transfer Impedance

The transfer impedance is a measure of the voltage induced on the inside surface of a shield for a current flowing on the outside surface of the shield. For a general situation, a transfer impedance and also an admittance were defined in Chapter 6, the former characterizing coupling to currents and the latter to voltages. For shields connected to ground by low-impedance paths, or when the shield is far from grounded structures and is of good quality (high braid coverage) then the transfer admittance is small and coupling to voltages may be neglected. Only in cases where there is a high impedance from shield to ground, as in cases of a bad or broken connection, effects due to Y_t need to be considered. In most cases of practical design, attention therefore focuses on reducing the transfer impedance to acceptable values. Screens can be made of solid material, braid, foil, or various combinations. A shield made out of solid high-conductivity material offers excellent isolation and a transfer impedance that decreases with increasing frequency. Its use, however, is restricted by its rigidity, which makes it unsuitable for many applications. A braided shield offers a good alternative.

The proportion of the shield area consisting of conducting material, referred to as the coverage, should be high — in excess of 80% for good shielding. Penetration of fields through the small holes in the shield and distortion of current paths limit the effectiveness of braided shields and, more importantly, cause the transfer impedance to increase with increasing frequency. A single braided shield offers a transfer impedance, typically of a few milliohms per meter up to about a few megahertz. If a lower transfer impedance is required, two shields may be used offering values in the range of 10^{-4} Ω/m at 10 MHz. Using four braids offers further improvements. Ferromagnetic materials may be used as part of a shield to reduce Z_t at low frequencies. Theoretical results have been reported of cables with two numetal shields placed between three copper braid shields giving values of Z_t in the 10^{-6} Ω/m range.[19] It should be borne in mind that the exact value of Z_t depends critically on the quality of manufacture and manner of use, and for this reason it is difficult to predict theoretically the behavior of cables. A number of results for a range of commercially available cables together with methods for measuring Z_t may be found in Reference 20.

In electrically long cables (length ≥ wavelength), resonances along the transmission lines formed by successive shields complicate matters and it is then difficult to predict performance without extensive study. Lossy materials (various ferrites) may be used to damp these resonances. In many systems, a large number of wires with a single ground return wire are placed in close proximity to each other inside a shield (e.g., in telephony). In such cases it is difficult to predict performance as it is affected by crosstalk and also coupling through the shield. The study of the latter case, which is addressed in this section, presents many difficulties. A single circuit consisting of a wire-pair inside the shield will be subject to common-mode interference (signal induced between the wire-pair and the ground return) and differential-mode interference (signal induced between the two wires). Hence, it is necessary to establish, effectively, a transfer impedance both for differential- and common-mode signals. Apart from saying that common-mode signals are in general higher than differential-mode signals it is difficult to offer any other general advice due to the complexity and variability of practical systems.

10.2.2 Earthing of Cable Screens

The grounding or otherwise of cable screens affects significantly the EMC performance of cables. As a general rule cable screens must be earthed at both ends to maximize magnetic field shielding. This is particularly true at high frequencies and for electrically long cable where earthing the shield at several points is necessary. The so-called "ground loops" are present but at high frequencies these are best described as transmission lines presenting a range of impedances depending on frequency and cable length. This point is discussed further in the next section. At low frequencies and in cases where the shield also serves as signal return, the formation of a ground loop causes interference, which appears as noise in the signal path. Various ways are available to reduce this noise by breaking the ground loop using, for example, optocouplers or transformers. Better performance may be achieved by introducing a second shield separate from the signal return path. Another issue that affects these arrangements is whether both sides of a circuit (source and load ends) are connected to ground. A discussion of these matters in connection with low-frequency instrumentation circuits may be found in Reference 13 (page 89) and in Reference 21.

In cables with multiple shields it is common practice to ground both ends of all shields. Studies have shown that in this case the induced voltage is proportional to cable length. If the lines formed by intermediate shields are matched or left on open circuit, then induced voltages vary as length squared and cubed, respectively.[18] As stressed before, grounding of shields must be made using a 360° connection avoiding breaches in conductor continuity. Entry of cables through equipment screens requires good connection between the cable and the screen. Attention must be given to the mechanical arrangement necessary to establish and maintain a good electrical connection, a suitable solution being to clamp the cables between a conducting block and bracket. This ensures that good electrical connection is made to the cable screen. The block should be bolted to the equipment screen, making sure that a good electrical connection is established. Consideration must also be given to the integrity of the entire ground path to avoid interference with and from other circuits.

10.2.3 Cable Connectors

It makes limited sense to select expensive cables with low transfer impedance, if the connectors used at the cable ends fail to meet the same high standards. It should be borne in mind that the total transfer impedance is the sum of the transfer impedance of the cable plus the two connectors at its ends. A standard BNC connector has a Z_t value of the order of 2 mΩ, whereas for a typical cable $Z_t \simeq 1$ mΩ/m. N-type connectors have a value of Z_t typically less than 1 mΩ up to frequencies of several hundred megahertz. High-performance connectors suitable for high-specification environments, such as EMP connectors, have a value of Z_t, less than a tenth of a milliohm. The comparison between typical values of Z_t for cables and connectors becomes more unfavorable at high frequencies. The connector can thus be the weak link in the chain and must be selected carefully. In selecting good connectors attention must be paid to the manner of connection to the cable screen (360°), compatibility of materials to avoid corrosion, and the quality of all other connections. Improvisations, such as pigtail connections, must be avoided, since even a short length of wire introduces an inductance of several nanohenry and thus considerably increases Z_t at high frequencies. The temptation to believe that a low DC resistance is a sufficient indication of low Z_t at all frequencies must be resisted.

Another factor that affects cable design in general is the selection of wires or PCB tracks to carry certain signals, i.e., digital, analogue, and power. This choice is sometimes determined by the ease of connection to circuit boards and by connector design. EMC considerations must influence these decisions at an early stage in the design. Common-sense advice is to keep as far apart as possible wire pairs carrying signals likely to cause interference and those connecting high-sensitivity circuits. It may be convenient and cheap to have one wire as a ground return for several circuits, but this approach may result in large loops and hence the risk of interference. These matters must be considered and sufficient connections allowed for coping with the special requirements imposed by EMC.

10.3 Grounding

Grounding is one of the most complex issues to confront the designer. Part of the difficulty associated with this topic is that the purpose of grounding is not simply EMC compliance, but it is also there to ensure the safety of personnel and equipment. Hence, any change in grounding arrangements to satisfy EMC requirements is always conditional on meeting electrical safety regulations. These two aspects of grounding do not sit comfortably together and care must be taken to meet all requirements as far as possible. For the power engineer, grounding is all to do with safety of personnel, tripping of protective devices, blowing of fuses, etc. For the electronic engineer, it has to do with signal referencing. In most cases, these professionals have little understanding of each other's problems, creating a tension that makes grounding such an emotive issue. The nomenclature used reflects the different attitudes and practices, e.g., protective earth, common, reference, chassis, etc. To avoid confusion the term grounding is used here to refer to arrangements at the low-voltage side of equipment. The primary purpose of these arrangements may be electrical safety (protective grounding) or signal reference (reference ground).

Understanding and improving grounding requires knowledge and control of current flow. This in turn implies detailed knowledge of the complete current path. Notwithstanding the difficulties of identifying this path, it is unfortunately true that in normal design practice no attempt is made to specify in detail grounding arrangements. In many drawings the grounding connection is shown by an arrow or a chassis symbol without any further explanation of where exactly this connection is to be made. Hence, the first step in helping to establish good grounding practices is to ensure that ground connections are explicitly shown in every detail in the layout of circuits and equipment. The practice of studying circuits using one-line diagrams does not do anything toward better grounding. Only the study of the complete path of grounding currents can help identify potential problems and point to solutions. The objective must be to ensure that the flow of these currents does not establish high-potential differences interfering with the normal operation of circuits (proper layout and small impedance of grounding conductors), and that induced fields cause the minimum disturbance in adjacent circuits (proper layout, shielding). It follows that keeping current flows as localized as possible minimizes earth loops and hence coupling, as does proper layout of circuits likely to produce or be particularly sensitive to interference. Particular aspects of grounding are examined below for different systems.

10.3.1 Grounding in Large-Scale Systems

Grounding in large electrical installations presents many complications. The physical separation between various parts of the installation and the variety of tasks that grounding is intended to perform present conceptual, practical, and management problems. The general advice is to ground at a single point the protective earth connection of the electrical supply to the installation. It is true, however, that in some cases more than one supply feed is present, and that, in any case, grounding at a single point requires several and long grounding conductors. This is impractical and may also be technically unsatisfactory. The following example helps to illustrate the difficulty of grounding in a large system.[22]

In Figure 10.5 a grounding connection is shown schematically. Of interest to the designer is the impedance Z_{in} seen across AA′. The grounding conductor AB has a diameter d, length l, and is at an average height h above the earth or reference plane CA′. The impedance Z_{in} depends entirely on the frequency and the parameters of the arrangement shown in Figure 10.5. At low frequencies (e.g., mains power) it may be acceptable to neglect all other contributions except the resistance of the conductor R. This may be kept low by choosing a high-conductivity, large-diameter conductor, and ensuring that connection at point C is made using appropriate electrodes properly connected to the plane. As the frequency increases, it is no longer possible to neglect the inductance L associated with this arrangement, since at some stage it becomes the dominant part of the total impedance. A low DC resistance does not guarantee a low

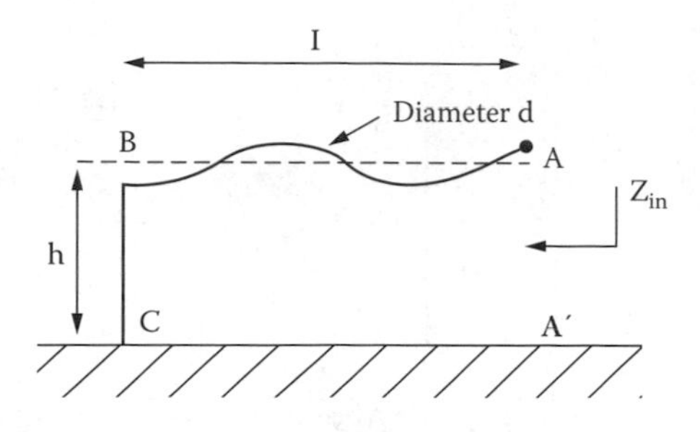

Figure 10.5 Schematic of a connection to ground.

impedance at high frequencies or for fast current pulses. As the frequency increases further, there comes a point when the length l is comparable to the wavelength. The impedance seen across AA′ is then obtained from transmission line theory and is

$$Z_{in} = j\, Z_0 \tan\left(2\pi l / \lambda\right) \tag{10.2}$$

where Z_0 is the characteristic impedance of the transmission line formed by conductor AB and the plane CA′. The important feature of this formula is that Z_{in} depends strongly on the ratio l/λ. For $l/\lambda = 0.25$, Z_{in} tends to infinity, suggesting very poor grounding arrangements. Although in various practical systems the values of parameters will be different, the main conclusion stands that Z_{in} varies very substantially from DC to frequencies at which the length of the conductor is comparable to the wavelength λ. A system in which $l \sim \lambda$ is called electrically large and may be found in large installations (large l but low frequency) or in a physically small circuit where the frequency is very high or the rise-time of pulses is very short. In such cases it is counterproductive to talk about or seek single-point earthing. The complexity indicated by Equation 10.2 and other phenomena not explicitly mentioned here, such as the conducting properties of the plane CA′, force the designer to a different way of thinking. Multiple earthing is then preferable, keeping the length of each conductor small, in the region between 1/10 and 1/20 of the wavelength. For large outdoor installations and buildings a ground grid is a useful means of providing a low-impedance path to which grounding connections are made.[23]

A further example, which illustrates the way the potential of a grounded object varies, is shown in Figure 10.6a. The structure consists of a steel tower 50 m in height, with a footing resistance $R_e = 10\ \Omega$. The latter quantity represents the resistance associated with the connection of the tower to earth. The problem considered here is the determination of the

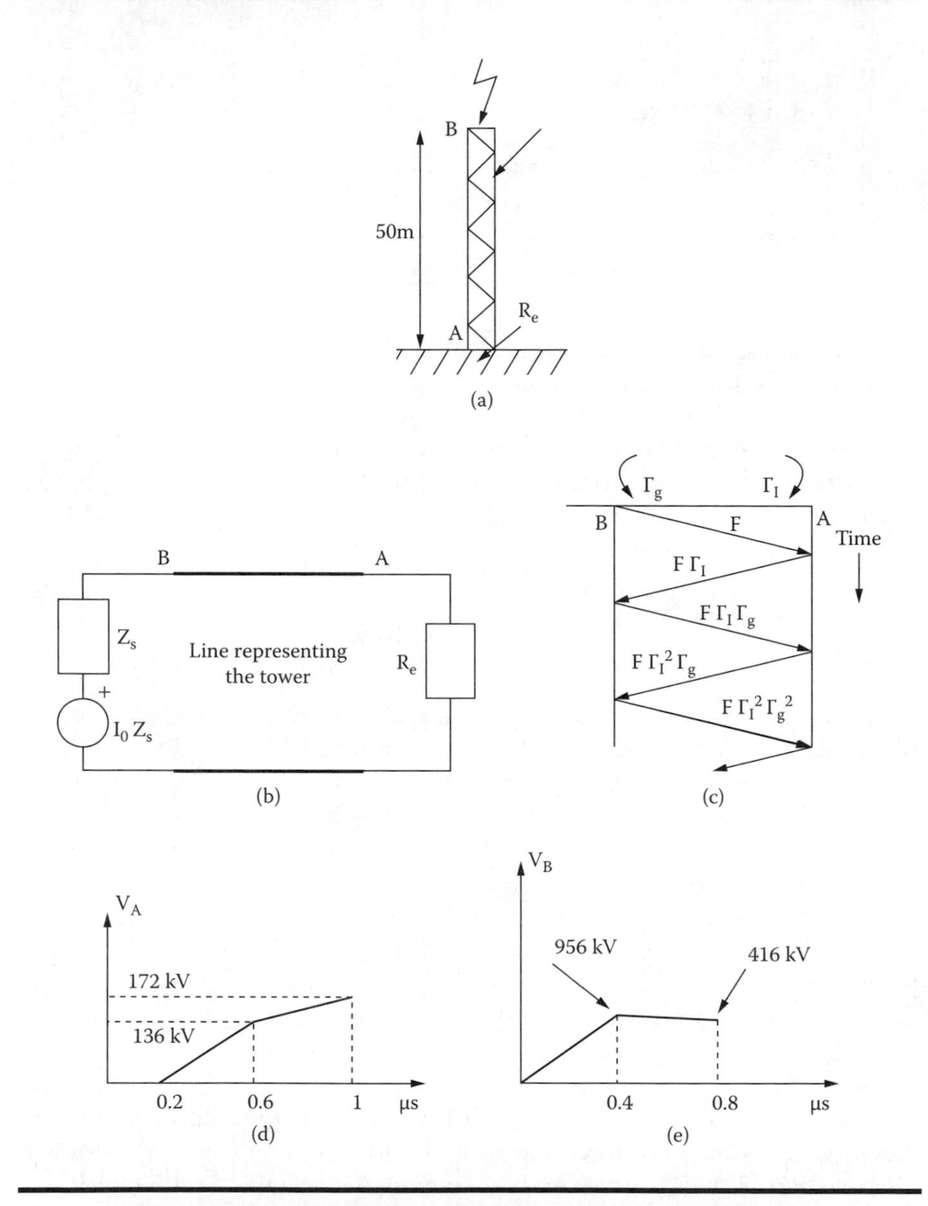

Figure 10.6 Schematic of a lightning strike on a tower (a), equivalent circuit (b), Bewley lattice (c), voltage at A (d), and at B (e).

potential at points A and B when a fast current pulse (e.g., lightning) is injected at point B. The tower may be regarded, approximately, as a transmission line of characteristic impedance $Z_T = 130\ \Omega$ and propagation velocity u = 240 m/μsec.[24] The line is driven at point B by a current source rising to 20 kA in 1 μsec and remaining constant thereafter. The electrical analogue of this problem is shown in Figure 10.6b. Although the exact value of all parameters used in this model is subject to uncertainties, the calculation nevertheless illustrates the basic behavior to be expected from such systems. The reflection coefficients at the load and source are

$$\Gamma_e = \frac{R_e - Z_T}{R_e + Z_T} = -0.857$$

$$\Gamma_g = \frac{Z_S - Z_T}{Z_S + Z_T} = 0.84$$

respectively, where the source impedance Z_s was chosen to be resistive and equal to 1500 Ω. The pattern of reflections is shown schematically in Figure 10.6c, by constructing the Bewley lattice (for details of this technique see section 15.2). The required voltages V_A and V_B are then shown in Figures 10.6d and e. Clearly, large voltages are induced at points that, on first examination, may be expected to be at a very low potential. The magnitude of these voltages is such as to risk a flashover to adjacent conductors that are at different potential, thus causing damage and further interference. In this example a very simple representation of the connection of the tower to earth was assumed (R_e = 10 Ω). In reality, the buried conductors at the foot of the tower constitute, at high frequencies, a transmission line and this introduces further complications to the model shown in Figure 10.6b. Connections to earth likely to experience high-frequency or impulsive currents, e.g., EMP installations, must be designed with these considerations in mind.[25,26]

It remains to stress that in extensive installations with several users, grounding principles must be established and adhered to over the lifetime of the installation and several changes in personnel, and that this requires that the grounding philosophy is explicitly stated and explained to all concerned.

10.3.2 Grounding in Self-Contained Equipment

Grounding in self-contained equipment where the only external connection is the mains supply presents the minimum number of complications. Protective earthing always adheres to the relevant electrical safety codes. Grounding inside the equipment is for referencing purposes and is designed to minimize impedance, eliminate as far as possible common grounding current paths between different circuits likely to interfere with each other, and to reduce coupling through radiated fields. The first requirement dictates that short conductors of adequate cross section are used. Whenever possible a ground plane is recommended to provide a low-impedance grounding connection. The typical arrangement is for the ground plane to be on the component side with through connections to signal tracks on the other side. This arrangement provides a well-defined transmission line between each track and the ground plane and also offers a certain amount of shielding. It is common to construct separate ground plane areas for analogue and digital signals that are joined at a point. This arrangement allows the option of adding further connections if it proves necessary, and controls the flow of current in the grounding plane so that the sharing of a common path between different circuits with different characteristics is minimized. Careful placement of components and circuits minimizes common paths and also the area of ground loops, especially for high-frequency or fast pulse circuits. In PCB construction, it is advisable to consider placing such circuits close to connectors (if fast communications are necessary) and, if it is not possible to provide an extensive ground plane, at least some form of grounding strip should be incorporated along a side of the PCB to provide a low-impedance grounding path. It is also customary to provide capacitors between supply rails as near as possible to circuits generating fast pulses so that high-frequency currents circulate in the smallest possible loop. Further advice as to how filtering should be used is given in the next chapter.

10.3.3 Grounding in an Environment of Interconnected Equipment

This situation is typically encountered in data processing rooms with several physically separate but interconnected units. Mains connection is to different wall-mounted sockets, sometimes supplied by different power phases, whereas these different units are interconnected via data cables. It is difficult in this environment to maintain any kind of control of grounding arrangements. It is recommended that ground potential variations between communicating devices should not exceed 0.25 V.[27] It is difficult to ensure this, especially if lightning currents are induced into the facility. The general grounding principles already discussed still apply, although changes to the working environment are necessary. Ideally, the equivalent of a ground plane should be established to provide a low-impedance path for grounding currents. Depending on the nature and extent of the installation, this plane could be the floor and or work bench and may involve already existing structural parts of the building, e.g., steel reinforcement bars. Equipment cabinets and any conducting penetrations, such as pipe work, should be bonded by the shortest possible route to the ground plane. Any external connections such as power and data cables should preferably be brought into a single distribution location and their respective ground connections bonded to the ground plane. Filtering and nonlinear protective devices may also be connected at this location to suppress external overvoltages and equalize ground potential fluctuations on various incoming connections. The objective must be to avoid making connections to equipment from outlets of an uncertain ground potential. Although difficulties may still arise, it is important that a well-thought-out grounding methodology exists and that it is documented and explained to all personnel likely to interfere with these arrangements. This applies not only to electrical contractors but also those responsible for other building services.

Further discussion of practical grounding arrangements may be found in the literature.[28,29]

References

1. Schulz, R B, Plantz, V C, and Brush, D R, "Shielding theory and practice,", IEEE Trans, EMC-30, pp 187–201, 1988.
2. Hemming, L H, "Architectural Electromagnetic Shielding Handbook — A Design and Specification Guide," IEEE Press, NY, 1992.
3. Scalora, M, Bloemer, M J, Manka, A S, Pethel, S D, Dowling, J P, and Bowden C M, "Transparent metallo-dielectric one-dimensional photonic bandgap structures," J. Appl. Phys. 83, pp 2377–2383, 1998.
4. Bloemer, M J and Scalora, M, "Transmissive properties of Ag/Mgf2 photonic bandgaps," Appl. Phys. Letts., 72(14), pp 1676–1678, 1998.
5. Sarto, M S, Sarto, F, Larciptere, M C, Scalora, M, D'Amore, M, Sibilia, C, and Bertolloti, M, "Nanotechnology of transparent metals for radio frequency electromagnetic shielding," IEEE Trans. on EMC, 45(4), pp 586–594, 2003.
6. Sievenpiper, D F, "High-Impedance Electromagnetic Surfaces," PhD Thesis, Dept. of Electrical Engineering, UCLA, CA, USA, 1999.
7. Kamgaing, T and Ramahi, O M, "A novel power plane with integrated simultaneous switching noise mitigation capability using high impedance surface," IEEE Microwave and Wireless Components Letters, 13 (1), pp 21–23, Jan. 2003.
8. Clavijo, S, Diaz, R E, and McKinzie, W E, "Design methodology for the Sievenpiper high-impedance surfaces: an artificial magnetic conductor for positive gain electrically small antennas," IEEE Trans. on AP, 51(10), pp 2678–2690, 2003.
9. Shahparnia, S and Ramahi, O M, "A simple and effective model for electromagnetic bandgap structures embedded in printed circuit boards," IEEE Microwave and Wireless Components Letters, 2005.
10. Mohajer-Iravani, B, Shahparnia, S, and Ramahi, O M, "Coupling reduction in enclosures and cavities using electromagnetic bandgap structures," IEEE Trans. on EMC, 48(2), pp 292–303, 2006.

11. Caloz, C and Itoh, T, "Electromagnetic Metamaterials-Transmission-Line Theory and Microwave Applications," Wiley, 2006.
12. Engheta, N and Ziolkowski, R W, editors "Metamaterials-Physics and Engineering Explorations," IEEE Press and Wiley-Interscience, 2006.
13. Ott, H W, "Noise Reduction Techniques in Electronic Systems," 2nd edition, John Wiley and Sons, NY, 1988.
14. Weston, D A, "Electromagnetic Compatibility — Principles and Applications," Marcel Dekker, NY, 1991.
15. Dawson, L and Marvin, A C, "Alternative methods of damping resonances in a screened room in the frequency range 30–200MHz," 6th Int Conf on EMC, York, IERE Publ No 81, pp 217–224, Sept 12–15 1988.
16. Catrysse, J, "Grounding, shielding and bonding," in AGARD Lecture Series 177 — Electromagnetic Interference and Electromagnetic Compatibility, pp 5.1–5.10, Neuilly sur Seine, France, 1991.
17. Wilson, P F, Mark M T, and Adams, J W, 'Techniques for measuring EM shielding effectiveness of materials: Part-I Far-field source," IEEE Trans, EMC-30, pp 239–249, 1988.
18. Williams, T, "EMC for Product Engineers," Newnes, Oxford, 1992.
19. Degauque, P and Hamelin, J (editors), "Electromagnetic Compatibility," Oxford University Press, Chapter 6, Oxford, 1993.
20. Benson, F A, Cudd, P A, and Tealby, J M, "Leakage from coaxial cables," IEE Proc — A 139, pp 285–303, 1992.
21. Morrison, R, "Grounding and Shielding Techniques in Instrumentation," 3rd edition, John Wiley and Sons, NY, 1986.
22. Christopoulos, C, "Electromagnetic Compatibility — Part 2, Design Principles," Power Engineering Journal, Vol 6, No 5, pp 239–247, 1992.
23. Simpson, J B K, Bensted, D J, Dawalibi, F, and Blix, E D, "Computer analysis of impedance effects in large grounding systems," IEEE Trans, IA-23, pp 490–497, 1987.
24. Diesendorf, W, "Insulation Coordinations in High-Voltage Electric Power Systems," Butterworth, London, 1974.
25. Verma, R and Mukhedkar, D, "Impulse impedance of buried ground wire," IEEE Trans, PAS-99, pp 2003–2007, 1980.
26. Post, C F, "Practical installation guidelines for EMC grounding," 20th Int Conf on Lightning Protection, Interlaken, Switzerland, pp 7.7p/1–7.7p/5, Sept 24–28, 1990.
27. St John, A N, "Grounds for signal referencing," IEEE Spectrum, pp 42–45, June 1992.
28. Lewis, W H, "Recommended power and signal grounding for control and computer room," IEEE Trans, IA-21, pp 1503–1516, 1985.
29. Goedbloed, J J, "Electromagnetic Compatibility," Prentice Hall, NY, Chapter 6, 1992.

Chapter 11

Filtering and Nonlinear Protective Devices

Proper design for EMC requires that every effort is made to reduce the generation of electromagnetic interference and to improve immunity by proper layout and shielding techniques. After all reasonable efforts have been expended to address these issues, it may still be necessary to take specific measures to influence undesired signal flows by incorporating special circuits in selected locations. Examples of such circuits are filters, isolators, clamps, etc. In this chapter the nature and application of these circuits is described. Filtering is interpreted in its widest sense, namely, as a general signal conditioner that is used to modify a particular attribute of a signal. Thus, in addition to conventional filters, components such as common-mode chokes, baluns, isolating transformers, opto-isolators, etc., will also be considered.

11.1 Power-Line Filters

Connection of equipment to a power supply is an almost universal feature and it is the source of both emission and susceptibility problems. It is therefore a common remedy to incorporate a power-line filter at the connection point of the power cable to the equipment cabinet. Filter design is well established in connection with communication circuits and many techniques exist to help design very sophisticated filters.[1] However, the design of EMC filters for power-line application presents special problems.

First, filters for EMC applications must work over a very wide frequency range extending to 1 GHz and beyond. This implies that all components used in filter design, including the physical layout of the filter and its connection to the equipment, must be fully characterized over the frequency range of interest. The so-called stray components associated with real components were discussed in Chapter 3 and are crucial in EMC filter design.

Second, filters on power lines may be subject to high-current pulses of considerable severity, causing, potentially, saturation of inductors wound on iron cores and deterioration of other components subject to high-energy pulses. These factors can cause reduced performance and long-term changes in filter behavior.

Third, components added on power-lines for EMC purposes must not interfere adversely with normal safety and power circuit protection arrangements.

Finally, the attenuation introduced by a filter depends on source- and load-side terminating impedances and in power circuits these two quantities, in contrast to the case for communication networks, are not normally known or entirely specified. This can lead to problems in predicting performance and specifying filters.

This list of special requirements for power-line filters is not meant to suggest that they are entirely different from other filters. It is simply meant to point out the care required in selecting a filter for a particular application. A typical filter consists of a combination of high-quality L and C components connected in such a manner as to reflect signals in a certain frequency range. The presence of an impedance discontinuity is therefore paramount in the functioning of such filters and it dictates their topology. For example, if the impedance at the source- and load-side is high, the configuration shown in Figure 11.1a must be chosen since it presents the maximum mismatch at high frequencies. In contrast, if the terminating

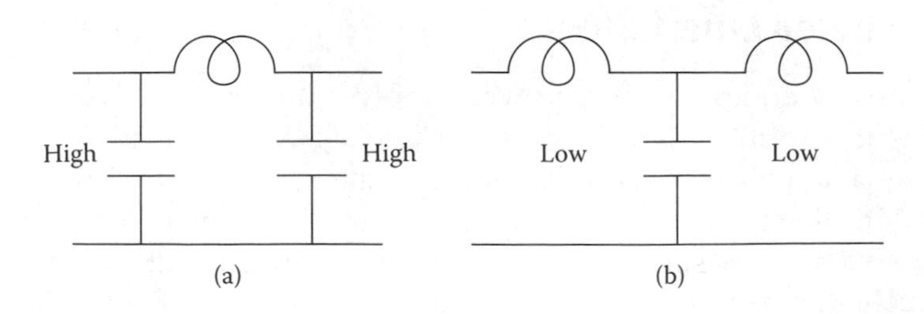

Figure 11.1 Optimum filter topology for (a) high- and (b) low-impedance networks.

impedances are low, the configuration shown in Figure 11.1b is chosen. Since in many cases these impedances are not known, design is a compromise and it is possible that the chosen filter may not in practice offer significant improvement. Another aspect of filters that must be apparent from this discussion is that interference signals are simply diverted by such filters — this energy is not dissipated — and hence care must be taken that this does not cause problems elsewhere. A filter design in which lossy components are added can help dissipate some of the energy associated with EMI and also provides some attenuation, irrespective of the value of the terminating impedance.[2]

A typical configuration used for filtering power signals is shown in Figure 11.2a. It consists of capacitors from line (L) and neutral (N) wires to the grounding (E) conductor (C_y), capacitor C_x between line and neutral, and the common-mode choke. The function of the capacitors is to present a low-impedance path for high-frequency currents. Capacitors C_y and C_x are intended for common- and differential-mode signals, respectively. Their selection must meet the required EMC specification and also safety regulations. Fairly high values for C_x are allowed (typically of the order of a microfarad) while C_y must not allow excessive current flow to ground at power frequencies. Depending on the type of equipment and safety requirements, total leakage current to ground must not exceed a milliampere or less. This requirement limits the value of C_y to a few thousand

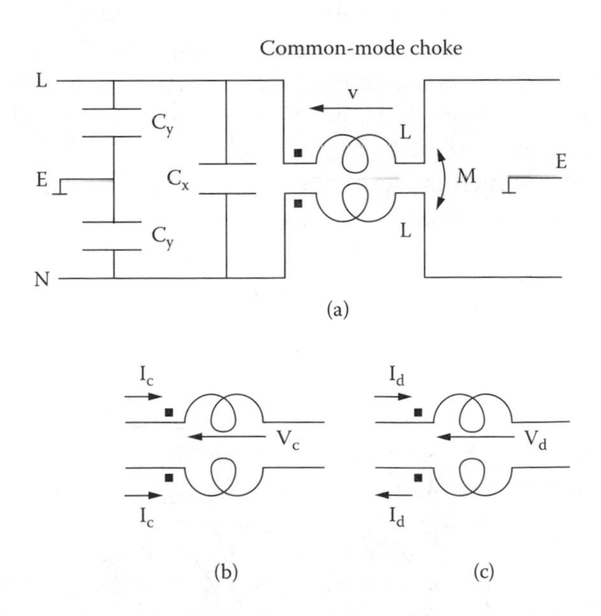

Figure 11.2 Mains filter topology (a) and flow of common- (b) and differential-mode (c) currents.

picofarad. High-quality components must be used for C_x and C_y since their failure may allow a follow-through current at power frequencies, which has implications for personnel and equipment safety. Of particular importance is the design of the common-mode choke. This consists of two identical coils wound on an iron core. Each coil has a self-inductance L and the mutual inductance between them is M. Since the two coils are closely coupled, the value of M is slightly smaller than L and for the ideal case of perfect coupling M = L. The purpose of the choke is to present a high impedance to common-mode current and a low impedance to differential-mode current. The voltage V_c across the top coil when common-mode currents flow, as shown in Figure 11.2b, is

$$\bar{V}_c = j\omega L\,\bar{I}_c + j\omega M\,\bar{I}_c \tag{11.1}$$

Hence, the impedance to common-mode currents is $j\omega(L + M)$ (high) and filtering of these currents is effectively due to C_y and L + M. Similarly, for differential-mode currents the voltage V_d across the top coil shown in Figure 11.2c is

$$\bar{V}_d = j\omega L\,\bar{I}_d - j\omega M\,\bar{I}_d \tag{11.2}$$

Hence, the impedance to differential-mode currents is $j\omega(L - M)$ (low) and filtering of these currents is effectively due to $2C_x$ and L – M. The power frequency currents are mainly differential mode and the flux they produce in the iron core is very small, thus minimizing saturation problems. This simple analysis neglects factors such as the parasitic capacitance across the choke, which are important at high frequencies. A discussion of these effects may be found in Reference 3. The design of the iron core is described in Reference 4. The basic configuration shown in Figure 11.2a may be modified to implement other filter topologies. As an example, additional capacitors may be incorporated on the other side of the choke to establish a filter similar to that shown in Figure 11.1a. Filters for three-phase power supplies may be designed following the same principles with a common-mode choke formed by winding the three lines and neutral on a common core. Filters where the required attenuation is larger than a few tens of decibels should be constructed inside metal enclosures and care must be taken to minimize stray components (short leads, minimization of capacitive coupling, individual shielded compartments for filter sections). Grounding connections must be checked for continuity and low impedance, especially when ready-made units with painted surfaces are installed. In systems with several supplies, such as switched-mode power supplies, it is probably best to provide some filtering as closely as possible to the source of interference in order to restrict the propagation paths of interfering signals and to allow scope for modular design. Ultimately, issues of cost, flexibility, and interchangeability determine the overall filtering strategy to be employed.

Example: Common-mode conducted emissions from a single-phase power electronic load are to be reduced using a common-mode choke. The Thevenin equivalent of the circuit driving common-mode currents at frequency f = 3 MHz consists of a voltage source V_{CM} in series with a stray capacitance C (100 pF). The CM voltage is measured across an LISN (section 14.2.5), which effectively presents a resistance of 50 Ω between live to ground and neutral to ground. The choke parameters are

$$C_x = 0.033\ \mu\text{F}, \quad C_y = 2200\ \text{pF}$$

$$L = 0.48\ \text{mH}, \quad M = 0.47\ \text{mH}$$

Solution: We repeat the CM choke circuit in Figure 11.3. At these frequencies the inductors on the live and neutral branches appear as in Figure 11.4a with the CM current as shown and the voltage V is

$$\begin{aligned} V &= L\frac{dI_{CM}}{dt} + M\frac{dI_{CM}}{dt} \\ &= (j\omega L + j\omega M)I_{CM} \\ &= j\omega(L + M)I_{CM} \end{aligned}$$

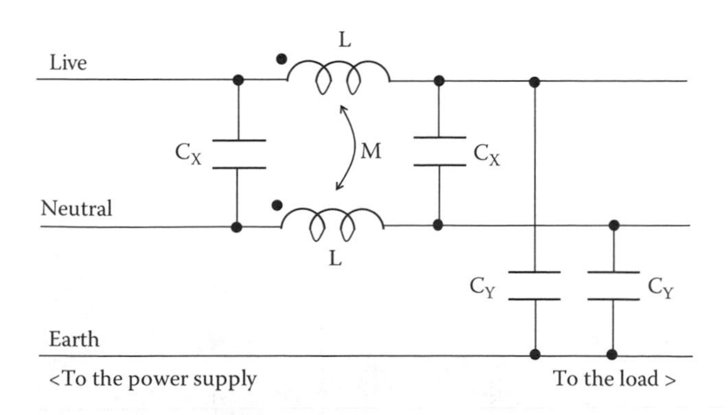

Figure 11.3 Schematic of a single-phase power line filter.

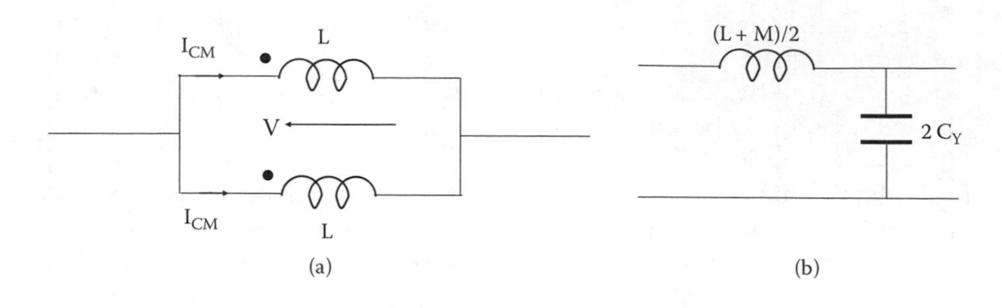

Figure 11.4 Common-mode current flow (a) and electrical equivalent (b).

Hence, the equivalent circuit to CM currents is as shown in Figure 11.4b. The complete circuit with the LISN and the driving source included is shown in Figure 11.5a. The required component impedances are

$$\frac{j\omega(L+M)}{2} = j8954$$

$$\frac{1}{j\omega 2C_Y} = -j12$$

$$\frac{1}{j\omega C_{stray}} = -j531$$

The two 50 Ω resistors of the LISN appear effectively in parallel, hence the complete circuit reduces to the one shown in Figure 11.5b. Reducing once more the circuit to its Thevenin equivalent, with the exception of the LISN impedance gives the circuit shown in Figure 11.5c. Hence the LISN voltage is

$$V_{LISN} = \frac{25}{\sqrt{25^2 + 8942^2}} 0.022 V_{CM}$$

$$= 61.5 \times 10^{-6} V_{CM}$$

11.2 Isolation

Isolators are used to prevent, as far as possible, the flow of common-mode currents while the flow of differential-mode currents remains unaffected. The common-mode choke described in the previous section provides

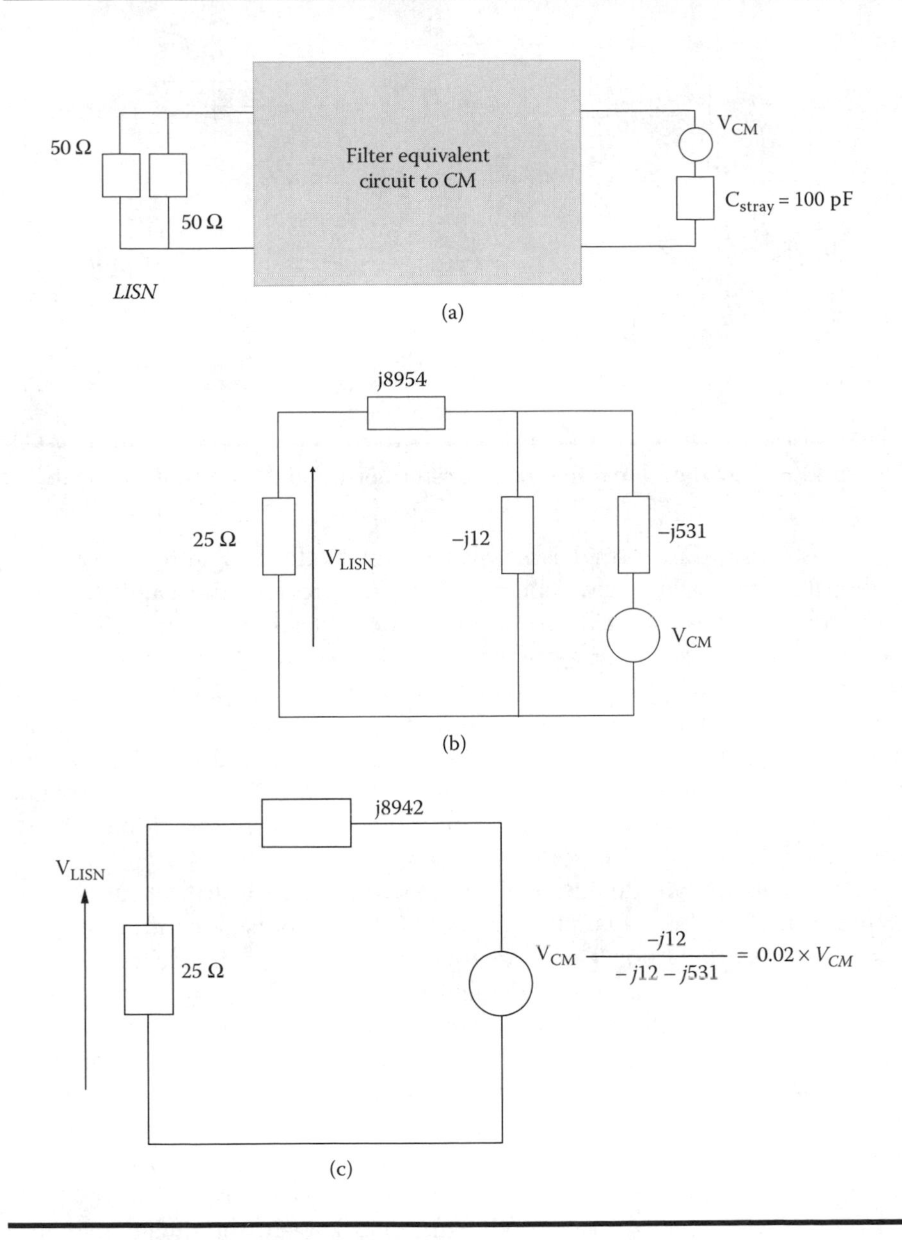

Figure 11.5 Complete system comprising of LISN, filter, and common-mode source (a), and subsequent circuit reduction (b) and (c).

a certain degree of isolation and can be simply used whenever necessary, by winding signal cables, including coaxial cables, on a magnetic core. Other isolating devices are also used extensively in signal conditioning. These are isolating transformers, opto-couplers, and fiber-optic links, and these are briefly described below.

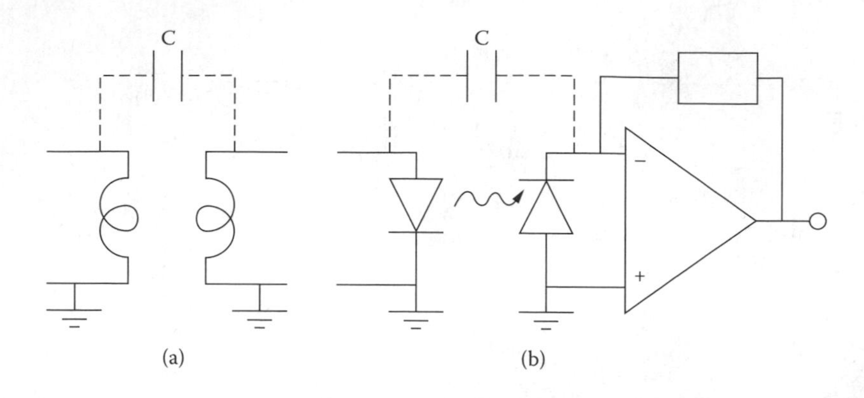

Figure 11.6 Isolation provided by a transformer (a) and an opto-isolator (b).

An isolating transformer is shown schematically in Figure 11.6a. Typically, the two coils are wound on an iron core and the capacitance C shown is the stray capacitance between the coils. Any voltage difference between the two grounding points shown will cause EMI to appear across the two coils, except if C is zero or the magnetic coupling between the two coils is perfect. These conclusions may be confirmed by using an analysis similar to that employed for the common-mode choke. Perfect isolation at high frequencies is difficult to achieve. The stray capacitance C may be reduced considerably by providing complete individual shields for the two coils and the core.

Opto-isolators are devices used to convert electrical to optical signals, which can then be transmitted without EMI. A typical configuration is shown in Figure 11.6b. It is clear that any stray capacitance C between the two circuits of the optical devices may cause the same problems as described earlier. Although these capacitances are small, in the fraction of a picofarad range, they may nevertheless contribute to loss of isolation.[5] Any other incidental wiring placed in the proximity of the opto-coupler may also contribute to C and thus the layout of the entire circuit must be carefully studied. The electrical circuit on both sides of the opto-coupler must also be designed with EMC considerations in mind, as any EMI introduced there may be converted to an optical signal and thus transmitted further.

Optical fiber links use an optical signal transmitted through a fiber and in principle are similar to opto-couplers, with the important extra feature that optical transmission over longer distances is possible. The same electrical considerations as regards EMC design apply, and it should also be pointed out that for both types of optical link it is only practical to transmit low-power-level signals.

11.3 Balancing

A two-conductor circuit is described as balanced if both conductors and all components connected to them have the same impedance with respect to ground and all other nearby conductors.[3] A balanced circuit tends to provide some immunity to common-mode currents and it is therefore desirable. A typical example of an unbalanced circuit is the antenna feed through a coaxial cable shown in Figure 11.7a. Although the dipole is perfectly symmetrical, the feed cable is strongly asymmetrical with respect to ground and other objects. Under these circumstances the normal distribution of currents in the coaxial cable (current on inner conductor returning along the inside surface of the outer conductor) is disturbed and an additional current flows on the outer surface of the outer conductor. The total current on the coaxial cable differs from zero by the current flowing on the outside surface (common-mode current). The cable therefore radiates and the two halves of the dipole do not carry symmetrical currents. The common-mode signal may be reduced by using techniques similar to those described in Section 11.1, namely, by winding part of the coaxial cable on a ferrite core, or by introducing ferrite clamps along the length of the cable. At high frequencies balancing of circuits such as the one shown in Figure 11.7a can be done by a special component known as a "balun" (BALanced UNbalanced), intended to connect a balanced system (dipole) to an unbalanced system (coaxial cable). An arrangement known as the

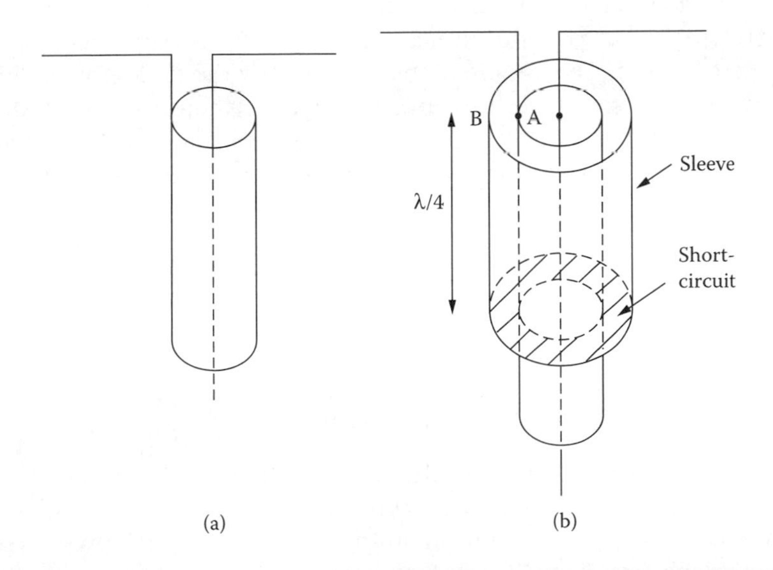

Figure 11.7 An unbalanced antenna feed (a) and an improved arrangement using a balun (b).

quarter-wavelength balun is shown in Figure 11.7b. The essence of this design is that a sleeve of length $\lambda/4$ is added as shown, thus forming a coaxial transmission line with the shield of the original coaxial cable. The input impedance to this line, as seen across AB, is that corresponding to a line $\lambda/4$ long with the far end shorted and it is therefore very large (Equation 3.30). Current flow is thus minimized on the outside coaxial line. This balun works well at a frequency having a quarter wavelength equal to the sleeve length. A number of other balun designs are available with improved characteristics to cope with various situations.[6]

11.4 Signal-Line Filters

Filtering on signal lines is a well-established technique and may be used to improve EMC when the spectrum of the EMI and the intentional signal are sufficiently separated. The design of filters is usually of the LC type and is well described in standard literature.[1] Automated and simplified design procedures are also available to ease calculations.[7] Signal-line filters operate over a very wide frequency range and care must be taken to place designs inside conducting enclosures, with separate compartments for each section, communicating through barriers by good-quality feed-through capacitors.

In addition to normal EMC requirements, it is necessary to provide additional isolation in certain communication systems to prevent unauthorized users from picking up signals emitted from badly screened and filtered equipment. This requirement is known by the codeword TEMPEST and although not strictly an EMC problem it nevertheless has considerable impact on EMC. TEMPEST requirements cover a very wide frequency range and thus demand high-quality shields and filtering of all power, control, and signal lines to reduce emissions to very low levels. In high-security military systems, very low electromagnetic emission must be maintained from all circuits and associated equipment carrying signals prior to encryption.

11.5 Nonlinear Protective Devices

Normal filtering arrangements as described in the previous sections cannot cope with severe electromagnetic threats, such as large overvoltages in the form of short pulses due to lightning, severe faults in power systems, or nuclear-generated EMP. In such cases additional components are required, intended to reduce the severity of the disturbance by clipping its amplitude, dissipating some of the energy, etc. These devices are invariably nonlinear so that under normal conditions signals are not affected. In contrast, to an abnormal pulse, the nonlinear component presents a high impedance if connected in series or a low impedance if it is a shunt connection. The simplest such component is a crowbar connected in parallel to the component to be protected, with the property that for voltages below a certain value V_0 the impedance of the crowbar is infinite, whereas when $V > V_0$ a current flows through the crowbar and the voltage across it is maintained at a value V_0 or smaller. In its simplest form, a crowbar may be a pair of sharp-pointed electrodes in close proximity to each other so that when $V > V_0$ breakdown takes place across the gap between the electrodes holding the potential difference to a low value corresponding to the arcing voltage. In practice, other components, such as nonlinear resistors, are connected in series to the gap to control the voltage-current characteristic during arcing. Such devices, especially those installed in power circuits, must be capable of self-healing, i.e., of interrupting the flow of power-frequency currents after the electromagnetic disturbance has ceased. In addition, it must be recognized that a sudden uncontrolled collapse of voltage due to the operation of such a device may generate its own interference, which, on occasion, may be more severe than that due to the original disturbance.

Thus, selection and application of protective devices must be done recognizing the problems that they may themselves generate. It is imperative that protective devices such as voltage clamps or crowbars are placed as closely as possible to the equipment to be protected. The rationale for this advice may be seen by examining the conditions on the system shown in Figure 11.8a. A pulse with a rate of rise of voltage equal to (a) travels along a transmission line O A B toward a device at B protected by a crowbar placed at A, a distance l from B, and set to operate at a voltage V_0. The propagation time from A to B is $\tau = l/u$ where u is the propagation velocity along the line. It is assumed for simplicity that the impedance of the device is much larger than the characteristic impedance of the line, so that pulses incident at B are reflected with the same sign. The total voltage at A is the sum of the original incident pulse ($dV/dt = a$) plus an identical pulse due to the reflection from B arriving after time 2τ. The crowbar operates after time $2\tau + t$ from the arrival of the pulse at A, as shown in Figure 11.8b, i.e., when

$$V_0 = a\left(2\tau + t\right) + at = 2a\left(\tau + t\right) \tag{11.3}$$

The firing of the protective device will be felt at B at time $(2\tau + t) + \tau$, when the pulse has been applied at B for a time interval $(2\tau + t) + \tau - \tau = 2\tau + t$.

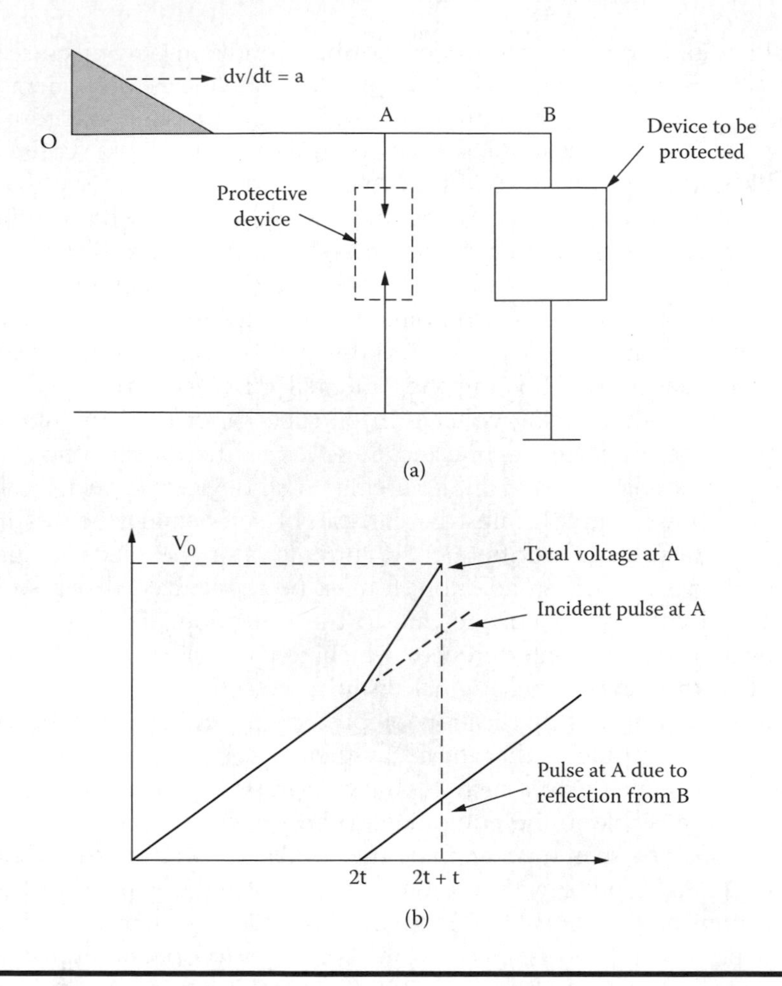

Figure 11.8 Circuit with a nonlinear protective device (a) and calculation of the voltage at A (b).

Hence the voltage at B across the device will have reached the value $V_B = 2a(2\tau + t)$ before the influence of the protective device was felt there. From this formula it is clear that V_B depends strongly on l since $\tau = l/u$. If, as an example, a = 100 V/nsec, l = 0.3 m, $u = 3 \times 10^8$ m/s, and $V_0 = 250$ V, the time interval t may be calculated from Equation 11.3 and is equal to 0.25 nsec. Hence

$$V_B = 2 \times 100(2 + 0.25) = 450 \text{ V}$$

This calculation shows that much higher voltages will develop across the protected device than those suggested by the firing voltage of the protective clamp. In practice there are additional delays in the operation of

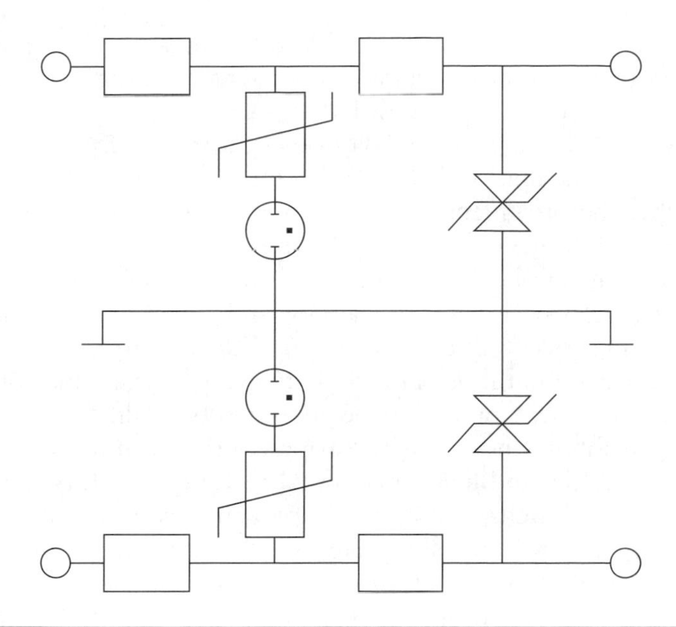

Figure 11.9 A voltage-clamping circuit for severe electromagnetic threats.

the protective device that have not been accounted for in this simple calculation.

A schematic diagram of an arrangement that can be used in association with the filter of Figure 11.2a is shown in Figure 11.9. The general philosophy of this circuit is that series components are used to slow down and dissipate some of the energy of fast pulses; nonlinear resistors (varistors) and gas discharge tubes are employed to limit overvoltages; and any remaining pulse energy is handled by solid-state devices (diodes), which although fast in operation cannot handle large amounts of energy. The remaining undesirable signals may then be dealt with by normal filtering.

A voltage crowbar cannot be used on its own to protect sensitive circuits and it is normally necessary to control the shape and level of the voltage across the protective device after firing. This is best done by a combination of components such as the gas discharge tube and nonlinear resistor shown in Figure 11.9. This combination is referred to as a voltage clamp in that a more controlled operation is achieved. All protective devices have an inherent operational delay and hence operation is dependent on the rate of rise of the disturbance. The advantage of spark gaps is that they can conduct large currents. They are, however, slow in operation. Advanced fast-operating designs have been produced that are suitable for EMP protection.[8] Solid-state devices such as thyristors may also be used as crowbar devices. Varistors (voltage-dependent resistors) of various kinds are available with fast operation and large current-handling capability and these can be used in isolation or in combination with other components for protection. The voltage current characteristic of these devices is of the form $I = kv^{\alpha}$ where α is of the order of 30.

Avalanche and zener diodes may be used to clip fast voltage pulses but it must be remembered that their current handling capability is limited. A comprehensive treatment of various nonlinear devices used for overvoltage protection may be found in Reference 9.

An issue of great practical significance, and which is difficult to address in a satisfactory way, is the survivability and deterioration of protective components in service. All components used to ensure EMC may at some stage be damaged but this is particularly relevant to front-line components such as the nonlinear protective devices described in this section. The difficulty is that it may not always be apparent that a particular component has failed in service and that it may need replacement. It is also the case that the performance characteristics of a particular component may change due to aging and/or the passage of several high-energy pulses through it. Although allowance may be made for aging and statistical variations, it is difficult to assess changes in characteristics due to voltage clamping operations since their frequency of occurrence is unknown. It is well known that the repeated passage of high-energy pulses through a varistor modifies its voltage-current characteristic. The general trend is for the voltage across the varistor at a fixed current to increase after exposure to large surges. For this reason some authors recommend that, in order to provide a safety margin, the value of α used in calculations for highly exposed circuits should be chosen to be equal to 4.[9] In any case, for a properly specified system some attempt must be made to assess and provide for aging and deterioration in component characteristics. The amount of protection necessary for each circuit is dependent on perceived risks, importance of maintaining operation, and costs.

The circuit shown in Figure 11.9 is suitable as the front end of a power-line filter. For signal lines used in control, instrumentation, and communications, various adaptations of this arrangement are used. As an illustration, arrangements suitable for protecting a balanced line and a general-purpose communication line are shown in Figure 11.10, a and b, respectively. In Figure 11.10a, the series components are typically resistors a few ohm in value and a three-electrode discharge tube is used as the front-line device. In Figure 11.10b component LA is typically a spark gap or low-voltage lightning arrester type to provide the first line of defence in hostile environments. The breakdown voltage of the diodes is selected to be above normal signal levels. A wide range of protective devices is produced commercially incorporating in a single package several components. Further details may be found in various technical publications and manufacturers' literature.[10,11]

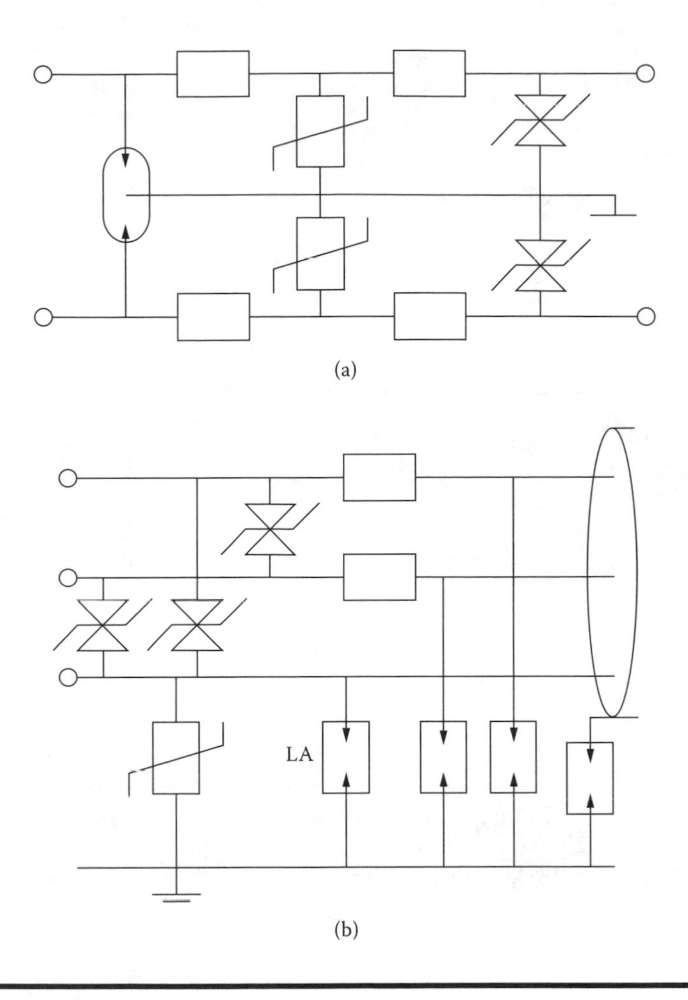

Figure 11.10 Protective arrangements for a balanced line (a) and a general-purpose communication line (b).

An interference problem with many similarities to EMC is the electrostatic discharge (ESD). Charge transfer takes place during contact and subsequent separation between different objects (triboelectricity).[3] When a person is charged up to high potential and subsequently comes into contact with a circuit or other object at a different potential, a rapid charge transfer takes place causing conducted and radiated interference. The basic feature of ESD is relatively low energy but very short rise-time. Normal protective measures for EMC, like shielding and grounding, offer some protection for the effects of ESD. Difficulties caused by handling circuit inputs by persons who are likely to be electrically charged may be overcome by connecting across the input to be protected capacitors,

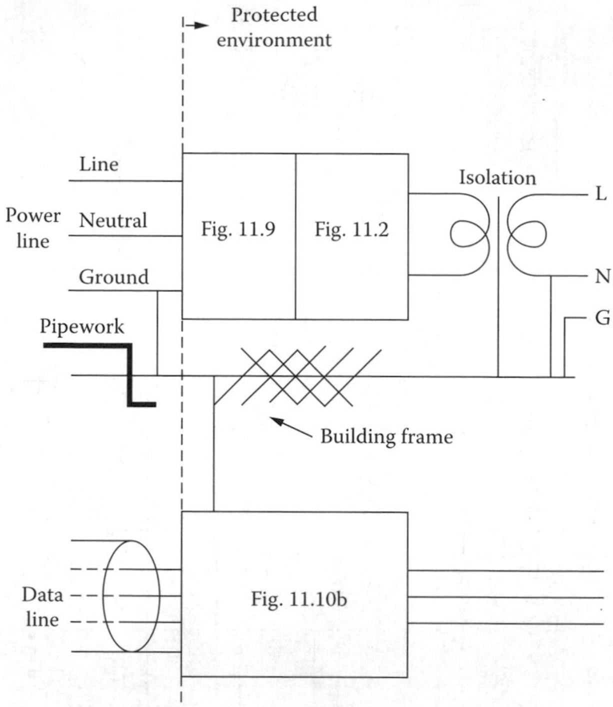

Figure 11.11 Schematic diagram of a general protective arrangement.

varistors, or avalanche diodes. Naturally, any measure that may be taken to prevent electric charging of persons or equipment in the first place (e.g., humidity control of rooms, conductivity control of working surfaces, and discharge paths through grounding straps) will help significantly to reduce ESD problems.

A schematic diagram of grounding, filtering, and protective arrangements that may be found in a typical installation is shown in Figure 11.11.

References

1. Lam, H Y-F, "Analog and Digital Filters: Design and Realization," Prentice-Hall, NY, 1979.
2. Broyde, F, "Dissipative filters for power electronics applications," Proc 7th Int Zurich EMC Symp, pp 393–398, 3–5 Mar 1987.
3. Ott, H W, "Noise Reduction Techniques in Electronic Systems," 2nd edition, John Wiley, NY, 1988.
4. Nave, M J, "On modelling the common mode inductor," EMC Symp, IEEE, Cherry Hill, NJ, 12–16 Aug 1991.

5. Goedbloed, J J, "Electromagnetic Compatibility," Prentice Hall, NY, 1992.
6. Burberry, R A, "VHF and UHF Antennas," Peter Peregrinus, London, 1992.
7. Wetherhold, E E, "Simplified LC filter design for the EMC engineer," Proc EMC Symp, IEEE, Wakefield, Mass, 20–22 Aug 1985.
8. Advanced gas-filled surge arresters — product catalogue, The M-O Valve Company Ltd., London, England, 1984.
9. Standler, R B, "Protection of Electronic Circuits from Overvoltages," John Wiley, NY, 1989.
10. Hasse, P, "Overvoltage Protection of Low-voltage Systems," Peter Peregrinus, London, 1992.
11. "Telecommunication Protection Circuits — Application and Data Manual," Texas Instruments, Bedford, England, 1992.

Chapter 12

General EMC Design Principles

In previous chapters, concepts underlying EMC and special techniques to characterize and control emissions, penetration, and coupling were described. Presentation of this material has inevitably resulted in the identification of good practices in EMC design. In this chapter it is not intended to repeat this material or to provide the reader with ready-made recipes for EMC design. The sole intention is to point designers in the right direction and help them establish a sound first design that can then be optimized to meet specific objectives. The opportunity will also be taken to address certain issues and techniques that did not fit comfortably with the theme of previous chapters.

Before embarking on more detailed aspects of design it is worth stressing from the very beginning the need to take a global, comprehensive, view of EMC. Design activity must be organized based on the understanding that all aspects, including mechanical and physical, have, potentially, an impact on EMC. Persons responsible for EMC must have access to and influence on all engineering activities — not simply electrical/electronic — from a very early stage and throughout the design and lifetime of a product. In order to impose some organization on this very broad activity, material in this chapter is presented under four subheadings dealing in turn with reduction of emissions at source, reduction of the number and efficiency of coupling paths, reduction of susceptibility, and, therefore, the construction of equipment that is more immune to EMC, and finally the management of EMC.

12.1 Reduction of Emissions at Source

Whenever possible, it always makes sense to endeavor to reduce the generation of interference at source. There are some electromagnetic threats, such as lightning, that are statistical events with a severity totally beyond the control of the designer. In contrast, interference generated within a system can be affected by design and a compromise can be reached that represents a balance between operational and EMC requirements. The task of reducing emissions at source must not be overlooked, as the majority of EMC problems are system generated and not externally induced. The single most important advice to reduce emissions in digital circuits is to limit the frequency and rise-time of signals to levels necessary to perform the intended function. The question must be asked: is such a high frequency necessary, or does the rise-time of a pulse need to be so short? Energy is conducted and radiated more effectively at high frequencies and the shorter the rise-time or the higher the repetition frequency, the more energy is present at high frequencies capable of causing undesirable interference. A high pulse repetition frequency is desirable in power electronic circuits to increase efficiency and reduce the weight of magnetic circuits. Against these benefits, the inevitable EMC problems must be weighed that lead to increases in costs and weight if shielding material needs to be provided. Results reported from a DC-to-DC converter design have shown interference from harmonics up to 490 MHz.[1] Results from the design of a pwm AC variable-speed motor drive are reported in Reference 2.

In digital clock circuits, emissions in excess of the 100th harmonic are not unknown. It is generally advisable to use the slowest logic family consistent with meeting operational objectives.

As mentioned before, arcing across electromechanical switches in inductive loads presents a rich source of interference and it is always worthwhile to provide some means of suppression. The general approach to such problems is to provide a combination of a capacitor and a flywheel diode in series with a resistor, connected in such a way as to provide an alternative path for the energy stored in the inductance at the instant of switching. Details of suitable designs may be found in specialist power electronics texts.[3] Other electromechanical components such as rotating machines, especially those with commutators and brushes, are likely to generate interference due to arcing at the brushes. The severity of interference is dependent on the eccentricity of the commutator and bearings, and the pressure on the brushes. Electrostatic discharges may also occur between moving parts (e.g., inner/outer cases of ball bearings). Although careful maintenance will minimize such problems, it is advisable to consider brushless motors in demanding applications.

12.2 Reduction of Coupling Paths

Under this broad heading, problems associated with the choice of operating frequency and pulse rise-times, reflection and matching, ground paths, component placement and segregation, and cable routing in general will be addressed.

12.2.1 Operating Frequency and Rise-Time

In choosing the operating frequency of circuits, operational requirements are predominant. From the EMC point of view, however, the operating frequency and, more generally, the rate of change of voltage and current have serious implications for coupling and shielding. Coupling can take place through shared conducting paths, capacitive and inductive near-field effects, and radiation.

Even the most carefully designed circuit will have some shared path with some inductance L and hence coupling through terms such as L di/dt will cause interference, especially at high rates of change of current. Elimination of shared paths and minimization of their inductance is therefore increasingly important as di/dt rises. Capacitive and inductive coupling are represented by terms of the type C dv/dt and dΦ/dt, respectively. High rates of change make this type of coupling more important and call for careful design to minimize such coupling. Radiative coupling becomes important when the physical length of connections becomes comparable to the wavelength. An estimate of the wavelength of radiation corresponding to a pulse of rise-time τ_r is $\lambda = u\tau_r$ where u is the speed of propagation (typically 2/3 or the speed of light for propagation in a PCB). Higher frequencies and rates of rise thus call for more compact designs. Small holes and gaps in shielding also become efficient paths for penetration of radiation when their largest physical dimension becomes comparable to the wavelength. It should be remembered that in the case of pulse waveforms typical of electronic power supplies and digital circuits, it is not only the fundamental frequency that matters but also its harmonics. In the case of power supplies, harmonics up to several megahertz should be expected, whereas for electronic circuits, depending on the clock frequency, harmonics into the gigahertz (i.e., 30th harmonic and beyond) range are possible. It must therefore be apparent that the designer should choose the lowest possible operating frequency and longest pulse rise-time consistent with meeting operational requirements. Similarly, the smallest possible change in voltage and current during switching reduces coupling effects and must be sought, provided that it does not compromise other operational requirements or circuit immunity to externally generated

Table 12.1 Typical Characteristics of Logic Families

	Voltage Swing V	*Rise-Time (τ_r) nsec*	*Bandwidth $1/(\pi\tau_r)$ MHz*	*DC Noise Margin mV*
CMOS	5 or 15	90–50	3.5–6	1000–4500
TTL	3.4	10	30	400
TTL (LS)	3.4	10	30	300
TTL (Schottky)	3.4	3	100	300
ECL	0.8	0.75–2	425–160	125–100

interference. The characteristics of some common logic families are listed in Table 12.1 and can be used to assess their suitability for different applications.

In systems where several frequencies are employed, it is a good practice to select these frequencies so that their values and those of their harmonics are sufficiently different from each other to avoid additive interference effects at certain frequencies. In practice, this implies that across the entire frequency range of interest, contributions from individual radiators must not be closer than the bandwidth of any likely victim system or test receiver. Particular care must be taken in the design of circuits operating with fast clock frequencies. The absence of high frequencies does not, however, guarantee EMC. In fact, very-low-frequency magnetic fields are very difficult to shield and hence a comprehensive look across the entire frequency spectrum must be taken.

A measure used to alleviate some of the problems caused by fast pulses is the placement of bypass or decoupling capacitors. The pulsed nature of the current drawn from power supplies during fast logic circuit transitions can cause large inductive voltage drops on supply rails and hence shared path interference, and also radiation if these essentially high-frequency currents are allowed to flow on substantial track lengths. A capacitor placed between the supply rails, at the nearest possible location to the component generating fast pulses, provides a reservoir of energy to cope with sudden current demands and a low-impedance path to high-frequency currents that do not now have to circulate around large areas of the board. It is imperative that the inherent inductance of these capacitors is very small and that the inductance of capacitor leads and any other connections is small so that the resonance frequency of each component and its connections is above the highest frequency in which substantial pulse energy is present. As this frequency may be in the gigahertz range, high-frequency capacitors, preferably in a surface-mounted package, connected as closely as possible to the component generating fast pulses, should be used. Such decoupling capacitors should certainly be placed near fast components and also elsewhere to avoid the spread of high-frequency current around the entire circuit.

The required capacitance may be estimated from the expression i = C dv/dt where i is the current required during the switching of the gate (typically tens of milliamperes), dt is the duration of the transition (typically several nanoseconds), and dv is the change in voltage that can be tolerated (typically 100 mV). In practice values of capacitance in the range 500 to 3000 pF are used. There is no merit in using too large values of a capacitance as it inevitably decreases the resonance frequency of the decoupling arrangement. Whenever possible, simultaneous switching of several gates should be avoided since this increases the peak high-frequency current demanded from supplies and thus impacts on EMC.

12.2.2 Reflections and Matching

Whenever the rise-time of a pulse τ_r is comparable to the transit time τ between two ends of a line, reflections and so-called transmission-line effects should be taken into account during design.[4] Depending on the geometry and material properties of the PCB the transit time is typically 2 n sec for propagation along a line 30 cm long. Transmission line effects are thus important for all logic families and especially those like ECL with very fast transitions. Reflections at line terminations may cause incorrect switching if they exceed the noise margin and also may cause interference through radiation. Reflections may be avoided if the transmission lines are matched, i.e., if they are terminated at the load and source ends by their own characteristic impedance. This implies that all transmission lines, including those using tracks on a PCB, are constructed in such a way that they have a well-defined characteristic impedance that is the same as the input and output impedance of circuits connected at their load and source ends, respectively. Microstrip-line-type construction, as used in high-frequency applications, should be used in circuit boards containing fast logic. The output and input impedance of logic gates vary widely between different families and also during transitions, typical values being of the order of tens of ohm and hundreds of ohm, respectively. Adding components across the load and in series with the source may adjust these values for better matching. The characteristic impedance of lines used in printed circuit boards is in the range 50 to 100 Ω for ECL and TTL logic gates, respectively. Some assessment of transmission-line effects should be made whenever $\tau_r < 10\ \tau$, i.e., for fast logic and/or long transmission lines. It is in general a good practice to terminate and if possible match all lines and to connect unused inputs of gates to the supply or ground as appropriate.

12.2.3 Ground Paths and Ground Planes

The single most important advice regarding ground paths is to recognize that, just as the details of connections at the high-potential part of signal lines are important to the operation of the circuit, so is the exact configuration of ground paths crucial to EMC. No discussion of any significance regarding ground effects on EMC can take place if all that the designer has to consider is a circuit drawing showing connections to "ground" in a schematic form. It is the responsibility of the designer to identify the complete current loop and thus to identify shared paths between different circuits and the scope for radiative coupling. Automatic PCB-routing CAD packages do not normally account for such effects. Shared paths between different circuits should be avoided, if possible, by proper design and placement. This is particularly important for sensitive circuits susceptible to interference and circuits that are known to generate substantial high-frequency currents. If complete elimination of shared paths is not possible, then the minimization of the impedance (not only the resistance) of these paths should be sought. This inevitably leads to the use of a substantial grounding structure in the form of wide strips, wire grids, or a complete ground plane, made out of copper, to form the ground connection. Whenever a complete ground plane is not necessary for technical and cost reasons and wire tracks are used for grounding, the temptation to economize on pin connections by using a single track as ground should be resisted, as it inevitably leads to shared paths and large-area ground loops that cause interference to spread and couple to other systems.

The presence of the ground plane makes a significant reduction to emission. This can be simply understood if a system consisting of a two-wire line above a ground plane is considered. The common-mode current I_c flowing in the two conductor line is, approximately, electromagnetically equivalent to a current I_c flowing on a single conductor above the ground plane. This system is in turn equivalent to the original conductor (current I_c) and its image conductor (current $-I_c$). Provided that the height above the ground plane is small, the emitted field is substantially reduced since the two opposing currents tend to cancel each other. A restricted ground plane is also effective in reducing emissions provided its width is larger than a tenth of the wavelength and the transmission line is not too near its edge.[5] Ground planes also have a partial shielding effect that contributes to higher immunity. As mentioned before, the long-term integrity of ground connections must be ensured by design and appropriate maintenance and inspection procedures.

12.2.4 Circuit Segregation and Placement

EMC design involves control of all electronic subcircuits, power supplies, interconnections, electromechanical components, and constructional details of a product. Only a broad EMC overview will ensure that the procedures adopted are consistent with the requirements of EMC. Conflicts between physical constructional details, which affect mechanical strength, appearance, and cost, and those regarding shielding, which affect EMC, should be resolved early on in the design, together with basic input/output and grounding arrangements. In regard to the manner of electronic construction, modularity is the norm. However, the minimum number of printed circuit boards should be used to avoid long connections between different boards. Careful thought must be given to external connections to a PCB. These should be the minimum number possible, should transmit the slowest signals possible, and pass the minimum amount of current necessary. These general principles imply that using the same clock between two boards is undesirable since a very fast signal must not be allowed to propagate long distances. Similarly, transmitting a signal between boards to drive a large number of gates is undesirable since this normally requires large currents. Some form of buffering can substantially reduce the magnitude of these currents. If there are unavoidable external connections, these should be shielded to reduce interference. The significance of the ground plane has already been emphasized.

In designs that contain analogue, digital, and input/output circuits on the same board it is advisable to provide separate ground planes so that control of shared paths is easier to exercise. These ground planes can be connected together at a location chosen to minimize interference. Grounded tracks may also be used between signal tracks to reduce coupling. The actual location of subcircuits in a board is a matter of some complexity, as it is affected by the general design philosophy and function of the circuit. In a well-partitioned system with identifiable interfaces, it is much easier to control interference. Circuits that generate fast signals (e.g., clocks, read/write) must be designed and placed in such a way as to minimize the length of connections. If fast signals are passed through connectors to other boards it is inevitable that the relevant circuits are placed near the connectors. However, it is best if fast signals are kept in the same board, in which case the relevant circuits should be placed centrally on the board with increasingly slower circuits constructed around the center. It is normally necessary for the designer to determine manually the route of critical circuits and grounding, before autorouting CAD packages are allowed to do the rest. Attention must be paid to connections adjacent to pins carrying fast signals. Some provision for placing suppressors (ferrites, capacitors) to the affected pins should be made at the design stage in case it is found to be necessary after testing the entire design. Circuits with different characteristics should be kept separate and the temptation to economize by using gates on the same package to service different circuits should be resisted. In high-performance boards, surface-mounted components and packages should be used to reduce emission and coupling from long wire connections. It should also be borne in mind that the EMC performance of nominally equivalent packages from different manufacturers can be quite different and therefore the design must have enough margin to cope with expected variability.

Advanced designs use multilayer board technology to increase component density and improve EMC performance.[6,7] A typical arrangement is to use a sandwich of different layers containing circuits-dielectric-voltage, supply plane-dielectric-ground, and plane-dielectric-circuits. In other designs a layer containing circuits is placed between dielectric layers and the voltage supply and ground planes. This arrangement considerably improves emission and immunity. The solution to be adopted depends on operational, EMC, and cost constraints and can only be arrived at after careful examination of alternatives.

Some basic PCB design guidelines for a better layout to improve EMC can be found in Reference 8.

12.2.5 Cable Routing

Cable positioning must be specified during design to avoid situations where cables are forced to occupy convenient gaps without regard to potential EMC problems. The basic principle of avoiding the transmission of fast signals over long distances still applies. Every effort must be made to reduce the necessary bandwidth to the lowest possible value. Cables must be classified according to their potential for causing or receiving interference signals. They must then be placed sufficiently apart from each other (several wire diameters) or in separate trunking to reduce coupling. Filtered and unfiltered portions of cables must not be placed close to each other or in the same trunking to avoid direct coupling and by-passing of the filters. In general, cables and trunking must be regarded as extensions of the equipment and the same principles regarding shielding and grounding must be applied.

12.3 Improvements in Immunity

A system operates satisfactorily if by careful design the level of interference (interference control level) is less than the level likely to cause an upset (susceptibility design level) by a safety margin as shown in Figure 12.1.

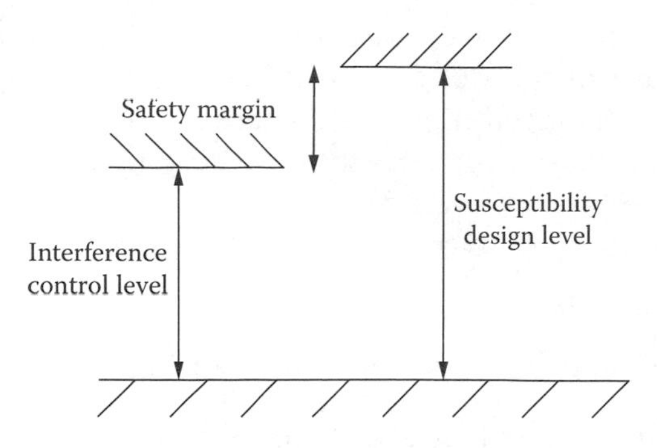

Figure 12.1 Safety margin in EMC design.

Naturally, establishing these levels and an adequate margin is a difficult task that can only be accomplished by a combination of experimentation, simulation, and design experience.

Many of the measures discussed earlier in connection with emission have, on the whole, a beneficial impact on immunity. It does not follow, however, that equipment with low emission is also immune to external interference. If, for example, low emission has been achieved by good shielding, it may follow that immunity will also be high. If, in contrast, low emission is the result of using low threshold logic, this does not necessarily offer any guarantee that immunity will be high. In achieving immunity from externally generated transients, proper grounding, filtering and nonlinear devices should be used at interfaces. The bandwidth of such circuits should be reduced to the minimum required by adding, if need be, simple R-C filters. The degree of immunity required depends on the application. Safety-critical and military systems are designed to a very high standard and are hardened to cope with very hostile environments.

A useful parameter in digital design is the "noise immunity" of components. The so-called DC noise margin is the difference in voltage between the worst-case output of a device and the input voltage required by the driven device to recognize correctly a logic state. For the case of TTL logic, the worst-case output is 2.4 V (high) and 0.4 V (low). Inputs interpret voltages below 0.8 V as low and above 2 V as high. Hence the DC noise margin for this logic family is 0.4 V. Data for other logic families are shown in Table 12.1. These are conservative limits and short-duration pulses or high-frequency signals may require somewhat higher voltages to cause upsets.[1] Information on the dynamic noise margin of various logic families showing the required duration vs. amplitude may be found in manufacturers' literature.[9] Logic families and protective devices must be selected according to the requirements of the application and the nature of the EM environment.

The techniques for enhancing immunity described above form the standard response of the designer to meeting an EMC specification and producing a well-engineered product. There are, however, specialist techniques that can be used, if necessary, to deal with particular problems and to enhance immunity even further. Two such methods, immunity by software design and spread spectrum, are described below.

12.3.1 Immunity by Software Design

Irrespective of the amount of care taken in the design of equipment, it must be accepted that under exceptional circumstances sufficient electromagnetic energy will penetrate to affect equipment operation adversely. It is the task of the designer to consider how best to deal with this situation in order to reduce malfunctions and damage. In computer-based equipment, the risk of upsetting program flow and causing the program to enter into interminable loops must be assessed and ways found for normal operation to be reestablished. A way of achieving this is by means of the so-called watchdog circuits.[10] This method consists of requiring the processor to execute a specific operation regularly. A dedicated timer seeks the outcome of this operation and if one is not forthcoming within a specified time period (normally a fraction of a second), it generates a reset to restart normal processor operation. Naturally, not all conceivable difficulties can be handled in this way and additional procedures must be implemented to trap errors and initiate recovery strategies. Examples of simple checks that can be useful in this respect are examining input data from sensors to check that they are in the correct range, inquiring about the same set of data several times to avoid acting on data corrupted by transient interference, reading and writing data into memory to check that they are the same, etc.

The integrity of data transmitted over a communication channel may be established in various ways. In slow operations, it is possible to confirm that a message is correctly received by requiring that it is echoed back for comparison with that originally dispatched. Systematic procedures are also available to check errors in the transmission of digital data over fast communication channels. These aim at error detection and, if possible, error correction. A common technique is parity check where one bit, the so called parity bit, is attached at the end of a group of bits so that the total number of those in the message is even (even parity) or odd (odd parity). A deviation from the chosen parity of the transmission indicates an error and a request for a retransmission of the message is initiated. Another version of this scheme is to use a parity bit for a group of characters in what is known as a longitudinal redundancy check to distinguish from the previous case, which is referred to as a vertical redundancy check. More sophisticated techniques are available, such as the cyclic redundancy check (CRC), which offer limited error correction capability.[11]

Most conventional schemes for dealing with error correction assume random errors (e.g., white noise) affecting the system. These are referred to as "random error correcting codes." However, under certain circumstances, which are not uncommon in EMC situations, errors occur in bursts, i.e., a sequence of bits is damaged. Special techniques referred to as "burst error correcting codes" are available that can detect and correct errors bursts. A particular example is the technique of interleaving. Here a sequence of bits is interleaved (mixed up) before transmission, so that a burst of EMI affects bits in several words not just a few. With this arrangement errors are few and randomly distributed in each word and hence can be corrected using random error correction codes. A de-interleaver then arranges bits in the right sequence after transmission. A detailed treatment of this specialist subject may be found in References 12 and 13.

The degree of sophistication incorporated into software to deal with errors reflects the importance attached to the correct operation of the system under design. However, some form of automatic or manual reset following errors should be incorporated in all but the most basic systems.

Further details of software-related measures to improve EMC may be found in References 14 and 15. The broad context in which the design of software for safety critical applications takes place is described in IEC Standard 61508 Part 2 (Requirements for electrical/electronic/programmable electronics safety-related systems).

12.3.2 Spread Spectrum Techniques

Developments in communications have resulted in heavy demands on the use of the frequency spectrum with the attendant problems of congestion and interference. In order to make the best use of the available spectrum, various common-sense measures are employed. These include using the minimum necessary transmitter power and bandwidth, directional antennas at minimum height, as high a carrier frequency as possible, and low harmonic emissions. A special transmission technique that was developed to make even better use of the available spectrum is the spread spectrum (SS) technique.[16,17] The impact of SS transmission methods is very wide in both military and civilian communications. The essence of the SS technique is described in Figure 12.2. This shows a simple direct sequence SS scheme whereby the data is used to amplitude modulate the carrier. The resulting signal is then amplitude modulated by a pseudorandom code sequence and the results of this process are sent to the transmission

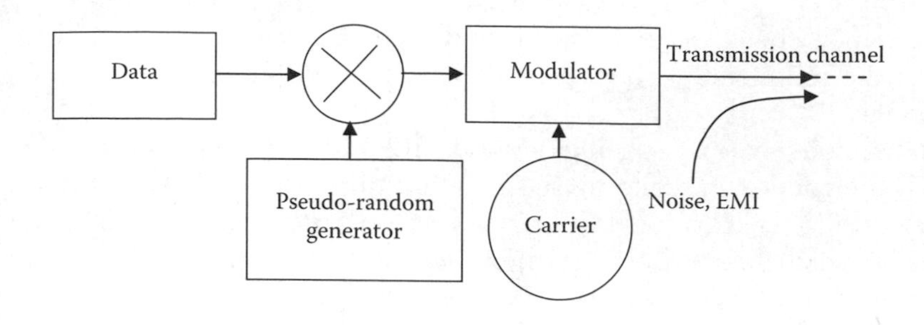

Figure 12.2 Schematic of a direct-sequence spread-spectrum arrangement.

channel. At any stage during these processes, noise and interference may be added to the transmitted message. In Figure 12.2 only the transmitter is shown. At the receiver, the reverse process of demodulation takes place, where it is necessary to have a synchronized copy of the pseudorandom code used by the transmitter. In practice all the signals referred to are binary sequences having values of ±1. The bit rate of the pseudorandom code is chosen to be much higher than the bit rate of the data. The result of this is that the frequency spectrum occupied by the message is much higher than that strictly required by the data. This may be regarded on first reflection as undesirable, but it must be borne in mind that although the message now occupies a very wide spectrum, the transmitted power per hertz is very low, typically below the thermal noise level of most receivers, which lack the correct code to reconstitute (despread) and recover the original data. Receivers with a correct copy of the code can implement despreading, demodulation, and filtering to obtain the transmitted data.

From this brief description of the SS technique, the following advantages are apparent: first, it provides a certain immunity from eavesdroppers, as receivers without a copy of the code cannot reconstruct the original data. Since the power spectral density is low, most users will not even be aware that there is a transmission in progress.

Second, several SS users may use the same frequency band but be addressed selectively using different code sequences. This feature maintains the secrecy of the communication and also increases spectrum utilization.

Third, the use of SS techniques increases interference rejection. This is particularly important in dealing with attempts to jam military and other communications.

It is outside the scope of this book to offer a comprehensive treatment of SS techniques. The interested reader may consult the references already cited where further details and other techniques, in additional to direct

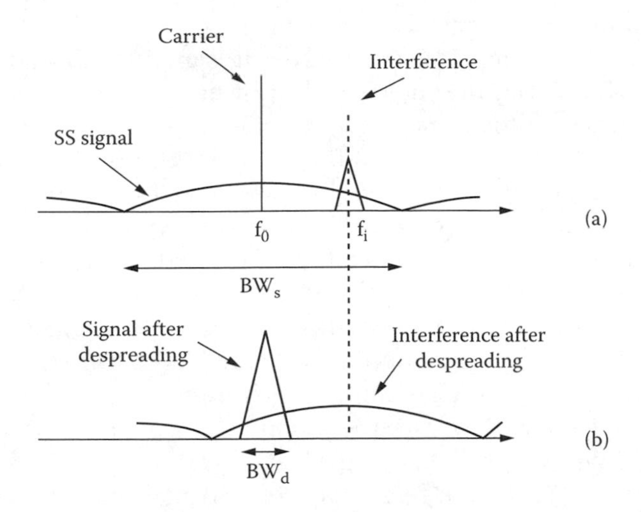

Figure 12.3 Spreading (a) and despreading (b) in SS systems.

sequence, such as frequency hopping are described. However, it is worth explaining briefly the mechanism of interference rejection in direct-sequence SS systems. Let a desired signal of power P_s at a carrier frequency f_0 and spread to a bandwidth BW_s be transmitted over a channel where an interfering signal at f_i of power P_i is superimposed, as shown in Figure 12.3a. The bandwidth BW_s is approximately equal to twice the clock frequency of the pseudorandom code. On reception, the desired signal after despreading appears as shown in Figure 12.3b. Spreading and despreading can be most simply done by implementing a modulo-2 addition of the data to the code, i.e., spreading = data ⊕ code. The despreading operations consists of another modulo-2 addition to the code, i.e., data ⊕ code ⊕ code = data, as can be easily confirmed from the properties of the modulo-2 addition. During the despreading operation, the modulo-2 addition of the interference signal to the code gives a broad spectrum of bandwidth BW_s as shown in Figure 12.3b. The subsequent passage of these signals through a bandpass filter of bandwidth BW_d, corresponding to the bandwidth of the data, results approximately in the following received power:

$$\text{Desired data power} = P_s$$

$$\text{Interference power} = (P_i/BW_s)BW_d$$

Hence, the signal-to-interference ratio is equal to $(P_s/P_i) \times (BW_s/BW_d)$. The quantity (BW_s/BW_d) is referred to as the process gain and it quantifies the improvement in signal-to-interference ratio as a result of the SS technique. As an example, if the interfering signal has a power P and a bandwidth of 10 kHz, the interference power at an SS receiver (BW_s = 1 MHz) operating at the same carrier frequency and with the same signal bandwidth will be P/100. Thus, worthwhile reductions to interference can be achieved by using SS systems. A more extensive discussion of the impact of SS on interference may be found in Reference 18. Issues related to spectrum management in general are explored in Reference 19.

Another area where SS have found application is in spread spectrum clock generators (SSCG), also known as dithered clock oscillators (DCO). Synchronous digital systems are driven by clocks. A perfect clock would have all radiated energy at its fundamental frequency and its harmonics. Receivers used for EMC tests have a bandwidth typically of 120 kHz. If all radiated energy were at a single frequency and this were within a single receiver BW, then a very large peak would be registered and the product would fail. A way around this is to modulate slightly the clock frequency around its nominal value and hence spread the radiated energy over a wider range of frequencies with the result that the peak values are reduced.[10,15] It is important to realize that the total radiated energy is not reduced, it is simply spread over a wider frequency range. It is debatable whether this strategy is ultimately beneficial. There is evidence that broadband systems may be more susceptible to radiation from equipment with dithered clocks. However, SSCG are widely used in many commercial systems. One obvious advantage is that the spreading function can be simply introduced without changing anything else in the design of the circuit — an obvious quick fix if EMC problems are encountered at a late stage.

The basic technique is locking a phase-locked loop to a crystal oscillator and adding a small modulating signal (±0.5%) as shown in Figure 12.4.[20] It is possible to provide center-spreading (output frequency around the nominal frequency) or down-spreading (output frequency below the nominal value). In the former case some clock periods are shorter, which may be a problem in systems with tight timing margins. In the latter, the average frequency is lower than its nominal value with some loss in performance. Normally, clock spreading can be set at different configurations or even turned off during the BIOS set-up.

12.4 The Management of EMC

Designing for EMC and ensuring compliance throughout the entire production cycle and in the field is a complex matter. A start should be made

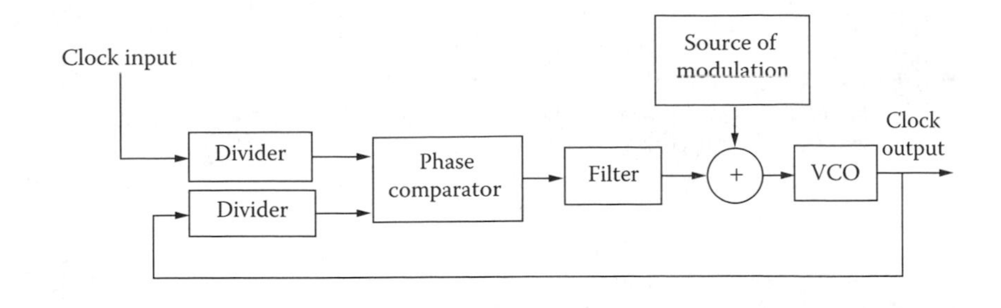

Figure 12.4 Schematic diagram of a dithered clock circuit.

by identifying the technical and legal constraints imposed by EMC and by making available to the design team the necessary resources to meet the desired objectives. Unfortunately, EMC is not a clearly identifiable aspect of a product, and in many cases spending money and time on EMC at an early stage feels like paying an excessive insurance premium. Questions are asked whether such an effort is justified and whether in fact the product would not have been acceptable without EMC-related expenses, which normally add little to functionality under normal conditions. It is also difficult to assess with confidence the safety margin required to meet EMC regulations comfortably, and to deal with the unexpected. Since EMC considerations can have considerable impact on aspects of design that do not at first sight seem relevant, it is essential to establish at the very beginning the ground rules and procedures necessary to ensure EMC. Most practitioners agree that a single person with responsibility for EMC should be appointed and must have the necessary status to see that the correct EMC procedures are established and maintained throughout the design. An important part of the EMC coordinator's task is to be aware of developments in regulatory aspects of EMC and to establish awareness of EMC at all levels of the organization.

Issues to consider at an early stage are the general level of EMC expertise within the company and the need to seek outside advice, the availability or not of test facilities within the organization and the need to use external facilities, and the timetable and procedures leading to compliance with EMC standards and regulations. Finally, proper EMC quality controls must be established and maintained during an extended production period when several modifications, redesigns, and sourcing of components from different suppliers may be necessary. It must be borne in mind that practically all aspects of design and manufacture of equipment can have an impact on EMC. Experience and problems with EMC must be carefully analyzed and the knowledge gained must be disseminated widely within the organization to improve future design for EMC.

References

1. Weston, D A, "Electromagnetic Compatibility — Principles and Applications," Marcel Dekker, NY, 1991.
2. Hargis, C, "Design of a pwm ac variable-speed motor drive for conducted emission compliance," in "Case Studies in EMC," IEE Colloquium, Digest No 1993/091, pp 3/1–312, 22 Apr 1993.
3. Lander, C W, "Power Electronics," McGraw-Hill, NY, 1981.
4. DeFalco, J A, "Reflection and crosstalk in logic circuit interconnections," IEEE Spectrum, pp 44–50, July 1970.
5. Jerse, T A, Paul, C R, and Whites, K W, "The effect of finite image plane width on the radiation from an electric line source," 10th Int Zurich EMC Symp, pp 201–206, 9–11 Mar 1993.
6. Montrose, M I, "Overview on design techniques for printed circuit board layout used in high technology products," Proc IEEE EMC Symp, Cherry Hill, NJ, pp 61–66, 12–16 Aug 1991.
7. Catrysse, J, "PCB design and EMC constraints," Proc 10th Int Zurich EMC Symp, pp 171–184,9–11 Mar 1993.
8. Hubing, T, "PCB EMC design guidelines: a brief annotated list," IEEE Symp. on EMC, Boston, pp 34–36, Aug. 18–22, 2003.
9. TTL Advanced Low-Power Schottky, Advanced Schottky, Data Book, Vol 2, Texas Instruments, Dallas, TX, 1989.
10. Williams, T, 'MC for Product Engineers," Newnes, Oxford, England, 1992.
11. Cegrell, T, "Power System Control Technology," Prentice Hall, Englewood Cliffs, NJ, 1986.
12. Lucky, R W, Salz, J, and Weldon, E J, "Principles of Data Communication," McGraw Hill, 1968.
13. Proakis, J G and Salehi, M, "Communication Systems Engineering," 2nd edition, Prentice Hall, 2002.
14. IEE Colloquium Digest 98/471, "Electromagnetic Compatibility of Software," London, Savoy Place, 12 Nov 1998.
15. Williams, T, "EMC for Product Engineers," 3rd ed., Newnes, Oxford, England, 2001.
16. Dixon, R C, "Spread Spectrum Systems," John Wiley, NY, 1984.
17. Taub, H and Schilling, D L, "Principles of Communication Systems," McGraw-Hill, NY, 1986.
18. Hopkins, P M and Cravey, D N, "Spread Spectrum communications-interference considerations — A tutorial overview," Proc 6th Int Zurich EMC Symp, pp 447–452, Mar 5–7 1985.
19. Darnell, M, "Spectrum management and conservation" in AGARD Lecture Series 177 — Electromagnetic Interference and Electromagnetic Compatibility," pp 7.1–7.24, 1991.
20. UK Radiocommunications Agency, Final Report, Contract AY 4092, "Further work into the potential effect of the use of Dithered Clock Oscillators on Wideband Digital Radio Services," 2002.

EMC STANDARDS AND TESTING

Following an understanding of concepts underlying EMC and coverage of the basic interference control techniques, it is now necessary to address the subject of EMC standards and the tests used to demonstrate compliance. However careful a design has been from the EMC point of view, it is always necessary to measure EMC performance against specific standards. The range of standards and associated measurements is very wide and no single person can have a complete grasp of every detail. However, the basic methodology employed in standards and testing should be familiar to all concerned with EMC. Demonstrating compliance with a particular standard involves measuring voltages (for conducted emission) and electric fields (for radiated emission) under formalized conditions. These are then compared with specific limits. There is no guarantee that a product that meets a standard will not experience interference problems. Hence, although complying with a standard is essential, it is no substitute for having a thorough grasp of the EMC design of the product. In the following two chapters both standards and testing are described in general terms to alert the designer to the difficulties that may be encountered.

Chapter 13

EMC Standards

In this chapter the international framework relevant to EMC is described. EMC standards and regulations cover in detail many aspects of EMC performance and testing and are continuously evolving. Neither the complete detail nor the current status of standards can therefore be reflected fully in a book. It is the responsibility of the designer to keep up to date with relevant standards in his or her field of activity and for the intended market. This can best be done through national standards and trade associations. The purpose of this chapter is to present the basic structure of EMC standards as it stands at the time of writing.

13.1 The Need for Standards

Progress in automation and the international nature of trade go hand-in-hand with standardization. International and national organizations have been set up with the aim of introducing standards affecting virtually all aspects of manufacture and trade. Good standards allow interchangeability and compatibility of equipment, relatively free trade, and satisfaction for the customer. In addition they should not be a barrier to technological progress. The setting of standards is a complex and lengthy process as it cuts across technical, trade, and national prestige issues. There are standards set by international organizations, national bodies, trade associations, insurance companies, etc. These standards are, in most cases, similar, but there are also many differences in the specific limits set by various organizations, the method of measurement and testing, and the route to compliance. Describing all standards that have an impact on EMC would require a whole book. The useful lifetime of such a book would be rather short. In addition, such coverage may generate the notion that there is nothing fundamental and no commonality among the bewildering variety of standards. For both these reasons, no attempt is made here to be comprehensive. Instead only the main standards are described, starting with the international framework and continuing with commercial, military, and company standards. In each case, typical examples are given to help the reader understand the nature of standards and prevalent limits.

It is worth pointing out that complying with an EMC standard is no guarantee that the product will not experience upsets due to electromagnetic interference. It is thus important to examine how far beyond meeting EMC regulations one has to go to produce a marketable product of high operational integrity at a reasonable cost.

13.2 The International Framework

The benefits of standardization are so significant that efforts at defining and maintaining standards are directed by major international bodies. Under the aegis of the United Nations Organization, various educational and scientific establishments have been set up, such as the International Organization for Standardization (ISO), the United Nations Educational Scientific and Cultural Organization (UNESCO), and the International Council for Scientific Unions (ICSU). Also, specialized agencies like the International Telecommunications Union (ITU) operate in areas where action at the international level is deemed necessary. Membership of these organizations is open to all nations and hence any agreements reached command, in general, widespread acceptance. The normal pattern is for such agreements and standards to be adopted by individual nations and embodied into national legislation. Hence, although major trading nations have their own standards organizations, they tend to reflect closely international practice whenever such practice has been established. From the EMC point of view, the International Electrotechnical Commission (IEC), which is a division of ISO, and ITU are the most important. The work of the IEC has impact on all EMC regulations and forms the model for most of them.[1] Within the technical committee structure of the IEC, the following three technical committees are the most important for EMC:

CISPR (International Special Committee on Radio Interference)
TC77 (Concerned mainly with EMC in electrical equipment and networks)
TC65 (Concerned mainly with immunity standards)

The output of the CISPR committee is in the form of standards with the designation CISPR10 to CISPR23 covering a wide range of EMC-related topics. These are mentioned in more detail when civilian standards are considered in the next section.

Technical Committee 77 is responsible for a series of publications with the designation 555 covering harmonics and flicker, and 1000 covering low-frequency phenomena in power networks. Technical Committee 65 has produced a series of standards with the designation 801 covering issues related to immunity. These form the background on which detailed immunity standards are developed. A discussion of IEC standards on EMC and related topics may be found in Reference 1.

13.3 Civilian EMC Standards

Civilian EMC standards have now been adopted by most countries and in most cases they have the force of law. In general, noncompliance with EMC standards can delay or prevent the marketing and sale of products and any infringements may lead to prosecution. It is therefore of the utmost importance to be aware of these standards and meet all detailed requirements for demonstrating compliance.

13.3.1 FCC Standards

Within the United States, the Federal Communications Commission (FCC) is responsible for radio communications and interference. No product can be sold in the United States if FCC requirements are not met. FCC Rules Part 15 Subpart J (RFI emissions) cover digital equipment and thus affect a wide range of products. The regulations concern conducted and radiated interference and refer to two classes of equipment. Class A equipment is for use in commercial, industrial, and business premises, whereas Class B equipment is intended for residential use. Regulations for class B equipment tend to be more stringent. Any device with timing (clock) pulses in excess of 10 kHz is covered by FCC regulations. The upper frequency tends to be of the order of ten times the highest clock frequency. Hence with clock frequencies now exceeding 1000 MHz, EMC regulations will cover emissions well into the gigahertz range.

Conducted interference covers frequencies ranging between 450 kHz and 30 MHz. Regulations aim at controlling interference current in power leads. Since specifications are given in terms of voltages measured on such leads, it is necessary to define more closely the impedance across which such measurements are made. This is done by interposing a line

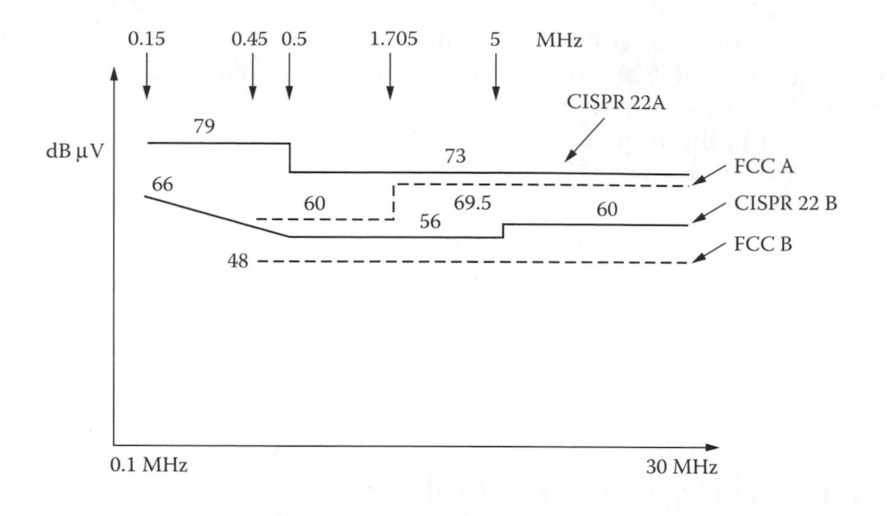

Figure 13.1 Typical conducted interference levels.

impedance stabilization network (LISN) between the equipment to be tested and the power supply network. This consists of a combination of capacitors, resistors, and inductors so that for frequencies up to 30 MHz the impedance to ground is essentially constant and equal to approximately 50 Ω. The conducted interference limits for classes A and B according to FCC rules are shown in Figure 13.1.

Radiated interference limits are set on measurements taken using resonant dipoles at a distance of 3 m for class B and 10 m for class A from the equipment under test (EUT). Measurements of the electric field (in dBμV/m) are taken in an open-field site with antennas scanned in horizontal and vertical polarizations and with the EUT rotated to obtain the maximum emission. The relevant limits are shown in Figure 13.2. It should be emphasized that the two limits shown are set for different measurement distances. It can be argued that class B limits are more severe and the limits for class A equipment may be referred to the 3-m distance for comparison. This can be done by assuming far-field conditions and adding 20 log (10/3) dBμV/m to the curve in Figure 13.2 for class A. This raises this curve by approximately 10 dB. However, it must be stressed that there is no firm scientific basis for assuming far-field conditions, especially at low frequencies.

Personal computers and peripherals are certified and may also be tested by the FCC. Other equipment is certified by the manufacturer. Separate FCC regulations (Part 18) apply for industrial, scientific, and medical (ISM) equipment, which specify operating frequencies and the amount of permitted radiation outside the specified frequency bands.

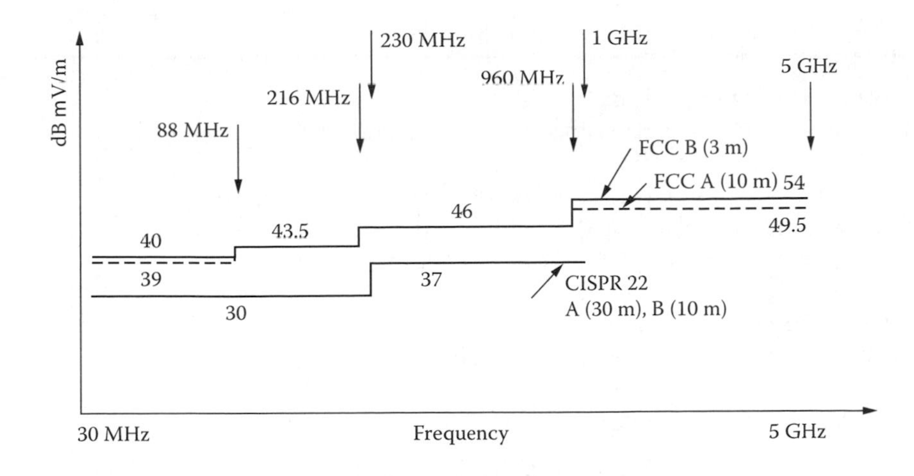

Figure 13.2 Typical radiated interference levels.

The coordinating organization for standardization in the United States is the American National Standards Institute (ANSI) to which other interested bodies, such as the IEEE EMC Society, contribute EMC-related standards.[2]

13.3.2 European Standards

The European Commission (EC) established, through its directive of May 3, 1989, the requirements for electromagnetic compatibility.[3] Subsequent amendments allowed for a transitional period, with the full force of the directive coming into effect on January 1, 1996. All member countries are required to enact legislation to implement the directive. The directive is all embracing, covering emission and susceptibility for a wide range of equipment and imposes duties on manufacturers whether or not appropriate standards exist. The regulations do not apply to equipment for export, specific large installations, spare parts, second-hand apparatus, amateur radio and military equipment, and those covered by other directives. The European Standards Committee (CEN) is responsible through the European Committee for Electrotechnical Standards (CENELEC) for the adoption of suitable standards. These are generic standards, covering in broad terms broad classes of equipment, i.e., class 1 (residential, commercial, and light industry) and class 2 (heavy industry); basic standards (setting out test procedures); and product standards (covering specific types of apparatus, e.g., information technology equipment). A partial list of harmonized standards is given in Table 13.1.[4,8] As an example, the limits set for conducted interference according to EN55022 (equivalent to CISPR22) are shown in Figure 13.1 assuming a quasipeak detector. Two

Table 13.1 European Harmonized Standards

Generic Emission Standards	
EN50081-1	Generic emission standard: residential, commercial and light industry
EN50081-2	Generic emission standard: industrial environment
Standards, EN 55011, 55014, 55022 although product standards are used widely. They are equivalent to CISPR 11, 14, 22 respectively.	
Generic Immunity Standards	
EN50082-1	Generic immunity standard: residential, commercial, and light industry
EN50082-2	Generic immunity standard: industrial environment
Basic Standards	
IEC61000-3	Electromagnetic compatibility: Limits (several sections)
IEC61000-4	Testing and measurement techniques for immunity of electrical and electronic equipment: Basic EMC standards (several sections)
Product Standards	
EN55013	Broadcast receivers and associated equipment: emission (equivalent to CISPR 13)
EN55020	Broadcast receivers and associated equipment: immunity (related to CISPR 20)
EN55014-1	Household appliances, electric tools, and similar apparatus: emission (equivalent to CISPR 14)
EN55014-2	Household appliances, electric tools, and similar apparatus: immunity
EN55015	Lighting equipment: emission (equivalent to CISPR 15)
EN61547	Lighting equipment: immunity
EN55022	Information technology equipment: emission (equivalent to CISPR 22)
EN55024	Information technology equipment: immunity (equivalent to CISPR 24)
EN55011	Industrial, scientific, and medical (ISM) devices: emission (equivalent to CISPR 11)
EN61800-3	Adjustable-speed electrical power drive systems: emission and immunity
...and several more.	

classes of equipment, A and B, are defined. Class B is equipment that is subject to no restrictions on its use.

Radiated emission limits are set for measurements taken on an open-field site of adequate quality. The EUT must be rotated and the antenna height varied between 1 and 4 m to obtain the maximum electric field strength. Broadband antennas in vertical and horizontal polarizations are used. The distance between the EUT and the measurement antenna is 10 m for class B and 30 m for class A equipment. The adopted limits, assuming a quasipeak detector, are shown in Figure 13.2. The attention of the reader is drawn to the fact that the limits shown refer to different measurement distances and that they may be referred to the same distance assuming far-field conditions (1/r decay). This approach, although used extensively, cannot be expected to give accurate results at low frequencies where far-field conditions cannot be guaranteed. Compliance with the requirements of the directive can be demonstrated by the manufacturer meeting EC approved standards, by producing a technical construction file prepared by a competent body (the only option available when an approved standard does not exist), or by meeting special type examinations (for radio transmission equipment only). A comprehensive discussion of the EC directive and routes to compliance may be found in References 5 and 6. A description of the process of implementation in one of the European countries may be found in Reference 7.

It is stressed again that Table 13.1 is a snapshot of some of the key European standards currently available, which are reflected in IEC standards usually with similar numbers (e.g., EN61547 and IEC61547). Revisions of standards are continuously produced and new ones are introduced, therefore frequent updates of the information in this table are necessary. These can be readily obtained from the EU Directorate on Enterprise and Industry[8] or from sites of commercial testing organizations who, in the process of advertising their services, provide regular updates.

In particular application areas, where safety is of paramount importance, special regulations apply. A particular example is the automotive area where EMC requirements are particularly severe and limits tend to be higher. Equipment used in road vehicles (ESA-electrical/electronic subassemblies such as radios, mobile phones) must comply with the Automotive EMC Directive (95/54/EC) and (2004/104/ED) (from 1 January 2006). Several test standards are involved in automotive EMC including, CISPR 12 and 25 and ISO 11451, 11452, and 7637.

In regard to the main EMC directive in Reference 3 this will be replaced from July 20, 2007, by a new directive 2004/108/EC.[9]

This new directive is aimed at simplifying regulatory procedures, reducing costs, and increasing documentation and information for inspection purposes. Manufacturers will be solely responsible for establishing the conformity of their products and for the "CE" marking. The simplified conformity procedures are accompanied by stricter requirements for information and documentation, a special regime for fixed installations (power plants, telecommunications networks, distribution networks), and the principle that strict regulatory harmonization is restricted to essential public interest requirements. Further details may be found in Reference 10.

13.3.3 Other EMC Standards

There is a variety of national standards covering EMC. Specific examples are those produced in the United Kingdom by the British Standards Institution (BSI) and in Germany by the Verband Deutscher Elektrotekniker (VDE). These are being rapidly brought into line with the provisions of the European Community EMC directive. Standardization in Japan is described in Reference 11. Canadian standards set by the Canadian Standards Association (CSA) are in general similar to those set by the FCC. EMC standards are undergoing a rapid evolution and it is hoped that greater commonality will be achieved through mutual recognition agreements between the major standards organizations.[12]

13.3.4 Sample Calculation for Conducted Emission

We give here a sample calculation of conducted emissions for a power electronic circuit, the type of a simple calculation that a designer may have made in an effort to ensure that the product meets CE standards. Because of the relatively low frequencies involved in CE problems it is far easier to do analytical calculations compared to RE problems. A more thorough discussion of conducted emission problems in three-phase power electronic drives may be found in References 13, 14, and 15.

The generic problem is the single-phase load shown in Figure 13.3. The parameters and basic assumptions are listed below:

- MOSFET duty cycle is 40%
- Switching frequency f = 50 kHz
- Fall- and rise-times 300 ns
- C_{stray} = 100 pF (only impedance between power circuit and earth)
- Neglect all impedances in CM path except C_{stray}
- At HF the DC capacitor is a short

The task is to calculate the CM voltage measured at the LISN at 3 MHz. The LISN interposed between the power supply on the left and the EUT

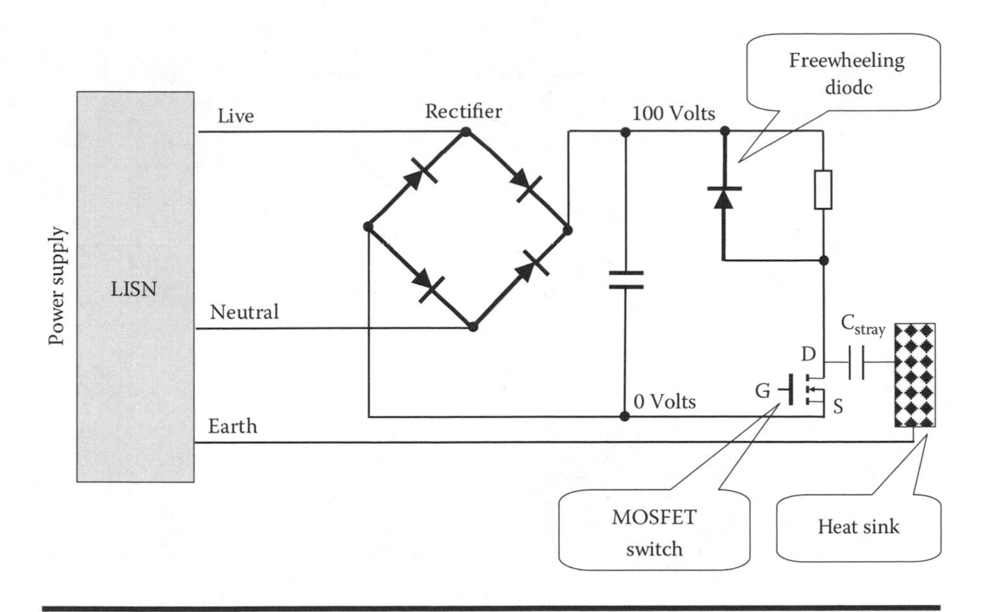

Figure 13.3 Schematic for studying common-mode interference from single-phase load.

is shown schematically in Figure 13.4. In this figure the CM and DM current paths are shown. Therefore, the current flows in the LISN resistors are as shown in Figure 13.5, Hence, the equivalent circuit for CM current has the form shown in Figure 13.6. It remains to establish the value of V_{CM} at 3 MHz. This we have done in the example of Section 5.3.3 and is 111.43 dBμV. At 3 MHz $1/(\omega C_{stray}) = 1/(2\pi 3 \times 10^6 \times 100 \times 10^{-12}) = 513\ \Omega$. Hence, the voltage across the LISN (parallel combination of the two 50Ω resistors) is

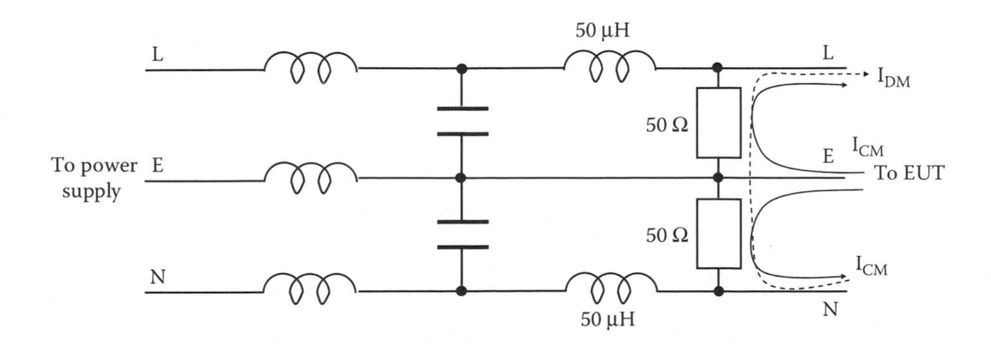

Figure 13.4 LISN, CM, and DM current flows.

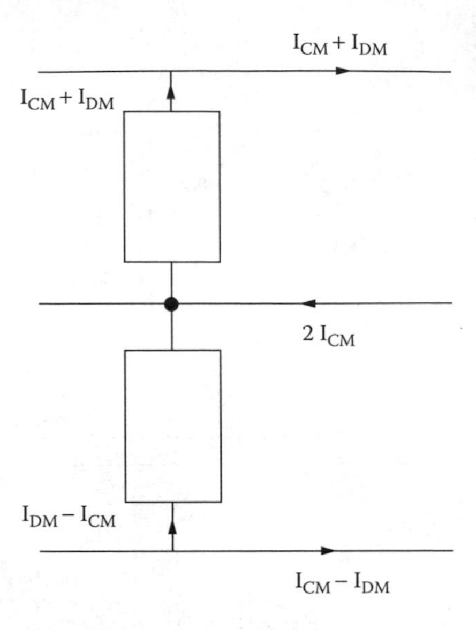

Figure 13.5 Details of current flow in the LISN.

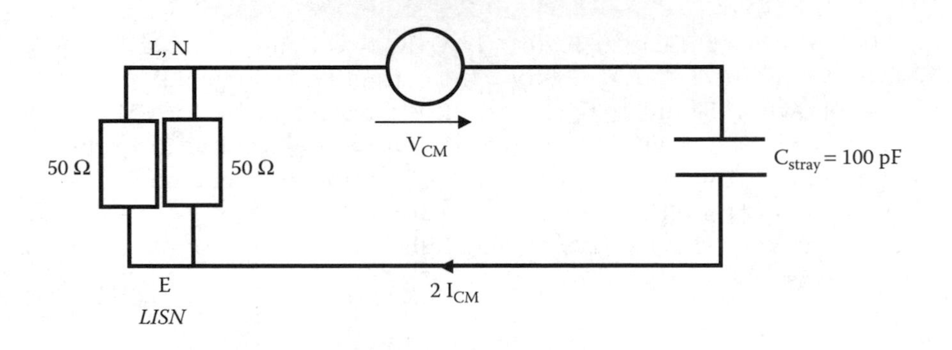

Figure 13.6 Equivalent circuit for calculating the LISN voltage.

$$V_{LISN} = \frac{25}{\sqrt{25^2 + 531^2}} V_{CM} = 0.047 V_{CM}$$

or, in decibels,

$$20 \log V_{LISN} = 20 \log(0.047) + 111.43 = 84.8\ \mu V$$

If we wish to reduce this level, we may introduce the CM choke, which we considered in the example of Section 11.1. We found in that section that the LISN voltage was 61.5×10^{-6} V_{CM}. Hence, the CM choke has introduced an attenuation of

$$20\log\left(\frac{61.8\times10^{-6}}{0.047}\right) = -57.6 \text{ dB}$$

bringing the LISN voltage within the limits shown in Figure 13.1.

13.4 Military Standards

Military systems present special problems to the EMC specialist. Increased use of electronic communications for integrated command, control, communications, and intelligence (C^3I) between the various services (land, air, and sea), the hostile nature of the battle environment, the need for portability and secrecy, and the deliberate use of electromagnetic energy for jamming, have created an extremely complex EMC system problem. A brief description of the relevant standards used in the United States and the United Kingdom is given below. NATO countries tend to follow similar procedures. Further details may be found in Reference 16.

13.4.1 Military Standard MIL-STD-461D[17]

The first successful attempts to establish a tri-service EMC standard in the United States date from the mid-1960s. The first documents issued were MIL-STD-461, -462, and -463 covering requirements, measurements, and definitions, respectively. These standards went through a process of evolution over the years with the intention of improving accuracy and repeatability. A feature of military standards is that they specify measurements in screened rooms in order to avoid difficulties with ambient electromagnetic noise. Measurements in unlined screened rooms are notoriously unrepeatable due to resonances. The introduction of radiation-absorbing material (RAM) can improve matters at high frequencies, but at low frequencies difficulties remain. Measurements are normally taken by placing the EUT on a conducting bench and using a variety of test antennas placed 1 m away from the EUT. Both electric and magnetic field measurements, the latter at low frequencies only, are taken. Loops antennas for LF magnetic field and dipole antennas for LF electric fields (≤ 30 MHz) are used. In the range 30 to 300 MHz, biconical antennas are generally

Table 13.2 Requirements According to MIL-STD-461D

Conducted Emissions	
CE101	Conducted Emissions, Power Leads, 30 Hz to 10 kHZ
CE102	Conducted Emissions, Power Leads, 10 kHz to 10 MHz
CE106	Conducted Emissions, Antenna Terminal, 10 kHz to 40 GHz
Conducted Susceptibility	
CS101	Conducted Susceptibility, Power Leads, 30 Hz to 50 kHz
CS103	Conducted Susceptibility, Antenna Port, Intermodulation, 15 kHz to 10 GHz
CS104	Conducted Susceptibility, Antenna Port, Rejection of Undesired Signals, 30 Hz to 20 GHz
CS105	Conducted Susceptibility, Antenna Port, Cross-Modulation, 30 Hz to 20 GHz
CS109	Conducted Susceptibility, Structure Current, 60 Hz to 100 kHz
CS114	Conducted Susceptibility, Bulk Cable Injection, 10 kHz to 400MHz
CS115	Conducted Susceptibility, Bulk Cable Injection, Impulse Excitation
CS116	Conducted Susceptibility, Damped Sinusoidal Transients, Cables and Power Leads, 10 kHz to 100 MHz
Radiated Emissions	
RE101	Radiated Emissions, Magnetic Field, 30 Hz to 100 kHz
RE102	Radiated Emissions, Electric Field, 10 kHz to 18 GHz
RE103	Radiated Emissions, Antenna Spurious and Harmonic Outputs, 10 kHz to 40 GHz
Radiated Susceptibility	
RS101	Radiated Susceptibility, Magnetic Field, 30 Hz to 100 kHz
RS103	Radiated Susceptibility, Electric Field, 10 kHz to 40 GHz
RS105	Radiated Susceptibility, Transient Electromagnetic Field

employed, while between 300 MHz and 1 GHz log-periodic antennas are the preferred choice. Above this frequency, horn antennas are generally used. The various requirements are shown in Table 13.2. It will be seen that the military EMC test requirements are more comprehensive than commercial EMC specifications. However, the test set-up used, with the test antenna in the near field of the EUT over a wide frequency range, makes direct comparisons between military and commercial limits difficult to make with any confidence. The number of tests and limits required are extensive and hence details cannot be given here. The reader is referred to the document in Reference 17 and also the companion document MTL-STD-461D covering the measurement procedures.

13.4.2 Defence Standard DEF-STAN 59-41[18]

In the late 1970s, a set of standards under the designation 59-41 was published in the United Kingdom covering the EMC performance of defence-related equipment. As in the case of similar standards in the United States, the great diversity between the needs of the different services has led to a complex set of requirements and limits. The main requirements are outlined in Table 13.3 and further details may be found in Reference 18. Emphasis is on testing inside screened rooms with a conducting bench and antennas placed 1 m away from the EUT. The standards are comprehensive in the sense that limits on both electric and magnetic field are specified covering a very wide frequency range. The difficulties associated with the repeatability of measurements inside an undamped or partially damped screened room in the near field of the EUT have led, as in the case of the MIL-STD, to a reexamination of the test procedures. The thrust

Table 13.3 Requirements According to DEF STAN 59-41 (PART 3)/3

Conducted Emission	
DCE01	Conducted Emission on Power Lines 20 Hz–150 MHz
DCE02	Conducted Emission on Control and Signal Lines 20 Hz–150 MHz
DCE03	Exported Transients Power Lines
Radiated Emission	
DRE01	Radiated Emissions E Field 14 kHz–18 GHz
DRE02	H Field Radiation 20 Hz–50 kHz
DRE03	Radiated Emissions Installed Antenna 1 MHz–76 MHz
Conducted Susceptibility	
DCS01	Conducted Susceptibility, Power Leads 20 Hz–50 kHz
DCS02	Conducted Susceptibility, Power, Control and Signal Leads 50kHz–400MHz
DCS03	Conducted Susceptibility, Control and Signal Lines 20 Hz–50 kHz
DCS04	Imported Transient Susceptibility (Aircraft)
DCS05	Externally Generated Transients (Ships)
Radiated Susceptibility	
DRS01	H Field Susceptibility 20 Hz–50 kHz
DRS02	Radiated Susceptibility 14 kHz–18 GHz
Magnetic Field Susceptibility	
DMFS01	Magnetic Field Susceptibility

of this effort is directed toward introducing limited amounts of radiation-absorbing material (RAM) and taking account in calibration of all instrument and proximity correction factors.

The intention is that measurements taken inside a screened room can be used to predict test results on an open-field site within an uncertainty of ±6 dB. The greatest difficulty in this respect is in introducing sufficient damping in a screened room at the first one or two resonance frequencies without incurring excessive costs and limiting the working volume to an unacceptable size. A summary of the work that affects the evolution of this standard can be found in References 19 to 22. Problems associated with the testing of large items such as aircraft are described in Reference 23.

13.5 Company Standards

Civilian standards have the force of law and noncompliance imposes severe limitations on trade and also other penalties. Military standards do not, in general, have a legal status and may be waived by permission from the relevant authorities. There are in addition a number of internal company standards and other specific requirements imposed by various agencies, insurance companies, etc. Company standards evolved over the years to fill gaps in national and international standards. They still have a role to play in setting limits and defining procedures that ensure that on completion of a design, the product will not have major difficulties in meeting international standards. In some countries, major utilities such as power supply or telecommunications may have their own standards that affect their purchasing. The situation must be ascertained in each case to avoid unexpected obstacles in the future. Similarly, major insurance companies and underwriters impose their own standards appropriate to their specific requirements.

In each country, national standards organizations and trade associations such as, for example, the National Electrical Manufacturers (NEMA) in the United States and the British Electrical and Allied Manufacturers' Association (BEAMA) in the United Kingdom, can advise on the current status of standards affecting their market sectors.

13.6 EMC at Frequencies above 1 GHz

The scope of EMC standards is very wide, extending from DC to 400 GHz, intended as it were to address a legal problem. But, in practice, testing is limited to frequencies below 1 GHz. In the United States, regulations require testing up to ten times the clock frequency and this appears a sensible approach. As, at present, clock frequencies are in the gigahertz

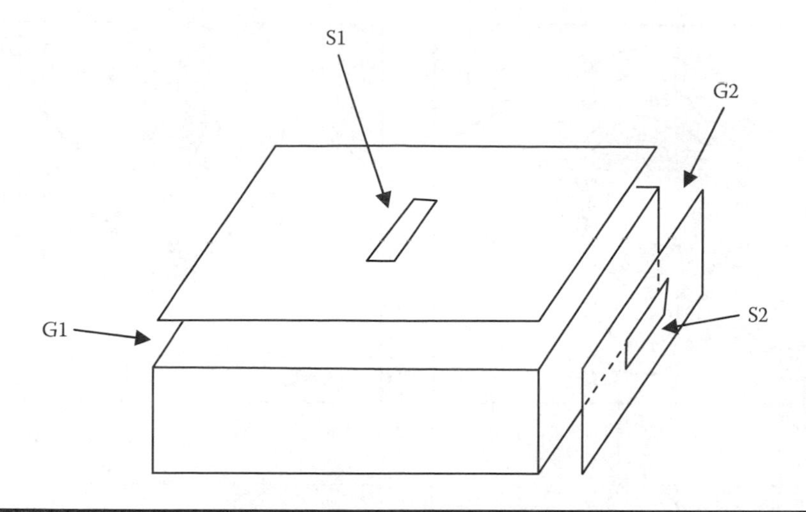

Figure 13.7 Device-under-test (DUT) showing several sources of emission.

range, it is clear that the current practice of limiting testing to 1 GHz is not a sensible one. Protection of the radio spectrum requires that testing to at least 2 GHz is done, for example, to deal with mobile services (GSM 1800). There is increasing evidence that man-made EM noise is rapidly increasing as indicated in Reference 24.

The difficulties of testing efficiently at frequencies above 1 GHz are considerable. Current testing practices, which are very time consuming, will become even more so at higher frequencies. It is not difficult to see why. As the frequency increases and the dimensions of the EUT become comparable to the wavelength, EUT behavior resembles that of an antenna or an array of antennas with highly directive radiation patterns, sometimes in the form of very narrow beams. A simple EUT is shown in Figure 13.7 to help illustrate the point. It consists of a base unit, a top lid, and a side panel, all good electrical conductors. The top lid can be offset from the base unit to exaggerate a narrow gap due to an ill-fitting top lid. Emissions from this gap are labeled as G1 in the calculations that follow. Similarly, an emission mode G2 applies if the side panel does not fit well to the base unit. Similarly, the top lid has a slot (S1) as does the Side panel (S2). The overall dimensions of the EUT are $48 \times 48 \times 12$ cm^3. The slots have dimensions 16×4 cm^2. The EUT was modeled numerically using the Nottingham TLM solver, where the gap modes were excited by impressing a voltage between the base unit and the lid (or panel) and the slot modes by impressing a voltage across the slot in the middle of the widest dimension. Figures 13.8 to 13.11 show results for the G1, G2, S1, S2, respectively. In each case only one mode is radiating, the other three having been suppressed. In this way it is easier to study how an EUT radiates. The plots

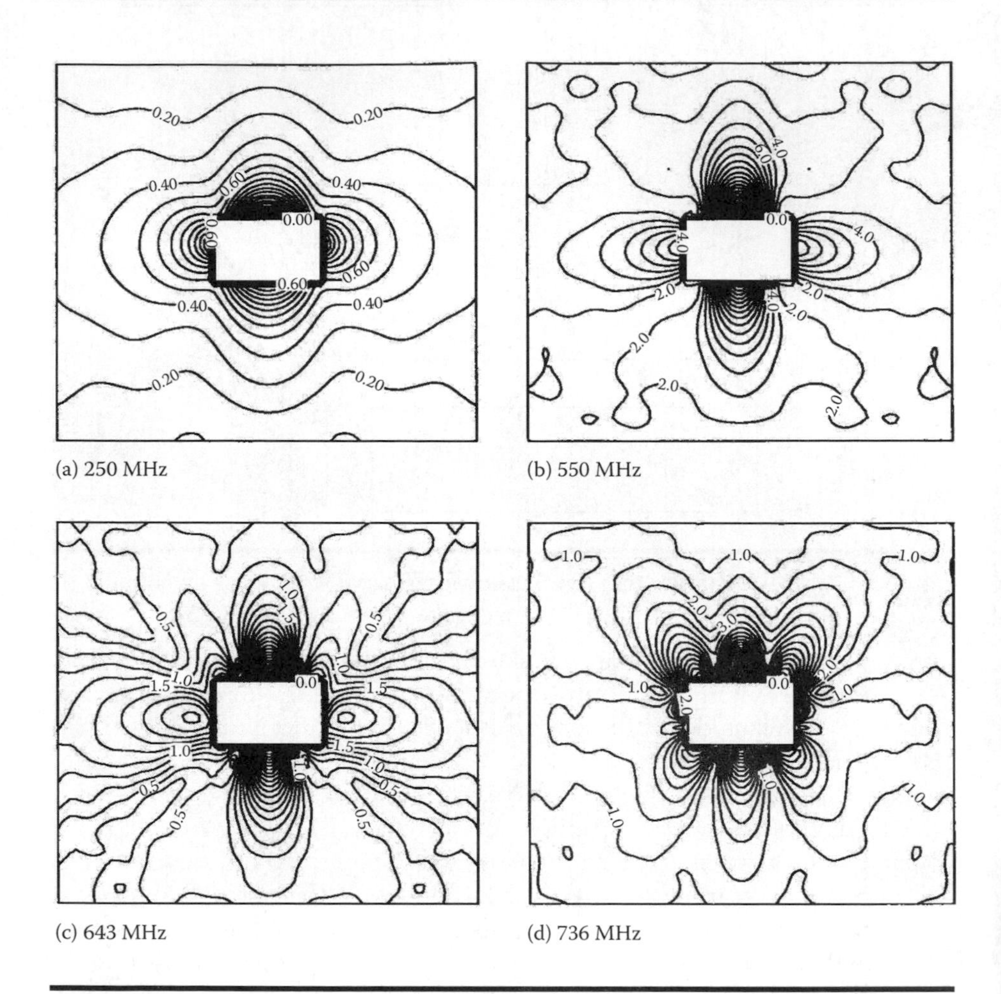

(a) 250 MHz

(b) 550 MHz

(c) 643 MHz

(d) 736 MHz

Figure 13.8 Electric field profile from gap G1.

show the vertical component of the electric field on a vertical plane passing through the EUT center. The outline of the EUT is clearly seen. Plots at four frequencies are shown corresponding to EUT internal resonances. It is clear that with increasing frequency the G-modes show an increasingly directive pattern with narrow beams. This is less pronounced for the S-modes as the size of the slots is small compared to the wavelength at these frequencies. Clearly, projecting the results to frequencies above 1 GHz one can see that practical EUTs become highly directive antennas. Under these circumstances, trying to identify the maximum radiated field will be difficult and time consuming. New ways of approaching EMC tests may be necessary and not merely an extension of existing ones.

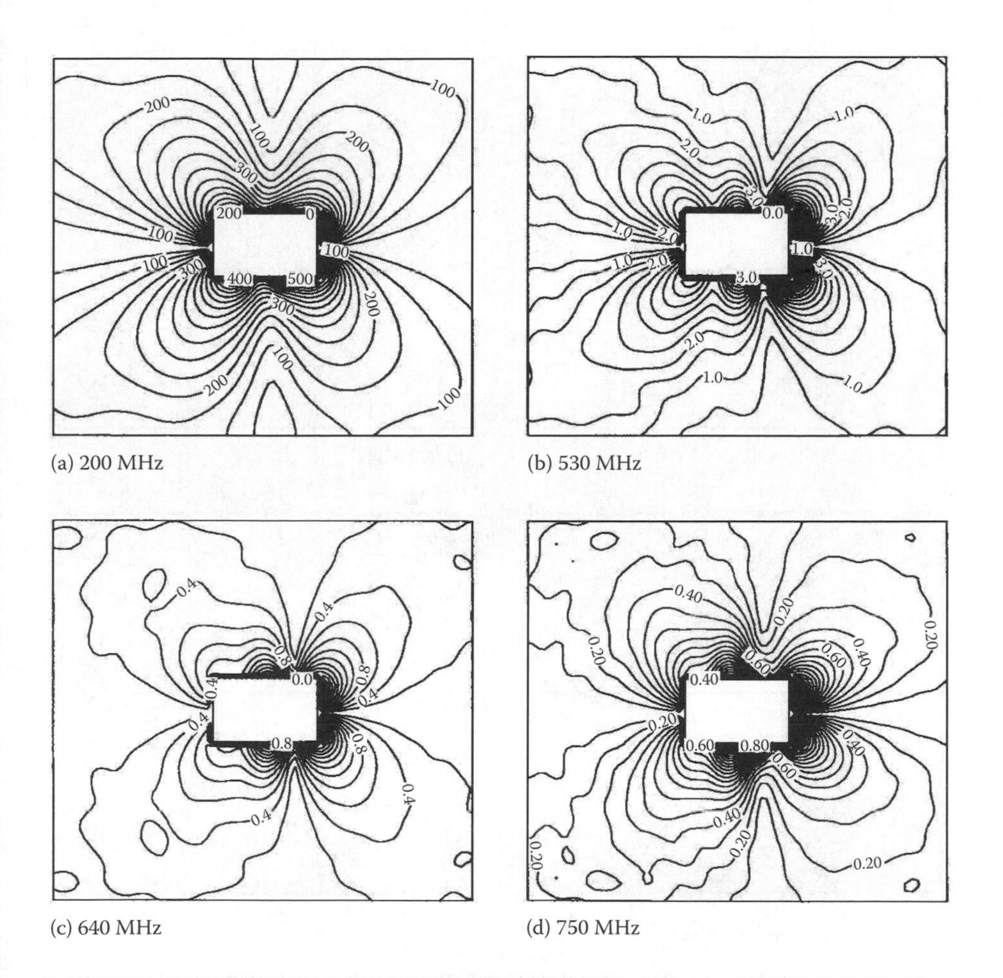

Figure 13.9 Electric field profile from gap G2.

At the time of writing, CISPR/A is developing standards for measurement methods above 1 GHz (CISPR 16-2-3) and for site validation above 1 GHz (CISPR 16-1-4). Several recommendations and amendments are in draft form. The arrangement envisaged is shown in Figure 13.12. The EUT is rotated in 15° increments. If the EUT is contained within the antenna beamwidth w then no height span is performed, but work is in progress to introduce a height span even for this case. Testing is likely to extend to 18 GHz. This is an area in EMC testing that will evolve rapidly in the years to come. For a discussion of the issues involved, reference is made to References 25 and 26.

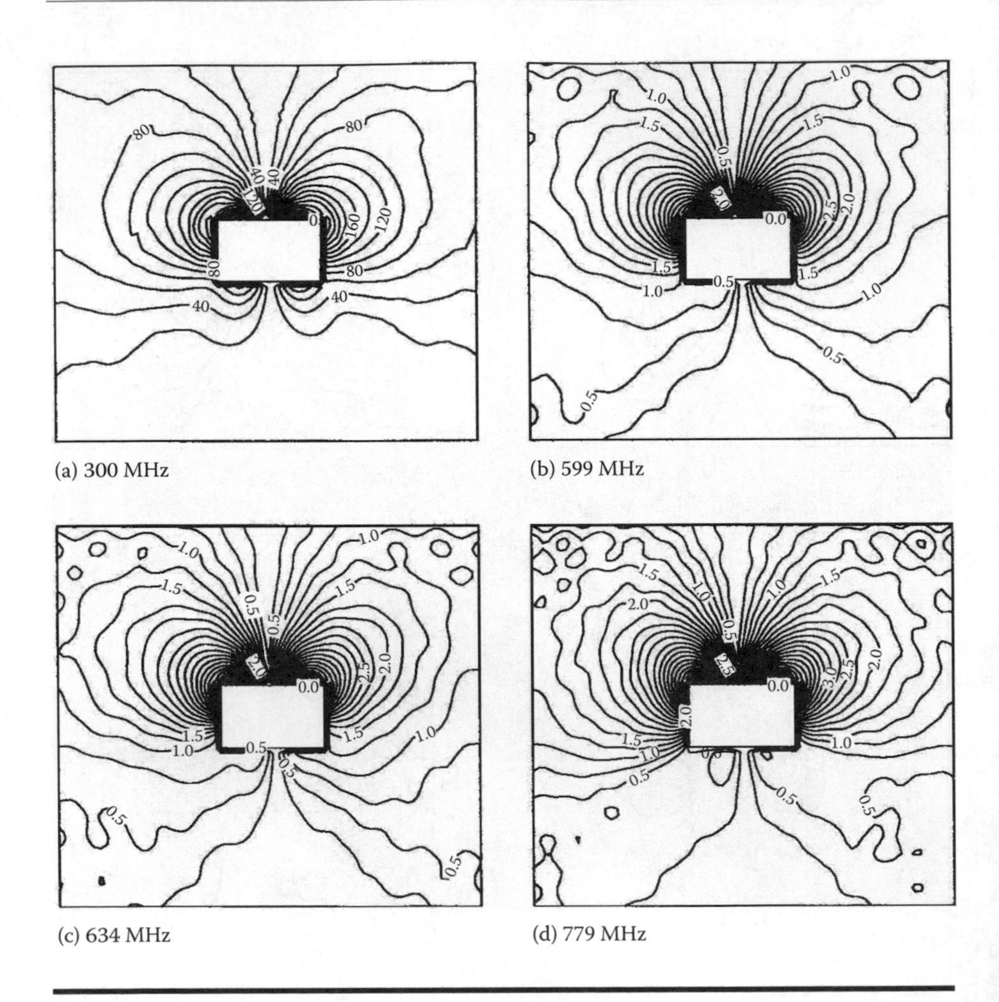

(a) 300 MHz

(b) 599 MHz

(c) 634 MHz

(d) 779 MHz

Figure 13.10 Electric field profile from aperture S1.

13.7 Human Exposure Limits to EM Fields

The setting of safe limits for human exposure to EM fields is not strictly an EMC matter. However, a brief description of these limits is given here to alert those involved with EMC measurements, especially susceptibility measurements, to the main issues. There is at present no complete understanding of the mechanisms responsible for the interaction of EM fields with human tissue. Traditionally, limits were set by assuming thermal effects and restricting the specific absorption rate (W/kg) in the body to a value giving a total amount of power absorbed comparable to the basic metabolic rate (~100 W). There is, however, concern that other nonthermal

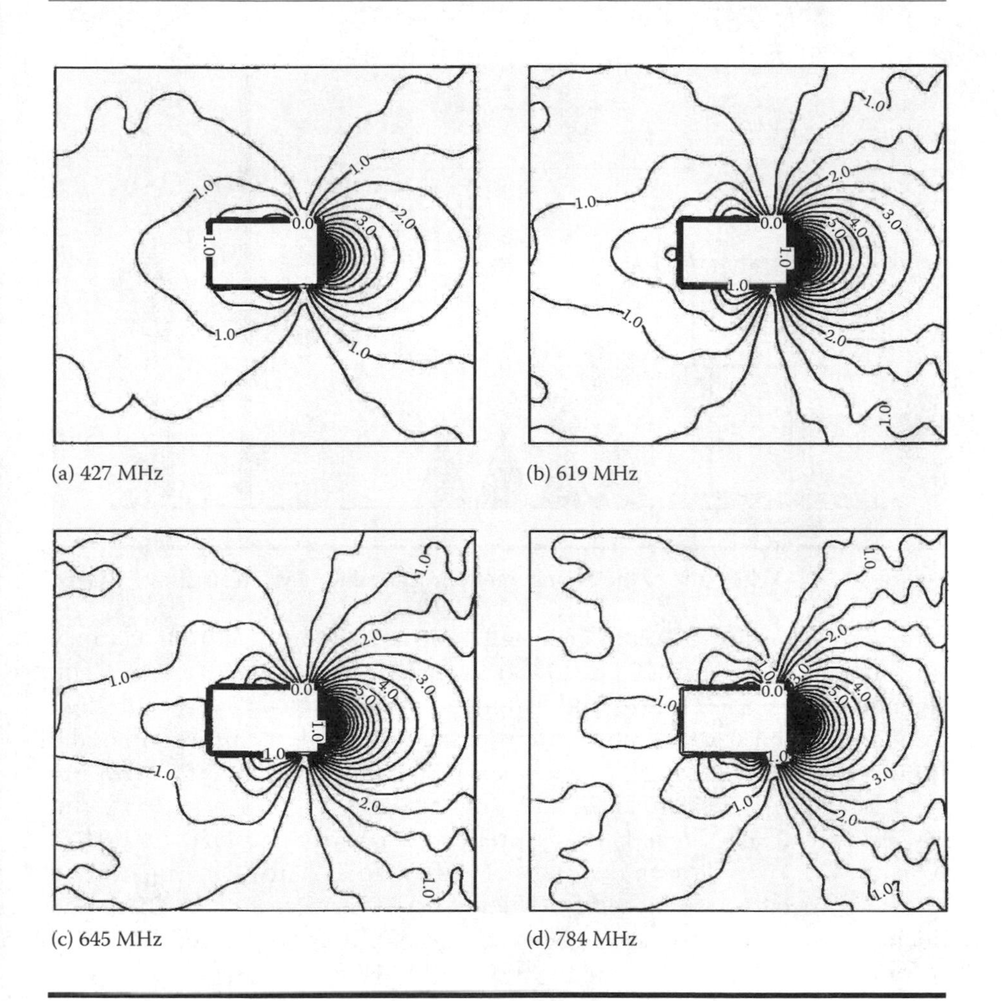

Figure 13.11 Electric field profile from aperture S2.

mechanisms are operating and that resonances (whole-body or in cavities such as the eye) can cause damage to humans.

The International Non-Ionizing Radiation Committee (INIRC) sets limits but in addition each country and each organization has its own ideas as to the most appropriate limits. The situation is unsatisfactory, as the basic interaction mechanisms at the cell level are not fully understood and the epidemiological evidence is confusing. The INIRC sets different occupational exposure limits and limits for the general public. For example, it is recommended by the INIRC that the absorption rate in the body over a 6-minute period should not exceed 0.4 W/kg for workers or 0.08 W/kg for members of the general public, at frequencies above 10 MHz. Guidance as to the exposure limits recommended in the United Kingdom may be

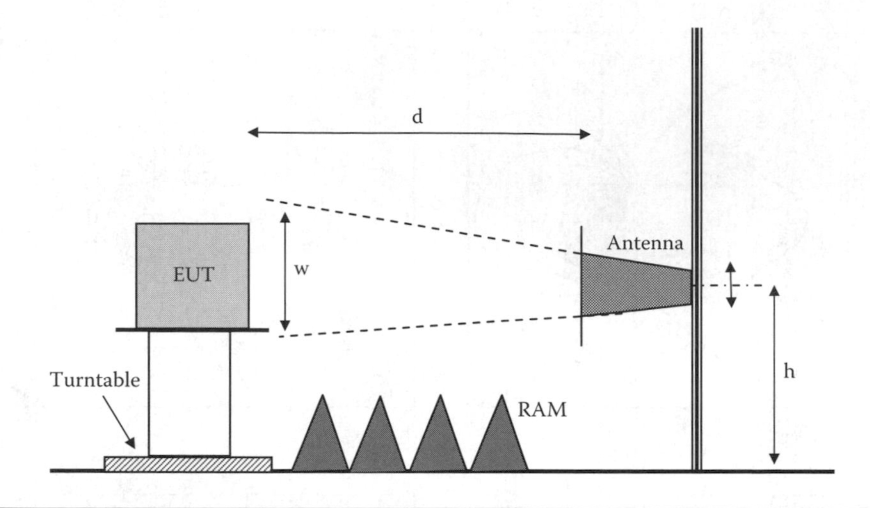

Figure 13.12 Schematic of the arrangement proposed for EMC tests above 1 GHz.

found in Reference 27. Further information regarding health effects may be found in Reference 28. At low frequencies (below 100 to 300 MHz) separate limits for electric and magnetic fields are set, while at high frequencies limits on the power density are specified. As an example, the IEEE recommends that at frequencies below approximately 1 MHz the electric and magnetic field should not exceed 614 V/m and 163 A/m, respectively. These limits fall 1/f to reach 27.5 V/m at 30 MHz and 0.073 A/m at 100 MHz. Power densities of less than 0.2 mW/cm^2 up to 300 MHz are specified, rising with frequency to 10 mW/cm^2 at 15 GHz. Full details may be found in Reference 29. The NRPB in the United Kingdom recommends 1 mW/cm^2 (30 to 400 MHz) rising with frequency to 5 mW/cm^2 at 2 GHz.[27] There is substantial variability in limits set in different countries. As an example, limits for exposure of the general public to power frequency (50/60 Hz) electric fields range between 1 and 30 kV/m. The interested reader is advised to consult national and international authorities and keep in touch with developments.

Detailed guidelines on exposure up to 300 GHz are provided by the ICNIRP in Reference 30.

The IEEE in 2005 updated its standard referred to in Reference 29 with guidelines that are not identical to those in Reference 30 or in Reference 29. For a summary of this standard and comparisons to other available guidelines the reader should consult Reference 31.

Regarding the impact of wireless technologies in this area, there have been numerous studies connected with possible adverse health effects due to mobile phones. A wide-ranging study commissioned by the U.K. government has looked at all the evidence and produced its recommendations

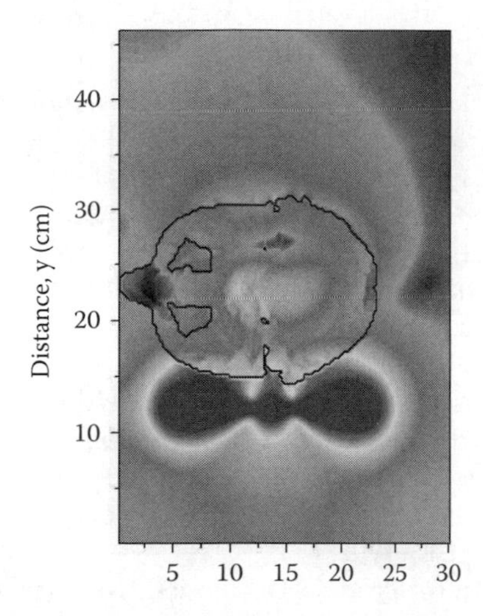

Figure 13.13 Electric field profile on a horizontal plane through the head for a horizontally polarized phone (900 MHz).

contained in the so called "Stewart Report."[32] While no specific problems were identified, the tone of the recommendations is to avoid unnecessary exposure, especially for children.

A recent review of standards related to mobile phones is presented in Reference 33, where the approach adopted in setting the specific absorption rate (SAR) below what is regarded as the threshold for adverse effects (4 W/kg) is explained. Occupational limits are set ten times lower (0.4 W/kg) and general public limits fifty times lower (0.08 W/kg).

Other studies focused on specific wireless technologies are reported in References 34 and 35.

Numerous studies of field penetration inside a model of a human head in the presence of a mobile phone have been done to help calculate the SAR in different parts. These numerical models can be very sophisticated and include the parameters of a great number of types of tissue. An example of such a calculation obtained with the Nottingham TLM solver is shown in Figure 13.13. The SAR is then obtained from

$$\mathrm{SAR} = \frac{|E|^2 \sigma_e}{2\rho}$$

where σ_e is the electrical conductivity of the tissue and ρ is its mass density.

Finally, attention is drawn to the EU Physical Agents Directive, which is all encompassing and aims at protecting workers against risks arising from physical agents such as electromagnetic fields.[36]

References

1. Showers, R M, "Influence of IEC work on national electromagnetic compatibility standards," Proc 10th Int Zurich EMC Symp — Supplement, pp 7–12, 9–11 Mar 1993.
2. Bronaugh, E L, "The EMC society contribution to EMC standards," Proc 10th Int Zurich EMC Symp — Supplement, pp 1–6, 9–11 Mar 1993.
3. "On the approximation of the laws of the Member States relating to electromagnetic compatibility 89/336/EEC," Official Journal of the European Communities, No L139/19–26, 25/5/89.
4. "Commission communication in the framework of the implementation of the 'new approach' directives 92/344/10," Official Journal of the European Communities, No C44/12 - 19/2/92 and No C90/2 - 10/4/92.
5. Marshman, C, "The Guide to the EMC Directive 89/336/EEC," EPA Press, Wendens Ambo, UK, 1992.
6. Williams, T, "EMC for Product Designers," Newnes, Oxford, England, 1992.
7. Bond, A E J, "Implementation of the electromagnetic compatibility (EMC) directive (89/336/EEC) in the UK," Proc 10th Int Zurich EMC Symp — Supplement, pp 169–177, 9–11 Mar 1993.
8. http://ec.europa.eu/enterprise/newapproach/standardization/harmstds/reflist/emc.html.
9. Directive 2004/108/EC of the European Parliament and of the Council of 15 December 2004, on the approximation of the Laws of Member States relating to electromagnetic compatibility, Official Journal EU, 31 Dec. 2004 (L390/24).
10. Info Day "Revised EMC Directive," Brussels 3 Feb. 2005, http://ec.europa.eu/enterprise/electr_equipment/emc/revision/workshop.htm.
11. Okamura, M, "Standardization in Japan in connection with CISPR Publication 11," Proc 10th Int Zurich EMC Symp — Supplement, pp 165–168, 9–11 Mar 1993.
12. Wall, A L, "Potential impact of changing international situation," Proc IEEE Int EMC Symp. Dallas, Texas, pp 1–4, 9–13 Aug 1993.
13. Ran, L, Clare, J, Bradley, K J, and Christopoulos, C, "Conducted electromagnetic emissions in induction motor drives systems Part I: Time domain analysis and identification of dominant modes," IEEE Trans. on PE, 13(4), pp 757–767, 1998.
14. Ran, L, Gokani, S, Clare, J, Bradley, K J, and Christopoulos, C, "Conducted electromagnetic emissions in induction motor drives systems Part II: Frequency domain models," IEEE Trans. on PE, 13(4), pp 768–776, 1998 (see also *erratum* in IEEE Trans. on PE, 13(6), Nov. 1998, page 1229).
15. Ran, L, Clare, J C, Bradley, K J, and Christopoulos, C, "Measurement of conducted electromagnetic emissions in PWM motor drive systems without the need for a LISN," IEEE Trans. on EMC, 41(1), pp 50–55, 1999.
16. Mertel, H K, "Military EMC Standardization," Proc 10th Int Zurich EMC Symp — Supplement, pp 13–30, 9–11 Mar 1993.

17. "Requirement for the control of electromagnetic interference emissions and susceptibility," MIL-STD-461D, Department of Defence, Washington, DC, 11 Jan 1993.
18. "Electromagnetic Compatibility," DEF-STAN 59-41 Parts 1 to 4, Ministry of Defence, Glasgow, U.K., 10 June 1988.
19. Marvin, A C and Simpson, G, "Screened room radiated emission measurement: A preliminary calibration procedure for the frequency range 1MHz to 30MHz," Conference Digest, Fifth British EM Measurement Conference, Organized by NPL, Winter Gardens, Malvern, pp 35/1–35/5, 11–14 Nov 1991.
20. Cook, R J, Bansal, P S, and Alexander, M J, "The measurement of field strength in a screened enclosure," Conference Digest, Fifth British EM Measurement Conference, Organized by NPL, Winter Gardens, Malvern, page 36/1, 11–14 Nov 1991.
21. Herring, J L and Christopoulos, C, "Numerical simulation for better calibration and measurements," Conference Digest, Fifth British EM Measurement Conference, Organized by NPL, Winter Gardens, Malvern, pp 37/1–37/4, 11–14 Nov 1999.
22. "Report on the Stage III research into the damping and characterization of screened rooms for radiated emission testing in Defence Standard 59-41," UK Ministry of Defence, 31 Jan 2005, http://www.dstan.mod.uk/59-41report.pdf.
23. Carter, N J, "The EMC testing of aircraft equipment," AGARD Lecture Series: Electromagnetic Compatibility, LS-116, pp 6.1–6.20, Aug. 1981.
24. Recommendation ITU-R P.372-8, "Radio Noise."
25. "Advances in Site validation techniques above 1GHz," Workshop MO-AM-WS-3, 2004 IEEE Int. Symp. On EMC, 9-13 Aug. 2004, Santa Clara, CA.
26. Van Dijk, N, Stenumgaard, P F, Beeckman, P A, Wiklundh, and Stecher, M, "Challenging research domains in future EMC basic standards for different applications," IEEE EMC Society Newsletter, Spring 2006, issue no 209, pp 80–86.
27. "Guidance as to restrictions on exposures to time-varying EM fields and the 1988 recommendations of the International Non-Ionizing Radiation Committee," NRPB-GS11, National Radiological Protection Board, UK, 1989.
28. Fischetti, M, "The cellular phone scare," IEEE Spectrum, pp 43–47, 30 June 1993.
29. "IEEE Standard for safety levels with respect to human exposure to radio frequency electromagnetic fields, 3kHz to 300GHz," IEEE C95.1–1991.
30. International Commission on Non-Ionizing Radiation Protection, "Guidelines for limiting exposure to time-varying electric, magnetic and electromagnetic fields (up to 300 GHz)," Health Physics, 74, pp 494–522, (1998), or, www.icnirp.de/documents/emfgdl.pdf.
31. Lin, J C, "A new IEEE Standard for safety levels with respect to human exposure to radio-frequency radiation," IEEE Antennas and Propagation Magazine, 48(1), pp 157–159, 2006.

32. Independent Expert Group on Mobile Phones, "Mobile phones and health," June 2000, www.iegmp.org.uk/report/text.htm.
33. Faraone, A and Chou, C-K, "Overview on standards related to the safety and compliance of mobile phone wireless transmitters," EMC Europe 2006, Sept. 4–8, 2006, Barcelona, pp 390–395.
34. Carrasco, A C, Gati, A, Wong, M-F, and Wiart, J, "Human exposure induced by operating wireless systems," EMC Europe 2006, Sept. 4–8, 2006, Barcelona, pp 468–472.
35. Ruddle, A R, "Influence of passengers on computed field exposure due to personal TETRA radio used inside a vehicle," EMC Europe 2006, Sept. 4–8, 2006, Barcelona, pp 592–597.
36. "Directive 2004/40/EC of the European Parliament and of the Council 29 April 2004, on the minimum health and safety requirements regarding exposure of workers to risks arising from physical agents (electromagnetic fields)," Official Journal of the European Union, 24.5.2004, L 184/1-L 184/9.

Chapter 14

EMC Measurements and Testing

The purpose of this chapter is to introduce the reader to the main experimental techniques used in EMC studies. A substantial part of published standards contains detailed experimental procedures for EMC testing. It is not possible or advisable to repeat here in detail all the relevant procedures. Instead, the basic experimental tools and test environments are described and emphasis is placed on pointing out difficulties that may lead to errors and poor repeatability of measurements. The reader is advised to consult the relevant standards prior to performing EMC measurements.

14.1 EMC Measurement Techniques

Measurements are performed to meet two requirements. First, measurements to ascertain the emission and susceptibility of equipment are necessary throughout the design phase. The purpose of these measurements is mainly diagnostic, in that it helps to identify likely problem areas and tests the effectiveness of various remedies. These measurements are under the complete control of the designer and test engineer and hence any number of a range of techniques may be used according to circumstances. Second, tests on complete equipment are prescribed by standards and these are mandatory in most cases. The measurement arrangement and receivers and transducers used are normally described in considerable detail. Hence, there are certain aspects of the measurement that are fully specified and there is little scope for variation. In many cases, formal tests required by certification authorities must be performed by accredited laboratories. If the equipment under test (EUT) fails to meet the standard, it is brought back to the design office for modification and it may thus undergo further diagnostic and standards testing. The purpose of EMC design is to reduce the need for retesting and modification to the bare minimum. Testing involves, depending on the particular standard, frequencies ranging from a few kilohertz to several gigahertz. Standards specify tests for emission and susceptibility both conducted and radiated as described in Chapter 13. It is obvious that testing on such a scale involves a grasp of many electromagnetic phenomena and it is subject to many uncertainties. It is fair to say that EMC measurements are, in general, less accurate when compared to other high-frequency measurements. The reasons for this are complex, as will be explained later. Recognition of this fact does not, however, absolve the experimenter from the responsibility of ensuring the lowest possible uncertainty in measurements. The purpose of this chapter is to present the main experimental techniques and outline areas where large uncertainties may be introduced. Detailed test arrangements and schedules may be found in the published standards.

A fully specified EMC test facility requires a substantial investment in buildings, equipment, and personnel and it is not within the reach of medium- to small-sized companies. However, useful diagnostic work may be done with modest resources and indeed it is essential that experience of measurements is part of the range of skills available to the design engineer. The material that follows should therefore be of value to all those involved with EMC.

A brief survey of EMC measurements may be found in Reference 1.

14.2 Measurement Tools

Irrespective of the formal standard involved, a number of basic tools are required to do EMC measurements. The issues involved in their selection, characterization, and use are briefly described below.

14.2.1 Sources

Sources are required for all susceptibility measurements. As is evident from the description of the various standards in Chapter 13, there is a lack of detailed commercial product standards for susceptibility compared to the corresponding situation for emission. A good start for ascertaining requirements is the IEC-801 standard.[2] The various parts of this standard are incorporated into national and European standards (e.g., BS6667: Part 1, BS EN60801: Part 2, BS6667: Part 3). Part 3 defines three severity levels over the frequency range 27 to 500 MHz, where the test field strength in V/m is 1, 3, and 10. No sources are capable of generating such fields directly. The usual configuration is for a low-power source to feed a broadband RF power amplifier capable of delivering a few watts of RF power. The selection of the amplifier is based on the required gain, linearity, and the practical requirement that it must be capable of feeding mismatched loads. A low-level power source covering frequencies at least up to 1 GHz is also necessary. It should be capable of feeding the power amplifier and it is also useful if a basic signal modulating facility is available.

Special source requirements apply to the testing for electrostatic discharge and fast transients. Part 3 of IEC801[2] describes a simplified ESD generator consisting of a high-voltage source charging a 150-pF capacitor via a high resistance (50 to 100 MΩ). The capacitor is discharged through a 330-Ω resistor and the load via a discharge switch. Using this basic arrangement, pulses of a very short rise-time (less than 1 nsec) can be produced. Charging voltages ranging from 2 to 8 kV and 2 to 15 kV for contact and air discharges, respectively, are specified. Further details of the required waveforms are given in the standard. Generators for this type of test are commercially available. Many factors, including environmental conditions, affect the rise-time of the injected current, as described in References 2 and 3.

IEC801 Part 4[2] describes tests for fast transients. These involve bursts of fast pulses with a rise-time typically of 5 nsec and repetition rate of 5 kHz. A burst lasts for 15 msec and is repeated every 300 msec. Severity levels ranging from 0.5 to 4 kV are suggested. Testing involves special generators capacitively coupled to equipment cables.[4]

14.2.2 Receivers

In EMC testing, interference is measured using receivers of a specified bandwidth and detector function. The most common instruments used for this purpose are measurement receivers and spectrum analyzers. Several units may be necessary to cover the entire frequency range and they represent a significant investment. Measurement receivers have better sensitivity, higher dynamic range, and are less susceptible to overloads compared to a basic spectrum analyzer. The latter feature is due to the presence of a tuned preselector in the measurement receiver, which limits the total amount of power in the input mixer. A preselector and tracking generator may be added to a spectrum analyzer to obtain a performance comparable to that of a measurement receiver at a comparable cost. Hence, in many ways the two instruments offer similar performance. Two issues determine the configuration of receivers for EMC measurements, namely, the bandwidth and the type of detector.

Clearly, the response of a receiver to a signal will be different depending on whether its bandwidth is narrower or wider than the signal bandwidth. For a narrowband signal, such as an analogue radio signal, the amount of energy reaching the detector is independent of the receiver bandwidth (provided the spectrum of the narrowband signal is narrower than the receiver bandwidth). If this last condition is not satisfied, the signal is classed as wideband, and the energy reaching the detector depends strongly on the receiver bandwidth. Thus, in order to make meaningful comparisons between various tests, it is necessary to specify the receiver bandwidth used in the measurements. Complications arise depending on the exact nature of the broadband signal being measured.[5] If the signal is noise-like (uncorrelated), an increase in bandwidth by a factor of 10 results in a tenfold increase in the energy reaching the detector, i.e., 10 dB. In contrast, in the case of a pulse with a spectrum consisting of a number of spectral lines phase-related to each other, increasing the bandwidth by a factor of ten increases the total voltage by the same factor and hence the energy reaching the detector by 20 dB. Receiver bandwidths are specified in international and national standards. As an example, CISPR Publication 16 specifies a 6-dB bandwidth in kilohertz of 0.2, 9, and 120 in the frequency ranges 10 to 150 kHz, 0.15 to 30 MHz, and 30 to 1000 MHz, respectively.[6]

The nature of the detector function was discussed in Section 4.4, where it was explained that, depending on the choice of the charge/discharge components, the detector may respond to the peak, average, or any other measure of the signal. There are three basic types of detector function, namely, peak, quasipeak, and average. The detailed nature of these detector functions is described in References 6 and 7. The various standards specify different limits depending on the detector type. The peak detector has a fast response and it is used extensively for diagnostic testing. It is clear that the response of each detector type will be different for a pulsed signal with different repetition frequency. Quasipeak and average type detectors require a long settling time (a substantial fraction of a second) at each frequency. Hence, when a wide range of frequencies must be covered, measurement time is quite long. The sweep rate must be selected so that the dwell time at each frequency is of the order of three time constants of the detector.

14.2.3 Field Sensors

In development and diagnostic work, it is often necessary to probe for the electric and magnetic field near transmission lines and individual circuit components. Similarly, it is often necessary to build up a picture of the field profile by numerous measurements using very small probes. Although such measurements are not specified by the various standards, they are nevertheless useful in giving an insight into the origin and mode of EM emissions. Several equipment manufacturers provide "sniffer" probes that are designed to respond to the near field from printed circuit tracks and other wiring. These give an indication of relative activity from various parts of a circuit, but cannot and should not be used to predict far-field performance. For the test engineer who wishes to develop in-house probes, the rudiments of design of small field sensors are described below.

Field sensors are small because a detailed mapping of the field may be required and also to avoid disturbing the field being measured. A sensor is electrically small if its largest physical dimension is much smaller (say 1/20) of the longest wavelength of interest. In its simplest form, an electric field sensor is a small piece of wire protruding above a ground plane (small monopole). If this sensor is placed in an electric field E, the open-circuit voltage is $V = l_e E$ where l_e is the effective length of the monopole. Normally, l_e is approximately equal to half the actual length of the monopole.[8] The short-circuit current is similarly equal to $I = C l_e \, dE/dt$, where C is the capacitance of the sensor. Thus, whether the sensor responds to electric field or its derivative depends on the impedance of the measuring instrument. If the time constant of the antenna capacitance with the measuring instrument input impedance is longer than the field characteristic time, then E is measured. Otherwise, the arrangement responds to dE/dt. Various modifications can be made to change the response as described in Reference 8, but at the expense of the already very low sensitivity of small sensors.

A small loop is the dual to the electric field sensor and provides a magnetic field sensor. If a loop of area A is immersed in a magnetic field H, the open-circuit voltage is $V = \mu_0 A \, dH/dt$ and the short circuit current is $I = \mu_0 AH/L$ where L is the inductance of the loop. It is clear, therefore, that either the magnetic field or its derivatives are measured depending on circumstances. If the time constant of the inductance with the input impedance to the measuring instrument is much larger than the characteristic time of the field, then the arrangement measures the magnetic field. Various field sensors have been developed to measure transient fields under hostile conditions and these may also be used as general instrumentation. Details may be found in References 8 and 9.

Another measuring arrangement that provides for very low distortion of the measured field by the measurement system is one based on a modulated scatterer. A very short piece of wire is used to probe the field. Incident fields are scattered by this wire and the scattered signal is received by a second antenna placed in the vicinity. This received signal contains complete information about the amplitude and phase of the original incident wave. The only difficulty is that the scattered signal is very weak in amplitude. In order to increase the chances of detection, a photosensitive component is placed at the center of the short probing wire and its impedance is modulated by a modulated light signal supplied through a fiber-optic link. In this way the scattered signal is also modulated and phase-sensitive detection techniques may be used to increase the chance of detection of even very weak signals. The sensitivity of this arrangement is low at low frequencies. Details of this technique may be found in Reference 10.

14.2.4 Antennas

Antennas are specified almost exclusively as the primary field measuring transducer in most standards. Understanding their behavior in actual test environments is thus of paramount importance. The basic theoretical background to the characterization of antennas in terms of their gain, effective aperture, radiation resistance, etc., was presented in Section 2.3.2. From the practical measurement point of view, the single most important parameter of an antenna is its antenna factor,[11] defined as

$$AF = \frac{\text{incident electric field } E_i \left(\text{in V/m}\right)}{\text{voltage measured at the receiver } V_L \left(\text{in V}\right)} \tag{14.1}$$

with the parameters defined shown in Figure 14.1a. Normally, the antenna factor is expressed in decibels, i.e.,

$$AF\ (\text{dB}) = E_i\ (\text{dB}\ \mu\text{V/m}) - V_L\ (\text{dB}\ \mu\text{V}) \tag{14.2}$$

Knowledge of the antenna factor and measurement of the receiver voltage permits the calculation of the electric field from Equation 14.2. Similar expressions apply for the antenna factor of antennas sensitive to the magnetic field.

The antenna factor may be calculated from first principles. It can be shown that for any antenna the effective aperture and the gain are related

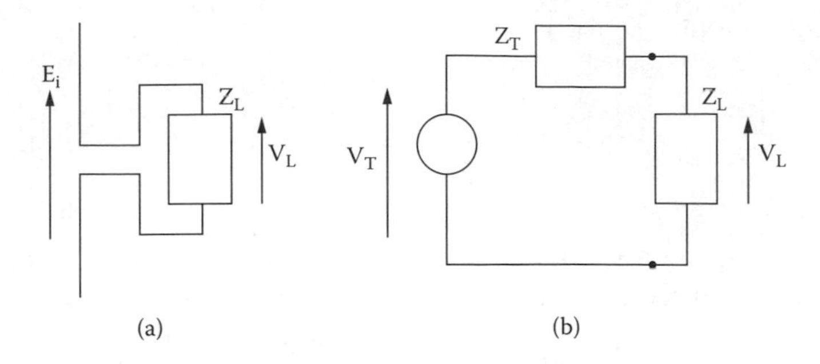

Figure 14.1 A receiving antenna (a) and its Thevenin equivalent (b).

by the expression $A_e = G\lambda^2/(4\pi)$. The antenna may be replaced by its Thevenin equivalent circuit as shown in Figure 14.1b. The maximum power available is that supplied to a conjugate load ($Z_L = Z_T^*$), and is therefore

$$P_{max} = \left|\bar{V}_T\right|^2 / 4R_T \tag{14.3}$$

where $|V_T|$ is an rms value and R_T the real part of the impedance Z_T. Assuming that the antenna is oriented for maximum response

$$P_{max} = \frac{\left|\bar{E}_i\right|^2}{Z_0} A_e \tag{14.4}$$

where $|E_i|$ is an rms value and Z_0 is the intrinsic impedance of the medium surrounding the antenna. Equating the right-hand side terms in Equations 14.3 and 14.4 and substituting for A_e gives

$$\left|\bar{V}_T\right| = \left|\bar{E}_i\right| \ell_e \tag{14.5}$$

where $l_e = \lambda\sqrt{GR_T/(\pi Z_o)}$ is described as the effective length of the antenna. In order to calculate the voltage at the receiver, Figure 14.1b is used to give

$$\left|\bar{V}_L\right| = \left|\bar{E}_i\right| l_e \left|\frac{Z_L}{Z_L + Z_T}\right|$$

Hence, the antenna factor is

$$AF = \frac{1}{l_e}\left|1 + \frac{Z_T}{Z_L}\right| \tag{14.6}$$

As an example, the AF of a half-wavelength resonant dipole in free space may be obtained from its gain and impedance $G \simeq 1.64$, $Z_T \simeq 73\ \Omega$.[12] Substituting in the expression for the effective length gives $l_e \simeq \lambda/\pi$. Substituting in Equation 14.6, assuming a receiver input impedance Z_L = 50 Ω and expressing the frequency f in MHz gives

$$AF\ (dB) = 31.78 + 20 \log f\ (MHz) \tag{14.7}$$

It should be noted that in this theoretical determination of the antenna factor it has been assumed that the antenna is oriented to give maximum response and also that the input impedance to the receiver has a specified value. It should be borne in mind that the AF will be different if any of these assumptions is invalid. The derivation leading to Equation 14.7 is based on the assumption that the resonant half-wavelength dipole is placed in free space and that the current distribution is sinusoidal. The latter assumption is good provided that the diameter of the antenna is very small. Inevitably, in practical measurements the antenna is not in free space as there is always, in close proximity, the ground and other structures. The proximity of the antenna to ground or other antennas modifies its impedance. In the case of two antennas 1 and 2 as shown in Figure 14.2, the voltage at the terminals of antenna 1 is

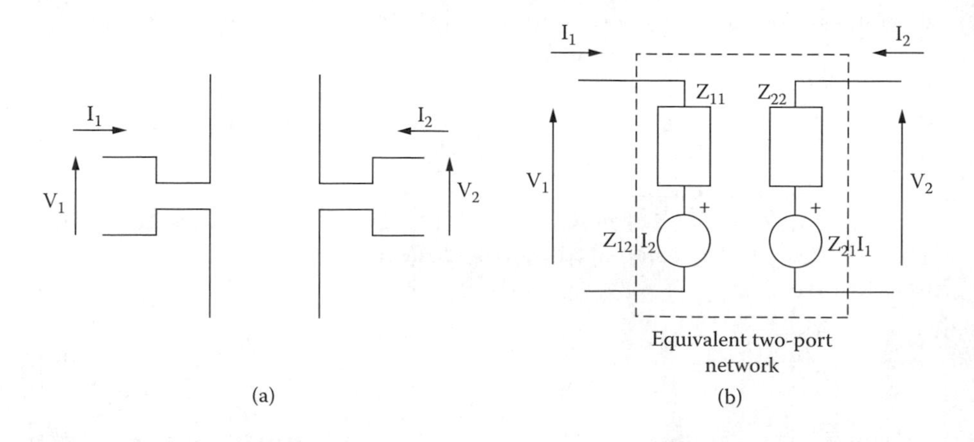

Figure 14.2 Coupled antennas (a) and their two-port equivalent (b).

$$\bar{V}_1 = Z_{11}\bar{I}_1 + Z_{12}\bar{I}_2 \tag{14.8}$$

where Z_{11} is the self-impedance of antenna 1 (impedance value under free-space conditions) and Z_{12} is the mutual impedance between the two antennas. The impedance of antenna 1 is thus

$$Z_1 = \frac{\bar{V}_1}{\bar{I}_1} = Z_{11} + \frac{\bar{I}_2}{\bar{I}_1} Z_{12} \tag{14.9}$$

Equation 14.9 makes clear that proximity effects must be taken into account in the calculation of Z_1.

The antenna factor in Equation 14.6 remains constant only if l_e and Z_T remain constant. Since in the case of close proximity between two antennas $Z_T = Z_1$, with Z_1 given by Equation 14.9, it is clear that the antenna factor will vary with the relative position between the two antennas and with any other factor affecting the current ratio $\bar{I}_2/\bar{I}_1$. A particular application of these ideas is in the case of an antenna above a ground plane. Two cases may be distinguished of antennas polarized in the horizontal and vertical directions. The effect of currents induced in the ground plane can be represented by an image antenna placed symmetrically with respect to the ground plane. The original antenna and its image form a two-antenna system in free space and can thus be described using Equations 14.8 and 14.9. The current in the image antenna is chosen to satisfy the boundary conditions at the position of the ground plane and this requires that $\bar{I}_2/\bar{I}_1 = +1$ for vertical polarization and $\bar{I}_2/\bar{I}_1 = -1$ for horizontal polarization. It is clear from Equation 14.9 that Z_1 is affected by the height of the antenna above the ground plane and by its polarization. It also follows from Equation 14.6 that the antenna factor also depends on these parameters. Another matter of practical significance, which affects the antenna factor, is the presence of other components, such as cable and/or a balun between the antenna and receiver. A full characterization of these components in terms of their S-parameters is necessary, as described in detail in Reference 13.

Antennas supplied by manufacturers are usually provided with a generic calibration, i.e., an antenna factor at different frequencies for this type of antenna. The antenna factor of the particular antenna supplied may deviate by several decibels from the generic value. For accurate measurements, it is thus necessary to calibrate antennas using either approved test houses or in-house facilities. There are basically three experimental techniques used for antenna calibration. In the standard field method a receiving antenna is calibrated against a calibrated transmitting antenna establishing a known field at a fixed observation point. Alternatively, an antenna of a known antenna factor may be used for calibrating other antennas in the standard antenna method. Finally, in the standard site method three uncalibrated antennas may be calibrated by doing three measurements between pairs of antennas. The accuracy of this calibration depends on the quality of the site, as discussed in detail in Reference 14. An alternative approach to antenna calibration is to use numerical simulation to establish the AF of an antenna-to-plane-wave illumination. This approach has been successfully applied, using the NEC computer code, to dipole and biconical antennas as described in Reference 13. It should always be borne in mind that the proximity of antennas to conducting structures and their use in screened rooms affects their calibration. A discussion of the scale of these affects may be found in References 15 and 16.

A number of different antennas is in use for EMC measurements depending on the frequency range. At low frequencies, below 30 MHz, it is common to use active antennas to achieve satisfactory matching. For electric field measurements, a monopole antenna known as the 41-in rod antenna is in common use. Because of the size and frequency of operation of this antenna it is difficult to calibrate. An approach based on the simulation techniques described in Reference 13 has been found the most acceptable. For low-frequency magnetic field measurements loop antennas are commonly used and are calibrated in standard TEM cells.

At frequencies above 30 MHz tunable half-wavelength dipoles may be used. Alternatively, broadband antennas may be employed to speed up measurements. In the range 30 to 300 MHz the biconical antenna is most commonly used. Normally, it has a total length of 1.3 m and an integral balun. Some designs exhibit a narrow resonance at approximately 278 MHz due to internal resonances in the cones. The addition of an extra wire inside each cone can reduce this resonance. For frequencies between 300 and 1000 MHz a log-periodic antenna may be used. Above frequencies of 1 GHz, broadband horn antennas are employed.

14.2.5 Assorted Instrumentation

A number of other components are necessary for EMC testing. As already mentioned, a line impedance stabilizing network (LISN) is necessary for conducted interference measurements.[6] The purpose of the LISN is twofold: first to prevent HF interference from the mains supply contaminating the measurement of the EM noise emanating from the equipment under test; second, to present an approximately constant impedance looking toward the mains to noise current emanating from the EUT. This impedance is 50 Ω over the range of frequencies 150 kHz to 30 MHz. The latter requirement is necessary since the impedance of the supply to HF currents

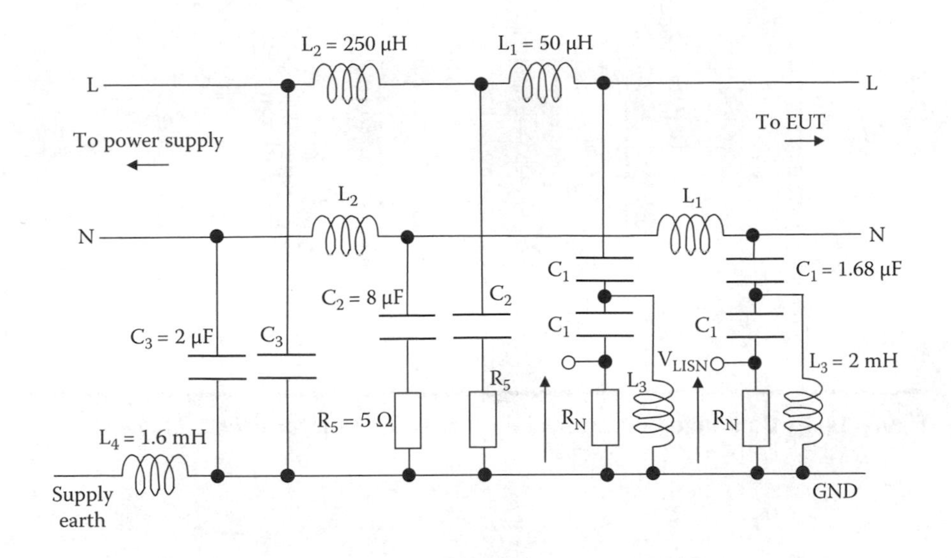

Figure 14.3 Schematic of a typical line impedance stabilizing network (LISN). Resistors R_N represent the input impedance (typically 50 Ω) of the measuring instrument.

varies widely depending on the location of the tests, and if this situation were allowed to persist it would result to poor repeatability of tests. A typical design of an LISN is shown in Figure 14.3. The resistors R_N are 50 Ω and represent the input impedance of the measuring instrument (spectrum analyzer, receiver). The isolation of the measurement set up from EMI originating from the power network is accomplished by components C_3, L_1 and L_2, C_2. The large inductor L_4 isolates at HF the reference of the LISN from the supply earth (which is not "clean" at HF). Capacitors C_1 are a DC block to protect the measuring instrument from overload and L_3 discharges these capacitors should the measuring instrument (and hence the 50-Ω resistor) be disconnected. With this arrangement the conducted emissions originating from the EUT are measured across 50 Ω. It is clear that inductors L_1 and L_2 must be rated to carry the full supply load at 50 or 60 Hz. This makes for a bulky and expensive design (especially for three-phase large loads). For this reason techniques were developed to achieve the same aims as the LISN but using a simpler and cheaper probe.[17]

Example: A three-phase motor drive is connected for testing as shown in Figure 14.4. The CM current on the reference conductor is 100 dBμA. Calculate the LISN voltage.

Solution: The CM is shared equally by the three branches of the three-phase LISN. Hence

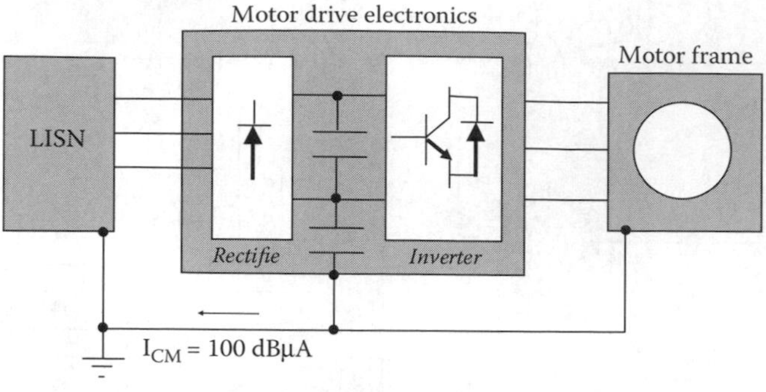

Figure 14.4 Conducted emissions in a three-phase motor drive.

$$V_{LISN} = \frac{I}{3}(50\ \Omega)$$

$$V_{LISN}(\text{dB}\mu\text{V}) = I(\text{dB}\mu\text{A}) - 20\log 3 + 20\log 50$$

$$= 100 - 9.54 + 33.98 = 124.44\ \text{dB}\mu\text{V}$$

Other items, such as ferrite clamps and current probes, can be placed around cables to measure currents (i.e., acting as current transformers) as specified in standards.[6]

Measurement of fast pulses requires broad bandwidth oscilloscopes. In Equation 4.15 we have seen that the description of a signal in the frequency and time domains results in a relationship between duration and bandwidth. In measurements, it is necessary to provide adequate bandwidth in order to measure accurately fast rise-time pulses. In traditional analogue instruments, where a number of circuit elements are joined together, we obtain a response that is Gaussian in nature. Under these circumstances the rise-time and bandwidth of the oscilloscope are related by the approximate formula

$$RT \simeq 0.35/BW$$

If several devices are cascaded together (e.g., scope and probe), then the overall bandwidth is given by the formula

$$\text{system bandwidth} = \frac{1}{\left[\left(\frac{1}{BW_{probe}}\right)^2 + \left(\frac{1}{BW_{scope}}\right)^2 + \ldots\right]^{1/2}}$$

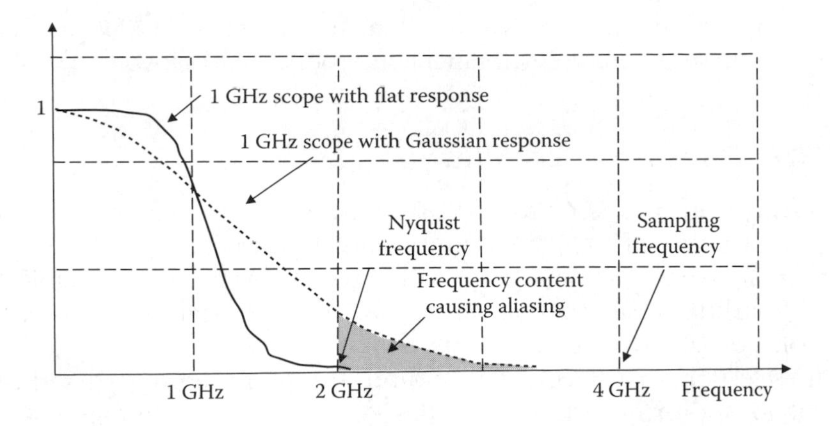

Figure 14.5 Frequency response of a 1-GHz bandwidth scope with a 4-GHz sampling rate.

When measuring a signal with a measurement system of finite rise-time, then the measured rise-time is

$$\text{measured risetime} = \left[\mathrm{RT}_{\text{signal}}^2 + \mathrm{RT}_{\text{system}}^2 \right]^{1/2}$$

Modern instruments contain fewer analogue amplifiers and use digital signal processing techniques to optimize response. Hence, the expressions above do not strictly hold. These instruments do not have a Gaussian response but instead have a maximally flat or brick-wall response as shown in Figure 14.5. Here the equivalent rise-time is

$$\mathrm{RT} = \frac{\mathrm{N}}{\mathrm{BW}}$$

where N ranges from 0.4 to 0.5, and the higher the value the flatter the response is. We also note in this figure the Nyquist frequency, which is half the sampling frequency. Any signal content above the Nyquist frequency causes aliasing errors (sampling theorem). A thorough discussion of the issues involved is given in Reference 18.

14.3 Test Environments

EMC measurements are normally performed in open-area test sites, in screened rooms, or in special test cells. Standards specify in some detail the basic requirements for each type of test environment and the experimental procedures to be followed. The problems encountered in EMC measurements in these different environments are described below.

14.3.1 Open-Area Test Sites

An ideal open-area test site (OATS) consists of a perfectly conducting ground plane of infinite extent, free from all obstructions, with very low levels of ambient electromagnetic noise. Actual OATS depart in significant ways from this ideal. They are nevertheless used extensively, especially for commercial EMC measurements.

In selecting an OATS, a flat ground area is necessary, of sufficient extent to approximate the properties of an ideal site (infinite plane). Assuming that a finite flat area is all that can in practice be available, it is required that the difference in path length between a ray leaving a transmitter and reaching a receiver, after reflections from the center of the flat area and from its edge, should be equal to at least half a wavelength. This is the so-called first Fresnel zone. Negligible contributions to the received signal are made from reflections outside this zone.[19] The lower the frequency and the higher the receiver and transmitter are placed above the ground plane, the larger the size of the first Fresnel zone will be and hence a large flat area is required. CISPR regulations specify that the flat area is at least an ellipse having major and minor axes equal to 2L and $\sqrt{3}L$, respectively, where L is the distance between receiver and transmitter. Typically, this distance is either 3, 10, or 30 m. Ideally, this flat area should also be perfectly conducting. In practice, a somewhat smaller metallized area constructed out of wire-mesh material is used. The mesh must be made out of material that does not corrode, of a welded loop construction, with a loop diameter smaller than one tenth of the smallest wavelength of interest. The mesh must make good electrical contact with the undersoil and surrounding area. The required flatness depends on the frequency of operation and the distance between transmitter and receiver and it is typically a few centimeters for test frequencies up to 1 GHz.

Understanding measurements in OATS requires a grasp of the fact that the signal at the observation point is the combination of signals arriving directly and after reflection from the ground, as shown in Figure 14.6a. Calculations are performed by replacing the ground plane by an image antenna, as shown in Figures 14.6b and c for vertical and horizontal polarizations, respectively. The direction of current flow in the image antenna is dictated by the boundary conditions on the ground plane and results in image currents that, in broad terms, strengthen (for vertical polarization) and weaken (for horizontal polarization) the original antenna current. The total field at the observation point is that due to two antennas

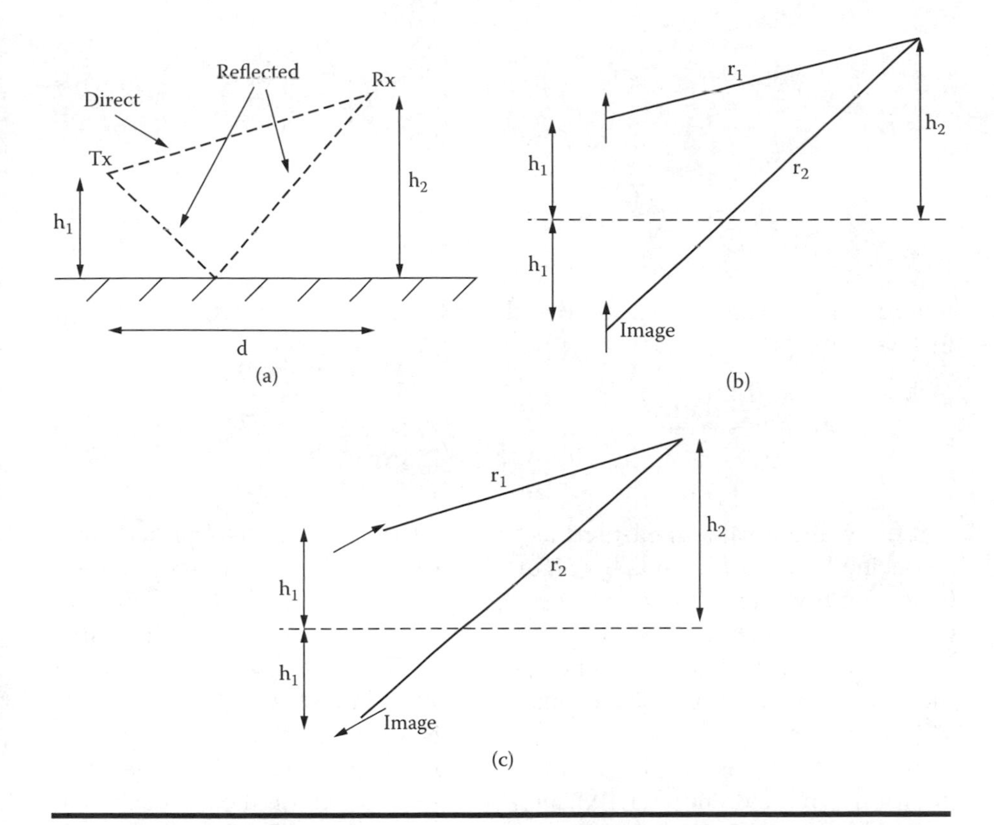

Figure 14.6 Antenna coupling in the presence of ground (a). Images for vertically (b) and horizontally (c) polarized antennas.

at distances r_1 and r_2 and may be calculated, for short dipoles, using formulae derived in Chapter 2. The two contributions differ in amplitude and in phase and it is clear that under particular circumstances the two signals may be added constructively or destructively. For a particular observation point and the same radiator strength, the response at low frequencies is higher for vertical polarization by approximately 10 dB. As r_1 and r_2 become comparable to the wavelength, phase differences become important and the total signals for vertical and horizontal polarizations become comparable in magnitude. At even higher frequencies a strong interference pattern is observed.

Before any OATS may be used to make EMC measurements, its suitability must be established by comparison to an ideal or standard test site. This is done by measuring the site attenuation (SA) and the normalized site attenuation (NSA) and comparing with values for an ideal site.[20,21] These two quantities are defined by the following expressions

$$SA = 20 \log \left| \frac{V_1}{V_2} \right|_{min} \tag{14.10}$$

$$NSA = \frac{SA}{AF_T AF_R} \tag{14.11}$$

where V_1 is the voltage measured at the receiver in the arrangement shown in Figure 14.7a,

$$\left| \frac{V_1}{V_2} \right|_{min}$$

is the minimum value measured as the receiving antenna is scanned over a specified range of heights h_z (Figure 14.7b), and AF_T, AF_R are the antenna factors of the transmitting and receiving antennas, respectively. Scanning the receiving antenna is necessary in order to take account of the interference between direct and reflected waves. The antennas used in these measurements are normally resonant dipoles connected to 50-Ω systems. The site is considered acceptable if its NSA is within ±4 dB of the theoretical NSA. Details of this measurement and tables of the theoretical NSA may be found in Reference 21. A detailed examination of the inaccuracies involved in the measurement of site attenuation and methods of improving these measurements may be found in Reference 13.

Once an OATS has been found to be acceptable, the test procedure for a radiated emission test follows well-defined lines. The EUT is placed

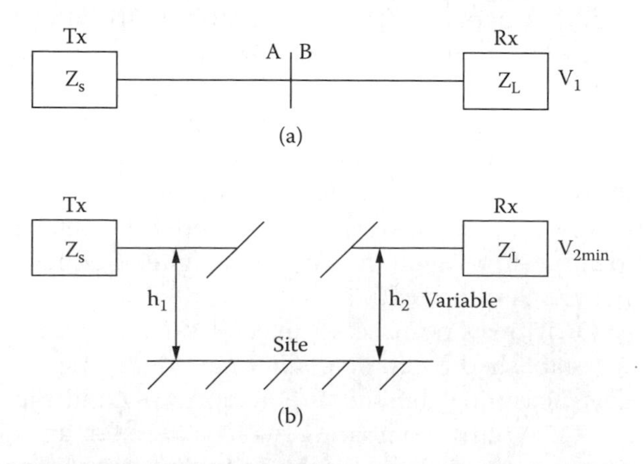

Figure 14.7 Tests used to measure site attenuation.

on a table 0.8 m in height and the measurement antenna is placed at a specified distance (3, 10, or 30 m) from it. The height of the antenna is varied between 1 and 4 m (or between 2 and 6 m for the 30-m range) and the EUT is rotated until maximum emission is observed. The test is repeated for horizontal and vertical polarizations. The tests are speeded up considerably if the EUT is placed on a turntable, the antenna is mounted on a mechanized mast, and the entire instrumentation, including the receiver, is under software control. Such systems are commercially available, but at a substantial cost.

A comparison of testing requirements to different standards (FCC, ANSI C63.4, EN55022/CISPR 22) may be found in Table 3 of Reference 22.

14.3.2 Screened Rooms

Open-area test sites have several disadvantages and this has led to a search for alternative EMC test environments. Among these disadvantages are the difficulty of ensuring a clean EM environment, dependence on the weather, and land costs. In the case of immunity tests, it is also difficult to avoid interfering with other users of the EM spectrum. Most military standards, immunity, or general diagnostic measurements, are therefore made in screened rooms. A screened room is an all-metal structure where all access for personnel, electrical, or mechanical services is designed in such a way as to provide a high degree of electromagnetic isolation up to very high frequencies. A well-designed screened room can be used for emission and immunity measurements without any EM interaction with the external environment. Screened rooms may in turn be distinguished as reverberating and anechoic types. In the former case all the internal surfaces of the room are unlined and highly conducting and the room is thus an electromagnetic cavity. Any EUT or antenna placed in this room interacts with the conducting surfaces in a complex manner. An empty rectangular cavity exhibits resonances at specific frequencies obtained from the formula given below:

$$f(\mathrm{MHz}) = 150\sqrt{\left(\frac{m}{a}\right)^2 + \left(\frac{n}{b}\right)^2 + \left(\frac{p}{c}\right)^2} \tag{14.12}$$

where a, b, and c are the internal room dimensions in meters and m, n, and p are integers with no more than one being zero. Typical room dimensions range from a few meters to a few tens of meters. The lowest resonant frequency depends on room dimensions and is generally of the order of a few tens of megahertz. Small rooms can be used to test small table-top products, whereas large rooms are used to test complete vehicles and may be equiped with dynamometers to exercise the vehicles. The electromagnetic field structure at each resonant frequency, obtained from Equation 14.12 (modes of the cavity), can be determined and depends on the numerical values of the indices m, n, and p. Thus, mode TE_{101} refers to a mode in which the vertical component of the electric field has a maximum on a line running from the center of the floor to the center of the ceiling (y-axis). Higher-order modes exhibit a more complex structure. The presence of resonances with pronounced field minima and maxima at particular frequencies make measurements in reverberating rooms difficult to interpret. In order to improve matters, the walls and ceiling may be lined with RAM to drastically reduce reflections. The resulting anechoic room has properties akin to those of an OATS. Various materials can be used as RAM, but some reflections remain, especially at low frequencies. Thus, most rooms, especially at low frequencies, can only be described as partially anechoic or highly damped. Each room type is described in more detail below.

Reverberating Rooms — A typical test arrangement specified in military standards is shown in Figure 14.8. The EUT is placed on a conducting bench, which is bonded to the room walls. A conducting extension to the bench may also be fitted to accommodate a rod antenna. Depending on the room height, the extension may be offset toward the ground to leave adequate clearance for the rod antenna. Biconical and log-periodic antennas are generally mounted on a mast. In military standards, a distance of 1 m is specified between the EUT and the measurement antenna. The

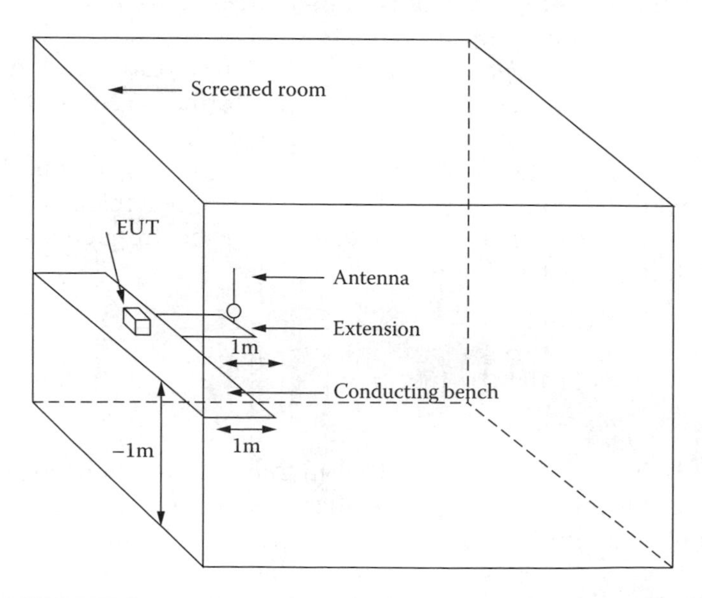

Figure 14.8 Typical test arrangement inside a screened room.

presence of the bench, extension, EUT, and antenna disturb the resonant pattern inside the room. Small shifts from the resonant frequency values obtained from Equation 14.12 are observed and the presence of the bench introduces new resonances. At low frequencies, understanding of the measurement may be achieved by resorting to lumped-circuit concepts. At frequencies for which the largest room dimension is smaller than $\lambda/10$, quasistatic concepts may be applied, whereby the coupling between EUT, antenna, and the room is essentially capacitive or inductive.[5] At higher frequencies and below the first resonant frequency of the room, coupling between the EUT and antenna is more complex. Apart from the direct coupling (capacitive or inductive) between the two, a further coupling TEM mode is established. This propagates on a coaxial line formed by the bench extension (inner conductor) and the room side walls (outer conductor). This line is terminated by a short circuit (back wall) and an open circuit (end of extension). The EUT couples capacitively and/or inductively with the line and the rod antenna forms a capacitive load near the open-circuit end. Further details of these models and comparisons with measurements may be found in References 23 and 24.

At even higher frequencies, the presence of cavity resonances makes the coupling mechanisms difficult to follow in detail and progress can only be made by resorting to sophisticated numerical modeling techniques. It is generally anticipated that measurement uncertainties in this complex environment can amount to up to ±40 dB. As a result there is little correlation between the same measurements taken inside different screened rooms or in an OATS.

The effect of resonances may be reduced by rotating an electrically large metal blade (size ~λ) in an arrangement known as a mode-stirred cavity as described in section 14.3.3.[25]

Anechoic Rooms — Improvements to the uncertainty of measurements in a reverberating chamber can be made by lining the walls and ceiling with absorbing material (RAM). Two types of material are used for this purpose, namely, carbon-loaded polyurethane foam in the shape of pyramidal cones and ferrite tiles. Several manufacturers provide such materials with low reflectivity, typically in the range –20 to –40 db. For the pyramidal RAM to be effective, its thickness must be a substantial fraction of a wavelength. At low frequencies this becomes impractical and uneconomical. In small rooms the reduction in working volume is unacceptable and in large rooms costs are excessive. Thus, below approximately 100 MHz, rooms cannot be regarded as anechoic and it must be accepted that a substantial amount of reflection will be present. An alternative is to use ferrite tiles, which are particularly effective at low frequencies. In broad terms, conventional RAM is associated with electric field losses whereas tiles are associated with magnetic field losses. Ferrite tiles remain effective from approximately 30 to 1000 MHz. In practice, a combination of ferrite tiles and pyramidal RAM is the most effective arrangement for a broadband anechoic room, but costs are very substantial. Antenna impedance measurements in an OATS and an anechoic room show little difference, but significant differences in site attenuation are observed at low frequencies where resonances persist in the anechoic room.[26] Significantly higher transmitter power is required inside an anechoic room compared to a reverberating room in order to establish the same electric field for immunity testing.[27]

The costs associated with a fully anechoic room and the difficulties that still persist at low frequencies have provided the impetus for examining other strategies for RAM placement and room damping. In essence, the difficulties are associated with the first one or two room resonances, which, depending on room size, occur below about 100 MHz. The structure of the first few modes is a rather simple one and it is possible to identify areas on the walls and ceiling where there are field maxima. RAM is then placed where the maxima occur, thus introducing substantial damping at low cost.[28] Other strategies may also be implemented where blocks of RAM are placed in the room away from the walls at the location of field maxima for particular modes. In practice it is only possible to damp in this way the first one or two resonances. A modest amount of RAM is then placed on the walls and ceiling in the normal way to deal with higher-order resonances. This approach has been found to be cost effective and offers excellent performance down to very low frequencies. Tests done in screened rooms damped in this way and in OATS have given comparable results within ±6 dB.[29]

The idea of damping resonances in screened rooms and in equipment cabinets is also explored in References 30 and 31. Systematic studies of damping screened rooms using numerical models and comparing with measurements were done by the Nottingham group and the National Physical Laboratory.[32,33] A selection from these results is shown here to illustrate the power and accuracy of these models. The studies were done in a room of dimensions $4.9 \times 7.1 \times 4.8$ m^3 for a number of different damping materials and antenna polarizations. In each case the normalized site attenuation (NSA) was measured or obtained from simulations. The results are shown in Figures 14.9 to 14.13. The first two figures show results for a fully ferrite tiled room for horizontal and vertical polarizations. Figure 14.11 shows results when the floor is not lined and the antenna is horizontally polarized. Figure 14.12 shows a fully lined room up to 1 GHz. Figure 14.13 shows results for a fully lined room with wood-backed tiles. In all cases we see that the numerical model predicts very well the response of the room. Such studies can be used to optimize room performance.

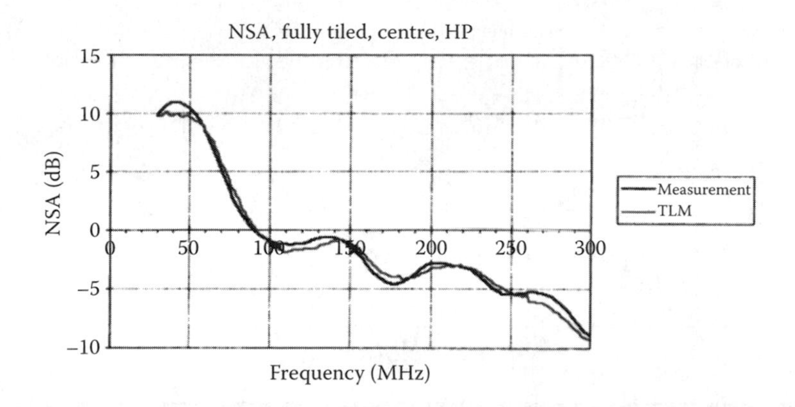

Figure 14.9 NSA in a fully tiled room with a horizontally polarized antenna. The two graphs are measurements and TLM simulations.

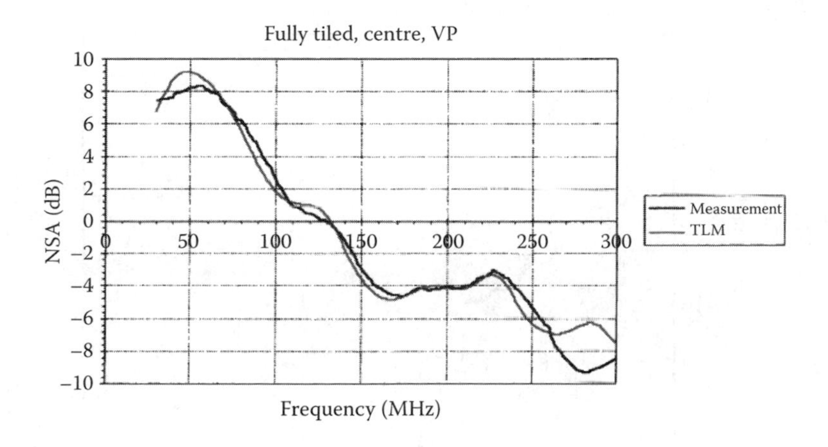

Figure 14.10 As above but for vertically polarized antenna.

14.3.3 Reverberating Chambers

In the previous section, we described screened rooms and the inevitable resonances that make accurate repeatable measurements difficult. By lining the conducting walls fully or partially, the room can be made anechoic and thus resemble, to a degree, free space. It may come as a surprise that the same resonances that we have previously found problematic can be suitably exploited to make a good test environment. The difficulty with resonances is that they introduce strong field nonuniformities at particular distinct frequencies and in particular locations in the room. Therefore, the

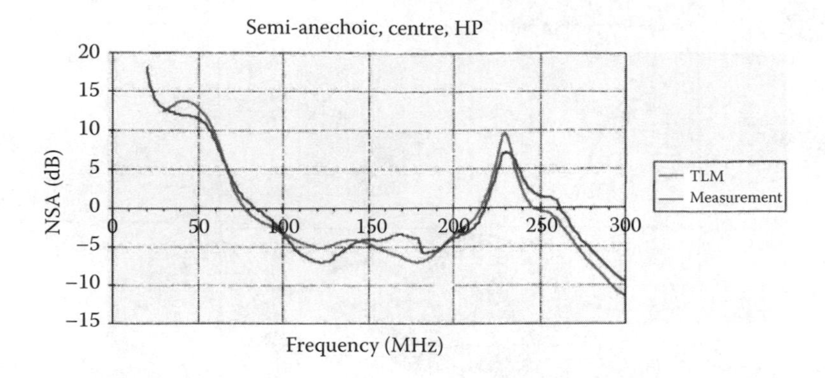

Figure 14.11 NSA in a room where all the surfaces are tiled except the floor, with the antenna horizontally polarized.

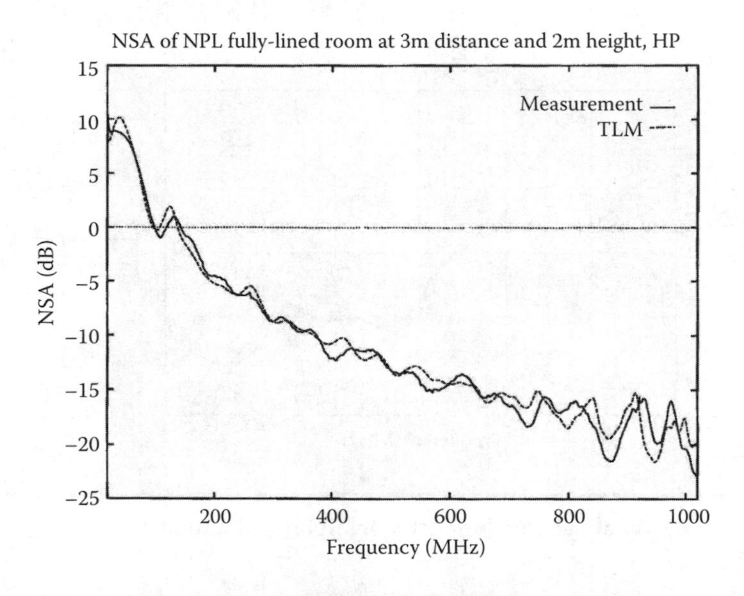

Figure 14.12 NSA in a fully tiled room. Comparisons between measurements and simulations (horizontal polarization).

location of the EUT and/or test antenna and the test frequency all have a strong influence on the quality of the measurement. If we were able to have a situation where there are many resonant frequencies close together and if the geometry of the room was varied in time to spread the location of field maxima and minima, then no frequencies or locations in the room would be particularly favored. This is the idea behind the reverberation chamber: we maximize the number of resonances and by a rotating

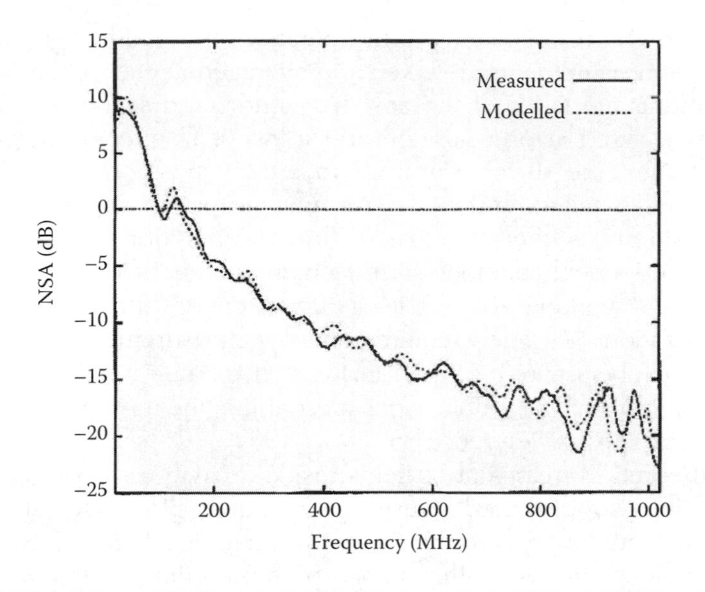

Figure 14.13 NSA in a fully lined room with wood-backed tiles (horizontal polarization).

"paddle" we perturb the chamber geometry and thus "stir" the resonant modes (mode-stirred chamber).

A schematic configuration of a mode-stirred chamber is shown in Figure 14.14. A shaft with a paddle attached to it rotates during the measurement so that the shape of the room changes and therefore resonances do not always occur at the same locations. In order for the stirrer to be effective, the size of the paddle must be considerable relative to the wavelengths of interest. A mode-stirred chamber cannot work effectively at low frequencies as there are few resonances there for effective stirring and mixing. As a rule of thumb, stirring becomes effective at frequencies above three

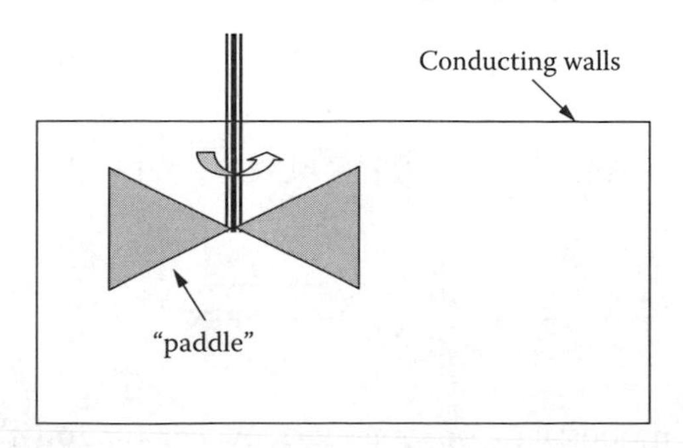

Figure 14.14 Schematic of a reverberating or mode-stirred chamber.

times the lowest resonance (f_{011}). In operation, the paddle rotates continuously and measurements are taken and averaged at each frequency, then measurements are taken at the next frequency, and so on. Alternatively, the stirrer is fixed at one position and a swept frequency measurement is taken. Then the stirrer is moved to a new position, another swept measurement is taken, and so on. An average of measurements over all stirrer positions is then calculated (tuned operation). In a properly designed and operated mode-stirred chamber the field is, on average, uniform in its working volume. It has equal energy flux in all directions and polarizations. Detailed requirements for measurements inside reverberation chambers may be found in Reference 34.

One of the earliest publications describing the use of mode-stirred chambers in EMC is Reference 35.

The theoretical treatment of fields inside a cavity is a complex matter as it involves sophisticated statistical techniques. The interested reader may consult publications in the literature, including References 36–40.

The basic elements of the theory of mode-stirred chambers are as follows. Consider that we have a one-dimensional structure of length ℓ representing one of the sides of a conducting chamber. Following Reference 39, at a frequency f, the number of half wavelengths that can fit in while satisfying the boundary conditions is

$$\frac{\ell}{\lambda/2} = \frac{\ell}{u/(2f)} = \frac{2f\ell}{u} \tag{14.13}$$

where u is the speed of propagation in the medium inside the room (normally air). The number of modes below a frequency f can be obtained from Equation 14.12

$$f = \frac{u}{2}\sqrt{\left(\frac{m}{a}\right)^2 + \left(\frac{n}{b}\right)^2 + \left(\frac{p}{c}\right)^2}$$

Assuming for simplicity that all dimensions are equal to ℓ and substituting we obtain

$$f = \frac{u}{2\ell}\sqrt{m^2 + n^2 + p^2}$$

The number of modes can be obtained by calculating the volume of a sphere in a coordinate system where the three coordinates are the possible values of the integers m, n, p. Only one octant of this spherical volume may be included (positive frequencies) and the number should be multiplied by 2 to account for two possible polarizations. The radius in this space is

$$r = \sqrt{m^2 + n^2 + p^2} = \frac{2\ell f}{u}$$

where we have used Equation 14.13.

Hence, for a 3D cavity with sides ℓ, taking account of two possible polarizations and positive frequencies only, we obtain as the number of modes below frequency f

$$N(f) = \frac{1}{8}\frac{4}{3}\pi r^3 2 = \frac{8\pi}{3}\frac{f^3\ell^3}{u^3}$$

The density of modes is then

$$\frac{dN(f)}{df} = \frac{8\pi}{u^3} f^2 \ell^3 \tag{14.14}$$

We see that the number and density of modes depend on the chamber volume (ℓ^3). As an example the number of modes in a 20 m^3 chamber below 200 and 300 MHz is 50 and 167, respectively. At 300 MHz the density of modes is 1.67×10^{-6} per Hz — there is at least one mode per MHz. This sets approximately the lowest usable frequency of this mode-stirred chamber. Larger chambers are necessary if a lower frequency of operation is required.

Another useful parameter is the Q-factor of the chamber. Wall losses are the main mechanism of lower Q but the presence of EUT and test equipment also contributes. For a given input power into the chamber the field strength is determined by the Q. Also, the sensitivity of the chamber in detecting emissions is affected by Q. For pulsed measurement it may be necessary to reduce Q (reduce the time constant of the chamber). Accurate calculation of Q is a complex matter.[36,37] Assuming wall losses only

$$Q = 1.5\left[\frac{V}{\mu S \delta}\right]\left[1 + \frac{3\pi}{8k}\left(\frac{1}{a} + \frac{1}{b} + \frac{1}{c}\right)\right]^{-1} \tag{14.15}$$

where V and S are the volume and total surface are of the chamber, k is the wavenumber, and δ is the skin depth. In general, due to wall imperfections, such formulae predict a much higher Q than actually observed.

In broad terms, in a small mode-stirred chamber an electric field of several tens of V/m can be generated at 1 GHz from 1 W input power. In contrast, in an anechoic chamber with considerable damping, 10 W would be needed to generate fields of the same order.

Reverberating chambers without a mode stirrer have also been described based on varying the angles between the walls, floor, and ceiling and with vibrating walls.[41,42]

14.3.4 Special EMC Test Cells

A design capable of producing a plane EM field is the so-called transverse electromagnetic (TEM) or Crawford cell.[43] The basic structure is shown in Figure 14.15 and it can be viewed as a modified stripline with sidewalls added. The dimensions of the cell and the two tapered sections are chosen so that a characteristic impedance of 50 Ω is maintained throughout. The cell may be used to establish a very uniform field as long as higher modes are not excited ($h < \lambda/2$). The maximum size of the EUT is limited to h/3. The TEM cell can be used for emission and susceptibility tests. In the former case the signal received at a port may be related directly to measurements obtained in an OATS, as described in Reference 43. The electric field established from a source of voltage V is $E = V/h$ and the power required is $(Eh)^2/50$. These cells are a very economical arrangement for immunity testing. TEM cells, which are large enough for whole aircraft testing, are described in Reference 44.

Another arrangement used for EMC testing is the gigahertz transverse electromagnetic (GTEM) cell, shown schematically in Figure 14.16.[45] It consists of a tapered transmission line that is terminated by a distributed 50-Ω load and RAM. Careful design ensures the setting up of a spherical wave without exciting higher-order modes up to very high frequencies. The maximum working size of the EUT is again limited to a third of the septum height. The septum is offset, as shown, to increase the working volume. The GTEM cell can be used in a similar way as the TEM cell.

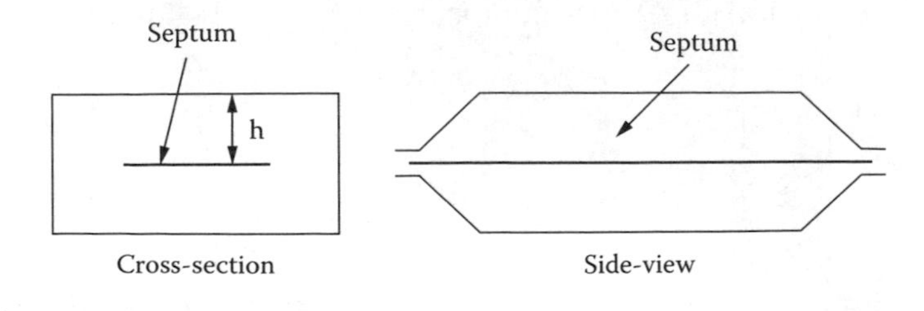

Figure 14.15 Schematic of a TEM or Crawford cell.

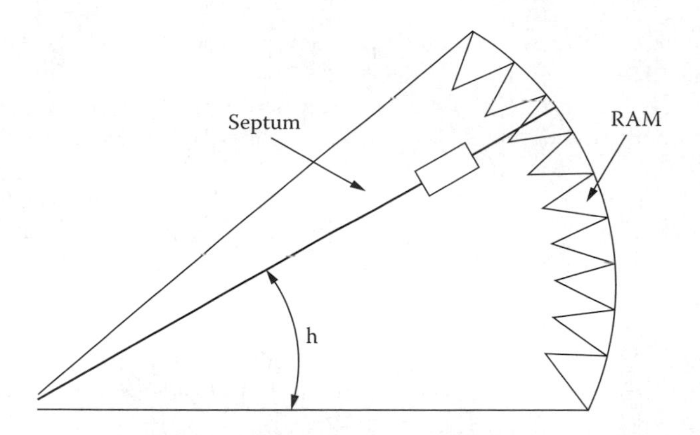

Figure 14.16 Schematic of a GTEM cell.

Both cells are useful for testing small- to medium-sized equipment and for accurate calibration of small transducers.

Requirements for the required field uniformity and for testing are presented in Reference 46.

A thorough discussion of issues related to the use of the GTEM cell for EMC testing is presented in References 47 and 48. The GTEM cell can be used for immunity studies by feeding the single port without the need for a transmit antenna. Compared to an anechoic room the same field can be achieved with less power. Similarly, for emission studies the field is inferred by measurements at the port of the cell; no receive antenna is required. For complete characterization of emissions, the EUT needs to be rotated around three orthogonal axes. The main field components coupling the EUT and the GTEM cell are the vertical electric field and the magnetic field normal to the paper (Figure 14.16). However, small cross-polarized field components are also present and in this respect the cell is inferior to an anechoic chamber.

References

1. Jackson, G A, "Survey of EMC measurement techniques," Electronics and Communication Engineering Journal, pp 61–70, Mar/Apr 1989.
2. IEC Publication 801, "Electromagnetic Compatibility for Industrial-Process Measurement and Control Equipment," Parts 1 to 6 (1990–1993).
3. Keenan, R R K and Rosi, L A, "Some fundamental aspects of EMC testing," Proc IEEE EMC Symp, Cherry Hill, NJ, pp 236–241, 12–16 Aug 1991.
4. Cormier, B and Boxleitner, W, "Electrical Fast Transient (EFT) testing — an overview," Proc IEEE EMC Symp, Cherry Hill, NJ, pp 291–296, 12–16 Aug 1991.

5. Marvin, A C, "Measurement environments and testing," AGARD Lecture Series, LS-177, pp 8-1–8-24.
6. CISPR Publication 16: "Specification for Radio Interference Measuring Apparatus and Measurement Methods," (see also British Standard BS727:1983 and American National Standard ANSI C63.2-1987).
7. CISPR Publication 22: "Limits and Methods of Measurement of Radio Interference Characteristics of Information Technology Equipment" (see also British Standard BS6527:1984 and Euro Norm EN55022:1987).
8. Miller, E K (editor), "Time-Domain Measurements in Electromagnetics," Van Nostrand Reinhold, NY, 1986.
9. Baum, C E, Breen, E L, Giles, J C, O'Neill, J, and Sower, G D, "Sensors for electromagnetic pulse measurements both inside and away from nuclear source regions," IEEE Trans, EMC-20, pp 22–35, 1978.
10. Hajnal, J V, "Compound modulated scatterer measuring system," Proc IEE-H, 134, pp 350–356, 1987.
11. Bennett, W S, "Properly applied antenna factors," IEEE Trans, EMC-28, pp 2–6, 1986.
12. Balanis, C A, "Antenna Theory — Analysis and Design," Harper and Row, NY, 1982.
13. Mann, S M, "Characterisation of antennas and test sites for the measurement of radiated emissions," D. Phil. Thesis, University of York, 1993.
14. Smith, A A, "Standard-site method for determining antenna factors," IEEE Trans, EMC-24, pp 316–322, 1982.
15. Tang, T and Gunn, M W, "Terminal voltage of a receiving antenna above a conducting plane," Proc 7th Int Zurich EMC Symp, pp 347–352, 3–5 Mar 1987.
16. Mishra, S R, Kashyap, S, and Balaberda, R, "Input impedance of antennas inside enclosures," IEEE EMC Symp, Wakefield, MA, pp 534–538, 20–22 Aug 1985.
17. Ran, L, Clare, J C, Bradley, K J, and Christopoulos, C, "Measurement of conducted electromagnetic emissions in PWM motor drive systems without the need for a LISN," IEEE Trans. on EMC, 41(1), pp 50–55, 1999.
18. Agilent Technologies, "Understanding Oscilloscope Frequency Response and its Effect on Rise-Time Accuracy," Application Note 1420, 2002.
19. Sander, K F and Reed, G A L, "Transmission and Propagation of Electromagnetic Waves," Second Edition, Cambridge University Press, Cambridge, 1986.
20. Smith, A A, German, R F, and Pate J B, "Calculation of site attenuation from antenna factors," IEEE Trans, EMC-24, pp 301–316, 1982.
21. American National Standard: Methods of Measurement of radio-noise emissions from low-voltage electrical and electronic equipment in the range 10kHz to 1GHz, ANSI C63.4–1988.
22. Heirman, D N, "Commercial EMC standards in the United States," Proc 10th Int Zurich EMC Symp, Supplement, pp 31–43, 9–11 Mar 1993.
23. Dawson, L and Marvin, A C, "New screened room techniques for the measurement of RFI," J Int Electron Radio Eng, 58, pp 28–32, 1988.

24. Goodwin, S, "Investigation of Emissions within a Screened Room," D. Phil. Thesis, University of York, 1989.
25. Wu, D I and Chang, D C, "The effect of an electrically large stirrer in a mode-stirred chamber," IEEE Trans, EMC–31, pp 164–169, 1989.
26. Kashyap, S, Mishra, S R, and Balaberda, R, "Comparison of absorber-lined chamber and open-field measurements," Proc IEEE Emc Symp, Wakefield, MA, pp 170–175, 20–22 Aug 1985.
27. Crawford, M L and Koepke, G H, "Comparing EM susceptibility measurement results between reverberation and anechoic chambers," Proc IEEE EMC Symp, Wakefield, MA, pp 152–160, 20–22 Aug 1985.
28. Dawson, L and Marvin, A C, "Alternative methods of damping resonances in a screened room in the frequency range 30–200 MHz," 6th Int Conf on EMC, York, IERE Publ No 81, pp 217–224, Sept 12–15 1988.
29. Cook, R J (editor), Consortium report "Requirements for the origination of a draft emission calibration procedure — Final report stage III," DRA/NPL/York and Nottingham Universities, National Physical Laboratory, U.K., Feb 1994.
30. Izzat, N, Hilton, G H, Railton, C J, and Meade, S, "Use of resistive sheets in damping cavity resonances," Electronics Letters, 32(8), pp 721–722, 1996.
31. Dixon, P, "Cavity-resonance dampening," IEEE Microwave Magazine, pp 74–84, June 2005.
32. Loader, B G, Alexander, M, Rycroft, R J, Rochard, O C, Jee, J, and Paul, J, "Partial lining of screened rooms: Validation of TLM model and assessment of room performance," NPL Report CEM S18, February 1998.
33. Christopoulos, C, Paul, J, and Thomas, D W P, "Absorbing materials and damping of screened rooms for EMC testing," EMC-99, June 1999, Tokyo, Japan, pp 504–507.
34. IEC 61000-4-21:2003, "Electromagnetic Compatibility (EMC): Testing and measurement techniques. Reverberation chamber test methods."
35. Corona, P, "Electromagnetic reverberating enclosures: Behavior and applications," Alta Frequ., 40, pp 154–158, 1980.
36. Chang, D C, Liu, B H, and Ma, M T, "Eigenmodes and composite quality factor for a reverberation chamber," NBS Technical Note 1066, Aug, 1983.
37. Ma, M T, "Understanding reverberating chambers as an alternate facility for EMC testing," Journal of Electromagnetic Waves and Appl., 2, pp 339–351, 1988.
38. Kostas, J G and Boverie, B, "Statistical model for the mode-stirred chamber," IEEE Trans. on EMC, 33(4), pp 366–370, 1991.
39. Holland, R and St John, R, "Statistical Electromagnetics," Taylor and Francis, 1999.
40. Arnaut, L R, "Effect of local stir and spatial averaging on measurement and testing in mode-tuned and mode-stirred reverberation chambers," IEEE Trans. on EMC, 43(3), pp 305–325.
41. Leferink, F B J and van Etten, W C, "Generating an EMC test field using a vibrating intrinsic reverberation chamber," IEEE EMC Society Newsletter Online, Spring 2001.

42. Kouveliotis, N K, Trakadas, P T, and Capsalis, C N, “Examination of field uniformity in vibrating intrinsic reverberation chamber using the FDTD method,” Electronics Letters, 38(3), pp 109–110, 2002.
43. Crawford, M L and Workman, J L, “Using a TEM cell for EMC measurements of electronic equipment,” National Bureau of Standards, Technical Note 1013. Boulder, CO, July 1981.
44. Walters, A J and Leat, C, “TEM cells for whole aircraft EMV testing,” IEEE EMC Society Newsletter Online, 2005.
45. Garbe, H and Hansen, D, “The GTEM cell concept: applications to this new EMC test environment to radiated emission and susceptibility measurements,” Proc 7th Int Conf on EMC, York, pp 152–156, 28–31 Aug 1990.
46. IEC Standard 61000-4-20 “Electromagnetic compatibility (EMC)-Part 4-20: Testing and measurement techniques-Emission and immunity testing in transverse electromagnetic (TEM) waveguides.”
47. Nothofer, A, Alexander, M, Bozec, D, Marvin A, and McCormack, L, “The use of GTEM cells for EMC measurements,” Measurement Good Practice Guide No. 65, The National Physical Laboratory, UK, 2003.
48. Ngu, X T I, Christopoulos, C, Thomas, D W P, and Nothofer, A, “The impact of phase in GTEM and TEM measurements,” EMC Europe 2006, Sept. 4–8, 2006, Barcelona, pp 11–16.

EMC IN SYSTEMS DESIGN

In Parts I to IV, we considered the traditional EMC topics concerned with the design and certification of commercial equipment to comply with EMC standards and regulations. Increasingly, EMC has to be seen in the context of recent developments in the design of information technology equipment and of the widespread use of wireless technologies for voice, data communications, and connectivity on the move of all kinds.

The very fast clock rates used routinely in equipment design make it now impossible to regard EMC issues as divorced from general signal integrity considerations. The issues of interoperability of systems and EMC are closely allied and special attention must be paid to developments in this area in connection with new wireless technologies. Traditional low-frequency networks (e.g., the telephone and power networks) are used to carry high-frequency signals and consideration must be given to the EMC impact of such practices. Many safety-critical controls are increasingly actuated electronically and issues of EMC and safety need to be brought into focus. Finally, the complexity of modern systems means that we need to consider alternative ways of characterization and modeling based on statistical techniques.

We address these issues in this final part of the book.

Chapter 15

EMC and Signal Integrity (SI)

15.1 Introduction

We have presented in previous chapters the main requirements for ensuring the electromagnetic compatibility of equipment and ways that compliance with standards and regulations may be achieved. Signal integrity is the discipline concerned with the maintenance of the shape, amplitude, and timing of signals as they propagate through complex networks, most notably in multilayer printed circuit boards (PCBs) or printed wire boards (PWBs) consisting of interconnects, passive components, and monolithic integrated circuits (ICs) or, as they are commonly known, chips. Integrated circuits convert DC power supplied to them to pulses, and therefore high-frequency signals, and are the main sources of interference in much electrical and electronic equipment. Boards are constructed from several layers of insulation and metalization to carry signals and DC power and provide a reference (ground planes). A typical configuration is shown in Figure 15.1.[1] Several similar stacks may be connected onto the same mother board. Each component in this figure is a complex electrical structure, especially the chip and the multichip module (MCM), which consists of several chips. It is clear that each component has complex electrical characteristics and it affects in a different but significant way the integrity of propagating signals and EMC. A full analysis of such a system at high frequencies is a forbidding task. A further insight into the inherent complexity of such systems is shown in Figure 15.2. Here, a multilayer

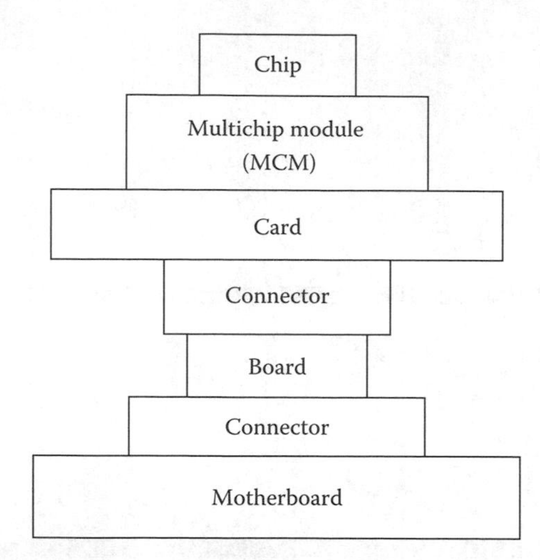

Figure 15.1 A typical stack of components onto a motherboard.

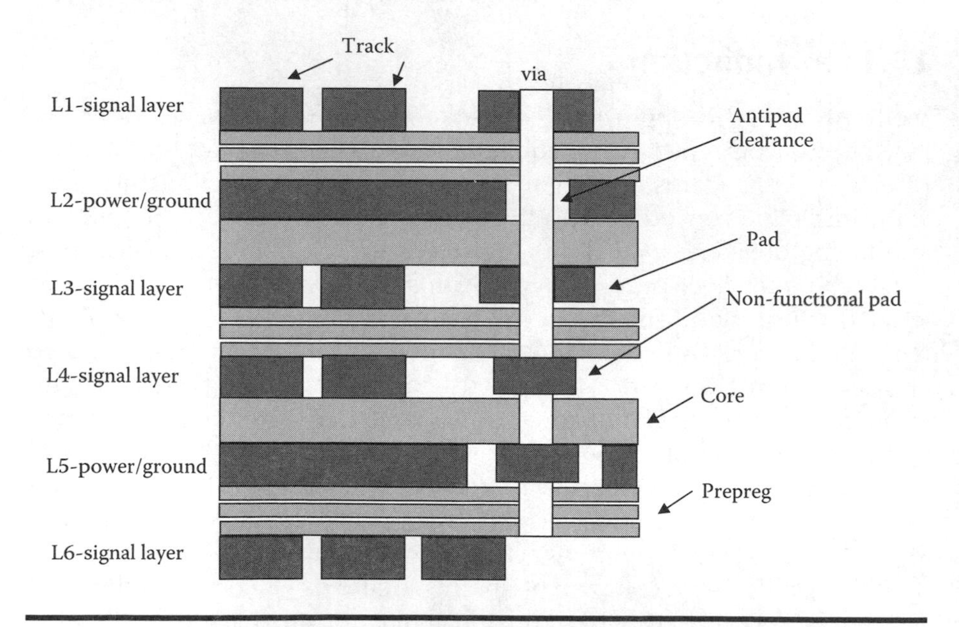

Figure 15.2 A typical arrangement of a multilayer board.

PCB is depicted[2] consisting of a number of pre-impregnated (prepreg) layers (a partly cured mix of resin and glass fiber) and fully cured prepreg layers with copper foil attached to form the cores (laminates) carrying signals and providing power and ground planes. The complete PCB is formed by heating the entire assembly under pressure so that the partially cured prepreg flows and bonds to the laminates. Vias are also provided to carry signals from one layer to another. Each of these components carries high-frequency signals and thus is effectively an unwanted antenna receiving and dispersing EM radiation to its neighbors and beyond. The problem is particularly acute with the current and projected clock rates of several gigahertz. One of the objectives of good EMC design is the minimization of emission at source, hence the design of chips must be subject to the same EMC discipline as any other circuit component. The reverse is also true, i.e., the EMC behavior affects signal integrity. A check list of the items to watch for in EMC and SI is shown below.[3]

In EMC one wishes to control:

- Emissions
- Penetration and crosstalk
- Susceptibility
- Cost, size, weight

In SI one wishes to control:

- Signal amplitude, shape, delays, timing
- Crosstalk
- Power dissipation
- Costs, silicon and board area, layout

We see that the two disciplines are related and strongly coupled, hence it is evident that EMC and SI have to be addressed in parallel so that efficient and effective designs may be developed. Crucial to SI is the issue of timing. This can be explained with the aid of Figure 15.3, which shows (a) a driver and a receiver and (b) the associated clock and data pulses. The driver is clocked and it takes a short delay T_{co} for the circuit to respond and the sent data to appear at the start of the interconnect. Due to the finite propagation velocity of waves on the interconnect (typically 2/3 of the speed of light), it takes the transit (or flight) time T_{flight} for data to reach the receiver. The flight time includes in addition to the actual propagation time any further delays caused by distortions to the signal's edge. The data at the receiver is sampled at the clock edge. For this operation to be performed properly the received data must be present and stable for a short time interval before this transition (setup time T_{su}) and after it (hold time T_h). Clearly, in order for the data to be clocked properly at the receiver, the clock period T_{clk} and the other time intervals in Figure 15.3 must satisfy the following inequalities:

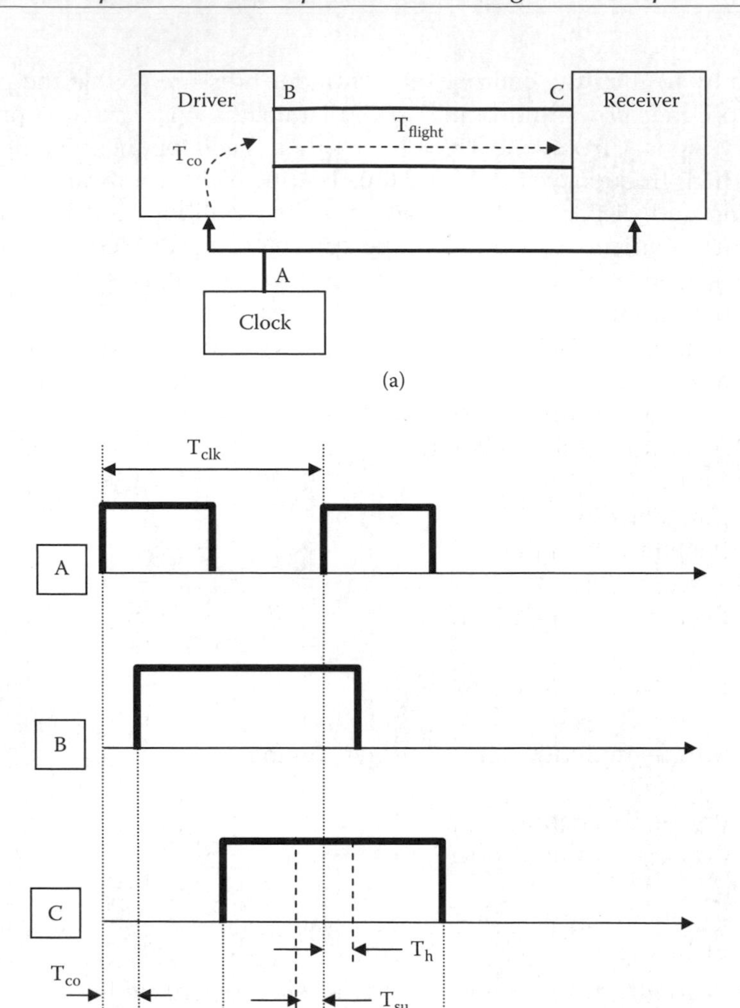

Figure 15.3 (a) Typical interconnect and (b) timing of signals at points A, B, C.

$$
\begin{aligned}
&T_{co} + T_{flight} < T_{clk} - T_{su} \\
&T_{co} + T_{flight} + T_{data} > T_{clk} + T_h
\end{aligned}
\tag{15.1}
$$

These inequalities can be converted to equalities if we introduce setup and hold margins to obtain

$$T_{co} + T_{flight} = T_{clk} - T_{su} - (\text{setup margin})$$
$$T_{co} + T_{flight} + T_{data} = T_{clk} + T_h - (\text{hold margin}) \tag{15.2}$$

The second expression in (15.2) may be simplified if we assume that T_{data} and T_{clk} are equal, to obtain

$$T_{co} + T_{flight} = T_h - (\text{hold margin}) \tag{15.3}$$

Equations 15.2 and 15.3 represent a somewhat ideal situation. We must also recognize that there are variations in the time of flight and in T_{co}, that all clock periods are not exactly the same (clock jitter), and that not all devices are clocked at exactly the same time (clock skew). Furthermore, the data are subject to alterations due to reflections, crosstalk, and coupling from external fields, in what we will refer collectively as "coupling" effects. The first expression in Equation 15.2 can thus be put in the following form:

$$T_{clk} = T_{co} + T_{flight} + T_{su} + (\text{setup margin}) + \text{skew} + \text{jitter} + \text{coupling} \tag{15.4}$$

The last three terms on the RHS may be positive or negative. Equation 15.4 shows the timing budget for the link shown in Figure 15.3. A reliable link with a low bit error rate (BER) demands that the three last terms in Equation 15.4 are well controlled. If we wish to increase the speed of the link (lower clock or cycle period T_{clk}) then every term on the RHS of Equation 15.4 must be reduced and controlled throughout the design. Jitter and skew are difficult to calculate and one has to rely on statistical estimates. The term described as "coupling" includes a multitude of phenomena that are in the domain of the EMC engineer. We see from this discussion that without an adequate EMC design, Equation 15.4 can only hold if we accept generous margins, thus sacrificing communication speed. The unpalatable alternative is to accept occasional malfunctions and thus an inferior BER. High-speed systems therefore require a good EMC design to satisfy strict timing requirements and a high level of reliability. SI and EMC go together at high data rates! We have focused here on the timing as it is critical in SI studies. If timing requirements are not met, nothing works. Coupling effects also affect the amplitude of signals and if they breach noise margins then malfunctions should be expected. We have discussed this topic in Chapter 9. A good EMC and SI design maintains noise and timing margins under most conditions. Any breaches that sometimes occur may be dealt with by error detection and correction codes as mentioned in Section 12.3.1.

EMC problems are caused due to emissions from interconnects carrying sharp-edged signals. Broadly, we describe these effects as board-level EMC. However, emissions also originate directly from ICs (drivers) and these are the subject of chip-level EMC. In the next sections we will introduce some of the main ideas and analysis tools used for SI and EMC analysis in this environment. More extensive coverage may be found in the references already given and also in Reference 4.

15.2 Transmission Lines as Interconnects

The importance of cable attachments to equipment (e.g., for supplying power, signaling, and control) as a means of EM emission and susceptibility is well known. It will come as no surprise, therefore, that interconnects in PCBs contribute significantly to emission and susceptibility problems, cause reflections and crosstalk, and are therefore responsible for many SI problems. Because of the crucial importance of interconnects we give here more details of the behavior of transmission lines to supplement the material presented already in Sections 3.2 and 7.2, and the example in Section 10.3.1.

We present two techniques for studying the pulsed response of lines and thus obtaining information on reflections for discontinuities, voltage, and current distributions, etc. The approaches described are the Bewley lattice (or bounce diagram) and Bergeron techniques.[5,6]

A systematic account of the propagation of pulse on a transmission line linking a driver and a load can be given using the Bewley lattice, which is constructed as follows. Let us consider the simple interconnect shown in Figure 15.4 where we have replaced the driver by its Thevenin equivalent and the load (or receiver) by its input impedance. In order to keep matters simple, and thus facilitate understanding of the basic ideas, we limit for the time being the source and load impedances to be purely resistive R_G and R_L respectively. The reflection coefficient experienced by pulses incident on the source and on the load are

$$\Gamma_G = \frac{R_G - Z_0}{R_G + Z_0}$$
$$\Gamma_L = \frac{R_L - Z_0}{R_L + Z_0} \tag{15.5}$$

where Z_0 is the characteristic or surge impedance of the line. We assume that the source is a unit step voltage of amplitude V_G and thus the Laplace transform of the source voltage is V_G/s where s is the Laplace operator. We also select a line that is not matched at either end so that we can

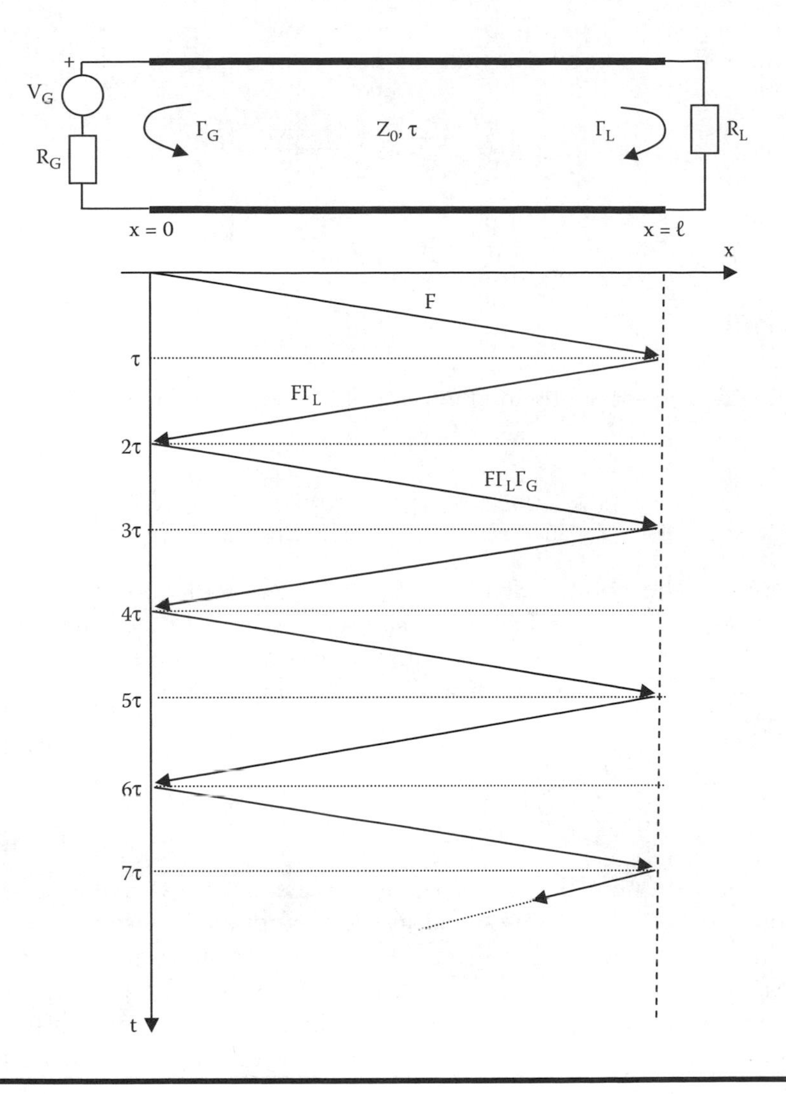

Figure 15.4 Bewley lattice (or bounce diagram) showing voltage pulses in the (x, t) diagram.

best observe the reflections. The line is lossless with a characteristic impedance R_0 and the source and load resistances are $2R_0$ and $R_0/3$, respectively

$$Z_0 = R_0, \quad R_G = 2R_0, \quad R_L = R_0/3 \tag{15.6}$$

The Laplace transform of the source voltage is

$$V_G(s) = V_G/s \tag{15.7}$$

and the reflection coefficients at the source and load are

$$\Gamma_G = \frac{R_G - Z_0}{R_G + R_0} = \frac{2R_0 - R_0}{2R_0 + R_0} = \frac{1}{3}$$

$$\Gamma_L = \frac{\frac{R_0}{3} - R_0}{\frac{R_0}{3} + R_0} = -\frac{1}{2} \tag{15.8}$$

The first wavefront penetrating into the line upon connection of the source is

$$F(s) = \frac{Z_0}{Z_0 + R_G} V_G(s) = \frac{R_0}{R_0 + 2R_0} \frac{V_G}{s} = \frac{V_G}{3s} \tag{15.9}$$

All terms in the bounce diagram of Figure 15.4 can now be calculated. The voltage at the termination is simply obtained by summing up the incident and reflected pulses, taking into account the time of their arrival. For example, the first incident and reflected pulse arriving at the load at time τ is

$$F(s)(1+\Gamma_L)e^{-s\tau} = \frac{V_G}{3s}\left(1 - \frac{1}{2}\right)e^{-s\tau} \tag{15.10}$$

where the exponential term accounts for the delay (τ) in the arrival of this pulse. The form of this signal in the time domain is simply obtained by taking the inverse Laplace transform of Equation 15.10 and since 1/s corresponds to a unit step pulse in the time domain and the exponential term represents a delay we see that the first pulse to arrive at the load has the form, $\frac{1}{6}u(t-\tau)$, i.e., it is a step pulse of magnitude 1/6 arriving at the load at time τ. The voltage level at the load will remain unchanged until the next set of pulses arrives at time 3τ as shown in Figure 15.4. The new disturbance is given by

$$F(s)\left[\Gamma_L\Gamma_G + \Gamma_L^2\Gamma_G\right]e^{-3s\tau} = \frac{V_G}{3s}\left[\frac{-1}{2}\frac{1}{3} + \left(\frac{-1}{2}\right)^2\frac{1}{3}\right]e^{-3s\tau}$$

$$= \frac{V_G}{3s}\left[\frac{-1}{6} + \frac{1}{12}\right]e^{-3s\tau} = \frac{V_G}{s}\frac{-1}{36}e^{-3s\tau} \tag{15.11}$$

In the time domain, this represent a step voltage of amplitude –1/36 arriving after time 3τ. This is superimposed with the previous pulse of

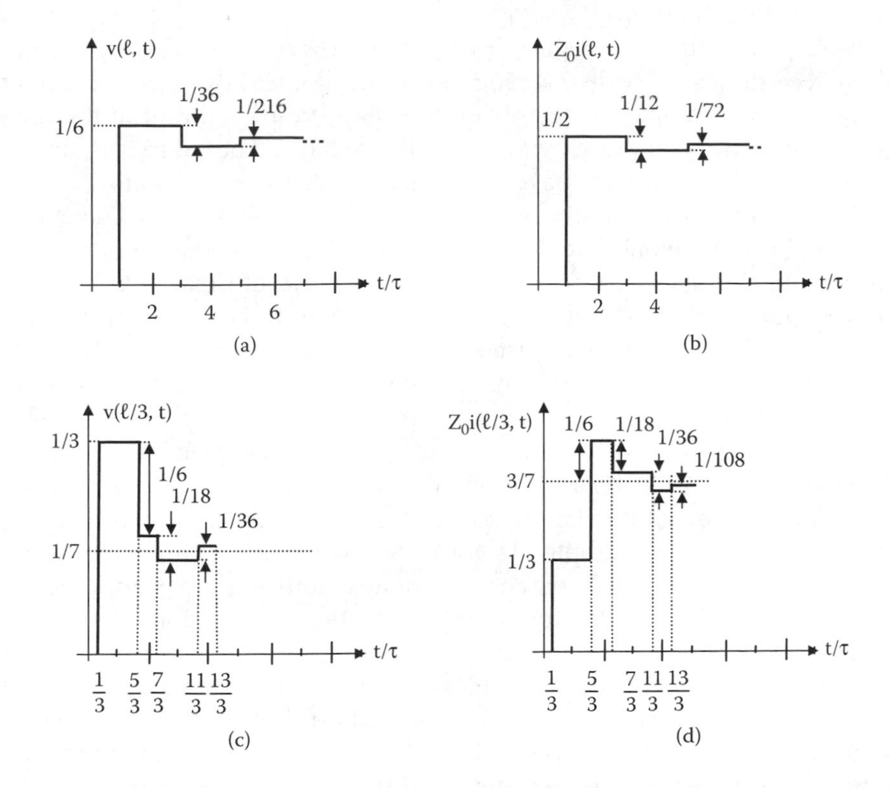

Figure 15.5 Voltage (a) and current multiplied by Z_0 (b) at the load for the configuration shown in Figure 15.4. Voltage (c) and current (d) at $x = \ell/3$.

amplitude 1/6 as shown in Figure 15.5a. In this way all pulses arriving at the load are accounted for to give the following expression:

$$v(\ell, t) = \frac{1}{6}u(t-\tau) - \frac{1}{36}u(t-3\tau) + \frac{1}{216}u(t-5\tau) + \ldots \quad (15.12)$$

This expression is shown in Figure 15.5a for the first three sets of reflections (for clarity we have set $V_G = 1$ in the graphs). We note that the steady state voltage at the load is

$$v(\ell, \infty) = V_G \frac{R_L}{R_L + R_G} = V_G \frac{R_0/3}{R_0/3 + 2R_0} = \frac{V_G}{7} \quad (15.13)$$

We see that at time equal to 5τ the voltage at the load has reached the value V_G (1/6 – 1/36 + 1/216) = V_G 0.143, which is very close to the value calculated in Equation 15.13. Therefore, at 5τ the voltage at the load is very close to steady state. Current pulses are calculated by dividing the forward pulses (those traveling from source toward the load) by Z_0 and the reflected pulses (those traveling toward the source) by $-Z_0$. The current terms are then summed up to obtain the shape of the current in the time domain in exactly the same way as for the voltage. The current is shown in Figure 15.5b where we have multiplied by Z_0 to make plotting easier.

If we wish to obtain the waveforms for the voltage and current in another location along the line, we proceed in a similar way. As an illustration, we show the voltage and current waveforms at $x = \ell/3$ in Figure 15.5c and d. We show this calculation for the first few voltage pulses. The first voltage pulse arrives at $\ell/3$ at time $\tau/3$ and it has amplitude F, i.e., $V_G/3$. The voltage remains constant until the first reflected pulse ($F\Gamma_L$) arrives at this location, which happens at time $2\tau - \tau/3 = 5\tau/3$. This pulse is equal to $-1/6$ and is superimposed on the previous pulse as shown in Figure 15.5c. The next time there is a change in voltage is when the first reflected pulse from the source end arrives there, which happens at $2\tau + \tau/3 = 7\tau/3$. Its amplitude is $-1/18$ and it is superimposed on the previous two pulses. The process continues until we have approached steady state sufficiently for further pulses to be of negligible magnitude.

In practical systems, impedances are not purely resistive and other components need to be taken into account. A common example is the case of a capacitive load, say C_L. The procedure described so far remains unchanged with the only proviso that the reflection coefficient at the load is not real but complex to account for the frequency dependence of the capacitive impedance:

$$\Gamma_L = \frac{\dfrac{1}{sC_L} - R_0}{\dfrac{1}{sC_L} + R_0} = \frac{-s + 1/T}{s + 1/T} \tag{15.14}$$

where $T = R_0 C_L$. Therefore, in the calculation of the Laplace transform of the signals and their inverse we need to take account, in addition to the source, the s-dependence of the load reflection coefficient in Equation 15.14. If we assume that the source is a step voltage as before and that the load is now a capacitor, the Laplace transform of the first set of pulse at the load is

$$\begin{aligned} 1^{\text{st}}\text{incident and reflected pulse at load} &= V_G\left(\frac{1}{s} + \frac{\Gamma_L}{s}\right)e^{-s\tau} \\ &= \frac{V_G}{s}\frac{2/T}{s + 1/T}e^{-s\tau} \end{aligned} \tag{15.15}$$

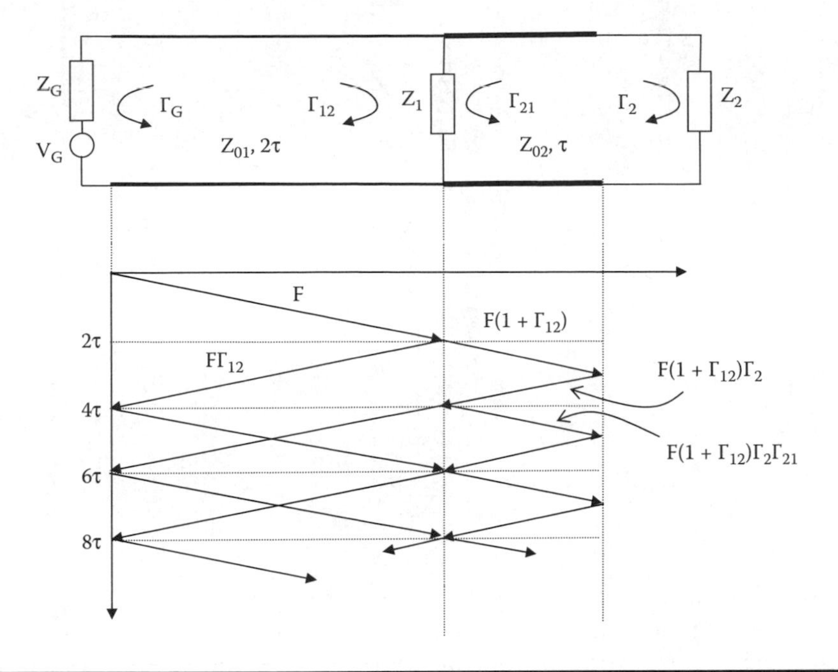

Figure 15.6 Bounce diagram for an interconnect with two segments of different characteristic impedance.

Taking the inverse Laplace transform of the expression in Equation 15.15 gives us the shape in the time domain of the first pulse experienced by the load,

$$2V_G\left(1-e^{\frac{t-\tau}{T}}\right)u(t-\tau).$$

A similar approach is applied for all the other pulses resulting inevitably in more complexity.

Another example is the junction between two lines with a different characteristic impedance as shown schematically in Figure 15.6. This occurs typically in a PCB where a track may change width, thus changing its characteristic impedance. We also introduce a lumped component at the junction for generality. The bounce diagram is shown in the same figure and the first few reflections are calculated from the first incident waveform F (obtained as before) and the corresponding reflection coefficients at all the discontinuities. We have assumed for simplicity that the transit time of the first line is 2τ and of the second τ. The reflection coefficients are

$$\Gamma_G = \frac{Z_G - Z_{01}}{Z_G + Z_{01}}$$

$$\Gamma_{12} = \frac{\frac{Z_1 Z_{02}}{Z_1 Z_{02}} - Z_{01}}{\frac{Z_1 Z_{02}}{Z_1 Z_{02}} + Z_{01}}$$

$$\Gamma_{21} = \frac{\frac{Z_1 Z_{01}}{Z_1 Z_{01}} - Z_{02}}{\frac{Z_1 Z_{01}}{Z_1 Z_{01}} + Z_{02}} \tag{15.16}$$

$$\Gamma_2 = \frac{Z_2 - Z_{02}}{Z_2 + Z_{02}}$$

We see that a very complex pattern of reflections will occur. At $t = 6\tau$ and at the junction of the two lines there are pulses arriving simultaneously from the left and from the right resulting in reflections from and transmission through the junctions. If the transit time of the second line is not exactly half of the first line, then very short pulses may results at this junction after superposition of the incident pulses coming from both sides.

A further example, which illustrates the complexity of propagation in this environment, is shown in Figure 15.7a. It illustrates a PCB with a ground plane, a substrate, and a track (characteristic impedance Z_{01}), which distributes signal V_G through a junction to two other lines Z_{02} and Z_{03}. This typically could be a clock pulse train distributed to two drivers. The model of this configuration is shown in Figure 15.7b. Naturally, reflections take place at the source and load ends if the lines are not matched to their characteristic impedance. In addition, a substantial pattern of reflections takes place at the junction of the three lines. As an illustration, the reflection coefficient seen by pulses traveling on line one and arriving at this junction is

$$\Gamma_1 = \frac{\frac{Z_{02} Z_{03}}{Z_{02} + Z_{03}} - Z_{01}}{\frac{Z_{02} Z_{03}}{Z_{02} + Z_{03}} + Z_{01}} \tag{15.17}$$

Similar expressions apply for the other two reflection coefficients shown in Figure 15.7b. A pulse originating from the source experiences a partial reflection at the junction and a part of it proportional to $(1 + \Gamma_1)$ penetrates

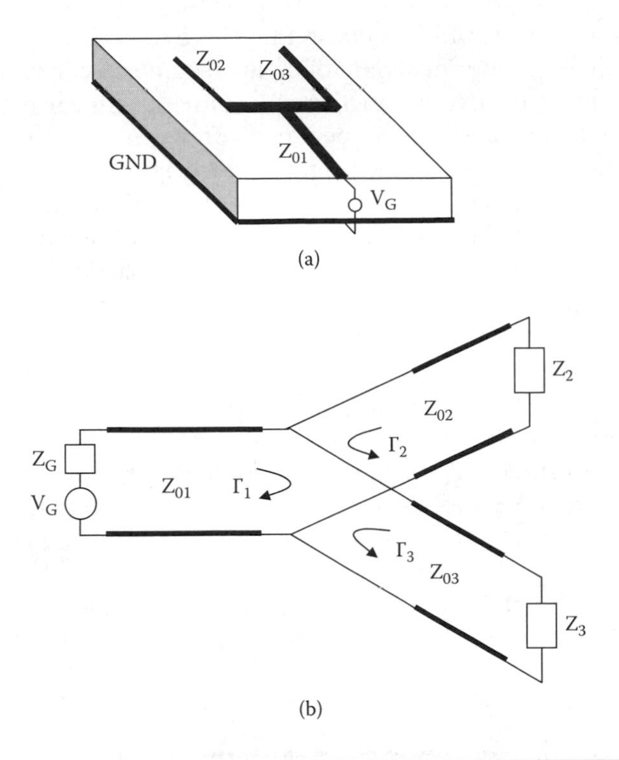

Figure 15.7 Typical PCB configuration (a) and transmission line equivalent (b).

into lines 1 and 2. When these signals reach the respective terminations they suffer further reflections to arrive after a delay back at the junction for further reflection and penetration into line 1. So does the first pulse, which reflected at the junction back into line 1 and encountered the source. Therefore, there are in general three pulses incident at the junction, at different instants in time (depending on the electrical length of the three line segments) generating a complex pattern of reflections.

From these examples we see that the propagation of fast pulses (where the duration of the pulse is comparable or shorter than the line transit time) generates complex patterns due to reflections that may alter radically the shape and amplitude of the original pulse train. This clearly has implication for SI in terms of timing and logic state recognition. In addition, the generation of so many sharp-edged signals traveling back and forth on interconnects of a length comparable to the wavelength does not bode well for control of emissions. It is inevitable that in this situation interconnects will radiate appreciable amounts of EM energy in addition to near-field coupling (crosstalk), addressed in Chapter 7 of this book. This creates EMC problems and also additional SI problems due to interference with other interconnects that are not part of the original transmission. It is therefore of considerable practical advantage to minimize reflections as far as possible by proper matching of all line segments carrying fast pulses. This is difficult to achieve in practice due to the nonlinear nature of loads (e.g., different high and low impedance) and the broadband nature of the termination required. Some examples are shown in Figure 15.8. In Figure 15.8a a resistive termination Z_R may be added when the input of the receiver is very high (e.g., capacitive). The disadvantage is that it draws DC current and that another component has to be added in an already very crowded board. An alternative is shown in Figure 15.8b where the effective terminating impedance is the parallel combination of the two resistive loads and the two values can be chosen individually to suit better the logic family properties. Finally, the arrangement in Figure 15.8c cuts down the DC current and hence power consumption, but it is difficult to implement in practice as it increases the component count. The capacitor may be implemented by parallel traces in a multilayer board and the resistor by a lower conductivity trace, but routing is difficult and expensive.

In PCB manufacture, several transmission line configurations are used: microstrip, embedded microstrip, stripline, etc. A stripline configuration

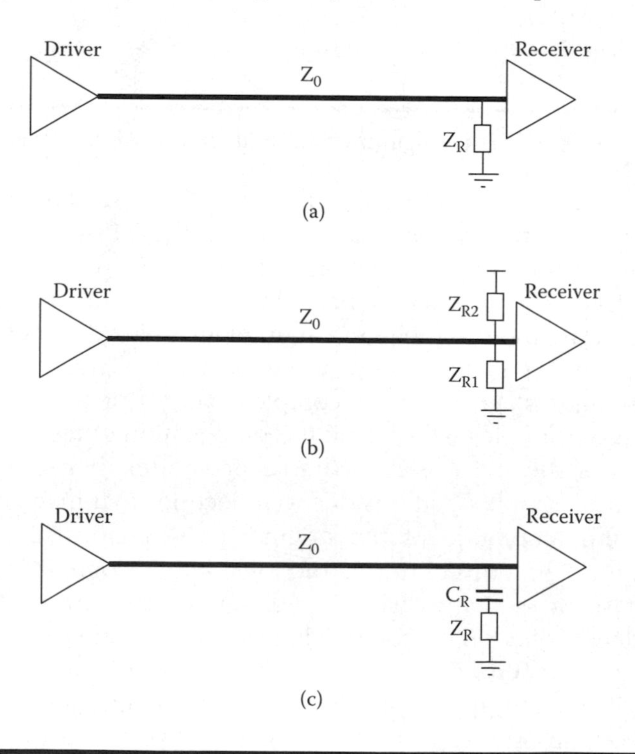

Figure 15.8 Ways to improve matching of a line at the load end.

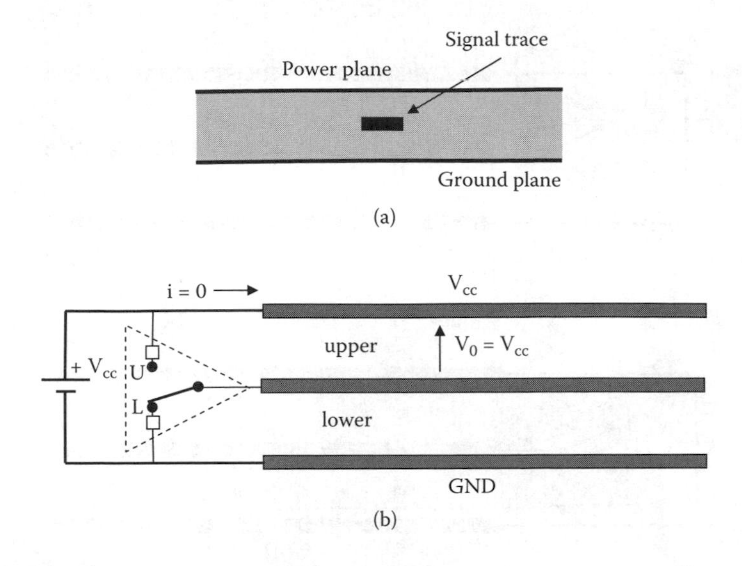

Figure 15.9 (a) A stripline configuration and (b) its electrical equivalent.

(track between power and ground planes) is shown in Figure 15.9a and the electrical configuration in Figure 15.9b.[2] We shall examine how the line is charged and discharged and this will give us the opportunity to see how the ideas presented above can be used in a situation where the line is initially charged. The starting point is the configuration shown in Figure 15.9b where the driver is shown schematically with an effective source resistance of 50 Ω (the square boxes shown but not labeled) irrespective of the position of the switch. The supply voltage is $V_{cc} = 2.5$ V. The switch is in the lower position, as shown, for some time so that steady-state conditions prevail. Hence, the lower segment of the line is completely discharged while the upper segment is charged to a voltage $V_0 = V_{cc} = 2.5$ V. No currents are flowing anywhere in the system. We wish to ascertain conditions in the circuit as the driver drives high, i.e., the switch moves from position L to position U. A full and accurate understanding of the redistribution of the field following switching needs solution of the full-field equations. However, we can gain a reasonable understanding of the prevailing conditions by making some bold assumptions, namely that the upper and lower segments of the stripline behave as two independent transmission lines, each with a characteristic impedance $Z_0 = 100$ Ω. Conditions following switching are the result of the superposition of two actions. First is the discharging of the top segment through the 50-Ω resistor (Figure 15.10a) and second is the charging of the lower segment through the 50-Ω resistor (Figure 15.10b). We examine

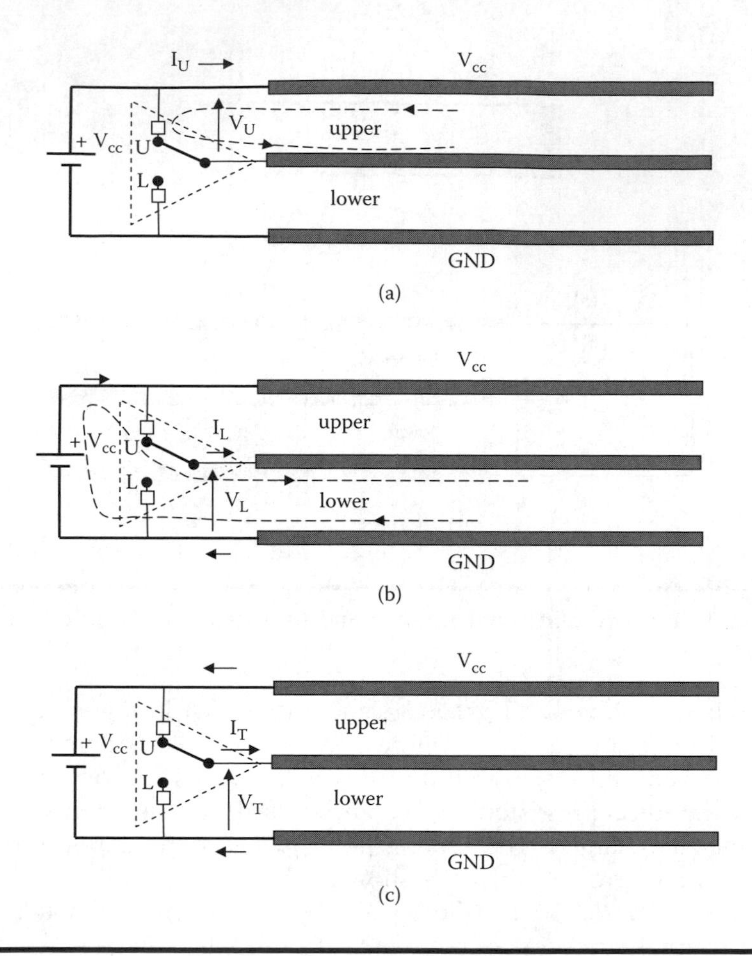

Figure 15.10 Conditions after moving the switch from L to U in Figure 15.9: (a) discharging of upper segment, (b) charging the lower segment, and (c) superposition of (a) and (b).

each action separately. At the start of the upper segment the voltage and current must comply with the boundary condition there, i.e.,

$$V_U = -I_U 50 \tag{15.18}$$

This requires that a pulse V^i is launched traveling toward the right together with an associated current pulse $I^i = V^i/100$ such that

$$V_0 + V^i = -\frac{V^i}{100}50 \tag{15.19}$$

where we have recognized that the line was initially charged to V_0 and that the initial value of the current was zero. Substituting for $V_0 = 2.5$ V and solving Equation 15.19 for the pulse voltage we obtain $V^i = -1.67$ V. Hence the current is $-1.67/100 = -16.7$ mA. This circulating current discharging the upper segment is shown schematically by a long-dash line in Figure 15.10a. Between the trace and the power plane is a displacement current (through the capacitance). We now turn our attention to the charging of the lower segment, which is depicted in Figure 15.10b. Since this line is initially uncharged, conditions are straightforward and the forward-propagating current pulse and associated voltage pulse are

$$\begin{aligned} I_L &= \frac{2.5}{50+100} = 16.67 \text{ mA} \\ V_L &= \frac{100}{100+50} 2.5 = 1.667 \text{ V} \end{aligned} \tag{15.20}$$

The charging current is shown by the long-dash line in Figure 15.10b. Combining the results from Figure 15.10a and Figure 15.10b gives the overall picture in Figure 15.10c where

$$\begin{aligned} I_T &= 2 \times 16.67 \text{ mA} = 33.34 \text{ mA} \\ V_T &= 1.67 \text{ V} \end{aligned} \tag{15.21}$$

The effective surge impedance is $V_T/I_T \approx 50\ \Omega$.We see that the charging of the line is the result of current flows from different parts of the circuit and therefore that careful design is required to minimize interference with other components sharing power and ground rails.

So far we have limited the study to terminations that are linear at each part of the switching cycle. Where a pronounced nonlinearity is present an alternative technique is available: the so called Bergeron method. It was originally developed to solve problems in hydraulics (the water hammer effect) but is readily applicable to EM problems. We know that signals on a line are a combination of forward- and backward-traveling waves, i.e., functions of $x \pm ut$. We can write therefore

$$\begin{aligned} V(x,t) &= f_1(x-ut) + f_2(x+ut) \\ Z_0 I(x,t) &= f_1(x-ut) - f_2(x+ut) \end{aligned} \tag{15.22}$$

where f_1, f_2 are arbitrary functions. Adding the two expressions in Equation 15.22 we obtain

$$V(x,t) + Z_0 I(x,t) = 2f_1(x - ut) \tag{15.23}$$

Similarly, subtracting we get

$$V(x,t) - Z_0 I(x,t) = 2f_2(x + ut) \tag{15.24}$$

The implication of Equation 15.23 is that if $x - ut$ is constant, then the value of the function f_1 is constant and therefore $V + Z_0 I$ is constant. The condition that $x - ut$ = const. means that $\Delta x - u\Delta t = 0$, or that $\Delta x/\Delta t = u$. An observer moving in the forward direction with the velocity of propagation u measures voltage and current such that $V + Z_0 I$ = const. Similarly, an observer moving with velocity u in the backward direction measures $V - Z_0 I$ = const. This important observation is the basis of the Bergeron method. Let us start with a simple example where we already know how to obtain the quantities on a bounce diagram, a line terminated by a resistive linear load as shown in Figure 15.11a. The Bergeron method is applied using a V-I diagram representing the conditions on the line as shown in Figure 15.11b. On the V-I diagram we draw the curves representing the boundary conditions (BC) at the source and load ends of the line. At the source end A the voltage is constant and equal to V_0, the source voltage, as the internal resistance of the source is zero. At the load

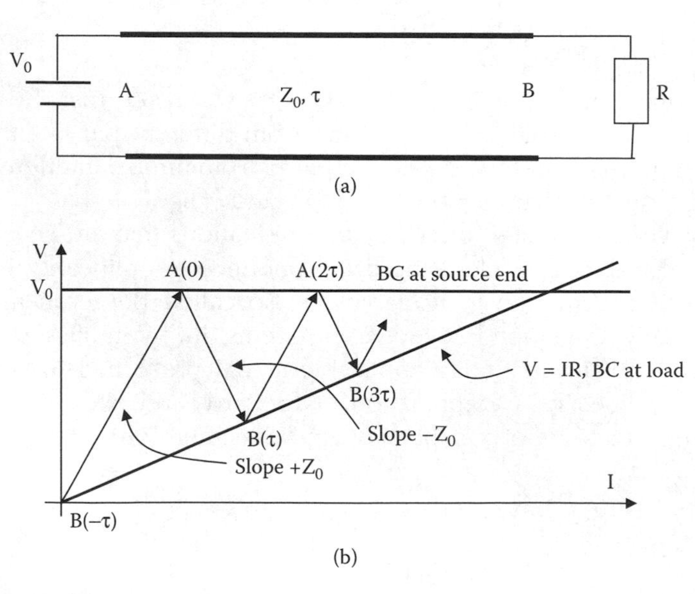

Figure 15.11 Example of the application of the Bergeron method (b) for the configuration shown in (a).

end B the BC is V = IR. The two BCs are shown in Figure 15.11b. The instant at which the source is connected to an initially uncharged line is t = 0. An observer traveling in the forward direction sees $V + IZ_0 = \text{const.}$ (see Equation 15.23), therefore $\Delta V/\Delta I = -Z_0$. Hence, a forward-traveling observer appears on the V-I diagram on lines of slope $-Z_0$. Similarly, a backward-traveling observer appears on lines of slope $+Z_0$. The essence of the technique is to start an observer from a point on the line where we already know V and I (from the initial conditions), which arrives at the source end at t = 0. Clearly, if we start a backward-traveling observer from the load end B at $t = -\tau$ it will have V = 0, I = 0, hence will start from the origin as shown in Figure 15.11b and trace a straight line of slope $+Z_0$. This line intersects the BC at the source end at the point marked as A(0). The coordinates of this point are the voltage and current at the start of the line at t = 0. The observer now travels in the forward direction and thus traces a line of slope $-Z_0$. This line intersects the BC at the load end at a point marked as $B(\tau)$, its coordinates giving the voltage and current at the load end at time $t = \tau$. Now, the observer travels backwards, thus tracing a line of slope $+Z_0$ to intersect the BC at the source end at point $A(2\tau)$, giving the conditions at the source end at time 2τ. The process continues as shown in the figure until steady state is reached at the intersection of the two BCs.

It is clear from this procedure that as long as we are able to plot the V-I conditions at the source and load ends we are can apply this technique. This is particularly useful for nonlinear loads. We illustrate this by an example where the configuration in Figure 15.11a is modified by adding a nonzero internal source resistance R_G and making the load resistance nonlinear with a V-I curve as shown in Figure 15.12a. In the same diagram we have plotted the BC at the source end of the line $V = V_0 - IR_G$. The calculation proceeds as before by starting a backward-traveling observer at B (the origin in the V-I diagram) tracing a line of slope $+Z_0$ until the intersection with the BC at A(0). The coordinates of this point give the voltage and current at the source end at t = 0. The forward-traveling observer and the intersection with the load boundary condition at $B(\tau)$ give the conditions at the load at time τ. A backward travel and a forward travel identify the intersection $B(3\tau)$ with coordinates, the voltage, and current at the load at time 3τ. The process continues for as long as required. A schematic of the load voltage is shown in Figure 15.12b.

The conclusion from the calculations shown in this section is that propagation of fast pulses in PCBs is a complex phenomenon that affects strongly EMC and SI in a manner that strongly couples these two disciplines especially at high frequencies. Other factors. especially losses, skin effect, and proximity effects, can also affect propagation especially at very high frequencies.

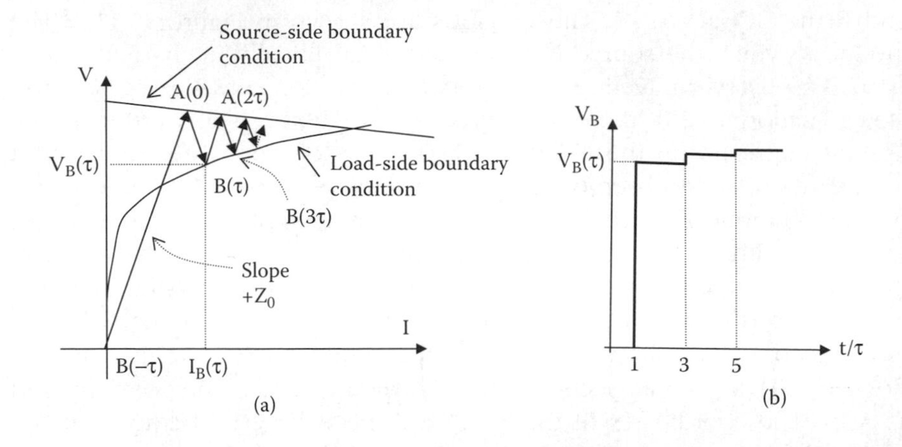

Figure 15.12 Bergeron diagram (a) and voltage at the load (b) for the configuration in Figure 15.11a where we have added a source internal resistance R_G and the load termination is now nonlinear.

Extensive literature is available covering transmission line theory, which may be consulted for more details.[7–10] Numerical techniques were described in Chapter 7.

15.3 Board and Chip Level EMC

The first systematic account of parasitic effects in PCBs and ICs was given in Reference 11. Since the publication of this book, substantial work has been published on a number of issues arising in this environment, especially over the last few years as increasingly high-speed systems are incorporated in a whole range of products. The purpose of this section is to highlight some of the major EMC and SI issues and techniques for tackling them. A full account of this rapidly expanding field cannot be given here and the reader should consult specialist texts on this subject.[2,4]

15.3.1 Simultaneous Switching Noise (SSN)

A typical problem arising in high-density high-speed circuits is shown schematically in Figure 15.13. This is an example of a single-ended signaling arrangement where the signal fed to the driver is $V_{in} - V_{ref}$ as shown. The difficulty with this arrangement is that V_{ref} is affected strongly by parasitic components in the ground track and by fast switching currents. It is clear from this diagram that $V_{ref} \approx L\ di/dt$ where L is the inductance associated with the ground track and di/dt is the rate of change of the

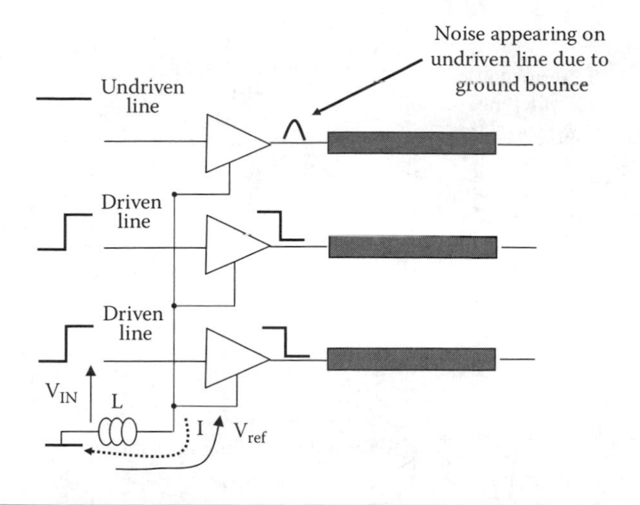

Figure 15.13 Ground bounce due to the voltage drop across L during the simultaneous switching of several drivers.

current. This current is due not only to the particular driver but also to all other driven lines. Moreover, other drivers that are not driven will experience a shift in their reference level. This is described as "ground bounce" and is the source of noise and interference in high-speed circuits. This situation is familiar to EMC engineers and was described in connection with grounding in Section 10.3. Another example is shown in Figure 15.14. A similar effect is observed in connection with the supply rails, the so-called "supply bounce." Since in modern circuits several gates switch simultaneously, the current flow and therefore the bounce in shared paths can be considerable. A more complex example is shown in Figure 15.15. Here gate 1 drives gate 2 through an interconnect. We have marked out the current path when gate 1 makes a high-to-low transition and also when it makes a low-to-high transition. We see that the parasitic inductance of the power, signal, and ground tracks will generate undesirable voltage drops at one or both of the transitions. Remedies to this situation are the reduction to the magnitude of the switched current (smaller di), increase in transition time (larger dt), reduction of the parasitic inductances (wider traces or a better transmission line configuration), or the addition of decoupling capacitors strategically placed to supply pulse current locally without the need for the pulse current to come from afar. Two such decoupling capacitors are shown in Figure 15.15. The placement of these capacitors must be chosen based on a clear understanding of the required current flows and paths so that pulsed current is supplied locally and thus voltage drops along long interconnects are minimized. It should also be borne in mind that capacitors are not ideal components and that they

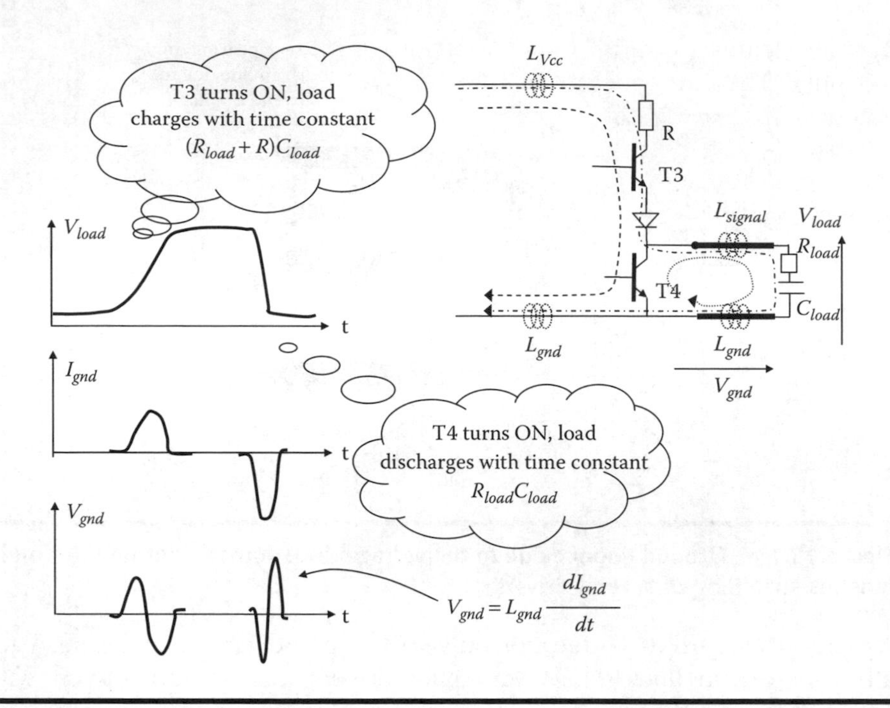

Figure 15.14 Current flows through L_{ground} and associated ground bounce.

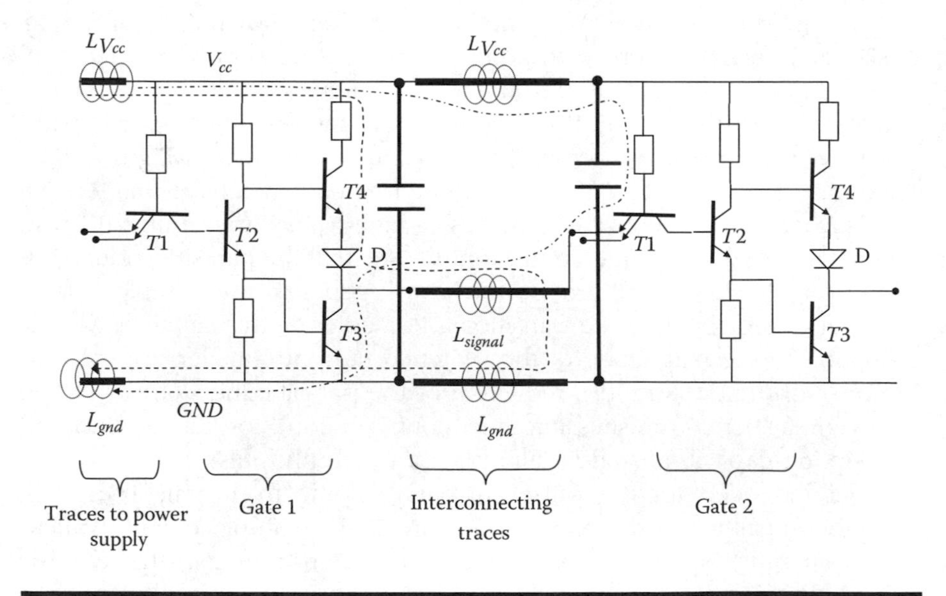

Figure 15.15 The short dashed line shows current flow during the high-to-low transition, and dash-dotted line for low-to-high transition. The two capacitors provide decoupling.

include a parasitic inductive component (mostly due to their lead connections). The presence of this inductance negates to some extent the beneficial impact of the capacitance in supplying pulse current while maintaining a constant voltage. At high frequencies there may also be resonances between these stray inductances and capacitances, further complicating the proper design of the circuit. The general rule of thumb is to select decoupling capacitors with low inductance value (high resonance frequency), very short lead connections to pins (preferably of surface mount technology), and typically of value 0.001 μF per module. Bulk tantalum electrolytic capacitors of a value ten times larger than the total value of decoupling capacitors are also placed around the board to provide an additional reservoir of energy for supplying current at a slower rate.

A more radical alternative is to send the signal in such a way that it provides its own reference. This is shown schematically in Figure 15.16a.[2] A differential transmitter sends the signal V_p and its complement Vn on two lines and therefore the differential receiver senses the voltage V_p-V_n as shown in Figure 15.16b. Any interference or common-mode signal couples on both lines in the interconnect and thus is canceled at the input to the differential receiver V_1-V_2. Because noise rejection is thus improved, the voltage swing can be dropped to a few hundred millivolts. Very fast transmission rates may be achieved in the GBits/s region. Several differential signaling standards are available including RS-422, RS-485, and low-voltage differential signaling (LVDS).[12] The disadvantage of these schemes is that an additional track is required.

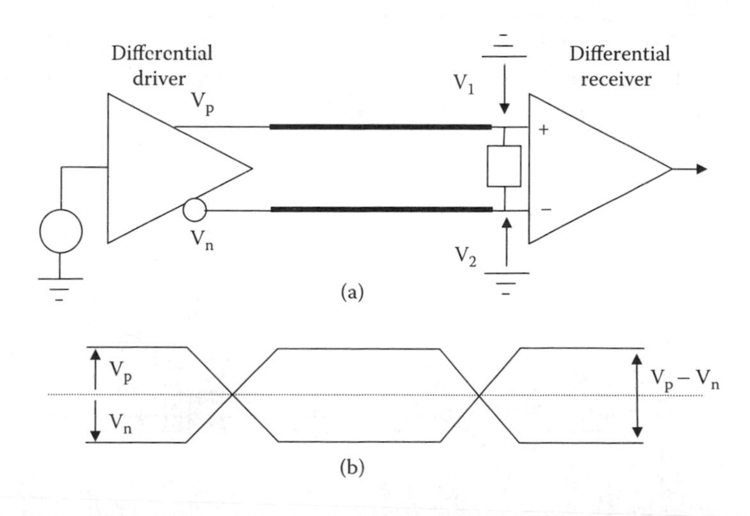

Figure 15.16 Differential signaling configuration (a) and signal levels on each line (b).

15.3.2 Physical Models

Physical models of ICs can be constructed in the normal way using lumped circuit components, distributed components, or full-field formulations based on Maxwell's equation. The challenge in doing this is handling the extreme complexity of practical systems. As the frequency increases, more sophisticated techniques are required to model particular ICs, their packaging, interconnects, and boards. Full rigorous system models and simulations are extremely complex and beyond current capabilities. The best approach is to develop a hierarchy of models starting from low frequency up to the highest frequencies. In Figure 15.1 the complexity in the assembly of modern integrated circuits was indicated. In Figure 15.17 two ways of attaching a chip to a package are shown. In Figure 15.17a the so-called wire bond is shown, which is easy to route and has good thermal and mechanical properties. Its inductance is variable and difficult to calculate and ranges between 1 and 5 nH. In Figure 15.17b an alternative connection is shown. The chip is mounted upside down (flip chip) with the pad at the bottom connected to the package through solder balls. This arrangement has smaller inductance (approximately 0.1 nH) and allows a denser interconnect pattern (lower minimum pitch, as low as 100 μm), but has poorer thermal behavior. More complex arrangements are in use such as stacking more chips on the same package, etc. Attempting to model this level of detail, in a situation with a large numbers of chips and packages populating a large multilayer board with a multitude of interconnects, in order to deal with the full range of EMC and SI issues is a daunting task. We give a brief introduction of the methodology appropriate in each case based on an example given in Reference 13.

We have seen that keeping the magnitude of the power distribution impedance low is highly desirable in minimizing ground and supply bounce and an important factor in keeping interference levels low. This is easy to see as the power distribution network in the PCB environment is extensive and all pervading. Figure 15.18 shows a schematic of this impedance versus frequency for a large processor. The dotted curve is

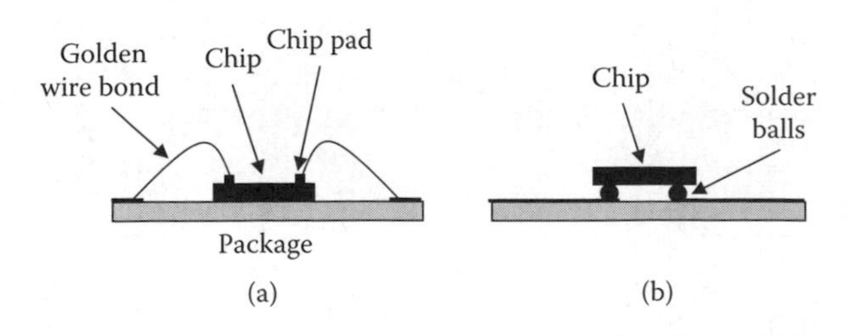

Figure 15.17 Wire bond (a) and flip-chip (b) configurations.

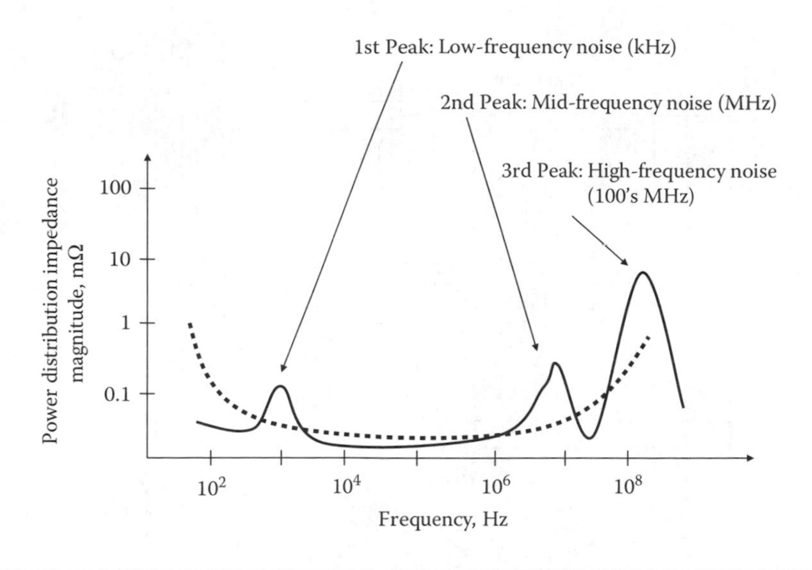

Figure 15.18 Power distribution impedance as a function of frequency.

what one would expect from theoretical considerations. At low frequencies inductance is not important and the dominant component is capacitance, hence impedance falls with increasing frequency. With rising frequency there comes a point when inductance cannot be neglected and therefore impedance rises with frequency giving the characteristic inverted bell-shaped curve. Measurements, however, show a behavior closer to the one depicted by the solid curve. Here we observe clear resonances, and the task of the designer and modeler is to establish the upper bounds for $|Z|$ and make sure that it is kept below the design limit. We will sketch out how this challenge may be met. There are three frequency regimes where we are able to make a different level of approximation. First, to predict behavior at low-frequencies (kilohertz range) we can neglect inductances and only account for resistances and capacitances associated with bulk and decoupling capacitors. Pulses that contribute energy in this part of the spectrum are relatively long in duration (100s μs) and therefore propagate over a large part of the physical extent of the circuit. Therefore, their parameters are affected by all parts of the circuit and as a consequence we must model the entire circuit. A schematic of such a model is shown in Figure 15.19. In a realistic problem a very large number of components will result, giving a cumbersome circuit to characterize. Similar problems arise as one tries to tackle problems at higher frequencies. In mathematical parlance, the problem possesses a large number of degrees of freedom (the order of the model is very high), making the solution computationally very expensive to achieve. However, not all degrees of freedom are important in determining system response with an acceptable

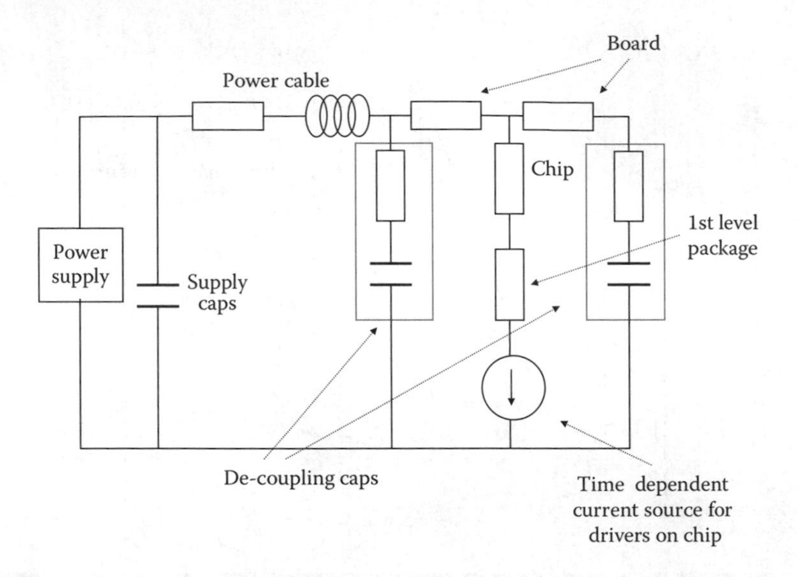

Figure 15.19 Low-frequency noise model.

accuracy. Mathematical techniques are available to reduce the order of a model without significant loss in accuracy. The approach is known under the collective name of model order reduction (MOR) technique. More details may be found in References 14 to 16.

As we move to higher frequencies in the MHz range the details of the packaging of the multichip module (MCM) and the board must be included. Inductance is important, but effects involving delays and time retardation can be neglected. Since pulses contributing to this part of the spectrum have a shorter duration (100s ns), conditions at a particular location are only affected by the circuit within a radius $u\tau$ (where u is the speed of propagation and τ the pulse duration), thus making it possible to make some simplification. A typical model is shown in Figure 15.20. The board may be modeled by a two-dimensional array of L and C component as shown in Figure 15.21. Each node represents a part of space typically $(\lambda/20) \times (\lambda/20)$ where λ is the wavelength at the highest frequency of interest. Models of decoupling capacitors, pins, and vias may be added to this model. The level of accuracy required depends on specific requirements and on frequency. A lumped circuit model of a via has the form shown in Figure 15.22a. A full-field model of a via is shown in Figure 15.23. The figure shows the magnetic field associated with the Gaussian shaped current pulse as it negotiates its way through the via. It shows clearly how the barrel of the via acts like a radiator causing emissions that couple to adjacent components. A track with a sharp bend can be modeled by a straight transmission line with a lumped capacitor

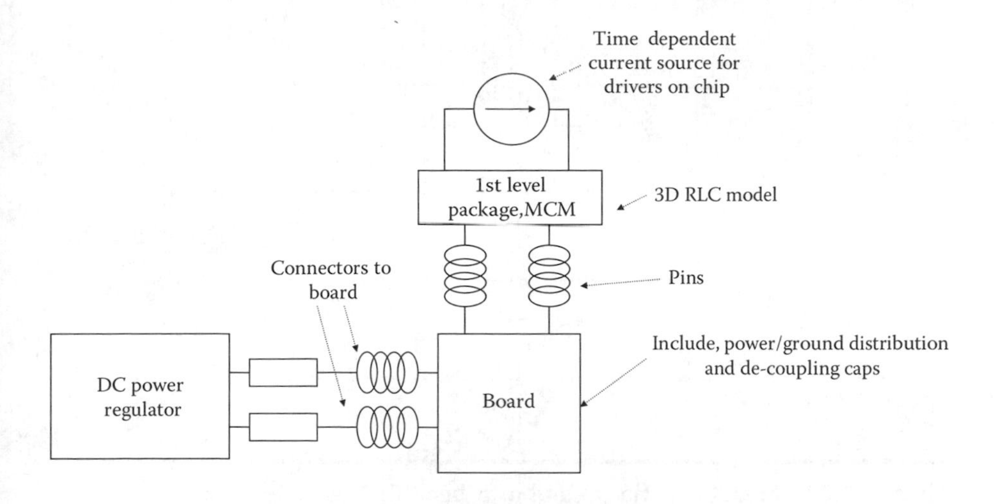

Figure 15.20 Mid-frequency noise model.

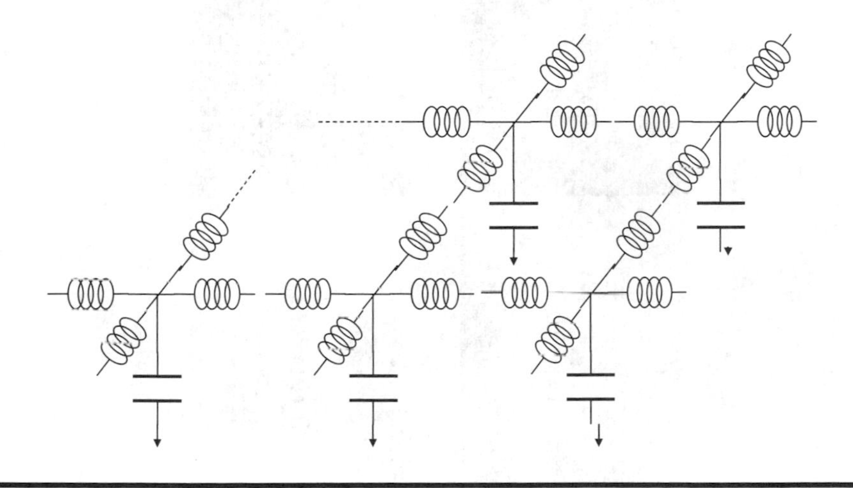

Figure 15.21 Basic board model consisting of a 2D lumped L-C configuration.

as shown in Figure 15.22b. More sophisticated models of bends that include emissions and signal distortion and yet maintain a measure of model simplicity are described in Reference 17.

For higher frequencies in the hundreds of MHz range, details of the chip and package must be included with full account of distributed R, L, C, and transmission line effects. Many of these parameters require three-dimensional calculations and are difficult to perform. One significant difficulty is that the details that may allow a full calculation to be made are not available as this would reveal proprietary material with substantial intellectual property loss. A schematic model is shown in Figure 15.24.

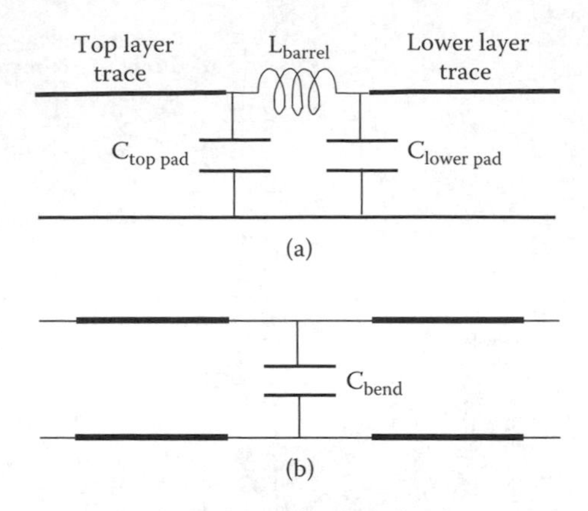

Figure 15.22 Model of a via (a) and of a bend (b).

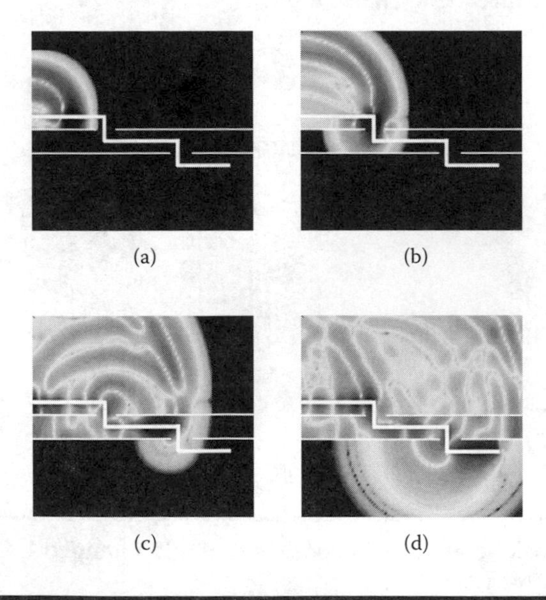

Figure 15.23 Schematic of pulse propagation through a cascade of vias. Magnetic field associated with a pulse prior to reaching the first via (a), going through the first via (b), through the second via (c), and emerging after the second via (d).

A model of the chip at very high frequencies must take account of different modes of propagation, which depend on the resistivity of the substrate and the importance of the skin effect in propagation. In order to give a taste of the complexity of these calculations, we consider the configuration shown in Figure 15.25. This is typical of semiconductor manufacture consisting of a metalized layer acting as ground, a Si substrate,

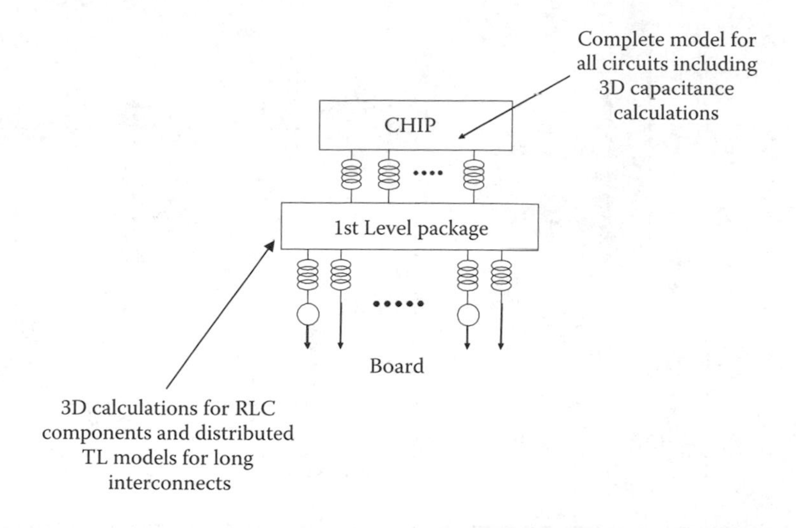

Figure 15.24 High-frequency model.

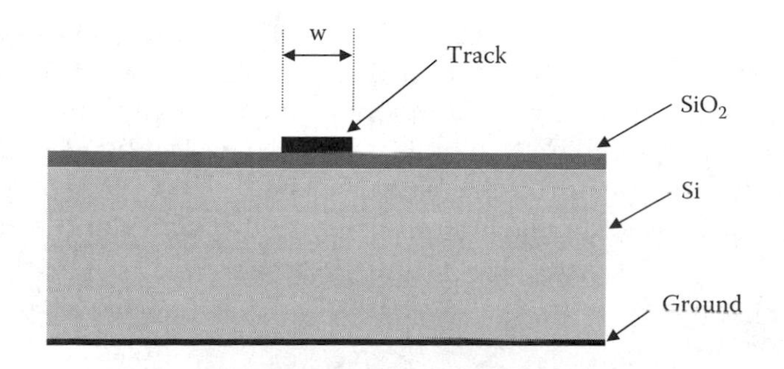

Figure 15.25 Microstrip line on Si-SiO_2 system.

a thin layer of SiO_2, and the conducting track of width w. The SiO_2 is a good insulator and acts as a passivation layer forming a hermetic seal to protect circuit elements. In a realistic situation several other tracks will be deposited on the same substrate. In considering crosstalk between adjacent tracks it is important to know how fields distribute in the vicinity of each track and to what extend they penetrate into the substrate and reach the ground plane. Clearly, a closely confined field between the track and the metalized ground layer will be very different as regards crosstalk compared to a situation where fields spread out and return is partially through the Si substrate. This can happen as the electrical conductivity of substrates varies widely depending on IC requirements. The problem cannot be solved analytically unless the track width is assumed infinite, resulting in a parallel plate waveguide configuration insulated with two layers, one

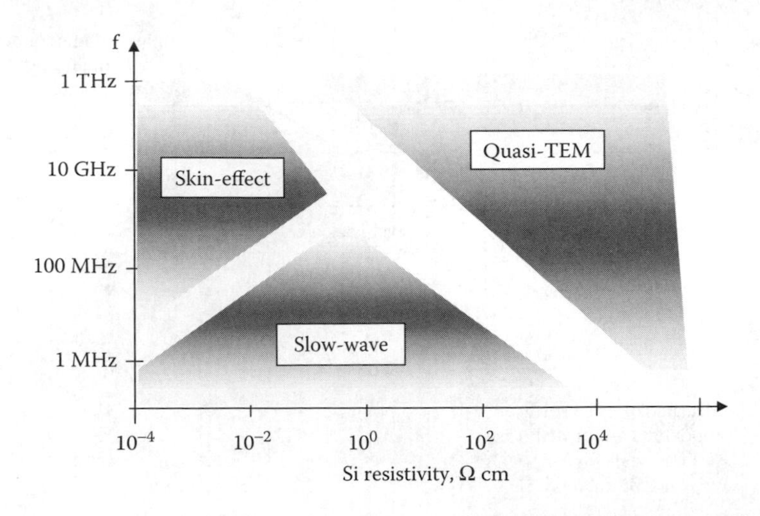

Figure 15.26 Resistivity-frequency graph showing the different propagation regimes.

of which (Si) has finite electrical conductivity. This calculation may be found in Reference 18. The graph in Figure 15.26 is based on this work and shows that, depending on frequency and resistivity, three modes of propagation are possible. Naturally, the boundaries between different modes are not clear-cut, but the basic ideas are valid. We know that the skin depth and the loss tangent are related to frequency and resistivity as follows:

$$\delta = \sqrt{\frac{\rho}{\pi \mu f}}$$
$$\tan \delta = \frac{1}{2\pi f \rho \varepsilon} \tag{15.25}$$

When the product of frequency and resistivity is large enough for the loss tangent to be small, the silicon substrate behaves as an insulator (dielectric) giving a quasi-TEM mode of propagation. When conditions are such that the depth of penetration into silicon is small, the substrate behaves as a conductor and effectively as an imperfect ground plane. Finally, for moderate frequencies and resistivities a slow surface wave mode prevails. For the practical case of finite width tracks numerical solutions are available showing field distribution in different regimes.[19]

The conclusion is that at high frequencies and in this complex environment it is very difficult to establish and operate numerical models that mimic the physical details of the circuit. Progress is being made in enhancing the capabilities of numerical methods so that they can cope with this challenge, but there are still formidable tasks ahead. This work is focused on multiscale methods to allow efficient treatment of fine features in electrically large structures and sophisticated techniques for the treatment of material behavior at high frequencies.[20,21] Reference should also be made to the MOR techniques already mentioned.

15.3.3 Behavioral Models

The complexity of current and projected systems[22] is such that physics-based models of the type described in the previous section are difficult to implement for system-level studies and therefore simpler models must be sought. Another factor, which pushes in the same direction, is the unwillingness of vendors of ICs to disclose circuit details that reveal proprietary and commercially valuable information. For these reasons ways are sought to allow a level of predictive capability at the system level using simplified models that mimic the behavior of components as seen at the pins without any details of the internal structure. They are described as behavioral models and have the advantage of efficiency, wide availability, and commercial confidentiality. On the negative side, they have a high sensitivity to loads and they account for a limited range of physics. The most common behavioral models are available under the label of IBIS (Input/output Buffer Interface Specification) and are widely available and accepted.[23] In Figure 15.27 a CMOS transistor output stage is shown. Block A shows the output transistors, which are modeled in IBIS by DC V-I tables as shown schematically in Figure 15.28 for the pullup and pulldown cases. Block B in Figure 15.27 shows the clamping diodes, which are also represented by V-I curves. The models for the power

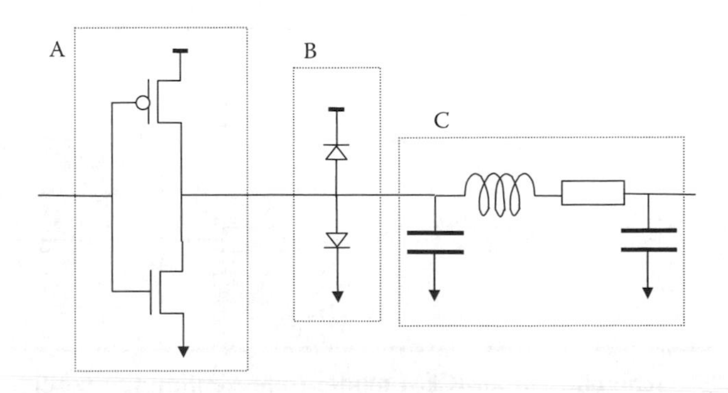

Figure 15.27 CMOS transistor output stage.

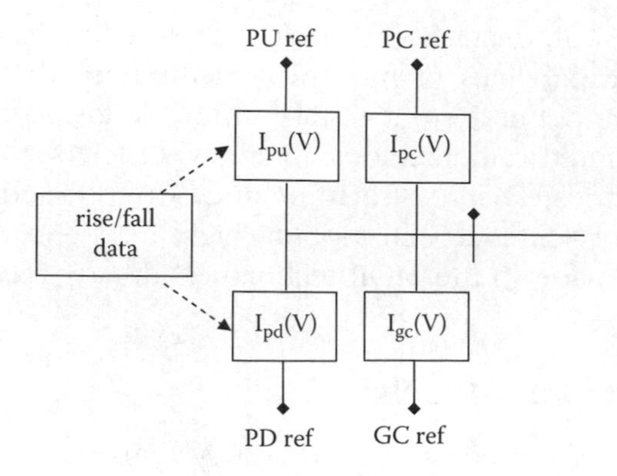

Figure 15.28 Schematic of an IBIS model.

clamp and ground clamp diodes are also shown schematically in Figure 15.28. These diodes clamp signals due to line reflections. Finally, block C represents the parasitics of the pad and package. The rise/fall data in Figure 15.28 relate to the transition time from one logic state to another. The IBIS models do not account for a range of effects that are important contributors to EMC. For example, the contribution of the internal activity in a component is not linked to the input/output. A more recent development are the ICEM (Integrated Circuit Electromagnetic Model),[24] models that complement the IBIS models to give a more complete picture of EMC behavior. The ICEM model introduces active elements that account for internal activities while IBIS addresses mainly interface issues. Thus it is proposed that ICEM models account for conducted emissions through supply and input/output lines, and direct radiated emissions. An ICEM model is shown schematically in Figure 15.29. The package model follows

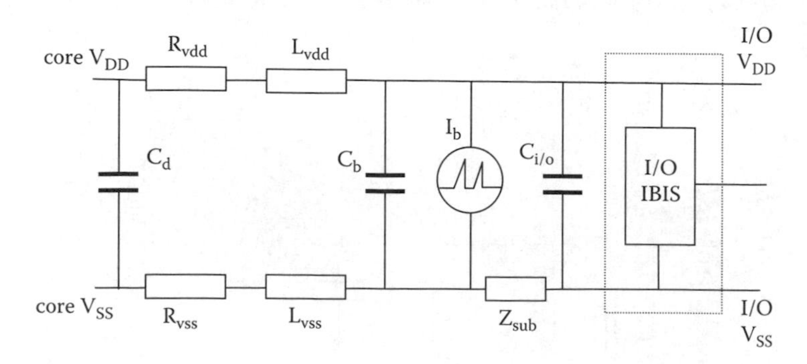

Figure 15.29 Schematic of an ICEM model (representing the core) and IBIS model (representing the input/output interface).

the IBIS input/output specification and the core noise model is shown on the left with an equivalent current generator I_b, which is a piece-wise linear approximation of the shape of the current in the time domain. Typically, this current ranges between 1 mA and 1 A with a period between 500 ps and 50 ns. The substrate resistance is typically a few Ω. The capacitor $C_{i/o}$ represents the decoupling capacitance for the input/output supply. Capacitors C_d and C_b stand for the decoupling and block capacitors. The other components shown account for the parasitics of the supply network. The necessary parameters for these models are obtained either by simulation or from measurements in a variety of experimental test configurations.[4,25]

A process of continuous development is in hand to enhance ICEM models so that they can be used for accurate EMC characterization of complex systems. The structure of these models is such as to fit in automatically with existing circuit solvers such as SPICE for ease of use by a wide range of vendors and developers.

References

1. Canavero, F, "Signal Integrity and EMC at Board Level," AUNP Tutorial: EMC and Signal Integrity for Electronic Design, Politecnico di Torino, Italy, 15–15 Dec. 2005.
2. Thierauf, S C, "High-Speed Circuit Board Signal Integrity," Artech House, 2004.
3. Christopoulos, C, "Basics of EMC and SI," AUNP Tutorial: EMC and Signal Integrity for Electronic Design, Politecnico di Torino, Italy, 15–15 Dec. 2005.
4. Ben Dhia, S, Ramdani, M, and Sicard, E, "Electromagnetic Compatibility of Integrated Circuits-Techniques for low emission and susceptibility," Springer-Verlag, Heidelberg, 2006.
5. Bewley, L V, "Travelling Waves on Transmission Systems," Wiley, NY, 2nd edition, 1951.
6. Bergeron, L, "Du Coup de Bélier en Hydraulique au Coup de Foudre en Electricité," Dunod, 1949, and english translation "Water Hammer in Hydraulics and Wave Surges in Electricity," ASME, 1961.
7. Hall, S H, Hall, G W, and McCall, J A, "High-Speed Digital System Design — A handbook of interconnect theory and design practices," Wiley 2000.
8. Granzow, K D, "Digital Transmission Lines-Computer modelling and analysis," Oxford University Press, Oxford, 1998.
9. Magnusson, P C, Alexander, G C, Tripathi, V K, and Weisshaar, A, "Transmission Lines and Wave Propagation," CRC Press, Boca Raton, FL, 2001.
10. Matick, R E, "Transmission Lines for Digital and Communication Networks," IEEE Press, 1995.
11. Bakoglu, H B, "Circuits Interconnections and Packaging for VLSI," Addison-Wesley, Reading, MA, 1990.

12. National Semiconductors: http://www.national.com/nationaledge/feb02/flavors.html.
13. Becker et al, "Mid-frequency simultaneous switching noise in computer systems," IEEE Trans. on Comp. Pack. And Man. Techn.," 21(2), 157–163, 1998.
14. Feldmann, B and Freund, R W, "Efficient linear circuit analysis by Padé approximation via the Lanczos process," IEEE Trans. on Computer Aided Design, 14, pp 639–649, 1995.
15. Cangellaris, A C, Celik, M, Pasha S, and Zhao, L, "Electromagnetic model order reduction for system level modelling," IEEE Trans. on Microwave Theory and Techniques, 47, pp 840–850, 1999.
16. Zhu, Y and Cangellaris, A C, "Finite element based model order reduction of electromagnetic devices," Int. J. Numer. Modelling, 15, pp 73–92, 2002.
17. Liu, X, Christopoulos, C, and Thomas, D W P, "Prediction of radiation losses and emission from a bent wire by a network model," IEEE Trans. on Electromagnetic Compatibility, EMC-48, 3, pp 476–484, Aug 2006.
18. Hasegawa, H, Furukawa, M, and Hisayoshi, Y, "Properties of microstrip line on a Si-SiO_2 system," IEEE Trans on Microwave Theory and Techniques, 19(11), pp 869–881, 1971.
19. Grabinski, H, "VLSI Interconnect Characterization," AUNP Tutorial: EMC and Signal Integrity for Electronic Design, Politecnico di Torino, Italy, 15–15 Dec. 2005.
20. Christopoulos, C, "The Transmission-Line (TLM) Method in Electromagnetics," Morgan and Claypool, 2006.
21. Christopoulos, C, "Multi-scale modelling in time-domain electromagnetics," Int. J. Eletcron. Commun. (AEÜ), 57(2), pp 100–110, 2003.
22. "International Technology Roadmap for Semiconductors," 2005 Edition, www.itrs.net.
23. IBIS, www.eigroup.org/ibis/.
24. ICEM, www.ic-emc.org.
25. Martens, L, "High-Frequency Characterization of Electronic Packaging," Kluwer, Dordrecht, 1998.

Chapter 16

EMC and Wireless Technologies

Traditionally, EMC has focused on the emission and immunity of commercial equipment primarily at frequencies below 1 GHz, and, at least in the EU, a lot of regulations were driven by competitive issues related to the free movement of goods. The digital revolution has had considerable impact on our electromagnetic environment but it has to be admitted that EMC still bears the imprint of its legacy in analogue radio. In recent years, the EM environment has had to accommodate a host of new wireless services, both licensed and unlicensed, that compete for the same part of the spectrum. EMC as traditionally conceived has therefore to expand to encompass broader issues related to the interoperability and coexistence of systems. The illustration in Figure 16.1 will be familiar to many readers. It shows a phone (mobile), which may encompass considerable multifunctionality both internal (e.g., camera, video, MP3 player) and external (e.g., GPS, Bluetooth). Some of the external wireless links are licensed (e.g., GSM) and others are unlicensed and nomadic (e.g., Bluetooth). The manner in which these different technologies use the frequency spectrum means that receivers and transmitters must be able to work at different frequency bands in an adaptive manner, adjust gains and power levels, and do so in as transparent and secure a manner as possible. Several ad hoc networks will be using the same part of the frequency spectrum, which are effectively heterogeneous.[1,2] A multitude of functional standards have been established to cope with this situation, but new standards are needed to meet radio regulatory and EMC requirements.[3,4]

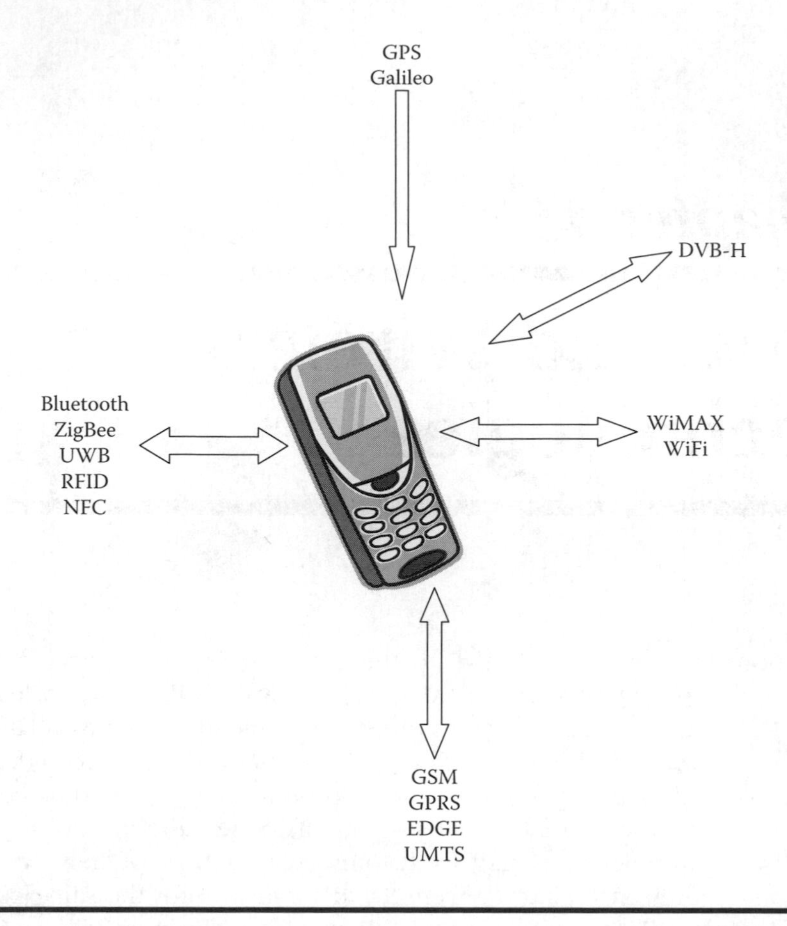

Figure 16.1 Schematic showing a mobile phone (with internal functionality) and external wireless links.

Rapid changes are taking place and functional standards are continuously under development to make it impossible to give an authoritative picture of the field at this stage. However, the impact of these technologies on EMC as currently understood is likely to be major and EMC engineers need to become familiar with the structure and future developments in wireless technologies. The purpose of this chapter is to sensitize EMC engineers to this challenging topic and thus facilitate interaction between professionals working in EMC and communications technologies. A more complete treatment of communications technologies is outside the scope of this book and must be sought in specialist texts.[5–8]

16.1 The Efficient Use of the Frequency Spectrum

In the early days of radio, national governments and international bodies would allocate parts of the spectrum to different services and users. After a century of this practice and with the current explosion of wireless applications more efficient ways must be sought to utilize this asset (the frequency spectrum). Some of the ingenious ways that communications engineers have sought to get the maximum out of the available spectrum are described below.

In conventional frequency division multiple access (FDMA), each user is allocated a certain amount of bandwidth spaced out as near as possible but with a small "guard band" to avoid interference between users as shown schematically in Figure 16.2a. This is the logical response to accommodating different users, and interference is not a problem provided that users keep within their allocated band. However, this is not the most efficient way, since a user has a dedicated band that, in most cases, is underused (e.g., during silent parts in a conversation). An alternative is to have several users using the same frequency band but allow them transmission and reception over short regular time intervals. This is shown schematically in Figure 16.2b and is known as time division multiple access (TDMA). In this scheme each user receives and transmits only part of the time. In practice, this does not represent a noticeable deterioration in the quality of service, as with fast digital signal processing techniques employed in modern communications the user is unaware of the interrupted nature of communication over the physical medium. Yet another way to use the spectrum is to allow users to communicate simultaneously over the same frequency band. At first, this may seem paradoxical as the chaotic nature of the aggregate transmitted signal would appear to make communication impossible. However, it should be possible to extract the signal we are interested in — after all, at a party there are several simultaneous conversations and yet our brain is able to listen to the one we are interested in. The engineering approach to this problem is for each transmitter (TX) to code each signal injected into the common channel and for the intended recipient (RX) of this signal to extract it. One requirement for this is that RX knows the code, in which case the transmitted signal can be extracted from the aggregate signal by correlating with the code. We have discussed signal correlation in Section 4.2. The other requirement is that, like doors and keys, each conversation must have a "different" code. In engineering parlance this means that the codes used by each user must be "orthogonal." This technique is known as code division multiple access (CDMA) and is shown schematically in Figure 16.2c. We have encountered a similar concept (a spreading code) in connection

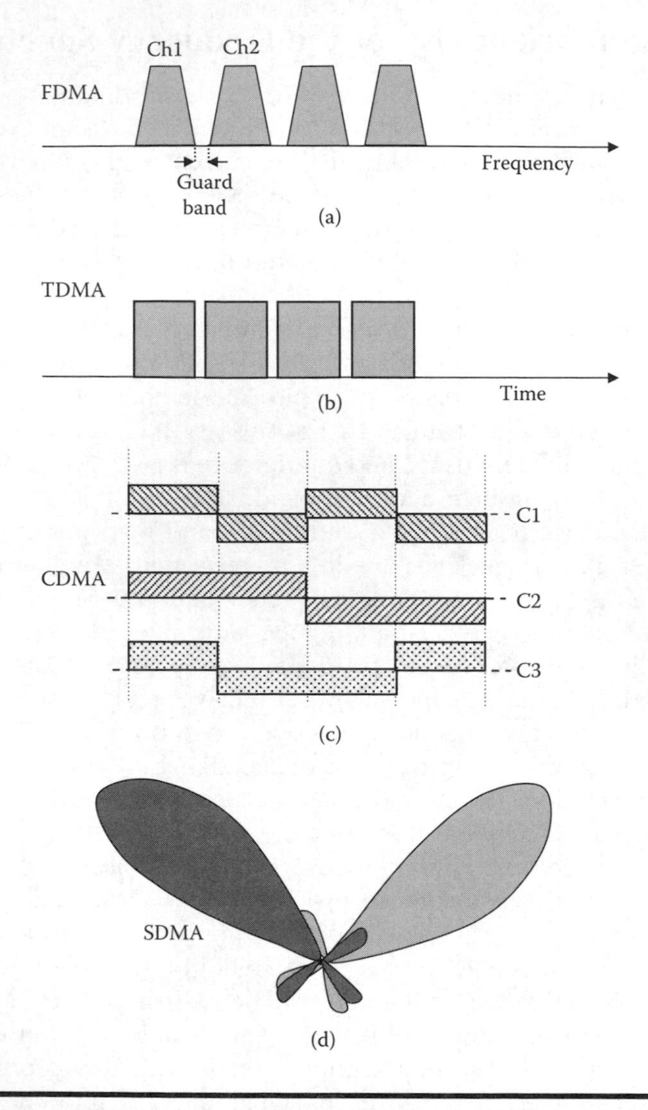

Figure 16.2 Examples of schemes for multiple access: (a) frequency division multiple access, (b) time division multiple access, (c) code division multiple access, and (d) space division multiple access.

with direct sequence spread spectrum techniques where the objective was security and spreading the energy in the signal over a wide band of frequencies and thus lowering the spectral density. The code operates at a higher rate than the data. Taking the three codes shown in Figure 16.2c as an example we see that they are orthogonal

$$C1 = (1, -1, 1, -1)$$
$$C2 = (1, 1, -1, -1)$$
$$C3 = (1, -1, -1, 1)$$

Hence, as expected, $C1 \times C2 = 1 \times 1 + (-1) \times 1 + 1 \times (-1) + (-1) \times (-1) = 0$. The same result is obtained for any pair combination of orthogonal signals. Many users can thus be placed on the same data stream provided each is first coded using a set of orthogonal codes as shown in Figure 16.3.

Yet another scheme is to use directive antennas to communicate with several users in the same frequency band as shown schematically in Figure 16.2d. This scheme is referred to as space division multiple access (SDMA). Cellular systems rely on the fact that each transmitter has a limited coverage area beyond which signal levels are too low, so other users can use the same frequency without appreciable risk of interference. A typical arrangement is of clusters of seven cells as shown in Figure 16.4. The size of a cell can range from 1 (in a city area) to 10 km (in a rural area) depending on the density of customers. Adjacent cells are allocated different frequencies so that there is no risk of interference since, in practice, the geographical boundary of each cell is uncertain. In Figure 16.4 the shaded cells can be allocated the same frequency with limited risk of interference. Clearly, the way the "handover" takes place as a user moves across the boundary between two adjacent cells so that the change is transparent to the user must be done carefully. This can be done in several ways. One way (used in GSM) is for the mobile to monitor and report to the network the received signal strength from several base stations. The network then

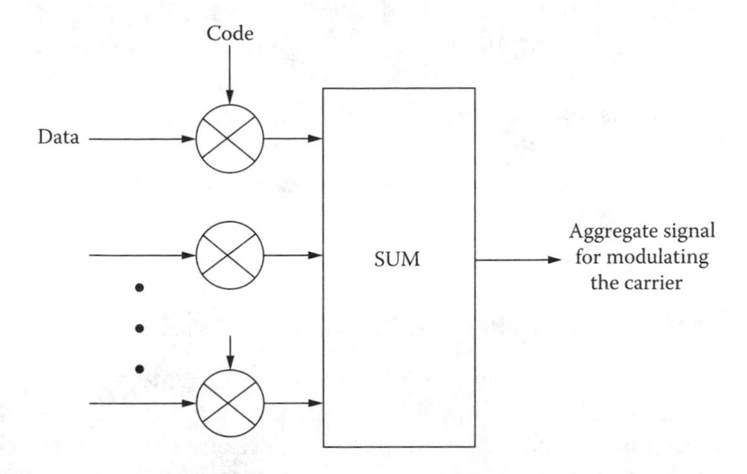

Figure 16.3 Data from several users are coded individually using orthogonal codes and then used to modulate a carrier.

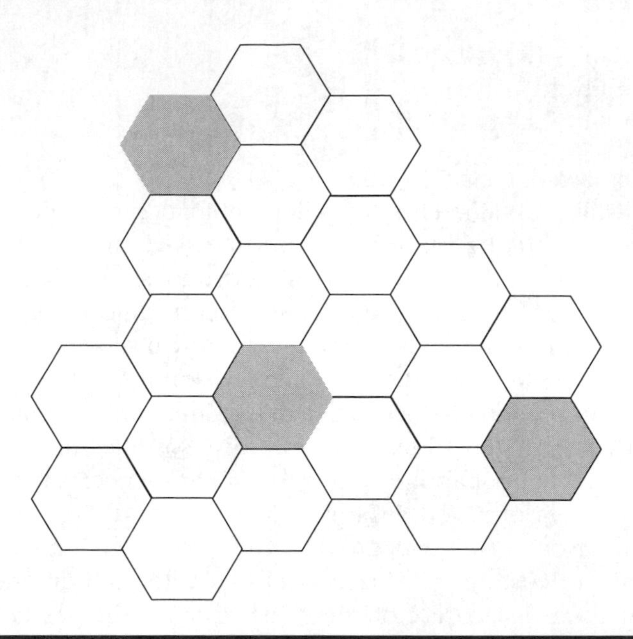

Figure 16.4 Schematic of a cellular system. Shaded cells transmit at the same frequency with minimal risk of interference.

decides on the basis of the available signal strength and the availability of spare channels in other cells when the handover must take place and informs the mobile. The mobile in turn then makes the necessary frequency adjustments to retune at the new channel. This is a complicated process and shows the degree of complexity involved in running a public service mobile network. In a telephone network duplex operation is required, i.e., it should be possible to speak simultaneously in both directions. This is done using either frequency division duplex (FDD) (Figure 16.5a) or time division duplex (TDD) (Figure 16.5b). The key to efficient and secure communications is diversity and all the approaches described above are focused on this aim.

16.2 EMC, Interoperability, and Coexistence

The data we wish to transmit are employed to modulate a high-frequency carrier that then feeds a transmitting antenna, propagates through free space and buildings, and suffers reflections and other distortions to arrive together with any other signals or noise to the receiver antenna for demodulation and decoding to recover, hopefully, the original data. The method of modulation clearly has an impact on how robust the transmission is and therefore on the quality of service. Modulation is a topic

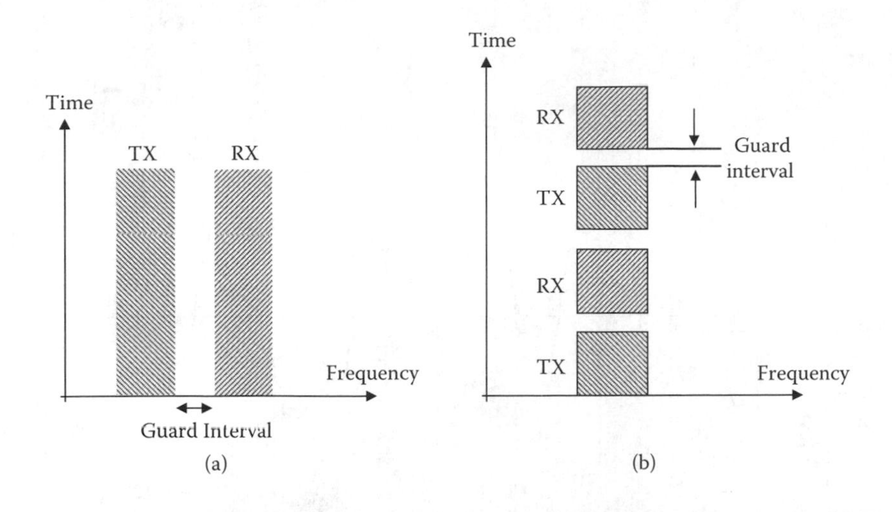

Figure 16.5 Duplex schemes: (a) FDD, (b) TDD, TX-transmit, RX-receive.

covered in many textbooks. We merely give here a couple of examples to show how it affects the immunity of the transmission to interference.

We start with the example of a modulation scheme known as binary phase shift keying (BPSK) where phase is used to transmit different symbols. Symbols (1 or 0) are transmitted in fixed intervals T. This modulation scheme inserts a phase shift 0 or π according to the rule

$$s(t) = h(t)A\cos(\omega_c t) \tag{16.1}$$

where A is the amplitude of the symbol and ω_c is the angular frequency of the carrier. The function h(t) is +1 when the symbol is "1" and –1 when the symbol is "0." This is shown schematically in Figure 16.6a for the sequence "1011." This scheme is best illustrated in terms of the signal constellation or phase transition diagram shown in Figure 16.6b. We see from this diagram that as long as the phase of the received signal can be recognized then detection is possible. All that is needed is to establish whether the signal lies on the right- or left-hand side of the plane. The amplitude of the received signal does not convey information. It needs to be large enough for detection. The error margin is of the order of the symbol amplitude, thus giving a degree of immunity to noise and interference. More sophisticated schemes such as the differential phase shift keying (DFSK) are also available.

It is also possible to expand further the idea of phase transitions as a means of transmitting more bits per symbol and also provide a degree of noise immunity. An example is the quaternary quadrature phase shift

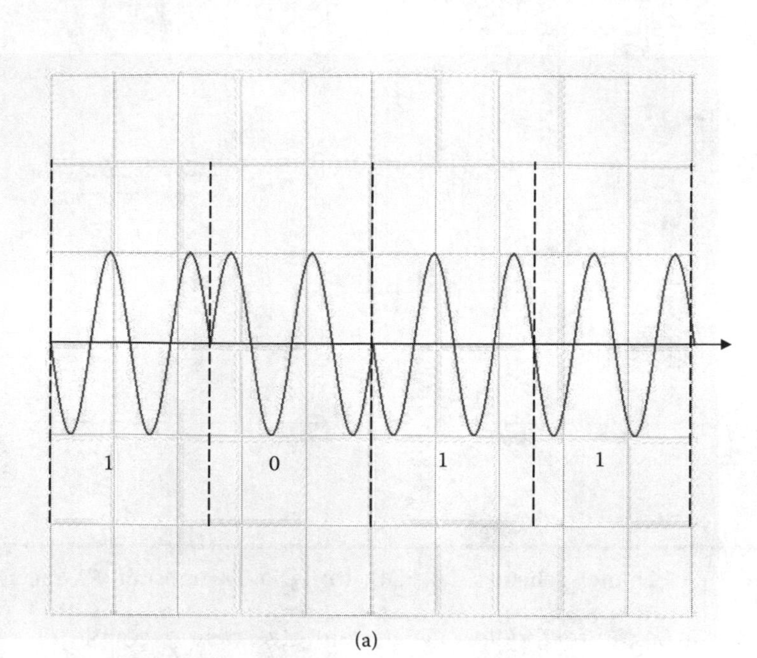

(a)

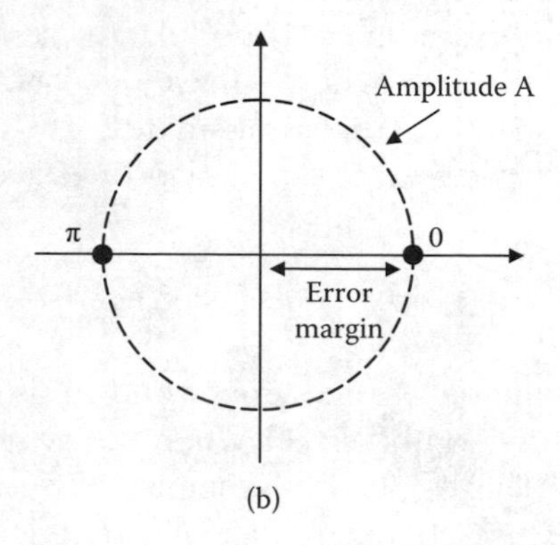

(b)

Figure 16.6 Principle of BPSK (a) and corresponding phase transition diagram (b).

keying (QPSK). The phase transition diagram for this scheme is shown in Figure 16.7. We see that at each of the four states it transmits two bits, the combinations 00, 01, 11, 10. We see, however, that the error margin is now smaller. By identifying more phase states we can transmit more bits per symbol (e.g., 8PSK), but at the cost of increasingly lower error margin.

More modulation schemes of increasing complexity are available to deal with a variety of situations. We give below a further example of a scheme to deal with multipath interference.

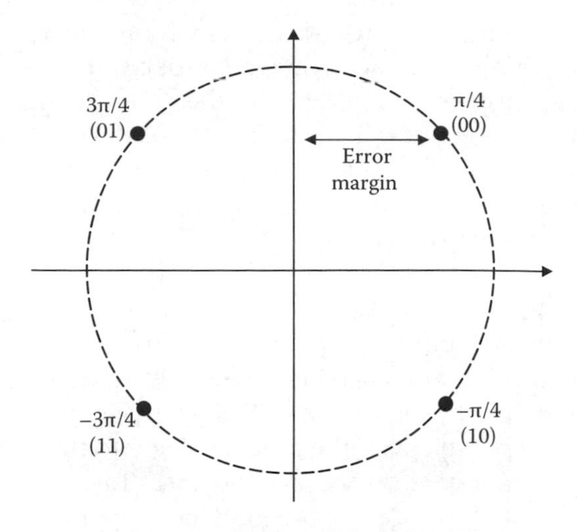

Figure 16.7 Phase transition diagram for the QPSK.

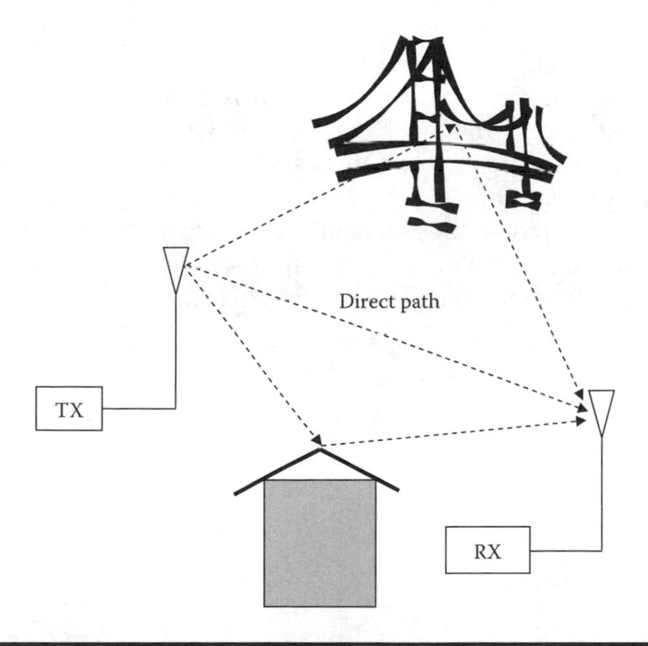

Figure 16.8 A typical multipath environment.

Consider the typical transmission situation depicted in Figure 16.8.[8] The transmitted signal from TX reaches the receiver RX through a number of possible paths. Three such paths are shown in the figure, the direct path and two other paths that are the result of reflections from obstructions from building and other features in the environment. In practice, many such reflections are present. The propagation delay along each path differs since the path length is different. Hence, the received signal at RX is the result of the combination of signal getting there through the direct paths and signals arriving later through indirect paths. If the delay of the indirect path signals is comparable with the symbol time and their strength is significant it is conceivable that the signal at RX will be severely distorted. Signals originating from one symbol and suffering a long delay due to reflections may arrive at the same time as the next direct path symbol. Depending on the circumstances it may not be possible to recognize correctly the second symbol (intersymbol interference). This so-called multipath interference becomes a problem when multipath propagation delays become comparable with the time allocated to transmit each symbol and is therefore dependent on the symbol rate. Taking as an example a relatively slow symbol rate of 0.1 mega symbols per second (Msps) this gives a symbol time T = 10 μs. With a propagation velocity of 3×10^8 m/s this corresponds to a path length of 3 km. Hence, at this slow symbol rate, significant multipath effects will become apparent if the coverage area is comparable or exceeds 3 km (e.g., a city area). If a higher symbol rate is desired, then the coverage area must be reduced accordingly to avoid multipath interference, e.g., for 1 Msps coverage < 300 m, 10 Msps coverage < 30 m, and 100 Msps coverage < 3 m. We see, therefore, that high symbol rates impose severe limitations in the area of coverage due to multipath effects.

In order to overcome these limitations, a way forward is to take a fast serial data stream and split it into, say, five parallel data streams as shown in Figure 16.9, where each of the parallel data streams uses a different

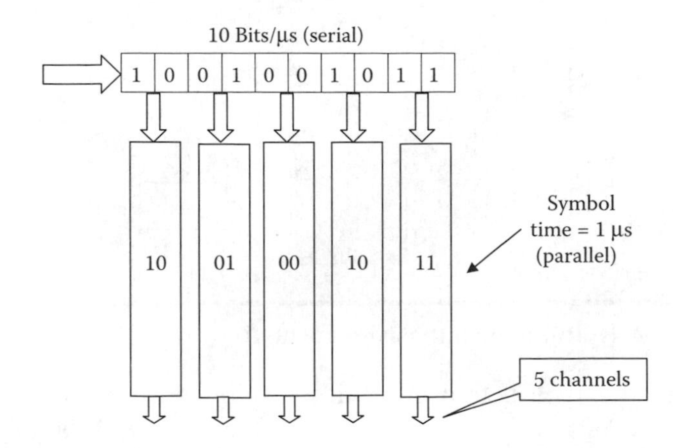

Figure 16.9 Splitting a fast serial stream into five parallel slower streams.

frequency channel. The symbol rate at each of the parallel streams is slower than the serial symbol rate (in this case 1/5). Hence, the reception of the parallel streams is multipath interference free over larger distances (five times larger in this example) compared with the overall serial symbol rate. The cost we pay, however, is needing more frequency channels (multicarrier transmission, five in this case). Referring to Figure 16.2a we see that five channels with their corresponding guard bands represent a considerable increase in spectrum use. A special modulation technique known as orthogonal frequency division multiplexing (OFDM) may be used to make spectrum use more efficient. The k-channel is

$$s_k(t) = \sin(2\pi f_k t) \tag{16.2}$$

where f_k is the center frequency of the k-channel. The separation between adjacent channels Δf is chosen to be equal to the symbol rate 1/T. If this is done, then adjacent channels are orthogonal over an integration time equal to the symbol time T. This follows from the expression[5]

$$\int_0^T \sin(2\pi f_j t + \varphi_j)\sin(2\pi f_k t + \varphi_k)dt = 0 \tag{16.3}$$

provided $f_j - f_k = \frac{n}{T}$, where n is an integer.

This expression implies that we can send symbols at different frequencies in parallel provided these frequencies differ by n/T and still recover each symbol by integrating over a time interval T. This is shown schematically in Figure 16.10. In Figure 16.10b the spectrum of two adjacent carrier-modulated symbols is shown. Since the carrier spacing is equal to 1/T, the two spectra overlap as shown — when one is maximum the other is zero. The implication of this is that we have no need of the guard band as shown schematically in Figure 16.11. This should be contrasted with the conventional FDM shown in Figure 16.2a and it confirms the higher spectrum efficiency of the OFDM.

We see that by increasing the diversity of the received signal we are able to use spectrum efficiently and at the same time provide a degree of immunity to noise and multipath effects.

Many other schemes are available to modulate and code signals, which the reader may access in specialist publications.[5–8]

Serious problems can be encountered in the coexistence of systems especially in the 2.4 GHz ISM band where many services are operating. Many of these services are unlicensed, nomadic, and operate on different

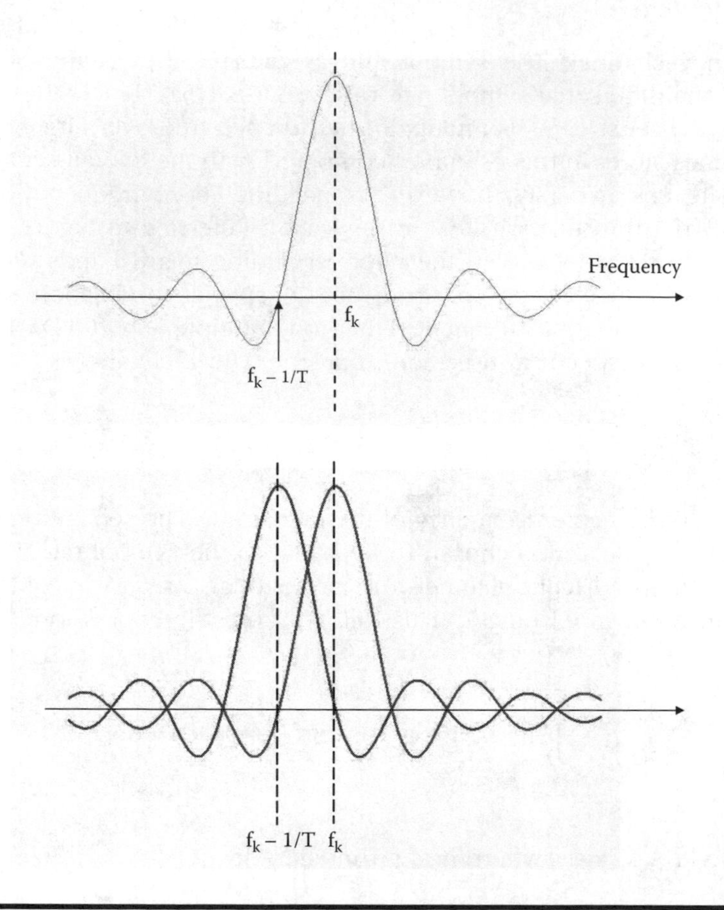

Figure 16.10 Spectrum of a single carrier modulated symbol (a), spectrum of two adjacent carrier modulated symbols (b).

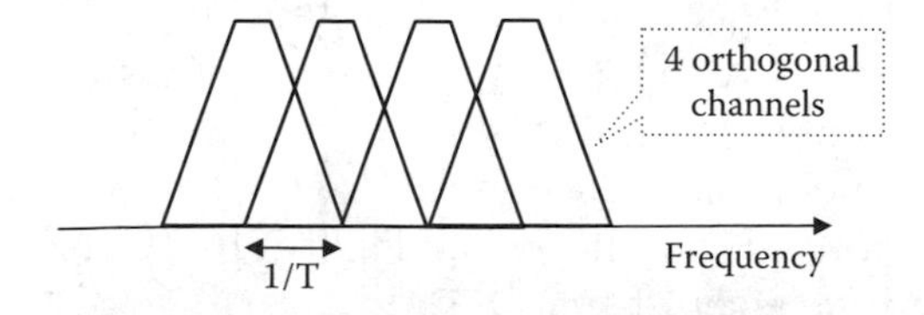

Figure 16.11 Schematic of power density for four orthogonal channels.

standards. Any attempt to coordinate the use of the spectrum between services is made difficult by the fact that communication in real time between users is hampered by the use of different communication protocols. We have in effect a number of heterogeneous networks. More details of the nature of these networks are given in the next section. Here we simply highlight some of the approaches to bring a degree of compatibility between different classes of users. Avoidance of interference may be achieved by classic approaches based on the listen-before-talk technique. Here a potential user listens to make sure that no other user is using a channel before transmitting. Other avoidance techniques are adaptive frequency hopping and dynamic frequency selection (DFS).[4] Suppression techniques are various error correction techniques whereby extra bits are added to transmissions to allow the receiver to detect errors and correct them as far as possible. Examples are the Reed-Solomon code, Hamming code, and convolutional codes.[5,6] Direct sequence spreading (DS) and frequency hopping (FH) are other approaches to suppressing interference.

In the case of heterogeneous networks, efforts are being made to establish a common narrowband control channel to which all potential users, irrespective of the particular transmission standards each adheres to, may contribute for coordination purposes. Such a shared broadcast channel (SBCH) is described in Reference 9 for the 2.4 GHz band.

16.3 Specifications and Alliances

Several wireless services have been established in recent years to implement a variety of functions from mobile telephony[6,7] to wire connectivity at home.[10] Several standards have been developed to regulate these services, together with commercial alliances to develop compatible products. It is impossible to give a complete picture of the entire field in a short chapter. However, presenting a broad outline to EMC engineers is a worthwhile goal as the impact of these services on the way the frequency spectrum is used and protected is likely to increase considerably in the years to come.

The first point to emphasize is the multitude of services using the spectrum; a partial list is shown in Figure 16.12. We will endeavor to make sense of the acronyms shown in this figure later in this section but it will suffice to point out how crowded the ISM band at 2.45 GHz has become. Some are highly regulated public mobile services (e.g., GSM), others private mobile networks for use by individuals (e.g., walkie-talkie) or the emergency services (e.g., TETRA). Another category encompasses ad hoc networks used to interconnect devices at home or at work without wires (e.g., Bluetooth). Many of these services compete for the use of the spectrum, both with each other and with long-established services such as radar (around 5.5 GHz) and navigation (GPS, Galileo). A further group of services using the spectrum are mobile broadcast (e.g., DVB-H). The

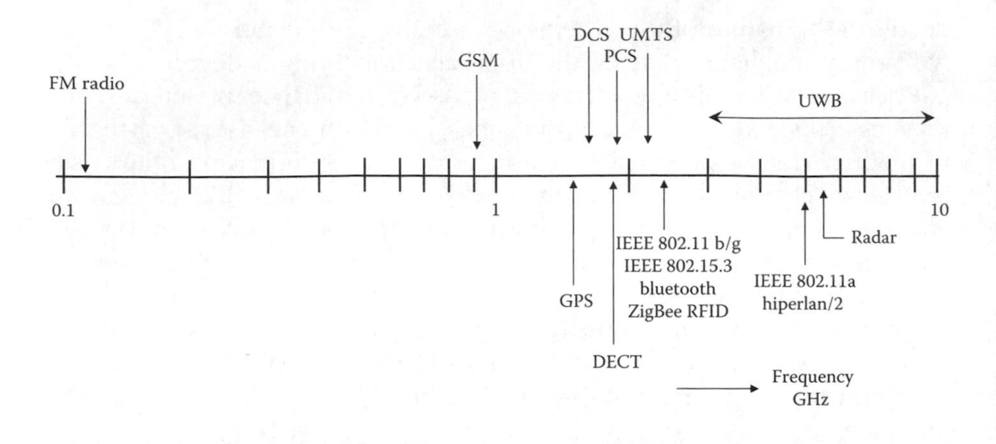

Figure 16.12 Some users of the electromagnetic spectrum.

exact frequency bands used by each service are set by the International Telecommunications Union (ITU) and depend on geographical location and local regulations. The appropriate national authorities should be consulted in each case for up-to-date and reliable information. For EU countries a comprehensive list may be found in Reference 11. The figures given here are indicative to help assess the range and scope for EMC and interoperability problems.

A convenient classification is in terms of the range of coverage of each technology from WPAN (Wireless Personal Area Network), WLAN (Wireless Local Area Network), WMAN (Wireless Metropolitan Area Network), to WWAN (Wireless Wide Area Network). These are depicted in Figure 16.13.

WPAN addresses connectivity around personal workspace, e.g., an individual's laptop computer, mobile phone, etc. Several examples are shown in the figure. Bluetooth[12] is a frequency hopping spread spectrum (FHSS) (1600 hops per second) wireless system with data rates typically 3 Mbits/s operating in the unlicensed 2.4 GHz band. It is based on the IEEE 802.15.1 standard. The modulation schemes used are GFSK (Gaussian Frequency Shift Keying), $\pi/4$ DQPSK (Differential Quadrature Phase Shift Keying), and 8PSK (Phase Shift Keying). Bluetooth uses a time division duplex system (TDD), the bandwidth is 1 MHz, and there are 79 1-MHz channels. Transmit power ranges from 20 dBm (100 mW) for class 1, 4 dBm for class 2, to 0 dBm for class 3, the most common option. Received signal of typically –70 dBm is required for acceptable bit error rate. The potential of coupling of Bluetooth signals into conductive structures is explored in Reference 13.

ZigBee[14] is a technology based on the IEEE 802.15.4 standard aimed at monitoring and control applications where low cost and long battery life are important. Multiple access is achieved by CSMA/CA (Carrier Sense

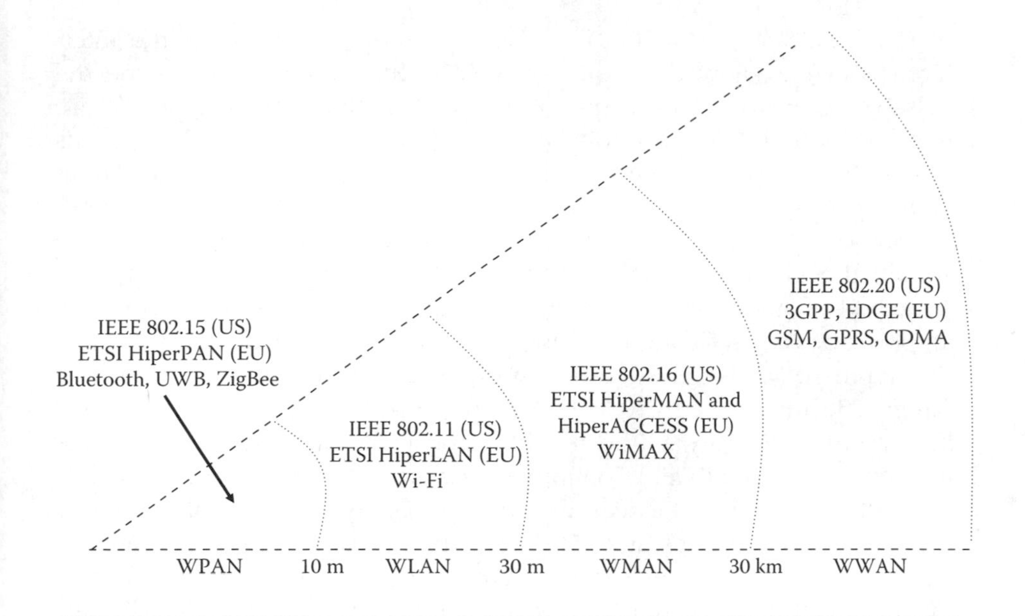

Figure 16.13 A classification of wireless technologies based on the coverage range.

Multiple Access/Collision Avoidance); channel bandwidth is 5 MHz and frequency ranges, number of channels, and peak data rate are 868 MHz (1, 20 kbits/s), 915 MHz (10, 40 kbits/s), and 2.4 GHz (16, 250 kbits/s). Modulation at the two lower frequency ranges is BPSK (Binary Phase Shift Keying) and at 2.4 GHz OQPSK (Offset Quadrature Phase Shift Keying).

Ultra Wideband (UWB) is a technique whereby rather than using a limited part of the spectrum we deliberately use a very wide band but at a low spectral density such that other users see a barely perceptible increase of the noise floor. These are schemes based on spread spectrum techniques, multiband OFDM, or sending very short pulses (pulse UWB radio). The receiver must be able to correlate with known spreading codes to recover the original signal. Spreading is typically in the range of 3.1 to 10.6 GHz and channel bandwidth is 528 MHz (for MB-OFDM) and 1.368 GHz, 2.736 GHz (for DSSS). The number of channels is 13 and 2, respectively, for the two schemes. Typical data rates are 480 Mbits/s, although higher rates are planned. Final standards are not yet available but the FCC specifies a power density below –41 dBm/MHz. EU harmonized standards are in preparation. For a snapshot of the deliberations on UWB of the European Conference of Postal and Telecommunications Administrations (CEPT) consult Reference 15. A useful reference on the impact of UWB on other services such as third-generation wireless services, radio astronomy, or radar is a consultation document by the UK Office of Communications (OFCOM). At issue is what proportion of the interference level is allowed to come from UWB and whether the US spectral mask is acceptable. One suggestion is that in-band emission levels are reduced by 6 dB (to –47.3 dBm/MHz), another that the out-of-band limits should be set to fall from –41 dBmW at 3.1 GHz to –85 dBm/MHz at 2.1 GHz. A discussion of these issues may be found in Reference 16. EMC aspects of UWB are further explored in References 17 to 19. Mention should also be made of Radio Frequency Identification (RFID),[20] which is a method of storing and retrieving data. An RFID tag may be attached to an object or person for identification purposes. Several standards are under development and several frequency ranges are used, the most prevalent being 125 to 134 kHz, 13.56 MHz, 868 to 928 MHz, and 2.45 GHz. Channel bandwidth is typically 250 to 500 kHz and peak data rate 848 kbits/s. A final example is Near Field Communication (NFC), which achieves wireless connectivity between devices in close proximity to each other (a few centimeters). It operates at 13.56 MHz and the data rate is 424 kbits/s.[21]

WLAN or Wi-Fi (Wireless Fidelity)[22] addresses wireless connectivity inside a plant or an office complex, an airport, etc., or at home to connect computers. It is based on the IEEE 802.11 family of standards. The first to emerge was IEEE 802.11b and operates at 2.4 GHz; 802.11g operates at the same frequency but is faster. 802.11a operates at 5 GHz. An even faster standard is the 802.11n, but is not yet in its final form. Details are given in Table 16.1. In Figure 16.14 a comparison of the spectral coverage of WLAN, Bluetooth, and RFID is shown.

WMAN is based on the IEEE 802.16 standard and extends the capabilities of WiFi and links up with WWAN applications. It competes with cables for “last mile” broadband access. WiMAX (Worldwide Interoperability for Microwave Access) is a forum of vendors of equipment based on this family of standards.[23] The 802.16-2004 covers the frequency range 2 to 66 GHz with commonly used frequencies in the range 2.45 to 5.825 GHz. A more recent standard, 802.16e, has a narrower frequency range (2 to 11 GHz) and more sophisticated modulation and access techniques based of OFDM for greater flexibility and lower power consumption. Bandwidth is in the range 1 to 28 MHz and peak data rate 134 Mbits/s (for 28 MHz BW), but typically 75 Mbits/s, and 15 Mbits/s for the 802.16e. WiBro is also based on the 802.16e standard and is supported by Korean manufacturers.

The communication effectiveness of different techniques and approaches to decreasing the transmit power of wireless systems are discussed in References 24 and 25.

WWAN cover the major public mobile phone networks and similar services. The first generation (1G), based on FDMA and relatively low data rates (less than 10 kbits/s), were followed by the second generation

Table 16.1 Summary of the IEEE 802.11 Specifications

Parameters	*802.11a*	*802.11b*	*802.11g*	*802.11n (provisional)*
Frequency range	5.15–5.35 GHz 5.47–5.825 GHz	2.4–2.485 GHz	2.4–2.485 GHz	5.15–5.35 GHz 5.47–5.825 GHz 2.4–2.485 GHz
Channel bandwidth	20 MHz	20 MHz	20 MHz	20 or 40 MHz
Max data rate	54 Mbits/s	11 Mbits/s	54 Mbits/s	600 Mbits/s
Transmitted power	40 mW	100 mW	100 mW	
Modulation	BPSK, QPSK, 16QAM, 64QAM	BPSK, QPSK	BPSK, QPSK, 16QAM, 64QAM	BPSK, QPSK, 16QAM, 64QAM
Multiple access	OFDM, CSMA/CA	OFDM, CSMA/CA	OFDM, CSMA/CA	OFDM, CSMA/CA
Duplex	TDD	TDD	TDD	TDD
Number of channels	12	3 nonoverlapping or 14 overlapping	3 nonoverlapping or 14 overlapping	At 5 GHz, 12 At 2.4 GHz 3 nonoverlapping or 14 overlapping

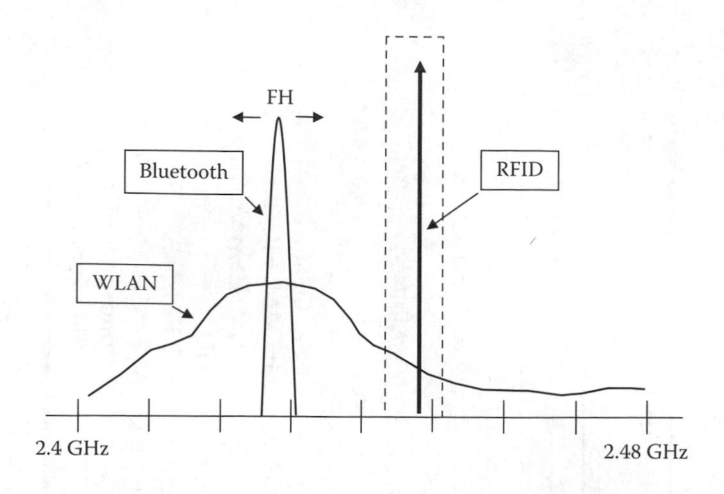

Figure 16.14 Schematic showing spectral occupancy of UHF RFID (bandwidth 350 kHz, transmitter power 4 W), WLAN (bandwidth 20 MHz, transmitter power 100 mW), and Bluetooth (bandwidth 1 MHz, frequency hopping, transmitter power 100 mW).

systems (2G) such as GSM (Global System for Mobile communications), CDMA One, and TDMA; 2.5G systems such as GPRS (General Packet Radio Service); and 3G systems such as CDMA2000 and UMTS (Universal Mobile Telecommunications System). The 3G developments are managed by a group known as the Third Generation Partnership Program (3GPP), which consists of six global partners and has been operating since 1998. UMTS is the development of GSM and is also known as 3GSM. The group dealing with the 3G development of CDMA2000 is known as 3GPP2. EDGE (Enhanced Data Global Evolution or Enhanced Data for GSM) is an upgrade of GPRS. GSM and GPRS use as modulation GMSK (Gaussian minimum shift keying) while EDGE uses 8PSK. The latter is a higher-order modulation that can accommodate more bits per slot and results in a threefold increase in the data rate but at the expense of noise immunity. The higher-order modulation scheme is more sensitive to noise, hence the high data rates are only attainable under strong signal conditions. Frequency ranges for GSM/GPRS/EDGE are generally in the 800 to 900 MHz and 1800 to 1900 GHz range depending on local radio regulations. Multiple access is through TDMA/FDMA, duplex operation is FDD, and the channel bandwidth is 200 kHz. The number of channels ranges from 124 for GSM900 to 374 for GSM 1800 with eight users per channel. Peak data rates are 14.4 kbits/s (GSM), 53.6 kbits/s (GPRS), and 384 kbits/s (EDGE). The 3G technologies are based on wideband CDMA (W-CDMA) and, in the case of UMTS, use a bandwidth of 5 MHz and CDMA2000

1.25 MHz. Typical data rates are 2 Mbits/s for UMTS and 3.1 Mbits/s for the forward link and to 1.8 Mbits/s for the reverse link of CDMA2000. A good general introduction to cellular systems may be found in Reference 7.

These are public networks but there are also networks used by organizations such as the police and other emergency areas. A particular example developed in Europe is TETRA (Terrestrial Trunked Radio System).[6] Various bands have been allocated in the range 350 to 1000 MHz with the range 410 to 430 MHz most commonly in use. Modulation is done using $\pi/4$ DQPSK (differential quadrature phase shift keying) and multiple access by TDMA, Duplex operation is through FDD, and the channel bandwidth is 25 kHz with four channels per carrier. Data rates are typically 7.2 kbits/s. Concerns have been expressed about the potential of TETRA to cause interference due to the high-transmit power and the low frame rate (18 frames/s).[26] Base station power ranges from 40 W to 0.6 W. The mobile stations have power in four classes: 30, 10, 3, and 1 W.

Other potential users of the spectrum are various broadcast services. An example is DVB-H (digital video broadcasting-handheld) based on the ETSI Standard EN 302 304.[27] There are three frequency ranges: 174 to 230 MHz, 470 to 862 MHz, and 1452 to 1492 MHz. Modulation is QPSK, 16QAM, and channel bandwidth from 5 to 8 MHz. The cell radius ranges from 17 to 67 km.

Finally we mention the major navigation systems, GPS (Global Positioning System) and Galileo.[28] The parameters for GPS are frequencies 1575.42 MHz, 1227.6 MHz, modulation is BPSK, channel bandwidth 20.46 MHz, 32 satellites (maximum), and data rate 50 bits/s. For Galileo, frequencies are 1575.42, 1176.45, 1207.14, and 1278.75 MHz, and the modulation is BOC (binary offset carrier).

16.4 Conclusions

The rapid evolution of wireless technologies will have an impact on the way the frequency spectrum is used and EMC is perceived. Many of the fundamental concepts of EMC are rooted to analogue radio system protection and adjustments must be made in the years to come in the way we understand and implement EMC measures. A trend evident from the description of the ingenious techniques used to code and transmit signals is that we can "spread" interference over a wide frequency range so that it becomes less noticeable. This, however, is bound to have an impact over the years on the noise floor as new technologies are rolled out and the number of users escalates. We need to look ahead and protect the radio spectrum from long-term contamination. The role of EMC engineers will be crucial in this very worthwhile aim.

References

1. Haartsen, J C, "Co-existence of licensed and unlicensed radio spectrum," COST 286 Meeting, Split, 13 Dec 2005.
2. Sterner, U and Linder, S, "Effects of intersystem interference in mobile ad hoc networks," Proc. EMC Europe, Barcelona, 4–8 Sept. 2006, ISBN 84-689-9438-3, pp 811–816.
3. OFCOM, "Electromagnetic Compatibility Aspects of Radio-based Mobile Telecommunications Systems," Final Report, 1999, www.ofcom.org.uk.
4. Del Mar Perez, M, Llamas, A, Soler, J C, Rojas, A, and Banos, J, "EMC challenges in emerging wireless technologies," EMC Europe, Barcelona, ISBN 84-689-9438-3, 4–8 Sept. 2006, pp 459–464.
5. Proakis, J G and Salehi, M, "Communications Systems Engineering," 2nd edition, Prentice Hall, NY, 2002.
6. Dunlop, J, Girma, D, and Irvine, J, "Digital Mobile Communications and the TETRA System," Wiley, NY, 1999.
7. Poole, I, "Cellular Communications Explained-From Basics to 3G," Elsevier-Newnes, 2006.
8. Dobkin, D M, "RF Engineering for Wireless Networks-Hardware, Antennas and Propagation," Elsevier-Newnes, 2005.
9. Bekkaoui, A and Haartsen, J C, "Shared broadcast channel for radio systems in the 2.4 GHz band," Proc. 12th Annual Symp. of the IEEE/SCVT, Nov. 2005. (wwwes.cs.utwente,nl/smartsurroundings/publications/Bekkaoui.pdf).
10. Zahariadis, Th, Pramataris, K, and Zervos, N, "A comparison of competing broadband in-home technologies," Electronics and Communication Engineering Journal, August 2002, pp 133–142.
11. IDATE, "Study on information on the allocation, availability and use of radio spectrum in the community," Final Report 30068/V5/FP, section 4.2, February 2005.
12. Bluetooth Official Site: www.bluetooth.com.
13. Schoof, A, Stadtler, T, and ter Haseborg, J L, "Two dimensional simulation and measurement of the coupling of Bluetooth signals into conductive structures," IEEE Int. Symp. on EMC, Boston, Aug 18–22, 2003, pp 258–262.
14. ZigBee Alliance Homepage: www.zigbee.org.
15. "Final CEPT Report in response to the Second EC Mandate to CEPT to harmonise radio spectrum use for the ultra-wideband systems in the European Union," 12th ECC Meeting, Cascais, 24–28 Oct. 2005.
16. Ofcom, "Ultra Wideband — consultation document," 13 Jan 2005, http://www.ofcom.org.uk.
17. Zeng, Q and Chubukjian, A, "Characterization of the cumulative effects of ultrawideband technology on the electromagnetic environment," IEEE Int. EMC Symp., Boston, Aug. 18–22, 2003, pp 252–257.
18. Buccella, C, Graziosi, F, Feliziani, M, Di Renzo, M, and Tiberio, R, "Ultra wide band propagation measurements in indoor working environment and through building materials," EMC Europe Workshop-EMC of Wireless Systems," Rome, 19–21 Sept. 2005, pp 106–109.

19. Munday, P and Tee, D, "Interference between UWB and conventional systems in theory and practice," EMC Europe Workshop-EMC of Wireless Systems," Rome, 19–21 Sept. 2005, pp 110–113.
20. www.aimglobal.org/technologies/rfid.
21. www.nfc-forum.org.
22. WiFi Alliance Site: www.wi-fi.org.
23. WiMAX Forum Site: www.wimaxforum.org.
24. Heddebaut, M, Gransart, Ch, Rioult, J, and Deniau, V, "WxAN communication effectiveness inside vehicle bodies in presence of passengers," EMC Europe Workshop-EMC of Wireless Systems," Rome, 19–21 Sept. 2005, pp 118–121.
25. Lienard, M and Degauge, P, "Use of MIMO techniques for decreasing the output transmitting power of wireless systems," EMC Europe Workshop-EMC of Wireless Systems," Rome, 19–21 Sept. 2005, pp 203–206.
26. Jarvis, T, "TETRA the noise source: preventing interference," in Compliance Engineering, www.ce-mag.com/archive/2000/mayjune/jarvis.html.
27. http://www.dvb-h-online.org/.
28. http://ec.europa.eu/dgs/energy_transport/galileo/index_en.htm.

Chapter 17

EMC and Broadband Technologies

To meet current needs for fast data transmission such as internet access into homes and businesses, new technologies are being developed that use the radio frequencies to achieve high speeds. One of the important commercial considerations in providing such services is what infrastructure to use and if such infrastructure is not available how best to provide it. Naturally, fiber-optic networks can be used with excellent results but they must be built afresh where they do not already exist, with consequent substantial costs. However, in the developed world virtually all domestic and commercial premises have telephone networks installed and in the less developed world substantial infrastructure exists in telephone networks and virtually all homes have a connection to the power network. The question therefore arises whether telephone and/or power networks could be used to transmit high-speed data. The commercial advantages are substantial since an existing widely available infrastructure can be used. In the case of the power network, not only is there a connection into every home but also into every room! Hence, high-speed data would be accessible from everywhere. The question relevant to EMC is what impact such services may have on the radio spectrum.

Two technologies, Digital Subscriber Line telecommunications (xDSL) and Power Line Telecommunications (PLT) are candidates for such schemes, the former for transmission over telephone lines and the latter for transmission over the power network. The frequencies proposed for

Table 17.1 Frequency Bands and Minimum Protected Field Levels

Band	*Frequency*	*Wavelength*	*Min Field Strength dBμV/m*
LF	30–300 kHz	10–1 km	66
MF	300–3000 kHz	1 km–100 m	60
HF	3–30 MHz	100–10 m	40

such applications cover the low-, medium-, and high-frequency bands that are traditionally used for emergency services, AM broadcasting, aeronautical, maritime and military, and by amateurs. Typically, protection is offered to broadcast services if they have the minimum field strength shown in Table 17.1. The requirement therefore is that if AM services are not to be disturbed, emission levels by xDSL/PLT must keep well below the minimum protected margins to provide an acceptable signal-to-noise ratio. Our interest in this chapter is to sensitize the reader to the issues involved. Since no EMC radiation limits are available below 30 MHz new standards need to emerge to balance the conflicting requirements of the radio and xDSL/PLT communities. This is a contentious issue but hopefully a widely acceptable solution will emerge in the years to come.

We first discuss the transmission of high-frequency signals over telephone and power networks, then we address the peculiarities of xSDL and PLT, and we finish with a presentation of the regulatory framework being currently debated.

17.1 Transmission of High-Frequency Signals over Telephone and Power Networks

Neither telephone nor power networks were designed to carry high-frequency signals. The consequence of this is that any attempts to transmit fast data streams on such networks will result in significant attenuation, encounter significant levels of noise, and cause the emission of radiation. Care is therefore required to address all these issues in a satisfactory way before these technologies are rolled out in the field. There is an insufficient number of studies done to characterize such networks at high frequencies. One of the difficulties is that every network is different and even the same geographical network changes during the day as consumers (loads) connect and disconnect. Any study therefore needs to be statistical in its methodology in order to collect sufficient data about network behavior. However, the issues to be addressed are not difficult to discern. Acceptable high-frequency transmission takes place if everywhere and at all times signals propagate in what can be termed as a transmission line, i.e., two good balanced conductors closely spaced without discontinuities. Under these conditions signals propagate without modification or appreciable radiation. There are no reflections and everywhere and at any time currents in the two conductors are equal and opposite. This is depicted schematically in Figure 17.1a. The last condition is profoundly important for EMC as it ensures that in the far-field the two elemental currents on any two conductor segments generate emitted fields that tend to cancel each other. Only in close proximity to the conductors (at distances comparable to the conductor spacing) field cancellation tends to become less effective as under such circumstances the conductor spacing has a significant impact on the distance between the observation point and each conductor. Balancing means that there are no CM currents. Reflections and unbalancing spoil matters as they can generate CM currents and can convert DM to CM currents (mode conversion). This is illustrated in Figure 17.1b, which shows a differential feed to a line. We assume that the characteristic impedance of Line1-gnd and Line 2-gnd is the same Z_0. Upon connection of the sources, two voltage pulses are injected into the line as shown in Figure 17.1c:

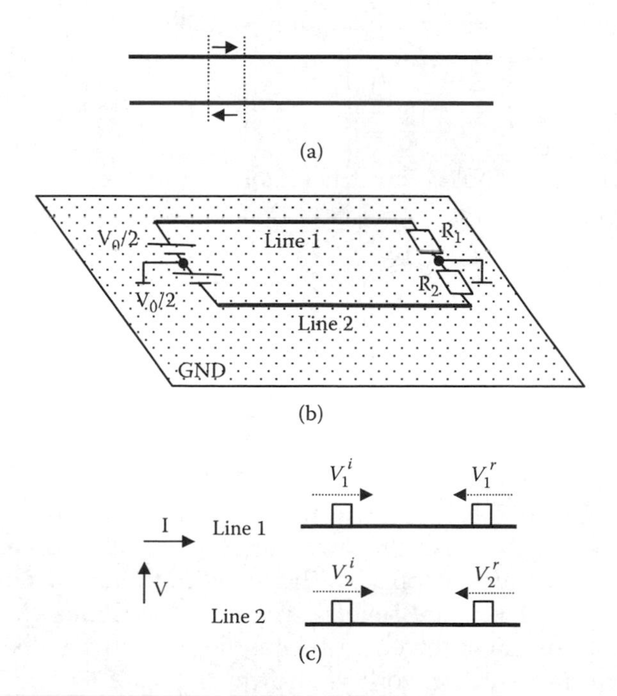

Figure 17.1 Schematic showing a two-wire line (a), (b), and incident/reflected pulse (c).

$$V_1^i = V_0/2, \quad V_2^i = -V_0/2 \tag{17.1}$$

The voltage and current positive reference directions are shown by the solid arrows to the left of the figure. The broken arrows indicate the direction of the propagation of the disturbance. We will neglect for simplicity any coupling between lines 1 and 2 so that we can approximately illustrate the principle. We now consider what happens when these pulses encounter the loads. Let us examine first the case where the line is balanced:

$$R_1 = R_2 = Z_0 \tag{17.2}$$

Then, since each line is matched, the reflected voltages will be zero:

$$V_1^r = V_2^r = 0 \tag{17.3}$$

This is a perfectly balanced line and we see that, throughout, the voltages and currents on the two arms of the line are equal and opposite. Let us now look at what happens when we deliberately unbalance the line loads:

$$R_1 = 2Z_0, \quad R_2 = Z_0/2 \tag{17.4}$$

The reflected voltage pulses are calculated as follows:

$$V_1^r = \frac{R_1 - Z_0}{R_1 + Z_0} V_1^i = \frac{2Z_0 - Z_0}{2Z_0 + Z_0} \frac{V_0}{2} = \frac{V_0}{6} \tag{17.5}$$

$$V_2^r = \frac{R_2 - Z_0}{R_2 + Z_0} V_2^i = \frac{Z_0/2 - Z_0}{Z_0/2 + Z_0} \left(-\frac{V_0}{2} \right) = \frac{V_0}{6} \tag{17.6}$$

We see that now not only do we have reflected voltage pulses (a bad situation for EMC), but also the two pulses are of the same polarity as are the associated current pulses. The radiated fields therefore do not cancel effectively. There are several ways that unbalances can occur in practical situations encountered in telephone and power networks. One example from a power network is shown in Figure 17.2. Although the main loop (live and neutral) follows good TL practice, the presence of the unswitched branch line introduces an unbalance as only the live line

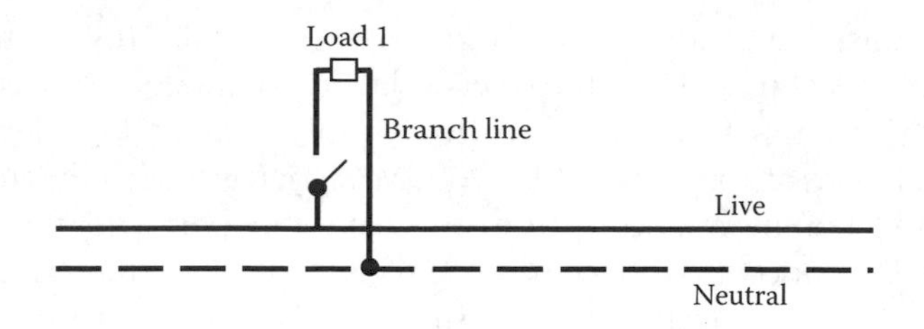

Figure 17.2 Power supply line with a switched branch.

is disconnected — the neutral remains connected. Hence, in total, live and neutral are not the same relative to their surroundings. Typically, in a distribution transformer (the local loop) there may be several hundred consumers nominally connected and all kinds of possibilities available for asymmetrical feed conditions.

Of the studies so far, several have been in connection with mains signaling or protection studies (at high voltage levels) but nevertheless they provide a background to the complexities of such networks as high-frequency propagation media.

Particular examples of models and simulations for EHV networks are given in References 1 and 2. An extensive study aimed at mainsborne signaling is reported in Reference 3. Data extend to a frequency of 200 kHz. The authors found high attenuation, which varies throughout the day by as much as 30 dB. At particular frequencies they found that attenuation is 20 or 30 dB greater than the average. They concluded that a single frequency carrier scheme would be unreliable under these conditions and that some form of spread spectrum approach would be preferable. A study of network models and performance specifically for high-data-rate transmission (2 to 10 Mbits/s) and therefore extending to frequencies of several megahertz is reported in Reference 4. They report an average noise power density of $1.6 \times 10^{-14} V^2 Hz^{-1}$ and strong signal attenuation. For an injected level of 65 dBmV the signal measured after 300 m of cable with a receiver bandwidth of 100 kHz was 45 dBmV at 1 MHz and –22 dBmV at 20 MHz. The conclusion is that some form of multicarrier technique would be necessary for reliable transmission.

17.2 EMC and Digital Subscriber Lines (xDSL)

The term xDSL is used to describe a generic class of technologies based on the use of copper wires in ordinary telephone connections to transmit internet in addition to the familiar voice telephone signals. Transmission thus takes place over unshielded twisted pair cable (UTP). In order to minimize emissions a good cable balance is essential. This means that the impedance of each wire in the pair relative to the ground is as far as possible the same. A measure of how well a cable is balanced is its longitudinal conversion loss (LCL). A DM signal is injected into the wire pair and the common-mode noise between the pair and the ground is measured. The ratio of the two in decibels is the LCL. Better balance means a higher LCL. ADSL stands for Asymmetric Digital Subscriber Line and operates between 25 kHz and 1.1 MHz. Bit rates as high as 8 Mbps are possible. Different data rates are provided downstream (to the home) and upstream (from the home) since in most cases the user does not need to communicate at a high rate with the provider. The downstream frequency band is 0.138 to 1.104 MHz and the power spectral density is –36.5 dBm/Hz. The upstream equivalent figures are 0.138 to 0.276 MHz and –34.5 dBm/Hz. VDSL stands for Very High Bit-rate Digital Subscriber Line and promises faster rates of the order of 50 Mbps. Its operation extends to higher frequencies, typically 10 MHz. A typical configuration is shown in Figure 17.3. Radiated emissions from such networks have been the subject of study by several researchers. The greatest emissions take place from the drop wires. The main concern is the deterioration of radio services due to the cumulative impact of a large number of xDSL users. Key studies are reported in References 5 to 7 and also in References 8 and 9. Radiation from such systems may propagate in several modes, i.e., via a sky wave (reflection from the ionosphere, frequency range 3 to 30 MHz), space wave (0.1 to 30 MHz), and the ground wave (0.1 to 3 MHz). Significant broadcasting services include digital AM radio (Digital Radio Mondiale or DRM), aeronautical radionavigation (nondirectional beacons or NDB), air traffic control, etc. In Reference 5 a typical city was taken as an example with an average fixed telephone line density of 3000/km^2, total area 50 km^2, market penetration of ADSL 10%, and cable balance of 30 dB typical of old copper wire

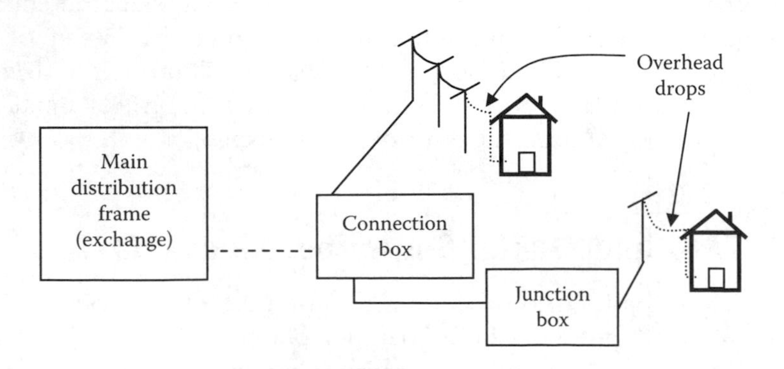

Figure 17.3 Configuration used in xDSL systems.

installation. On this basis and models developed in References 5 to 7, the cumulative electric field was calculated at different distances from the main distribution frame of the exchange for a number of frequencies. As an example, at 1 km distance and at 1 MHz (downstream ADSL) a value of 48.9 dBμV/m was obtained. At a distance of 10 km this drops to 27.6 dBμV/m. If better wiring is used (balance 35 to 45 dB, typical of new copper installations) then the electric field falls by an amount in decibels equal to the improvement in balance. According to ITU-R the minimum sensitivity of an average receiver is 60 dBμV/m based on a SNR of 30 dB at its input. On this basis, the maximum interfering signal that a receiver can sustain at its input is approximately, 30 dBμV/m. The final conclusion is that the cumulative field is such as to increase the noise floor up to a distance of 7 km from the city and thus impair analogue reception of MF broadcasting. In regard to practical remedies, a very high degree (better than 40 dB) of cable/connector balance and also good shielding of the wiring inside customer premises are recommended.

17.3 EMC and Power Line Telecommunications (PLT)

This system is similar in concept to the xDSL but exploiting a different transmission medium — the power network outside and inside the home. The issues and concerns remain the same but more accentuated due to the imperfect nature of power networks as high-frequency transmission media. While with telephone systems a twisted cable pair is employed, thus affording a certain degree of balance, this is far from the case for power networks. The concern is therefore that significant emissions will result leading to an increase in the noise floor and degradation of other services. Another concern is that since the power network is by its very nature a noisy environment it may be difficult to establish reliable transmission without significant levels of injected HF signal power. A schematic configuration is shown in Figure 17.4. Sophisticated coding and modulation

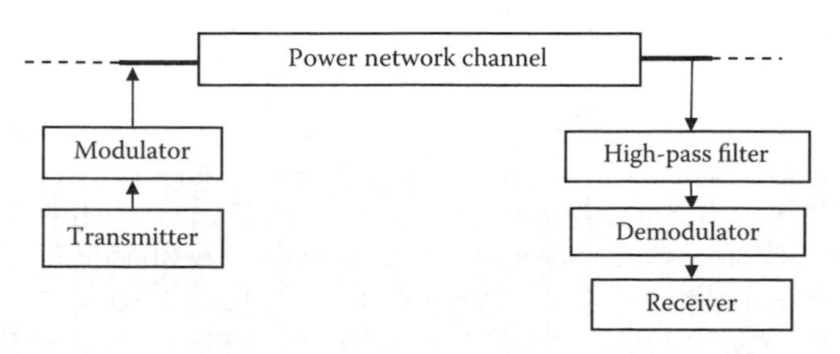

Figure 17.4 Schematic of a PLT configuration.

techniques can be used to deal with the noise and imperfections of the transmission medium.[10] Several investigators have looked into the emissions from such systems.[11–17] The references mentioned earlier in connection with xDSL also contain assessments of emission from PLT systems. Field trials performed by the British Broadcasting Corporation[18] came to the conclusion that there is demonstrable potential for PLT to cause interference to indoor reception. It appears therefore that in situations where a telephone network is widely available, there are limited commercial opportunities for PLT with its associated technical difficulties; however, this will depend on the final form of the regulatory environment, which is at present unclear. This is addressed in the next section.

17.4 Regulatory Framework for Emissions from xDSL/PLT and Related Technologies

The development of international standards is a lengthy business, especially when there are potential users of the spectrum with conflicting demands. Radio users feel that insufficient consideration is given to their requirements and xDSL/PLT vendors feel that spectrum protection is a form of trade restriction. A comprehensive discussion with a range of views is reported in Reference 19. The point of view of broadcasters and radio amateurs in put forward in References 20 and 22.

In terms of standards there is considerable delay and uncertainty in establishing appropriate limits and test procedures. A good survey of the current situation is given in Reference 20 on which Figure 17.5 is based. For comparison, proposed field values are converted to 1-m distance assuming far-field (1/r) conditions (although this is not always justified). Many standards in this area are given in terms of the magnetic field and whenever this is the case they are converted to electric field as follows:

$$H = \frac{E}{377} \qquad (17.7)$$

$$20 \log E = 20 \log H + 51.5$$

Again, Equation 17.7 is only justified under far-field conditions. In Figure 17.5 we plot the proposed electric field limits in dBμV/m versus frequency according to different requirements. The baseline should be the minimum protected AM broadcast field, which is set at 40 dBμV/m. The curve labeled K.60 originates from ITU-T (the telecommunication sector of ITU). In contrast, the curve labeled ITU-R is a recommendation by the radiocommunication sector of the ITU and represents, in their view, the

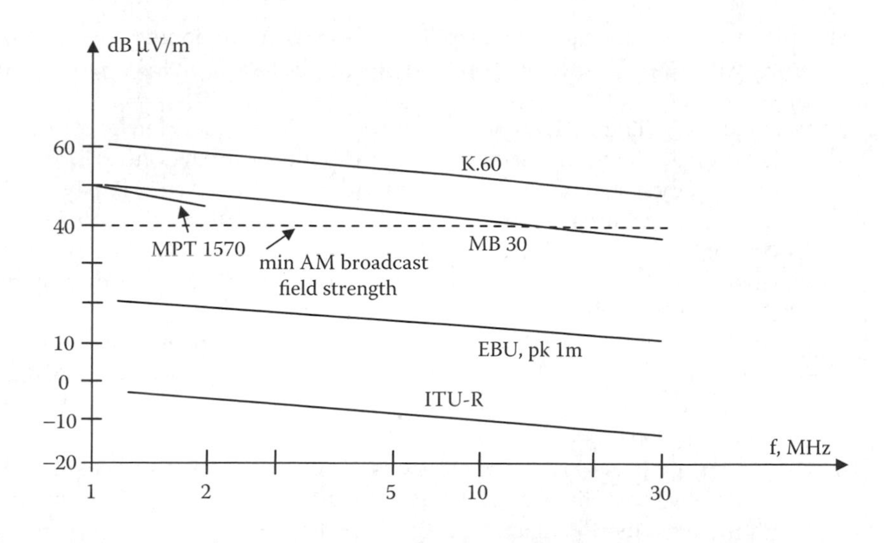

Figure 17.5 Electric field limits for various standards under development in the xDSL/PLT area.

level necessary to protect reception everywhere. The difference is striking and typifies the inherent difficulties in setting widely acceptable limits. The German NB30 and UK MPT 1570 recommendations are also shown. A compromise worked out by the BBC and the EBU is also shown.[23] Another way forward is to provide for a number of notches — frequencies where the PLC transmission is 30 dB lower. The issue then is whether these notches are fixed, who selects them and how, how many there are, etc.[20] A possibility is dynamic notching, which adapts to reception conditions. Work that addresses EMC issues and also the co-existence between PLC and VDSL is reported in Reference 24.

Interested readers need to follow developments as, hopefully, a consensus emerges that balances the needs of all spectrum users.

References

1. Cristina, S and D'Amore, M, "A new efficient method to evaluate carrier channel performances on long EHV power lines with transpositions," IEEE Trans. on PAS, 101(9), pp 3053–3060, 1982.
2. D'Amore, M and Sarto, M S, "A new formulation of lossy ground return parameters for transient analysis of multiconductor dissipative lines," IEEE Trans. on Power Delivery, 12(1), pp 303–311, 1997.
3. Burr, A G and Burbidge, R F, "Computer modelling of the LV distribution network for mainsborne signalling," IEE Proc. Part C, 133(1), pp 83–96, 1987.

4. Dostert, K, "RF-models of the electrical power distribution grid," Int. Symp. on Power Line Communication and its Applications, Tokyo, Japan, pp 105–114, March 1998.
5. Welsh, D W, Flintoft, I D, and Papatsoris, A D, " Cumulative effect of radiated emissions from metallic data distribution systems on radio based services," Report for the Radiocommunications Agency AY3525, May 2002, www.ofcom.org.uk/static/archive/ra/topics/research/topics/emc/ay3525/report2.pdf.
6. Papatsoris, A D, "Spectrum management of ADSL," Electronics Letters, 38(17), pp 1001–1003, 2002.
7. Papatsoris, A D, Flintoft, I D, Welsh, D W, and Marvin, A C, "Modelling of cumulative emission field of unstructured telecommunication transmission networks," IEE Proc, Science, Engineering and Technology, 151(4), pp 244–252, 2004.
8. Carpender, D, Macdonald, J M, Morsma, T, Standley, D, and Foster, K T, "The EM environment and xDSL," Proc. 8th Int. Conf. on High-frequency Radio Systems and Techniques, IEE, 10–13 July 2000, pp 379–383.
9. Bullivant, A R, "Continuation of investigations into the possible effects of DSL related systems on radio services," Report for the Radiocommunications Agency, AY3949, ERA Technology Ltd, May 2001.
10. Degardin, V, Lienard, M, and Degauque, P, "Optimization of an equalization algorithm for a power line communication channel," Electronics Letters, 39(5), pp 483–485, 2003.
11. Burrascano, P, Cristina, S, D'Amore, M, and Salerno, M, "Digital signal transmission on power line carrier channels: an introduction," IEEE Trans. on Power Delivery, PWRD-2, No 1, pp 50–55, 1986.
12. D'Amore, M and Sarto, M S, "Digital transmission performance of carrier channels on distribution power line networks," IEEE Trans. on Power Delivery, 12(2), pp 616–623, 1997.
13. D'Amore, M and Sarto, M S, "Electromagnetic field radiated from broad band signal transmission of power line carrier channels," IEEE Trans. on Power Delivery, 12(2), pp 624–631, 1997.
14. Issa, F, Chaffanjon, D, and Pacaud, A, "Outdoor radiated emission associated with power line communication systems," IEEE Int. EMC Symp., 13–17 Aug, 2001, Montreal, pp 521–526.
15. Marthe, E, Rachidi, F, Ianoz, M, and Zweiacker, P, "Indoor radiated emission associated with power line communication systems," IEEE Int. EMC Symp., 13–17 Aug, 2001, Montreal, pp 517–520.
16. Vick, R, "Radiated emission of domestic main wiring caused by power-line communication systems," Int. Wroclaw Symp. on EMC, pp 111–115, 2000.
17. Goldberg, G, "Evaluation of power line communication systems," Int. Wroclaw Symp. on EMC, pp 103–106, 2000.
18. Stott, J H, "The effects of power-line telecommunications on broadcast reception: brief trial in Crieff," BBC R&D White paper, WHP 067, Sept. 2003.
19. IEE Seminar on "EMC and Broadband for the last mile," Savoy Place, London, 17 May 2005, Digest No 2005/11037.

20. Stott, J, "Potential threats to radio services from PLT systems," EBU Technical Review, 15 pages, July 2006.
21. The European Broadcasting Union (EBU), www.ebu.ch contains several technical articles in this area.
22. The Radio Society of Great Britain, www.rsgb.org/emc/.
23. Stott, J H, "Emission limits," BBC R&D White Paper, WHP 013, Nov. 2001.
24. Zeddam, A, Moulin, F, and Gauthier, T F, "EMC and co-existence issues of broadband communications over copper," URSI General Assembly, New Delhi, October, 2005, 4 pages.

Chapter 18

EMC and Safety

EMC standards and the associated design effort aim in the first instance to comply with certain requirements that have legal force. It may be assumed that equipment that satisfies EMC regulations is "safe." This would be a bold statement and in many cases it would be untrue. It is quite possible to comply with EMC standards and yet find that in practice a product may fail in an unsafe manner due to EM disturbances. This should not come as a surprise as EMC standards are not formulated with safety in mind. Their background, imposed limits, and way of thinking come from the need to protect the radio spectrum and not explicitly to make equipment safe. Although safety is a legal requirement, EMC standards do not offer any explicit guidance as to how to proceed. There are also a few cases where designers may perceive a conflict between making a product EMC compliant and safe. If one looks at safety standards for guidance on EMC matters the situation is similar — there is very little coverage on this issue. It therefore appears that there is a divide between the way engineers practice EMC and functional safety. This divide is rooted in the different approach of the two communities. Traditionally in EMC, verification is done by testing one or a few samples without regard to the design and construction practices employed in producing equipment.[1] This process is described as "black box testing." In contrast, the verification process for safety follows a different approach. This is based on a hazards and risks assessment. Designs have to follow well established design criteria that reflect their intended environment as regards all environmental factors such as temperature, humidity, vibration, etc. Reasonably foreseeable faults are introduced to ensure that safety is not compromised even under these circumstances. All equipment leaving the factory has to undergo some basic safety tests to ensure that safety-related features are in place.

The regulatory framework in this area is based on IEC standard 61508 and Technical Specification 61000-1-2.[2,3] The Institution of Electrical Engineers (now the IET) in London has produced guidance on EMC and Safety that contains much of the background to the issues involved.[4] We provide here a summary of the difficulties encountered in this area based on References 1 and 4.

EMC tests are done under ideal conditions with equipment prepared, optimized, and checked for testing. This is far from real-life situations. An example of this is that in immunity tests EM threats are applied one at a time. In practice, however, more than one threat may be applied simultaneously (time coincident threats), thus eroding whatever noise margin has been allowed for. This may not be a serious problem in terms of radio spectrum protection but may have serious implications in safety-critical equipment. The environment inside an anechoic chamber or an open-area test site is very remote from real-life situations (realistic operating environments). As an example a car encounters a dynamic environment, e.g., may be boxed in by trucks and 18-wheelers, thus effectively confined in a resonant cavity. These are not rare situations and yet they are not accounted for in EMC standards. The time-history/frequency spectrum, amplitude, and modulation of threats vary widely. Yet standards specify a very limited formalized range of waveforms. In particular, if the modulation frequency of a threat coincides with key equipment operating frequencies, the risk of failure increases considerably. While standards cannot specify all conceivable modulation frequencies, manufacturers are aware of the particularities of their equipment (e.g., key clock frequencies, transfer rates, etc.) and could and should test that safety-critical functions are not impaired when a threat modulation frequency coincides with operational frequencies. Yet such procedures are not specified in standards and are left to the discretion of manufacturers. Naturally, responsible manufacturers will have developed over the years procedures to deal with such matters, but nevertheless no guidance is offered in standards and there is, in particular, no formalized procedure for dealing with retrofits, equipment subsequently installed, etc.

The word "safe" has a different meaning to different people. For engineers it is a relative term that can be defined in a statistical sense — there is no absolute safety. Yet the notion of absolute safety is very common in the law and the media. Engineers are asked to "make sure that this never happens again." The problem is in the word "never." EM disturbances follow statistical distributions with a mean value and standard deviation. Events two or three standard deviations away from the mean are very rare but they do happen. Questions that need answering are formulated as follows: "If 1% of this type of EM threat exceeds the designed safety limit, would this be acceptable?," or "How much money should be spent to reduce this 1% to say 0.5%?" Naturally, there are no simple answers to these questions and our response depends on the severity of the consequences of failure, attitudes to risk, and cultural traits. We return to the topic of statistical EMC in Chapter 19.

EMC tests are done on brand-new equipment. However, every-day use of equipment over its lifetime means that its EMC performance will change and, more specifically, will deteriorate. The reasons are manifold, including aging of protective devices, replacement of some components with new parts that have different EMC characteristics, dry joints, general corrosion, ill-fitting panels and gaskets, etc. We have discussed the issue of EMC management in Section 12.4. The assessment of EMC-associated risks in a statistical sense is discussed in Chapter 19. From the practical point of view there is no requirement for lifetime testing. The conclusion to be drawn is that since currently only very few representative samples are tested at the time of manufacture, no definite statement can be made regarding the condition of most new equipment leaving the factory or of equipment after several years of use.

This is a snapshot of the situation regarding EMC and functional safety. Naturally, industries where safety is paramount, e.g., aerospace and automotive, are particularly active in addressing these issues in a systematic way and further regulatory developments are likely to emerge over the years to cover the entire range of industrial activities. Reference 4 should be consulted for more details. The final conclusion to be kept in mind is that adherence to current EMC regulations on its own is no guarantee of functional safety or adequate protection in product liability litigation.

References

1. Armstrong, K, "Why EMC immunity testing is inadequate for functional safety," IEEE Int. Symp. On EMC, Santa Clara, CA, 9–13 Aug. 2004.
2. IEC 61508: 1998, "Functional safety of electrical/electronic/programmable electronic safety-related systems."
3. IEC 61000-1-2: 2001, "Methodology for the achievement of functional safety of electrical and electronic equipment."
4. "IEE Guidance on EMC and Functional Safety," 2000, http://www.iee.org/policy/areas/electro/core.pdf.

Chapter 19

Statistical EMC

19.1 Introduction

Current emphasis in standards, measurements, and simulations is in characterizing the response of systems under fixed, well-defined conditions. In Chapter 4 we have given an introduction to the random nature of signals. It should come as no surprise that practical systems also exhibit statistical behavior in their parameters and response. No two components are ever the same. Transistors of the same type have different parameters depending on the manufacturer, and even if the manufacturer is the same there are significant differences if components are coming from a different batch. The assembly of equipment cabinets, even of the same type, does not result in identical EMC performance as at high frequencies control of electrical performance is not easy. Two cars out of the assembly line would not give identical EMC performance. Manufacturers are aware of these variations and strive to establish bounds and worst-case scenarios to meet EMC requirements. Substantial effort was expended on the design and analysis of deterministic systems, systems where all parameters have a fixed known value, and rightly so. If one cannot analyze deterministic systems, what chance is there to analyze random systems? However, the endeavor is well worth the efforts for several reasons:

- Sources of EMI are statistical in nature. The amplitude, frequency, propagation direction, and polarization of an EM threat are not known precisely in advance but vary randomly around a mean value. Can we design with the mean value and also the expected spread in mind?

- Nominally identical systems as they actually materialize in practice off a production line exhibit random variations around a mean value, e.g., the gap between two ill-fitting panels is small and its value is a random variable. This can have a profound effect on EMC behavior as it would affect the shielding effectiveness in disproportionate ways. How can we account for this while avoiding overdesign as an expedient to mask our poor understanding of the issues involved?
- The immunity of components is a statistical phenomenon as already alluded to in Chapter 9. Components do not fail at a fixed magnitude of a threat but with a certain probability at different levels. Rather than asserting that a certain component never fails (a dubious statement), would it not be better to be able to say that the component at this level of threat has a 99% probability of survival?
- Systems are increasingly dominated by complexity. Analysts and designers must be able to do their jobs in an environment characterized by extreme complexity where the co-existence of different subsystems is the order of the day. One example will suffice to illustrate the point — the mobile phone. This is no longer a device for audio conversation but includes exchange of text, video images, movie and still camera, email, GPS, MP3 player, notebook, etc. Increasingly more functionality is introduced and complexity increases by orders of magnitude. Might it not be more sensible to design such systems recognizing from the beginning the statistical nature of their parameters and their responses?
- A system tested today is not the same system after 5, 10, or 20 years. Meeting an EMC test today says something about the quality of engineering that has gone into its design. But what does it say about its EMC behavior in ten years' time? The issues are the aging of various components, deterioration in service of contacts, etc. Although it is not difficult to quantify these changes for individual components, the interesting question (for which there are no answers today) is how all these random changes combine together to affect overall system response. What is the sensitivity of the design to certain of its components or features?
- A large system today will suffer several modifications through its lifetime. These can be very severe, as for example in the case where electromechanical systems were partially replaced by solid-state systems. Here we have two different technologies with very different EMC characteristics combined. Without prior knowledge of the response of each technology in the statistical sense (i.e., in terms of probability of failure) it is difficult to design efficiently. For reasons

of cost, legacy systems are increasingly upgraded by introducing new technologies that are required to co-exist with older technologies. Design would be faster and more efficient if we understood system behavior in the statistical sense. The same applies to the increasing use of commercial equipment as components in military systems (commercial off-the-shelf equipment or COTS). Here civilian technology with its own EMC characteristics is combined with military technology with very different characteristics.

- Although at present, by and large, EMC standards are expressed in deterministic fashion, the push to high frequency is likely to require a fundamental re-examination of the basic approach and the introduction of statistical measures of failures, an approach that is common in the safety industry.
- Material properties, component values, and construction details all vary from sample to sample. A good predictive model should take account of the "fuzziness" of these parameters to represent in a more meaningful way system variability and uncertainty.

This list highlights the impetus for examining more deeply the response of systems not just simply through deterministic models but through stochastic models. A general discussion on statistics and EMC can be found in Reference 1. We give in this chapter a brief introduction into emerging techniques.

19.2 The Basic Stochastic Problem

A random variable X assumes a value x with a probability $f_x(x)$. This so-called probability density function (pdf) is in general bell-shaped as shown in Figure 19.1, although other distributions are also possible. The random

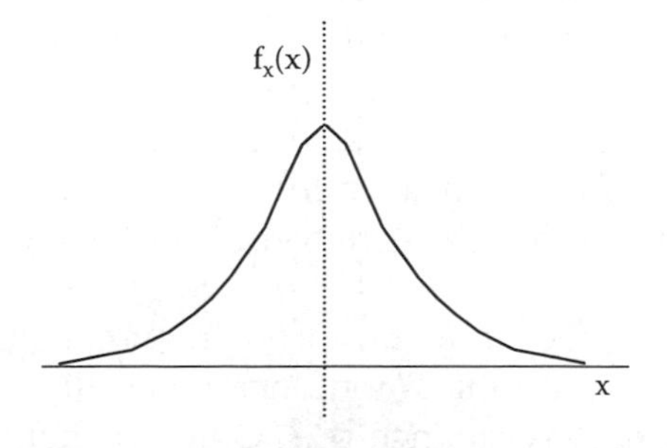

Figure 19.1 A general bell-shaped function representing a typical probability density function (pdf).

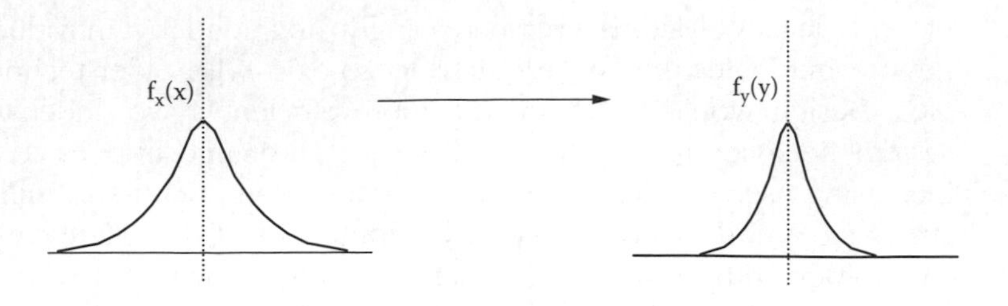

Figure 19.2 The probability density function (pdf) of the input random variable X and the resulting pdf of the output random variable Y.

variable X may represent the resistance of a component, the gap of an aperture, etc. Naturally, in an engineering application we are interested in the impact the random nature of X has on some other important output parameter, e.g., the current in the case of resistance or the shielding effectiveness in the case of aperture size. The output parameter, let us call it Y, is in itself a random variable that depends on X. The key engineering question is, given the functional dependence of the output Y on the input X and the pdf of X, to find the pdf of Y. All the interesting problems we have outlined in the previous section can be formulated in this way. The mapping is shown schematically in Figure 19.2. It goes without saying that establishing the pdf of an input variable can be time consuming. Even more so, the functional relationship between input and output variables can be very difficult to establish except in simple cases where a straightforward parametric expression links the two. However, if we somehow obtain this information, statistical theory offers guidance as to how to proceed to find $f_y(y)$. We will give a simple example of this further on in this section. It must be pointed out that in practical problems we have more than one random input variable. Hence the problem is how to work out the pdf of the output quantity when the pdf of several input random variables are known. This adds considerable complexity to the task. In regard to the functional dependence, this cannot always be in the form of a simple differentiable function — much beloved of mathematicians — but more often than not it is the result of a numerical simulation and it is implicit in a collection of data stored in computer memory. The difficulties therefore are formidable but the potential rewards are considerable.

In many situations, the complete information contained in a pdf is not required and instead a few of its "moments" will suffice. I mean specifically its mean value $\bar{x}$ and its variance σ_x^2.[2] Other higher moments may be desirable for more unusual distributions such as the "skew," which measures how symmetric the distribution is and the "kurtosis," which relates to how peaky the distribution is.

Consider the situation where we have random variable X (the input) for which we know its pdf $f_x(x)$, mean value $\bar{x}$, and variance σ_x^2. Another quantity Y (the output) is related to X through the function g:

$$Y = g(X) \tag{19.1}$$

Since X is a random variable, so is Y. The theory of functions of a random variable provides the following formulae for the pdf, mean, and variance of Y[2]:

$$f_y(y) = \frac{f_x(x)}{\left|\frac{dg(x)}{dx}\right|} \tag{19.2}$$

$$\bar{y} \simeq g(\bar{x}) + \frac{\sigma_x^2}{2}\left[\frac{d^2g(x)}{dx^2}\right]_{x=\bar{x}} \tag{19.3}$$

$$\sigma_y^2 \simeq \sigma_x^2 \left\{\left[\frac{dg(x)}{dx}\right]_{x=\bar{x}}\right\}^2 \tag{19.4}$$

where we have only kept the lowest-order terms in the expansions. If more than one random variable affects the output, then similar expressions are available in the literature. Complications arise when the variables are not statistically independent. We provide a simple example here of an application of Equations 19.1 to 19.4 to illustrate how these techniques are applied.[2]

Consider the situation of a resistor with a resistance value in the interval 900 to 1100 Ω, where the probability of having any particular value in this interval is the same (uniform distribution). The resistance is represented by the random variable X in Equation 19.1. This resistor is connected to a voltage source of E = 120 V. The pdf, mean value, and variance of the current are required.

The current is represented by the random variable Y in Equation 19.1 and clearly the functional relationship is g(x) = E/x. The normalized pdf of X is

$$f_x(x) = \frac{1}{1100 - 900} = \frac{1}{200} \tag{19.5}$$

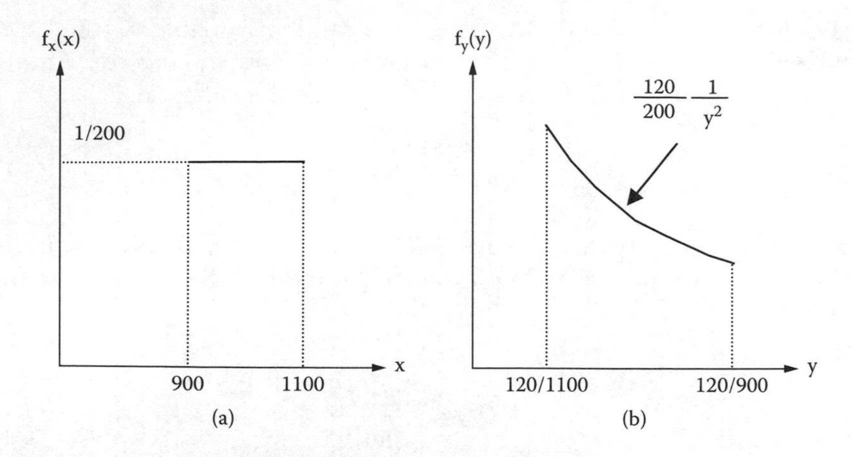

Figure 19.3 Probability density function of the input variable (a) and of the output (b).

This pdf is shown in Figure 19.3a. We now calculate the mean value (first moment)

$$\bar{x} \simeq \int x f_x(x)dx = \frac{1}{200}\int_{x=900}^{1100} x dx = 10^3\Omega \tag{19.6}$$

Similarly, the variance (second moment) is

$$\sigma_x^2 \simeq \int \left(x-\bar{x}\right)^2 f_x(x)dx = \frac{1}{200}\int_{x=900}^{1100}\left(x-\bar{x}\right)^2 d\left(x-\bar{x}\right) = \frac{100^2}{3} \tag{19.7}$$

We now calculate the derivatives of g(x) as required in Equations 19.2 to 19.4.

$$\frac{dg}{dx} = -\frac{E}{x^2}, \quad \frac{d^2g}{dx^2} = \frac{2E}{x^3} \tag{19.8}$$

These derivatives evaluated at the mean value are

$$\frac{dg}{dx} = -\frac{120}{10^6} = -12\times10^{-5}, \quad \frac{d^2g}{dx^2} = \frac{2\times120}{10^9} = 24\times10^{-8} \tag{19.9}$$

Using the value calculated in Equations 19.6 to 19.9 and substituting into Equations 19.3 and 19.4 we obtain

$$\bar{y} \simeq \frac{120}{1000} + \frac{100^2/3}{2} 24 \times 10^{-8} = 0.12 + 0.0004 \text{ A}$$
$$\sigma_y^2 \simeq \frac{100^2}{3}\left(-12 \times 10^{-5}\right)^2 = 48 \times 10^{-6} \text{ A}^2 \tag{19.10}$$

We now focus on obtaining the pdf of the current. Using Equation 19.2 we obtain

$$f_y(y) = \frac{1}{200} \frac{1}{120/x^2} = \frac{120}{200} \frac{1}{(120/x)^2} = \frac{120}{200} \frac{1}{y^2} \tag{19.11}$$

The pdf of the current $f_y(y)$ is plotted in Figure 19.3b. At x = 900 Ω we get y = 120/900 A and at x = 1100 Ω, y = 120/1100 A. We confirm that the pdf in Figure 19.3b is normalized

$$\int f_y(y)dy = \frac{120}{200} \int_{y=120/1100}^{120/900} y^{-2}dy = 1 \quad \text{QED} \tag{19.12}$$

This simple example describes the rudiments of stochastic calculations. The difficulty is in extending these ideas to more complex practical problems. We survey progress in this direction in the next section.

19.3 A Selection of Statistical Approaches to Complex EMC Problems

The analytical approach described in the previous section is not the only way to collect statistical information about the variability of a system. We outline some of the advantages and disadvantages of available techniques.

It is possible to explore the parameter space by repeated calculations, simulations, or experimental trials. This, however, is very time consuming and expensive due to the large number of parameters (multidimensionality) of systems. A very large number of trials would be necessary, which would be prohibitive for any but the simplest systems. The alternative is to judiciously sample the parameter space so as to minimize the number of samples or trials and yet extract meaningful statistical information. There is a substantial amount of work in the design of experiments so that a subset of the complete set of experiments can give good quality information but at a fraction of the cost and on the Taguchi method.[3] There is also work in social and organizational research that may help in some situations.[4] Nevertheless, this trial and error approach, even with the accelerators mentioned, is very time consuming and is not widely used in engineering applications.

A method that has been used for engineering problems is the Monte Carlo method.[5,6] In this method, essentially, a probabilistic model of a mathematical problem is set up and the problem is solved by using a sequence of random numbers. It is not difficult to see how the method works for the solution of a simple potential problem. Consider the configuration shown in Figure 19.4. The four boundaries of the problem are at potentials V_1 to V_4, respectively. We wish to calculate the potential at point (x,y). We set up a number generator to generate random numbers from zero to one. We pick a random number and if it is less that 0.25 we move right one grid point, if it is between 0.25 and 0.5 we move left, if between 0.5 and 0.75 we move up, otherwise we move down. We do this until we reach a boundary where the potential is known. For random walk A this happened after six throws of the dice and we reached potential V_3. Now, we start another random walk B, which after five trials gives V_2. If we were to stop at this point with only two random walks, we would conclude that the potential at (x,y) is $(V_3 + V_1)/2$. In practice many random walks must be done before an average is worked out that

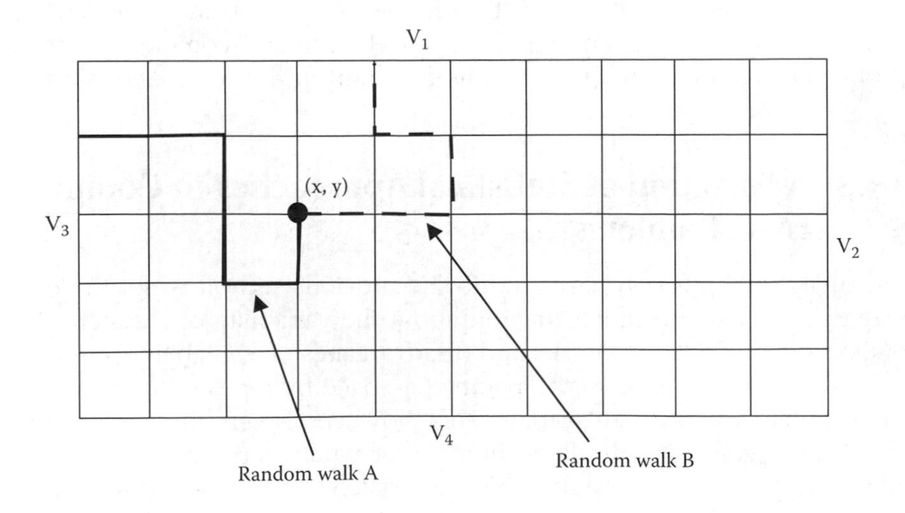

Figure 19.4 Arrangement to illustrate application of the Monte Carlo method.

converges to the true value of the potential. An example of application of the Monte Carlo method in EMC may be found in Reference 7.

More direct methods similar in spirit to those described in the previous section are increasingly explored in efforts to address more complex problems.

One way is to introduce a small perturbation to input parameters and then study how this perturbation propagates to the output. As an example, if the dielectric permittivity is the quantity with an uncertainty $\varepsilon = \bar{\varepsilon} + \delta$, where δ is a small variation, then all the relevant quantities are expanded in terms of δ (up to say second order). All calculations thereafter operate on the expanded quantities to give eventually the expansion coefficient for the required output quantity. The algebra involved can be handled using operator overloading available in some programming languages.[8]

A thorough study of the response of an electrically short two-conductor transmission line to an incident random wave is presented in Reference 9. This is done using techniques that are similar to those described in Section 19.2.

Progress in attempts to couple propagation uncertainty in large-scale numerical codes is described in Reference 10. Here techniques based on the Monte Carlo method, local linearization, and the unscented transform (UT) were compared. The UT transform approximates the pdf using a small number of carefully selected test points that then propagate through the nonlinear system.

Further work based on the techniques in Section 19.2 is described in Reference 11. A general model in lumped circuit component form with five impedances Z_n, which are random variables, is solved by the mesh current method to obtain

$$\mathbf{I} = [Y]\mathbf{V} \tag{19.13}$$

where **I** and **V** are column vectors for the mesh current and source voltages respectively. The elements of the mean current are given by

$$\bar{I}_i \simeq \sum_{n=1}^{5} \left\{ I_i(\bar{Z}_n) + \frac{\sigma_{Z_n}^2}{2} \left[\frac{d^2 I_i}{dZ_n^2} \right]_i \right\} \tag{19.14}$$

where

$$\frac{d^2 I_i}{dZ_n^2} = -\left\{ [Y]\frac{d^2[Z]}{dZ_n^2}[Y] - 2\left([Y]\frac{d[Z]}{dZ_n}[Y] \right)\frac{d[Z]}{dZ_n}[Y] \right\}\mathbf{V} \tag{19.15}$$

The current variances may be similarly obtained:

$$\sigma_{Ii}^2 \simeq \sum_{n=1}^{5} \sigma_{Z_n}^2 \left\{ \left(\frac{d[I]}{dZ_n} \right)_{\bar{Z}_n}^2 \right\}_i \tag{19.16}$$

where

$$\left\{ \left(\frac{d[I]}{dZ_n} \right)_{\bar{Z}_n}^2 \right\}_i = \mathbf{V}^T \left\{ [Y] \frac{d[Z]}{dZ_n} [Y] \right\}_i \left\{ [Y] \frac{d[Z]}{dZ_n} [Y] \right\}_i^T \mathbf{V} \tag{19.17}$$

where [Y] and [Z] are calculated at their mean values. The complexity of these equations is evident and further work is required to develop efficient techniques suitable for the statistical evaluation and design of systems.

References

1. Pignari, S A, "Statistics and EMC," Radio Science Bulletin, 316, pp 13–26, March 2006.
2. Papoulis, A and Pillai, S U, "Probability, Random Variables and Stochastic Processes," McGraw Hill, NY, 4th edition, 2002.
3. Abraham, B, "Quality Improvement Through Statistical Methods," Birkhäuser, Basel, 1998.
4. Bartlett, J E, Kortlik, J W, and Higgins, C C, "Organizational research: Determining appropriate sample size in survey research," Information Techn., Learning and Performance Journal, 19(1), pp 43–50, 2001.
5. McCracken, D D, "The Monte Carlo Method," Scientific American, 192, pp 90–96, 1955.
6. Sadiku, M N O, "Numerical Techniques in Electromagnetics," Chapter 8, CRC Press, Boca Raton, FL, 1992.
7. Goudos, S K, Vafiadis, E E, and Sahalos, J N, "Monte Carlo simulation for the prediction of the emission level from multiple sources inside shielded enclosures," IEEE Trans. on EMC, 44(2), pp 291–308, 2002.
8. Smartt, C J, Private communication, 2006.
9. Bellan, D and Pignari, S, "A probabilistic model of the response of an electrically short two-conductor transmission line driven by a random plane wave field," IEEE Trans. on EMC, 43(2), pp 130–139, 2001.
10. Borges, G A and de Menezes, L R A X, "Uncertainty propagation in transmission line modelling method (TLM)," FACE 2006, 19–20 June 2006, Victoria BC, Canada.
11. Ajayi, A, Sewell, P, and Christopoulos, C, "Direct computation of statistical variations in electromagnetic problems," in Workshop on "Statistics in EMC Measurements and Predictions," Europe EMC, 2006, Sept. 4–8, 2006, Barcelona.

Appendix A

Useful Vector Formulae

The following operations are described on a scalar and a vector function and A, respectively.

Cartesian coordinate system (x, y, z):

$$\text{Gradient } \nabla\Phi = \hat{x}\frac{\partial\Phi}{\partial x} + \hat{y}\frac{\partial\Phi}{\partial y} + \hat{z}\frac{\partial\Phi}{\partial z}$$

$$\text{Divergence } \nabla\cdot A = \frac{\partial A_x}{\partial x} + \frac{\partial A_y}{\partial y} + \frac{\partial A_z}{\partial z}$$

$$\text{Curl } \nabla\times A = \hat{x}\left(\frac{\partial A_Z}{\partial y} - \frac{\partial A_y}{\partial z}\right) + \hat{y}\left(\frac{\partial A_x}{\partial z} - \frac{\partial A_z}{\partial x}\right) + \hat{z}\left(\frac{\partial A_y}{\partial x} - \frac{\partial A_x}{\partial y}\right)$$

$$\text{Laplacian } \nabla^2\Phi = \frac{\partial^2\Phi}{\partial x^2} + \frac{\partial^2\Phi}{\partial y^2} + \frac{\partial^2\Phi}{\partial z^2}$$

$$\nabla^2 A = \hat{x}\nabla^2 A_X + \hat{y}\,\nabla^2 A_y + \hat{z}\,\nabla^2 A_z$$

Cylindrical coordinate system (r, ϕ, z):

$$\text{Gradient } \nabla_{\phi} = \hat{r}\,\frac{\partial \Phi}{\partial r} + \hat{\phi}\,\frac{1}{r}\,\frac{\partial \Phi}{\partial \phi} + \hat{z}\,\frac{\partial \Phi}{\partial z}$$

$$\text{Divergence } \nabla \cdot A = \frac{1}{r}\,\frac{\partial}{\partial r}\left(rA_r\right) + \frac{1}{r}\,\frac{\partial A_{\phi}}{\partial \phi} + \frac{\partial A_z}{\partial z}$$

$$\text{Curl } \nabla \times A = \hat{r}\left(\frac{1}{r}\,\frac{\partial A_Z}{\partial \phi} - \frac{\partial A_{\phi}}{\partial z}\right) + \hat{\phi}\left(\frac{\partial A_r}{\partial z} - \frac{\partial A_Z}{\partial r}\right)$$

$$+ \hat{z}\left(\frac{1}{r}\,\frac{\partial\left(rA_{\phi}\right)}{\partial r} - \frac{1}{r}\,\frac{\partial A_r}{\partial \phi}\right)$$

Laplacian:

$$\nabla^2 \Phi = \frac{1}{r}\frac{\partial}{\partial r}\left(r\frac{\partial \Phi}{\partial r}\right) + \frac{1}{r^2}\frac{\partial^2 \Phi}{\partial \phi^2} + \frac{\partial^2 \Phi}{\partial z^2}$$

$$\nabla^2 A = \hat{r}\left(\nabla^2 A_r - \frac{2}{r^2}\,\frac{\partial A_{\phi}}{\partial \phi} - \frac{A_{\phi}}{r^2}\right) + \hat{\Phi}\left(\nabla^2 A_{\phi} + \frac{2}{r^2}\,\frac{\partial A_r}{\partial \phi} - \frac{A_{\phi}}{r^2}\right) + \hat{z}\,\nabla^2 A_z$$

Spherical coordinate system (r, θ, ϕ):

$$\text{Gradient } \nabla \Phi = \hat{r}\frac{\partial \Phi}{\partial r} + \hat{\theta}\frac{1}{r}\,\frac{\partial \Phi}{\partial \theta} + \hat{\phi}\frac{1}{r \sin\theta}\,\frac{\partial \Phi}{\partial \rho}$$

$$\text{Divergence } \nabla \cdot A = \frac{1}{r^2}\,\frac{\partial}{\partial r}\left(r^2 A_r\right) + \frac{1}{r \sin\theta}\,\frac{\partial}{\partial \theta}\left(\sin\theta\, A_{\theta}\right) + \frac{1}{r \sin\theta}\,\frac{\partial A_{\phi}}{\partial \phi}$$

$$\text{Curl } \nabla \times A = \frac{\hat{r}}{r \sin\theta}\left(\frac{\partial}{\partial \theta}\left(A_{\theta}\,\sin\theta\right) - \frac{\partial a_{\theta}}{\partial \phi}\right)$$

$$+ \frac{\hat{\theta}}{r}\left(\frac{1}{\sin\theta}\,\frac{\partial A_r}{\partial \phi} - \frac{\partial}{\partial r}\left(rA_{\phi}\right)\right) + \frac{\hat{\phi}}{r}\left(\frac{\partial}{\partial r}\left(rA_{\theta}\right) - \frac{\partial A_r}{\partial \theta}\right)$$

Laplacian:

$$\nabla^2\Phi = \frac{1}{r^2}\frac{\partial}{\partial r}\left(r^2\frac{\partial\Phi}{\partial r}\right) + \frac{1}{r^2\sin\theta}\frac{\partial}{\partial\theta}\left(\sin\theta\frac{\partial\Phi}{\partial\phi}\right) + \frac{1}{r^2\sin^2\theta}\frac{\partial^2\phi}{\partial\phi^2}$$

$$\nabla^2 A = \hat{r}\left[\nabla^2 A_r - \frac{2}{r^2}\left(A_r + \cot\theta\, A_\theta + \operatorname{cosec}\theta\frac{\partial A_\rho}{\partial\phi} + \frac{\partial A_\theta}{\partial\theta}\right)\right]$$

$$+ \hat{\theta}\left[\nabla^2 A_\theta - \frac{1}{r^2}\left(\operatorname{cosec}^2\theta\, A_\theta - 2\frac{\partial A_r}{\partial\theta} + 2\cot\theta\operatorname{cosec}\theta\frac{\partial A_\theta}{\partial\phi}\right)\right]$$

$$+ \hat{\phi}\left[\nabla^2 A_\phi - \frac{1}{r^2}\left(\operatorname{cosec}^2\theta\, A_\phi - 2\operatorname{cosec}\theta\frac{\partial A_r}{\partial\phi} - 2\cot\theta\operatorname{cosec}\theta\frac{\partial A_\theta}{\partial\phi}\right)\right]$$

Appendix B

Circuit Parameters of Some Conductor Configurations

The capacitance and inductance per unit length and the characteristic impedance of some simple conductor configurations are given below. The formulae may also be used to obtain approximate values of these parameters even for configurations that do not conform strictly to the ideal conditions assumed in these calculations.

Two-wire line (Figure B1) — It is assumed that d >> a and that the two conductors have the same radius.

$$\text{Capacitance (in F/m)}: C = \frac{\pi\varepsilon}{\ln(d/a)}$$

$$\text{Inductance (in H/m)}: L = \frac{\mu}{\pi}\ln(d/a)$$

$$\text{Characteristic Impedance (in }\Omega\text{)}: Z = 120\sqrt{\frac{\mu_r}{\varepsilon_r}}\ln(d/a)$$

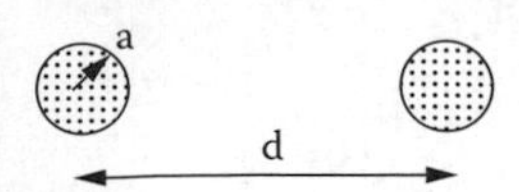

Figure B1 A two-wire line.

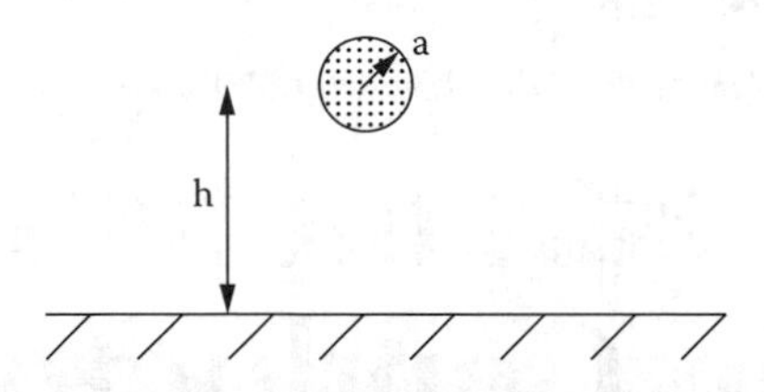

Figure B2 A wire-above-ground line.

Wire above ground (Figure B2) — It is assumed that h >> a.

$$\text{Capacitance (in F/m)}: C = \frac{2\pi\varepsilon}{\ln(2h/a)}$$

$$\text{Inductance (in H/m)}: L = \frac{\mu}{2\pi}\ln(2h/a)$$

$$\text{Characteristic Impedance (in }\Omega): Z = 60\sqrt{\frac{\mu_r}{\varepsilon_r}}\ln(2h/a)$$

Two-wire line above ground (Figure B3) — It is assumed that d, h >> a.

$$[q] = [c][v] = \begin{bmatrix} c_{11} & c_{12} \\ c_{21} & c_{22} \end{bmatrix}\begin{bmatrix} V_1 \\ V_2 \end{bmatrix}$$

The capacitance coefficients are

$$c_{11} = c_{22} = \frac{\ln(2h/d)}{A}$$

$$c_{12} = c_{21} = \frac{-\ln(D/d)}{A}$$

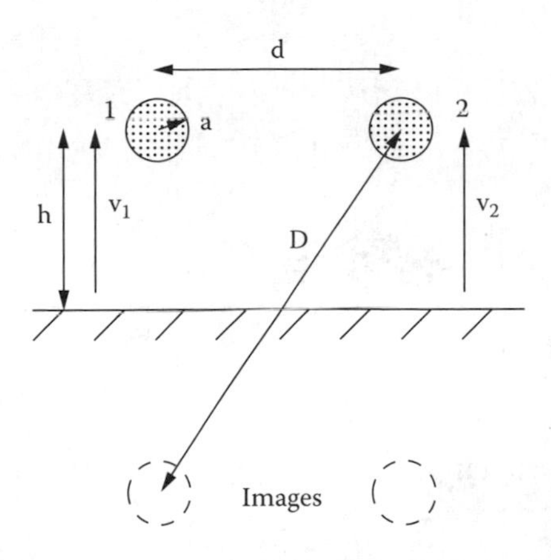

Figure B3 A two-wire line above ground.

where

$$A^{-1} = \frac{2\pi\varepsilon}{\left(\ln\frac{2h}{a}\right)^2 - \ln\left(\frac{D}{d}\right)^2}$$

The inductance matrix is

$$\left[L\right] = \frac{\mu}{2\pi}\begin{bmatrix} \ln\frac{2h}{a} & \ln\frac{D}{d} \\ \ln\frac{D}{d} & \ln\frac{2h}{a} \end{bmatrix}$$

The characteristic impedance matrix is

$$\left[Z\right] = u\left[L\right] = 60\sqrt{\frac{\mu_r}{\varepsilon_r}}\begin{bmatrix} \ln\frac{2h}{a} & \ln\frac{D}{d} \\ \ln\frac{D}{d} & \ln\frac{2h}{a} \end{bmatrix}$$

Coaxial line (Figure B4)

$$\text{Capacitance }\left(\text{in F/m}\right):\ C = \frac{2\pi\varepsilon}{\ln\left(r_2/r_1\right)}$$

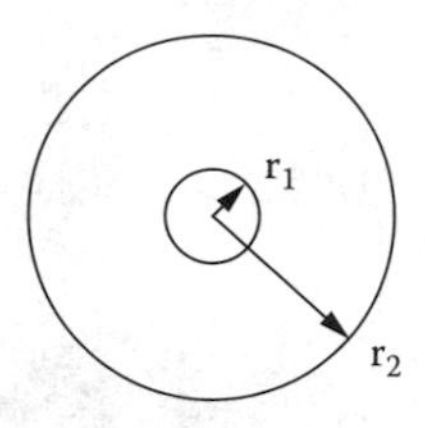

Figure B4 A coaxial line.

$$\text{Inductance (in H/m): } L = \frac{\mu}{2\pi} \ln(r_2/r_1)$$

$$\text{Characteristic Impedance (in } \Omega\text{): } Z = 60 \sqrt{\frac{\mu_r}{\varepsilon_r}} \ln(r_2/r_1)$$

Parallel-plate line (Figure B5) — Fringing of field is neglected. This is acceptable provided $u >> v$. If, as an example, $u = v$ then the inductance value is half that given by the formula for L below.

$$\text{Capacitance (in F/m): } C = \varepsilon u/v$$

$$\text{Inductance (in H/m): } L = \mu v/u$$

$$\text{Characteristic Impedance (in } \Omega\text{): } Z = \frac{v}{u} \sqrt{\frac{\mu}{\varepsilon}}$$

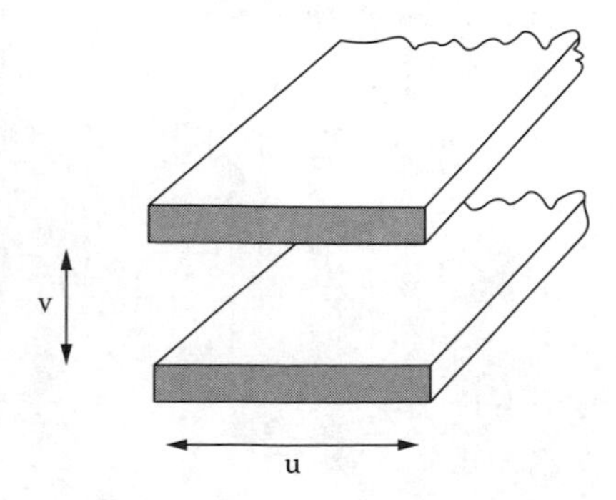

Figure B5 A parallel-plate line.

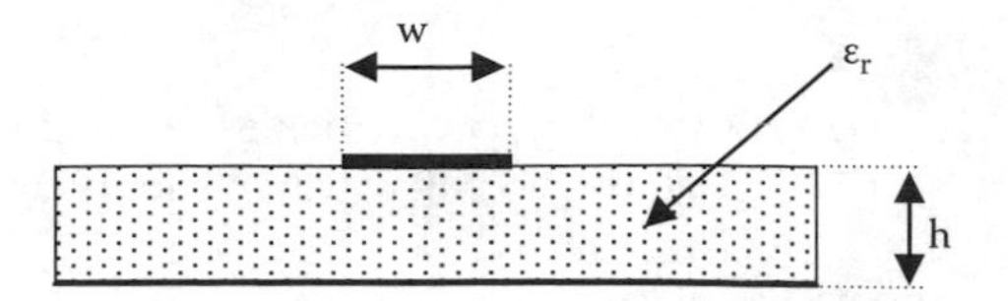

Figure B6 Schematic of the cross-section of a microstrip line.

Microstrip Line (Figure B6) — Since the conductors are immersed in a nonuniform dielectric (substrate and air) the best approach is to define an effective dielectric constant ε_{eff} and an effective width w_{eff} such that the wave propagation velocity in the actual microstrip line is the same as in an equivalent waveguide for which parameters can be calculated.[1,2] The effective quantities are

$$\varepsilon_{eff} = \left\{ \sqrt{\varepsilon_r} + \frac{(\varepsilon_r - 1)\left[\ln(\pi/4)^2 + 1 - \varepsilon_r \ln(\pi e/2)(w/2h + 0.94)\right]}{2\sqrt{\varepsilon_r}\,\varepsilon_r \left\{ w\pi/2h + \ln\left[2\pi e(w/2h + 0.94)\right]\right\}} \right\}^2$$

$$w_{eff} = \left\{ \frac{w}{h} + \frac{2}{\pi} \ln\left[2\pi e \left(\frac{w}{2h} + 0.92 \right) \right] \right\} h$$

The velocity of propagation and the characteristic impedance are then given by

$$u = \frac{u_0}{\sqrt{\varepsilon_{eff}}}, \quad Z_0 = \sqrt{\frac{\mu_0}{\varepsilon_0}}\, \frac{h}{\sqrt{\varepsilon_{eff}}\, w_{eff}}$$

Simpler formulae are also available in the literature for restricted ranges of the parameter space. Capacitance and inductance formulae for interconnect wiring in VLSI circuits may be found in Reference 3. Frequency-dependent line parameters are discussed in Reference 4. Extensive analytical treatment of microstrip lines and discontinuities may be found in References 2 and 5.

References

1 Wheeler, H A, "Transmission line properties of parallel strips separated by a dielectric sheet," IEEE Trans. on Microwave Theory and Techniques, 13, pp 175–185, 1965.
2. Itoh, T, editor, "Numerical Techniques for Microwave and Millimeter-Wave Passive Structures," Wiley, 1989.
3. Delorme, N, Belleville, M, and Chilo, J, "Inductance and capacitance analytic formulas for VLSI interconnects," Electronics Letters, 32(11), pp 996–997, 1996.
4. Williams, D F and Holloway, C L, "Transmission-line parameter approximation for digital simulation," IEEE Trans. on Electromagnetic Compatibility, 43(4), pp 466–470, 2001.
5. Collin, R E, "Field Theory of Guided Waves", 2nd edition, IEEE Press, 1991.

Appendix C

The sinx/x Function

The sinx/x function is used quite often in the calculation of the spectra of pulsed waveforms. This function is shown in graphical form up to an argument value of 12 radians in Figure C1.

The amplitude of this function $|\sin x/x|$ is also shown in Figure C2. More specifically, the quantity $20\log|\sin x/x|$ is plotted vs. x with x ranging from 0.1 to over 20,000. Consecutive points on the x-scale differ by 5%. It is clear from this graph that the maxima are bound asymptotically by a line at 0 dB up to $x = 1$ and by a line with a slope equal to –20 dB per decade of x for $x > 1$.

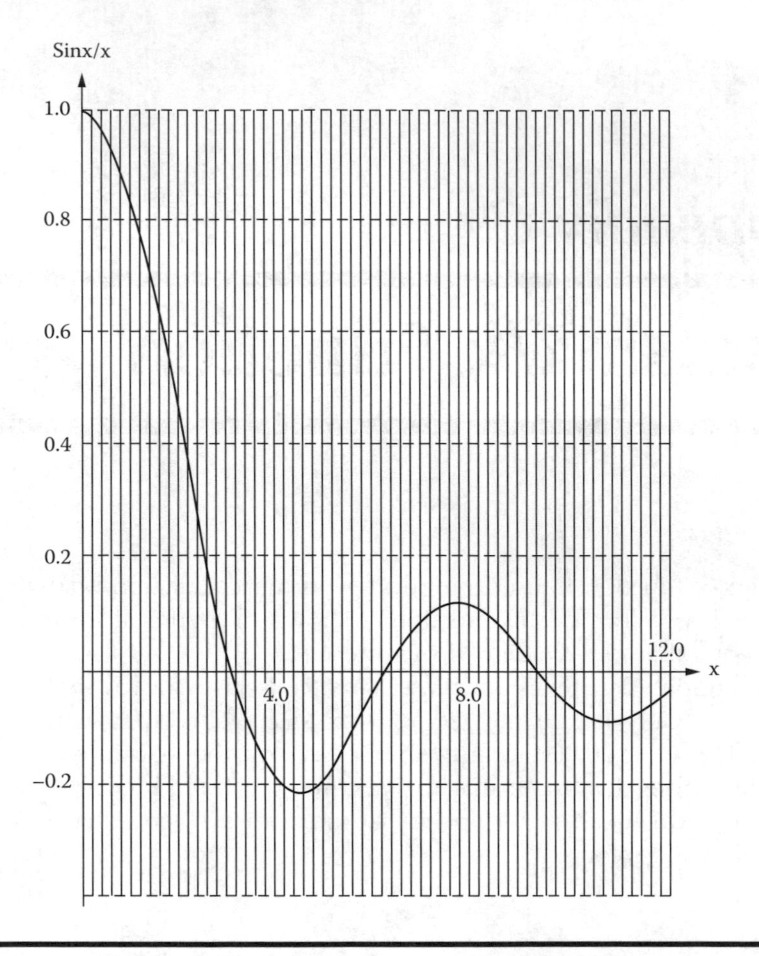

Figure C1 The sinx/x function.

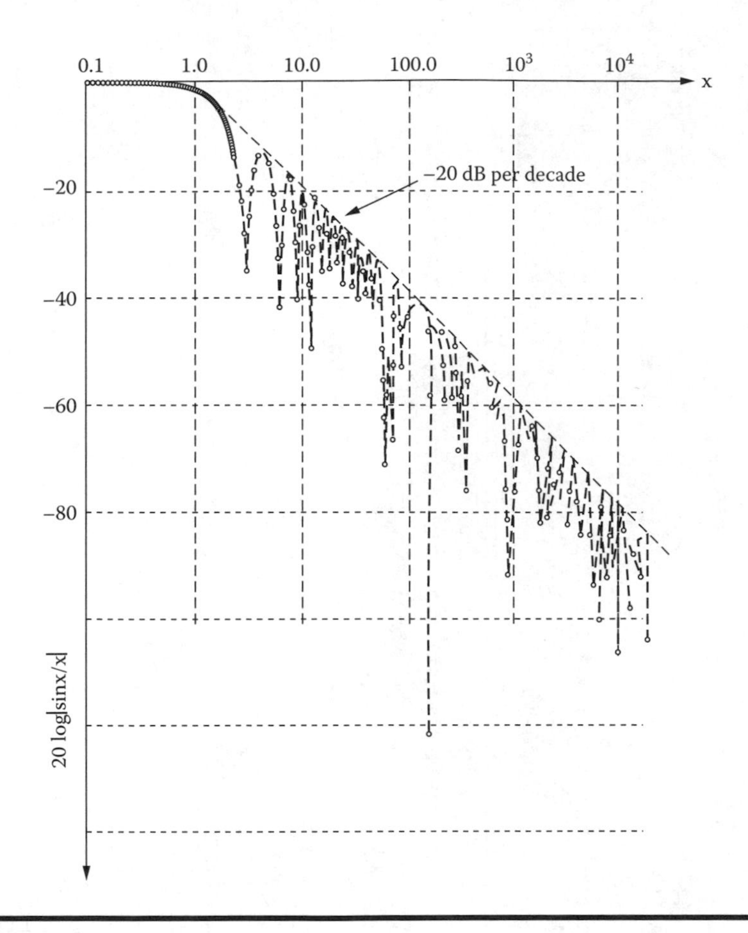

Figure C2 The |sinx/x| function.

Appendix D

Spectra of Trapezoidal Waveforms

In practical systems, the pulse rise-time and decay-time is always finite and it is useful, therefore, to study the spectral content of a periodic signal consisting of a succession of trapezoidal pulses, as shown in Figure D1.

The Fourier series coefficients may be obtained by using Equation 4.7 and are

$$C_n = V_0 \frac{\tau - \tau_r}{T} \frac{\sin \frac{\pi n(\tau - \tau_r)}{T}}{\frac{\pi n(\tau - \tau_r)}{T}} \frac{\sin \frac{\pi n \tau_r}{T}}{\frac{\pi n \tau_r}{T}} e^{-jn\pi\tau/T} \tag{D.1}$$

where n = 0, 1, 2, …

The magnitude spectrum of this signal is thus

$$2|C_n| = 2V_0 \frac{\tau - \tau_r}{T} \left| \frac{\sin \frac{\pi n(\tau - \tau_r)}{T}}{\frac{\pi n(\tau - \tau_r)}{T}} \right| \left| \frac{\sin \frac{\pi n \, \tau_r}{T}}{\frac{\pi n \, \tau_r}{T}} \right| \tag{D.2}$$

where n = 1, 2, … and $C_0 = V_0 (\tau - \tau_r)/T$. It consists of frequency components multiples of 1/T with amplitude determined by |sinx/x|

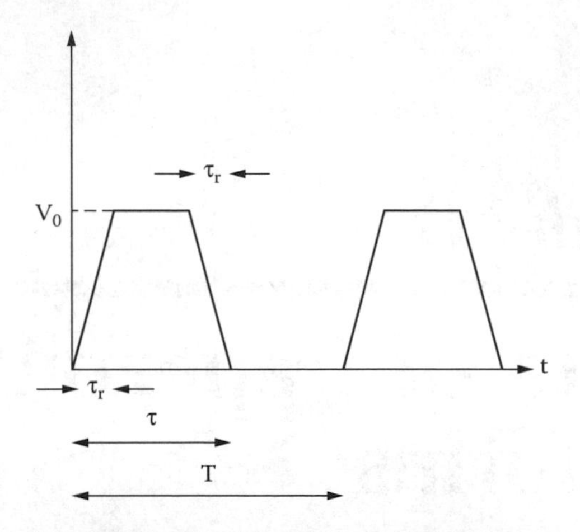

Figure D1 A pulse waveform with finite rise- and fall-times.

terms. Its envelope may easily be sketched by recognizing that $|\sin x/x|$ tends to one for x small and to $1/|x|$ for x large, with the breakpoint of the two asymptotes at $x = 1$. The expression in Equation D.2 consists of three terms. The first has magnitude $2V_0(\tau - \tau_r)/T$ and remains constant at all frequencies (zero slope). The second term has zero magnitude up to a frequency $1/[\pi(\tau - \tau_r)]$, corresponding to the argument of sin(.) equal to one and then diminishes with a slope equal to –20 dB per decade of frequency. The third term is similar but with a breakpoint at $1/(\pi\tau_r)$. These terms are shown in Figure D2. The envelope of all terms in Equation D.2

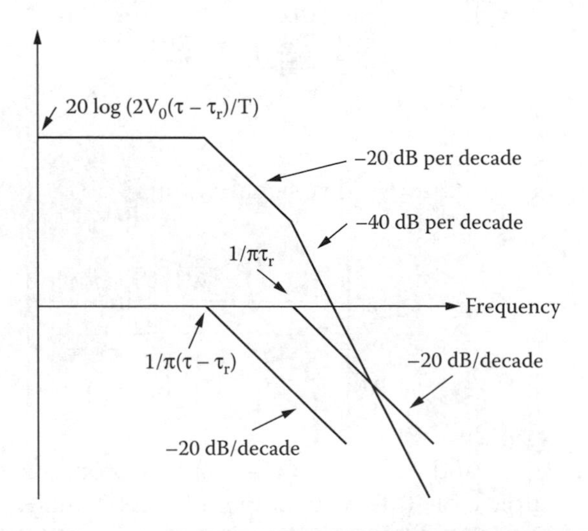

Figure D2 Envelope of the frequency spectrum of the waveform in Figure D1.

expressed in decibels is the sum of the three contributions just described and is shown in heavy outline in Figure D2.

It is clear that at high frequencies the spectrum is particularly sensitive to the pulse rise- and fall-times. If the rise-time is long, the spectral content starts to decay rapidly (–40 dB per decade of frequency) at relatively low frequencies and high-frequency components, which are particularly troublesome, become negligible in magnitude.

When the duty cycle is such that $\tau - \tau_r = T/2$, it is dear from Equation D.2 that all even harmonics are zero. However, even a modest deviation from this value of the duty cycle is likely to have significant effects on the magnitude of the even harmonics.

Appendix E

Calculation of the Electric Field Received by a Short Electric Dipole

The electric field received by a short electric dipole can be calculated if the effective antenna noise factor is known. The power associated with atmospheric noise is

$$P_n = f_a \, k_B \, T_0 \left(BW\right)$$

where

P_n is the power in W
f_a is the effective antenna noise factor (F_a in dB)
k_B is Boltzmann's constant (1.38×10^{-23} J/K)
T_0 is the temperature (288 K)
BW is the receiver equivalent bandwidth in Hz

Taking the logarithm of both sides of this equation gives

$$P_n\left(dBW\right) = F_a\left(dB\right) + 10\log\left(BW\right) - 204 \tag{E.1}$$

But

$$P_n = \frac{\left(E/\sqrt{2}\right)^2}{120} A_e \tag{E.2}$$

where E is the peak noise electric field received by a short electrical dipole and A_e is its effective aperture (= $3\lambda^2/8\pi$). Substituting in Equation E.2 and taking the logarithm of both sides gives:

$$P_n\left(\text{dBW}\right) = 20\log E + 131.5 - 20\log f \tag{E.3}$$

where f is the frequency. Expressing in dBµV/m and f in MHz in Equation E.3 gives:

$$P_n\left(\text{dBW}\right) = E\left(\text{dB}\mu\text{V / m}\right) - 20\log f\ \left(\text{MHz}\right) - 108.5 \tag{E.4}$$

Equating the left-hand side of Equation (E.1) and Equation (E.4) gives:

$$E\left(\text{dB}\mu\text{V/m}\right) = F_a\left(\text{dB}\right) + 10\log\left(\text{BW}\right) + 20\log f\left(\text{MHz}\right) - 95.5 \tag{E.5}$$

Equation E.5 relates the electric field received by a short vertical dipole to the effective antenna noise factor.

Index

A

Absorption rate, specific, 364–367
ADSL, 464
Ampere's Law, 17, 24, 25
  and circuit laws, 28
Analogue circuits, immunity, 283
Anechoic room, 389
Antennas, 376–380
  biconical, 380
  coupled, 378
  directivity, 47
  effective aperture, 376
  effective length, 377
  factor, 376
  log periodic, 380
  loop, 48
  noise temperature, 101
  proximity effects, 378
Apertures, 143, 167–174
Aperture
  dipole moment, 171
  effective, 47
  with thick walls, 172
Arc reignition and recovery, 134, 135
Arc voltage, influence on switching, 131–135
Arcing, 330
Arcing ground, 135
Attenuation, practical levels, 292
Atmospheric noise, 503–504
Autocorrelation, 89, 91

B

Balancing, 319
Balun, 319
Bandwidth, 86
Bandwidth, system, 382
Basis function, 80
BBC, 467
Behavioral model, 433–435
Bewley lattice, 307, 408–415
BER, 407
Bergeron technique, 408, 419–422
Binary phase shift keying, 443
Bit error rate (BER), 407
Black box testing, 471
Bluetooth, 449, 450
Bounce diagram, see Bewley lattice
Boundary element method (BEM), 188
Boundary integral method (BIM), 188
BPSK, 443
Broadband technologies, 459
Brushes, 330
Buffering, 335

C

Cable screens, 300
Cables, screening of electrically long, 301
  earthing of screens, 301
  shields, 163

Capacitance, 12
  per unit length, 224
  coefficients, 226
  requirements for decoupling, 333
Capacitors, 59
  bypass, 332
  decoupling, 332, 423
  feed-through, 262
Cavity modes, 386
CDMA, 439, 454
CDMA2000, 455
Cellular systems, 441
CE marking, 354
CENELEC, 351
Chip-to-package attachment, 426
Circuit parameters. 489–493
Circuits and fields, 26, 30
Circuits, nonlinear, 98
Clamp voltage, 323
Clock, dithered, 342
  jitter, 407
  skew, 407
Clocking of data, 405
CM currents, 461
CMOS output stage, 433
Coaxial line parameters, 491, 492
Code division multiple access, 439
Coexistence, 442–449
Common-mode, 222, 247
Common-mode
  choke, 313
  current, 319
  emissions example, 315
  radiation, 248, 250
Communications secrecy, 340
Components
  distributed, 64
  ideal, 56
  real, 57
Compton effect. 137
Conducted emission, example, 354
Conducted interference levels, 350
Connectors, 302
Constitutive expression, 25
Continuity equation, 25
Convolutional code, 449
Correlation, 88, 439
Coupling, 259–270
  capacitive, 208
  external fields, 216, 232–246
  to electrically short lines, 237
  far-end, 217
  far-field, 210
  frequency-domain solutions, 235
  high-frequencies, 240
  inductive, 209
  low-frequency, 221
  medium frequency, 239
  near-field, 210, 217
  paths, reduction of, 331–336
  time-domain solution, 240–246
  wire-to-wire inside cabinet, 264–270
Cross-correlation, 90
Crosstalk, 207–231, 270
Crowbar, 321
Curl, 17, 485, 486
Current
  conduction, 24
  displacement, 24
  interruption, 129
Cyclic redundancy check, 339

D

Damping of resonances, 390
Decibel, 15
DFSK, 443
Delay, gate propagation, 286
Delta function, 79
Despreading, 340
Detector
  average, 374
  circuit, 98
  peak, 374
  quasi-peak, 374
  time-constant, 374
Device
  impact of interference, 281
  susceptibility, 273, 281
Dielectric constant, effective, 493
Differential
  formulation, 186, 187
  mode, 222, 247
  signalling, 425
  phase shift keying, 443
Diffusion term, 37
Digital circuits, immunity of, 284
Digital
  filter interface (DFI), 197
  radio mondiale (DRM), 464

video broadcasting –handheld (DVB-H), 455
Dipole
electric, 41
half wavelength, 51, 380
magnetic, 49
moment, 42
Direct sequence spread spectrum, 339, 440, 449
Directivity, 46
Directive gain, 46
Divergence, 9, 486, 487
DM, 248, 250
DSL, 459, 463–465
Duty cycle, 501
DVB-H, 449
Dynamic frequency selection, 449

E

Earthing, 301, 303–309
Eavesdropping, 340
EBU, 467
EDGE, 454
Eigenvalues, 213
Electrostatic discharge, 135
Electric field, 6
Electric field integral equation (EFIE), 190
Electric flux density, 7
Electromagnetic bandgap (EBG) structures, 293
EM environment surveys, 139
EMC
board level, 422
chip level, 422
management, 342
and signal integrity, 405
simulation requirements, 275
statistical, 473, 475–484
Emission
mechanisms, 250–252
reduction at source, 330
End effects, 231
Energy storage
electric, 13–14
magnetic, 20
Error correction, 339, 449
Error margin, 443
ESD, 135, 325, 330
ESD equivalent circuit, 136,137
Excitation
broadside, 238
endfire, 238
impulse, 85
sidefire, 238
Expected value, 90
Expert systems, 272

F

Faraday's law, 23, 25
Faraday's law and circuit laws, 27
Far-field extrapolation, 353
FDD, 442
FDMA, 439
FDTD, 275
FE, 276
Fields and circuits, 26,30
Field
electric, 6
high-frequency, 33
magnetic, 15
quasistatic, 22
radiated from transmission lines, 246–252
static, 6
Filters
power line, 311
signal line, 320
Finite Difference Time Domain (FDTD), 190
Finite Element Method, (FE), 187
Fourier transform, 84
Fourier series, 80
Fresnel zone, 384
Frequency allocation, 118
Frequency bands, 460
Frequency division duplex, 442
Frequency division multiple access, 439
Frequency-domain, 86, 186
Frequency hoping, 341, 449
Frequency response, 93
Frequency spectrum, envelope, 500

G

Gain, antenna, 46
Galerkin's method, 188

Galileo, 449, 455
Gaskets, 263, 297
Gauss's law, 7–9, 25
Gaussian probability distribution, 102
Gaussian, pulse, 85
Gibbs phenomenon, 82
GPRS, 454
GPS, 449, 455
Gradient, 485, 486
Ground
  bounce, 423
  paths, 334
  planes, 334
  reflections, 384
Grounding, 301, 303–309
  and inteconnects, 308
  in large systems, 304–307
  in self-contained equipment, 307
  single point, 305
GSM, 441, 449, 454
GTEM cell, 396
Guard band, 439, 447

H

Hamming code, 449
Hardening procedure, 283
Harmonics, 122
Harmonic distortion, 122
High-frequency propagation in power networks, 460
High-impedance surfaces (HIS), 294–296
High-power electromagnetics, 137
HPEM, 137
Human exposure to EM fields, 364
Hybrid models, 273

I

IBIS model, 433–434
ICs, physical models, 426–433
ICEM model, 434
IEC, 348
Impedance
  characteristic, 68
  to common-mode currents, 229
  to differential-mode currents, 228
  intrinsic, 36
  power distribution, 427
  surge, 68
Impulse
  excitation, 85
  function, 84
  response, 92
Immunity
  analogue circuits, 283
  digital circuits, 284
  improvements, 336
  by software design, 338
Inductance, 18, 32, 33
  internal, 20
  mutual, 227, 230
  partial, 227
  per unit length, 225
  self, 227,230
Inductors, 61
Interconnects, 408
Interface conditions
  electric, 11
  magnetic, 18
Integral formulation, 186, 187
Interference
  conducted, 122, 124, 207
  control level, 336
  electroheat applications, 119
  intersymbol, 446
  multipath, 444
  radiated, 123,124, 210
  rejection, 340
  to signal ratio, 341
Interference sources
  classification, 111
  digital signal processing, 119
  natural, 112
  man-made, 118
Interleaving, 339
Intermediate models, 272
Interoperability, 442–449
International non-ionizing radiation commission (INIRC), 365
International telecommunications union (ITU), 348
Intersymbol interference, 446
Ionosphere, 117
ISM band, 447, 449
Isolation, 316
ITU-R, 466

J

Joints, effectiveness of, 293

K

Kirchhoff's laws, 27,30
Kurtosis, 478

L

Laminate, 404
Laplace's equation, 11
Laplacian, 485–487
Lightning, 113
Lightning
  NEMP, 138
  test waveform, 115, 116
Line parameter calculation, 223–227
Line impedance stabilising circuit (LISN), 317
LISN, 355, 380
Logic family, impact on emissions, 330, 332
Longitudinal conversion loss (LCL), 464
Loop, small, 375
Lorentz gauge, 32
Loss tangent, 60, 432
Lumped components, 55

M

MCM, 403
Magnetic cores, inductance, 63
Magnetic field,
  integral equation (MFIE), 189
  intensity, 15,16
Magnetic flux conservation, 25
Magnetic flux density, 15
Magnetosphere, 117
Mains signalling, 123
Margins, timing, 407
Matching, 333, 416
Mean value, 90, 478
Mean square value, 90
Measurements
  EMC, 373–397
  sources, 372
Method of moments, 188
Metamaterials, 296
Microstrip
  line parameters, 493
  propagation regimes, 431
MM, 276
MPT 1570, 467
Modes
  cavity, 386
  common, 214
  density of, 395
  differential, 214
  of propagation, 214
Mode-stirred chamber, 393–396
Modal voltages, 212
Modulation, 442
Modulated scatterer, 376
Moments, statistical, 478
Multichip module, 403
Multiconductor systems, 210
Multilayer board, 404
Multipath interference, 444–446
Monopole, small, 375
Monte Carlo method, 482

N

Nanotechnology, 293
NB30, 467
Near-field, 45
NEMP, 137, 320
NEMP and lightning, 138
NFC, 452
Networks
  ad hoc, 437
  heterogeneous, 449
Noise, 95
  atmospheric, 503–504
  bandwidth equivalent, 105, 106
  characterization, 100
  cosmic, 117
  factor, 103
  figure, 104
  immunity, 337
Noise margins, 285, 337
  DC, 337
  dynamic, 337
Noise temperature, 101
Noise temperature, effective, 104
Noise, white, 92

Nondirectional beacons (NDB), 464
Nonlinear circuits, 98
Nuclear EM pulse, 115, 137

O

OATS, 384
OFDM, 447
Ohm's law, 25
Open area test site, 384
Operator overloading, 483
Optical fibre, 318
Opto-isolator, 318

P

Package, 426
Parallel data stream, 446
Parallel plate line parameters, 492
Parity check, 338
PCB, 403
PCB routing, 334
Penetration, 143, 259–270
Phase constant, 35
Phase-locked loop, 342
Phase velocity, 35
Physical agents directive, 367
Pigtail connection, 263
Placements of circuits, 335
Planck's formula, 100
PLT, 459, 465–466
PLT, regulatory framework, 466–467
Poisson's equation, 11
Polarizability, 171
Potential
  electric, 10
  retarded, 40
  scalar, 31
  vector, 31
Power flux, 36
Power line telecommunications, 459, 465–466
Power spectrum, 86, 91
Power supply model
  HF, 430
  LF, 428
  MF, 429
Poynting's vector, 36
Prepreg, 404
Probability density function (pdf), 477
Prony's method, 197
Propagation, 207–231, 270
Propagation constant, 38, 72
Propagation delay, gate, 286
Propagation, multimode, 245
Protected field strength, minimum, 460
Protection of circuits, 326
Protective devices
  discharge tubes, 323
  nonlinear, 311, 320
  proximity to protected equipment, 321, 322
  solid state, 323
  survivability and deterioration, 324
  varistors, 323
  zener diodes, 324
Pseudorandom code, 339
Pulse duration and bandwidth, 86
Pulse repetition frequency, 330
PWB, 403

Q

Q-factor, 299, 395
QPSK, 444
Quasi-TEM propagation, 432
Quaternary quadrature phase shift keying, 443

R

Radiating systems, 39
Radiation
  absorbing materials, 389
  from CM currents, 248
  from dipole, 50
  from DM currents, 248
  from monopole, 50
Radiation resistance, 44
  small loop, 49
Radiated field from transmission line, 246–252
Radiated interference levels, 351
RAM, 389
Random variable, 477, 478
Random variable, function of, 478

Rayleigh's law, 101
Receiver (RX) , 373, 439
  and narrowband signals, 374
  and wideband signals, 374
Reed-Solomon code, 449
Reflection, 333
  coefficient, 69, 408
  ground, 384
  minimization, 416
  nonlinear loads, 419
Resistors, nonlinear, 321
RFID, 452
Resonances
  damping, 299
  room, 387
Return stroke, 114
Reverberating chamber, 391
Reverberating room, 386
Rise-time, 330, 331, 333
Routing of cables, 336
Routing on PCBs, 335

S

Safety and EMC, 471–473
Safety limits, human exposure, 364–67
Schwarz-Christoffel transform, 169
Screening effectiveness, measurement, 299
Screening materials, 292
Screened room, 387
SDMA, 440,441
Seams, 297
Segregation of circuits, 335
Semiconductor failure modes, 282
Sensors, field, 375
Serial data stream, 446
Sferics, 115, 117
Shared broadcast channel (SBCH), 449
Shielding, 143–201, 291–303
  attenuation, 292
  circuit analogies, 152
  analytical approaches, 161
  for cables, circuit equivalent, 167
  circuit approach, 146
  and eddy currents, 152
  general remarks, 154,161
  with high-temperature superconductors, 163
  materials, 292
  multiple apertures, 194
  two-layered, 162
  wave approach, 156
Shielding effectiveness, 145
  calculation using numerical techniques, 185–201
  of equipment cabinets, 174–185
  impact of loading, 183
SI, 403–435
SI and EMC, 405
Signal
  aperiodic, 83
  detection in noise, 95
  deterministic, 77
  ergodic, 91
  minimum detectable, 107
  orthogonal, 439
  periodic, 80
  random, 77, 90
  representation, 78
  square wave, 81
  stochastic, 77
  timing, 406
  differential, 425
Signal integrity, 403–435
Signal-to-noise ratio, 97, 103
Site attenuation, 385
  measurement, 386
  normalised, 385
Skew, 478
Skin depth, 39, 59, 297, 432
Sniffer probe, 375
Solar radiation, 117
Solder balls, 426
Sources
  ESD testing, 373
  of interference, 111
Source region, 257, 258
Space division multiple access, 440,441
Spark gap, 323
Spectrum
  amplitude, 83
  analyzer, 373
  content, 86
  phase, 83
  use, 439
Simultaneous switching noise, 422–425
sinx/x function, 495–497
Spread spectrum techniques, 339–342
Spreading code, 439
SSN, 422–425
Standards, 150, 348

above 1GHz, 360–364
CISPR, 348
class A, 350
class B, 350
company, 360
EMC, 347
European, 351–352
FCC, 349
International, 348
Military, 357–360
other, 354
Standard deviation, 91
State equation, 199
Stepped leader, 113
Stoke's theorem, 33
Stripline configuration, 417
Statistical EMC, 475–484
Stochastic problems, 477
Superposition, 7
Supply bounce, 423
Suppressors, 335
Susceptibility, 257
design level, 336
Symbol rate, 446
Switch
arcing, 132
idealised, 126
realistic, 131

T

TDD, 442
TDMA, 439
TEM cell, 396
Testing, EMC, 371–397
Test environments, 383–397
TETRA, 449, 455
Third generation (3G), 454
Time constant, 99
Time division duplex. 442
Time division multiple access, 439
Time-domain, 86
Time-domain methods, 186
Timer, dedicated, 338
Timing diagram, 287
Timing margins, 407
TLM, 275
Transfer admittance of shields, 167
Transfer characteristic, inverter, 285
Transfer impedance, 163
cable, 300
connectors, 302
Transient
closing on inductor, 127
nature and origins, 125
opening on capacitor, 131
opening on inductor, 128
switching, 125
Transmission lines
ABCD parameters, 73
balanced, 324
characteristic impedance, 72, 333
electrical length, 74
electrically small, 75
equations, 267
frequency-domain response, 70
input impedance, 74
propagation constant. 72
time-domain response, 67
transmission matrix, 73
Transmission-Line Modelling or Matrix method (TLM), 192
Transmitter (TX), 439
Trapezoidal waveform, spectra, 499–501
Transit time, 333
Triboelectric series, 136
Trunking, 336
Two-wire line above ground, parameters, 490–491
Two-wire line, parameters, 489

U

Unbalance, 123
UMTS, 454
Unit step function, 78
Unscented transform (UT), 483
UWB, 451

V

Variance, 91, 97
VDSL, 464
Vector formulae, 485–487
Velocity of propagation, 68
Via model, 430
Voltage dips, 123

Voltage fluctuations, 122
Voltage induced, 23

W

Watchdog circuits, 338
Waves, electromagnetic, 34
Wave
- equation, 32
- impedance, 45
- term, 37

Wavelength, 36
White noise, 92
Wiener-Khinchine theorem, 89, 91
Wi-Fi, 452
WiMAX, 452
Wire-above-ground, parameters, 490
Wire-to-wire coupling inside cabinet, example, 264–270
Wireless
- alliances, 449–455
- links, 438
- specifications, 449–455
- technologies, 437–455

WLAN, 450, 452
WMAN, 450, 452
WPAN, 450
WWAN, 450, 452

X

xDSL, 459, 463–465
- regulatory framework, 466–467

Z

ZigBee, 450